VOLUME ONE HUNDRED AND FORTY NINE

CURRENT TOPICS IN DEVELOPMENTAL BIOLOGY

Cell Signaling Pathways in Development

CURRENT TOPICS IN DEVELOPMENTAL BIOLOGY

"A meeting-ground for critical review and discussion of developmental processes"
A.A. Moscona and Alberto Monroy (Volume 1, 1966)

VOLUME ONE HUNDRED AND FORTY NINE

CURRENT TOPICS IN DEVELOPMENTAL BIOLOGY

Cell Signaling Pathways in Development

Edited by

PHILIPPE M. SORIANO

Department of Cell, Developmental, and Regenerative Biology, Icahn School of Medicine at Mount Sinai, New York, NY, United States

ELSEVIER

ACADEMIC PRESS

An imprint of Elsevier

Academic Press is an imprint of Elsevier
50 Hampshire Street, 5th Floor, Cambridge, MA 02139, United States
525 B Street, Suite 1650, San Diego, CA 92101, United States
The Boulevard, Langford Lane, Kidlington, Oxford OX5 1GB, United Kingdom
125 London Wall, London, EC2Y 5AS, United Kingdom

First edition 2022

ISBN: 978-0-12-817097-7
ISSN: 0070-2153

For information on all Academic Press publications
visit our website at https://www.elsevier.com/books-and-journals

Publisher: Zoe Kruze
Developmental Editor: Naiza Mendoza
Production Project Manager: Vijayaraj Purushothaman
Cover Designer: Vicky Pearson Esser

Typeset by STRAIVE, India

Contents

Contributors

Sabine L. Bosman
Hubrecht Institute—KNAW (Royal Netherlands Academy of Arts and Sciences) and University Medical Center Utrecht, Utrecht, The Netherlands

Jeffrey O. Bush
Department of Cell and Tissue Biology; Program in Craniofacial Biology; Institute for Human Genetics; Eli and Edythe Broad Center of Regeneration Medicine and Stem Cell Research, University of California San Francisco, San Francisco, CA, United States

James F. Clark
Department of Cell, Developmental, and Regenerative Biology, Icahn School of Medicine at Mount Sinai, New York, NY, United States

Mercedes Costell
Departamento de Bioquímica y Biología Molecular, Facultad de Ciencias Biológicas; Institut Universitari de Biotecnologia i Biomedicina, Universitat de València, Burjassot, Spain

Reinhard Fässler
Department of Molecular Medicine, Max Planck Institute of Biochemistry, Martinsried, Germany

Ameera Haque
Program in Craniofacial Biology and Department of Orofacial Sciences, University of California, San Francisco, CA, United States

Caroline S. Hill
Developmental Signalling Laboratory, The Francis Crick Institute, London, United Kingdom

Philip William Ingham
Lee Kong Chian School of Medicine, Nanyang Technological University, Singapore, Singapore

Ophir D. Klein
Program in Craniofacial Biology and Department of Orofacial Sciences; Department of Pediatrics and Institute for Human Genetics, University of California, San Francisco, CA, United States

Peter W. Krenn
Department of Molecular Medicine, Max Planck Institute of Biochemistry, Martinsried, Germany; Department of Biosciences and Medical Biology, Cancer Cluster Salzburg, Paris-Lodron University of Salzburg, Salzburg, Austria

Pauline Marangoni
Program in Craniofacial Biology and Department of Orofacial Sciences, University of California, San Francisco, CA, United States

James F. Martin
Department of Molecular Physiology and Biophysics, Baylor College of Medicine; Cardiomyocyte Renewal Laboratory, Texas Heart Institute, Houston, TX, United States

Sarah E. McFann
Department of Chemical and Biological Engineering; Lewis Sigler Institute for Integrative Genomics, Princeton University, Princeton, NJ, United States

Marek Mlodzik
Department of Cell, Developmental, & Regenerative Biology, Graduate School of Biomedical Sciences, Icahn School of Medicine at Mount Sinai, New York, NY, United States

Eloi Montanez
Department of Physiological Sciences, Faculty of Medicine and Health Sciences, University of Barcelona and Bellvitge Biomedical Research Institute, L'Hospitalet del Llobregat, Barcelona, Spain

Neha Pincha
Program in Craniofacial Biology and Department of Orofacial Sciences, University of California, San Francisco, CA, United States

Stanislav Y. Shvartsman
Lewis Sigler Institute for Integrative Genomics; Department of Molecular Biology, Princeton University, Princeton, NJ; Center for Computational Biology, Flatiron Institute, Simons Foundation, New York, NY, United States

Katharina F. Sonnen
Hubrecht Institute—KNAW (Royal Netherlands Academy of Arts and Sciences) and University Medical Center Utrecht, Utrecht, The Netherlands

Philippe M. Soriano
Department of Cell, Developmental, and Regenerative Biology, Icahn School of Medicine at Mount Sinai, New York, NY, United States

Jared E. Toettcher
Department of Molecular Biology, Princeton University, Princeton, NJ, United States

Chang-Ru Tsai
Department of Molecular Physiology and Biophysics, Baylor College of Medicine, Houston, TX, United States

Linh T. Vuong
Department of Cell, Developmental, & Regenerative Biology, Graduate School of Biomedical Sciences, Icahn School of Medicine at Mount Sinai, New York, NY, United States

Preface

At first thought, one may think of cell signaling as if it was abstract art. One sees lines, curves, and association with objects, but the directionality is not clear and the general impression is complex. If one thinks about cell signaling in more depth, one may realize, however, that lines and curves as shown on this book cover represent signaling waves or gradients that instruct developmental pathways and processes. This volume of *Current Topics in Developmental Biology* focuses on the topic and does much to clarify many of the aspects that may feel blurry at first.

Many aspects of cell signaling have been worked out in established cell lines, which by their nature lack an *in vivo* perspective. Thus, we have come to believe that any biochemical interaction is essential for functional circuitry; however, the analysis of the pathway in the organism often provides a more nuanced perspective. One example that comes to my mind, dear to my heart, is how various effectors were found to associate with receptor tyrosine kinases (RTKs) in the early 1990s, but at the same time genetic screens conducted in the Rubin lab identified a role for many of the same components that interacted genetically with *sevenless*. Additionally, many concepts of genetic redundancy and interaction with parallel pathways have come to light from developmental genetics studies.

The first three chapters in this volume provide numerous insights into the roles of individual signaling pathways in development, in various contexts. With Philip Ingham covering the Hedgehog pathway, Linh Vuong and Marek Mlodzik discussing both canonical and noncanonical Wnt signaling, and Chan-Ru Tsai and James Martin providing an in-depth perspective on Hippo signaling, we can garner a historical perspective, how these signaling pathways operate in development and disease, and perspectives on future research.

This is then followed by three chapters on RTKs, their downstream pathways, and their cross talk with cell adhesion mechanisms. A discussion by James Clark and Philippe Soriano focuses on both canonical RTK signaling, as well as on how these RTK can impact other process such as cell adhesion beyond their classic signaling properties. Following closely on this theme, Jeffrey Bush covers the interactions between Ephrins and the RTK Eph receptors, and how they impact cellular organization by impacting on the actin cytoskeleton or cell adhesion behaviors. Last, Peter Krenn,

Eloi Montanez, Mercedes Costell, and Reinhard Fässler cover integrin signaling. Focusing in on hematopoietic cells (HSCs) and their niche, the authors discuss integrin-mediated sensing for HSC-niche interactions, the functional properties of the integrins and their activators, and how these pathways can be targeted in specific leukemias.

Moving on to how signaling dynamics regulate development, Sarah McFann, Stanislav Shvartsman, and Jared Toettcher focus on how transient expression of a critical RTK target, ERK1/2, regulates mesoderm development. They also discuss optogenetic methods for manipulating this pathway during development, and the role of ERK1/2 dynamics in the developing embryo. Next, Caroline Hill discusses TGF-β superfamily members, namely Nodal and bone morphogenetic proteins, focusing on how opposing gradients are formed, how the ligands can signal as heterodimers with related factors, and how the gradients are interpreted. She also discusses buffering mechanisms and how the gradients can be further shaped by feedback loops. Following on this general theme, Sabine Bosman and Katharina Sonnen describe signaling oscillations in embryonic development, at the level of single cells and whole tissues, in multiple contexts including neurogenesis and somitogenesis.

Wrapping it all together, Neha Pincha, Pauline Marangoni, Ameera Haque, and Ophir Klein draw parallels in signaling between embryonic development and regeneration, in several ectodermal organs. They focus on four signaling pathways, several of which have been discussed elsewhere in this volume, to draw differences and similarities during regenerative processes.

With numerous technological advances in the ease of manipulating signaling pathways in various organisms, ranging from gene-editing approaches to optogenetics, I feel that the future for this field is very bright indeed. Deciphering signaling pathways, much like examining lines, curves, or the patterns of abstract art, provides insight and greater understanding of the embryo and development as a whole. In closing, I want to state how rewarding it has been to work with this group of talented and experienced investigators who have contributed chapters to this volume. I also want to thank them for having made such meaningful contributions, at a time when our scientific and personal lives have all been upended by the present pandemic.

Philippe M. Soriano
Professor, Department of Cell,
Developmental, and Regenerative Biology
Icahn School of Medicine at Mt. Sinai

CHAPTER ONE

Hedgehog signaling

Philip William Ingham*
Lee Kong Chian School of Medicine, Nanyang Technological University, Singapore, Singapore
*Corresponding author: e-mail address: pingham@ntu.edu.sg

Contents

Abstract

Hedgehog (Hh) proteins constitute one family of a small number of secreted signaling proteins that together regulate multiple aspects of animal development, tissue homeostasis and regeneration. Originally uncovered through genetic analyses in *Drosophila*, their subsequent discovery in vertebrates has provided a paradigm for the role of morphogens in positional specification. Most strikingly, the Sonic hedgehog protein was shown to mediate the activity of two classic embryonic organizing centers in vertebrates and subsequent studies have implicated it and its paralogs in a myriad of processes. Moreover, dysfunction of the signaling pathway has been shown to underlie numerous human congenital abnormalities and diseases, especially certain types of cancer. This review focusses on the genetic studies that uncovered the key components of the Hh signaling system and the subsequent, biochemical, cell and structural biology analyses of their functions. These studies have revealed several novel processes and principles, shedding new light on the cellular and molecular mechanisms underlying cell-cell communication. Notable amongst these are the involvement of cholesterol both in modifying the Hh proteins and in activating its transduction pathway, the role of cytonemes, filipodia-like extensions, in conveying Hh signals between cells; and the central importance of the Primary Cilium as a cellular compartment within which the components of the signaling pathway are sequestered and interact.

Current Topics in Developmental Biology, Volume 149
ISSN 0070-2153
https://doi.org/10.1016/bs.ctdb.2022.04.003

1. Introduction

Hedgehog (Hh) proteins constitute one family of a small number of secreted signaling proteins that together regulate multiple aspects of animal development, tissue homeostasis and regeneration. Genetic analysis in model organisms has played a major part in unraveling the processes regulated by these signals and the pathways through which cells respond to them. The elucidation of the mechanisms and functions of Hh signaling is a prime example: first discovered through its role in controlling segmental patterning of the epidermis during *Drosophila* embryogenesis, it has subsequently been implicated in a host of other processes in both the larval and adult fly. Moreover, identification of vertebrate homologs, initially in chicken, mouse and zebrafish, has led to the discovery of its fundamental role in virtually every organ of the body, as well as its involvement in numerous human congenital abnormalities and diseases.

Along with the archetypal processes of neural tube and limb patterning, Hh proteins play major roles in the development, maintenance or repair of many other tissue and organs, such as the vasculature, lungs, intestine, kidney, liver, heart, muscle, mammary gland, ovary, testes, skin, bones, teeth, cartilage and more. Clearly a comprehensive survey of the many roles of this signaling system is well beyond the scope of a single review and will certainly not be attempted here. Instead, what I aim to do is present an overview of our current understanding of how this signaling system operates as well as a compendium of the fundamental discoveries upon which this understanding has been elaborated. I will begin by summarizing the genetic analyses that uncovered the various components of the pathway, principally in *Drosophila* but also in mouse, zebrafish and human. In doing so, my intention is to provide the reader with an understanding of how we know which proteins are involved in Hh signaling and the developmental processes through which they were identified. These processes have formed the basis of the *in vivo* assays underpinning much of the subsequent analysis of Hh signaling.

2. The developmental genetics of Hh signaling in *Drosophila*

The *hh* gene owes its name to the phenotype of *Drosophila* larvae homozygous for loss of function mutations. These were isolated in a systematic screen designed to identify all of the zygotically expressed genes required for the normal patterning of the larval epidermis, which secretes a cuticle with an easily observable segmented pattern (Nusslein-Volhard & Wieschaus, 1980). The screen, which was carried out at the EMBL in Heidelberg, yielded three distinct phenotypic classes that disrupt this pattern, known as the gap, pair-rule and segment polarity mutations. Amongst the latter class, the predominant phenotype was a loss of the smooth or "naked" cuticle, that characterizes the posterior half of each thoracic and abdominal segment, and its replacement with belts of chitinous processes known as denticles in mirror image polarity to those normally decorating the anterior half of each segment (see Fig. 1). This phenotype—a continuous "lawn" of spikey denticles covering the ventral surface of the cuticle—led Nüsslein-Volhard and Wieschaus to choose the name *hedgehog*, after the nocturnal mammals well known for the short spines that cover their backs. Lethal alleles of *wingless (wg), the Drosophila* ortholog of the murine oncogene *Wnt1* (Rijsewijk et al., 1987), isolated in the same screens display a similar phenotype. A key early finding was that *hh* is required to maintain the spatially restricted pattern of *wg* transcription in the developing embryo (Hidalgo & Ingham, 1990), a finding which in part explains the similarity between the *hh* and *wg* mutant phenotypes (Fig. 1). The *wg* gene is thus considered a Hh-target and its expression has been widely used as a standard assay for Hh activity in the *Drosophila* embryo.

The *hh* gene was cloned independently and contemporaneously in four different laboratories, three of which exploited the isolation of enhancer traps in which the *E.coli lacZ* gene (encoding beta galactosidase) had randomly inserted into the locus (Lee, von Kessler, Parks, & Beachy, 1992; Tabata, Eaton, & Kornberg, 1992; Tashiro et al., 1993). These lines were identified through their pattern of β-galactosidase expression in larval imaginal discs, epithelial sheets of progenitor cells that give rise to the adult appendages such as the wings, antennae and legs. Strikingly, expression was restricted to the posterior lineage compartment of each disc, consistent with the earlier finding that clones of cells mutant for *hh* in the posterior but not the anterior compartment of the legs and antennae disrupt the patterning of these appendages (Mohler, 1988). Intriguingly, these posterior clones exerted an effect on non-mutant anterior compartment cells, suggesting that *hh* functions to mediate interactions between cells in each compartment (Mohler, 1988). The molecular cloning and sequence analysis of *hh* revealed that it encodes a protein containing a signal sequence, a characteristic of secreted proteins (Lee et al., 1992). Consistent with this, Hh was shown to be distributed in distinct puncta in cells extending beyond its

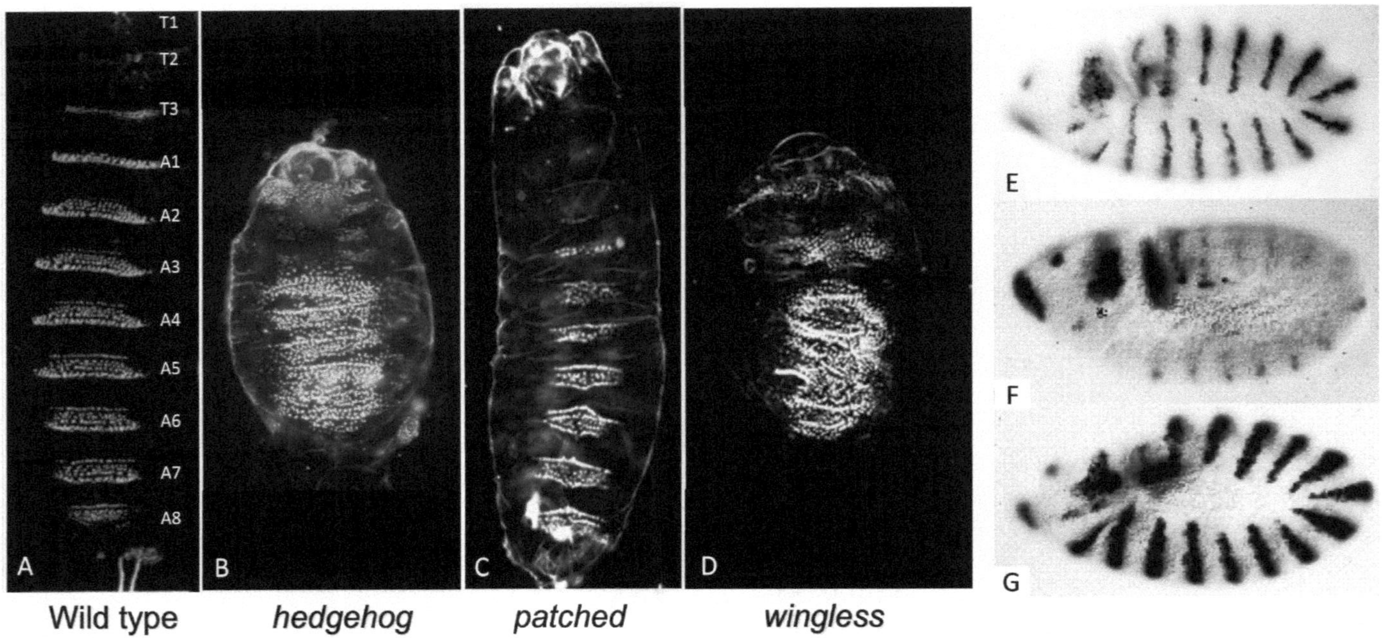

Fig. 1 (A–D) Dark-field photomicrographs of the ventral cuticles of wild-type and segment polarity mutant *Drosophila* larvae. (A) The ventral surface of the wild-type cuticle is decorated with 11 belts of denticles that mark the three thoracic (T1–3) and eight abdominal (A1–8) segments. In segment polarity mutants all the segmental units are present, but parts of each segment are deleted; both *hedgehog* (B) and *wingless* (D) mutants lack the posterior "naked" cuticle in each segment and instead have duplicated belts of denticles with reversed polarity, giving rise to a continuous "lawn" of denticles in the abdominal region. By contrast in *patched* mutants (C), the middle region of each segment is deleted and replaced by mirror image duplications of the anterior half of each denticle belt and adjacent naked cuticle: this results in a series of belts that are more rectangular than the trapezoidal arrangement typical of wild-type. (E) The generation of the cuticle pattern is presaged by the segmentally restricted pattern of *wingless* transcription in wild type embryos. This expression is absent in *hh* mutant embryos at the same developmental stage (F), whereas in *ptc* mutant embryos, each expression domain is expanded (G).

expression domain, both in the embryo and the imaginal discs (Tabata & Kornberg, 1994; Taylor, Nakano, Mohler, & Ingham, 1993).

Several other mutations with a phenotype similar to *hh* were isolated in the Heidelberg screens (Jürgens, Wieschaus, Nüsslein-Volhard, & Kluding, 1984; Nüsslein-Volhard, Wieschaus, & Kluding, 1984; Wieschaus, Nusslein-Volhard, & Jurgens, 1984); two of these turned out to be allelic to previously described mutations, *cubitus interruptus (ci)* (Tiniakow & Terentieva, 1933) and *fused (fu)* (Lynch, 1919), while *smoothened (smo)* was novel. Gene expression analyses revealed that *fu, ci* and *smo* are also required for normal *wg* transcription, implicating them with *hh* as components of a common gene regulatory network or pathway (Forbes, Nakano, Taylor, & Ingham, 1993; Hooper, 1994; Limbourg-Bouchon, Busson, & Lamour-Isnard, 1991; van den Heuvel & Ingham, 1996). Additional genes that generate a *hh*-like phenotype when mutated were identified in screens of the progeny of females carrying germ line clones of cells homozygous for mutagenized chromosomes, meaning that the contributions of maternal as well as zygotic activity of the gene are simultaneously eliminated. Genes identified in this way include *dispatched* (*disp*) (Burke et al., 1999) which encodes a transporter like protein, *skinny hedgehog (aka sightless, central missing and rasp)* (Amanai & Jiang, 2001; Chamoun et al., 2001; Lee & Treisman, 2001; Micchelli, The, Selva, Mogila, & Perrimon, 2002) the product of which is an acyl transferase and *tout velu* (*ttv*) (Bellaiche, The, & Perrimon, 1998) encoding a glycosyltransferase. Screens for mutations that disrupt wing patterning when homozygous in clones of imaginal disc cells led to the isolation of two additional genes with homology to *ttv*, accordingly named *brother of tout velu (botv)* and *sister of tout velu (sotv)* (Takei, Ozawa, Sato, Watanabe, & Tabata, 2004). A complete list of these and the other *Drosophila* Hh pathway genes described below can be found in Table 1.

Overexpression of *hh* in the wing imaginal disc results in the duplication of anterior compartment structures (Basler & Struhl, 1994; Capdevila, Estrada, Sanchez-Herrero, & Guerrero, 1994; Fietz et al., 1994; Ingham & Fietz, 1995; Kojima, Michiue, Orihara, & Saigo, 1994) (Fig. 2). These duplicate structures form in response to the activation of various Hh target genes such as *collier (col), engrailed (en)* and *patched (ptc).* In normal development, these genes are expressed in precise domains at varying distances from the A/P boundary in response to the graded distribution of Hh protein emanating from the posterior compartment. In this way, Hh specifies the pattern of the central region of the wing in a dose dependent manner, as well as driving proliferation and further patterning of the imaginal disc by

Table 1 Mutation in Hedgehog pathway genes isolated in Drosophila and their vertebrate counterparts.

Species	*Drosophila*	Isolation method (in *Drosophila*)	Mouse	Zebrafish
Process				
Signal	*hedgehog (hh)*	Forward genetics Zygotic	*Dhh* *Ihh* *Shh*	*dhh* *ihha, ihhb* *sonic you/shha, shhb*
Processing and release	*skinny hedgehog (ski)*	Forward genetics glc	*Hhat*	*hhat*
	dispatched (disp)	Forward genetics glc	*Disp1*	*chameleon/disp1*
Distribution and reception	*interference hedgehog (ihog)*	RNAi	*Cdon*	*cdon*
	brother of ihog (boi)	Sequence homology	*Boc*	*umleitung/boc*
	division abnormally delayed (dally)	Forward genetics glc		
	dally like protein (dlp)	Forward genetics glc		
	tout velu (ttv)	Forward genetics glc	*Ext1*	*ext1a, ext1b, ext1c*
	brother of tout velu (botv)	Forward genetics somatic clone	*ExtL3*	*boxer/extl3*
	sister of tout velu (sotv)	Forward genetics somatic clone	*Ext2*	*dackel/ext2*
	shifted	spontaneous	*WIF-1*	*WIF-1*
		N/A	*Gas1*	*gas1a, gas1b*
		N/A	*Scube2*	*you/scube2*

Reception and transduction	*patched (ptc)*	Spontaneous/forward genetics zygotic	*Ptch1* *Ptch2*	*leprechaun/ptch1 blowout/* *ptch2*
	smoothened (smo)	Forward genetics Zygotic	*Smo*	*slow muscle omitted/smu*
	costal-2 (cos-2)	Forward genetics Zygotic	*Kif7*	*kif7*
	fused (fu)	Spontaneous/forward genetics zygotic	*Stk36*	*stk36*
	protein kinase A catalytic subunit C1 (pka-C1)	Forward genetics somatic clone	*Pka-C*	*pka-c*
	glycogen synthase kinase 3β (gsk3)	Forward genetics somatic clone	*Gsk3β*	*gsk3β*
	casein kinase 1 (ck1)	RNAi	*CK1*	*ck1*
	g-protein coupled receptor kinase 1 (gprk1)	Sequence homology	*Grk2/3*	*grk2*
	Suppressor of fused (Sufu)	Forward genetics	*SuFu*	*dreumes/SuFu*
	supernumerary limbs (slimb)	Forward genetics somatic clone		
	cubitus interruptus (ci)	Spontaneous/forward genetics zygotic	*Gli1* *Gli2* *Gli3*	*detour/gli1* *you-too/gli2a gli2b, gli3*

glc = germ line clone. For zebrafish, homologous genes and corresponding mutant names are listed. For mouse, only gene names are shown.

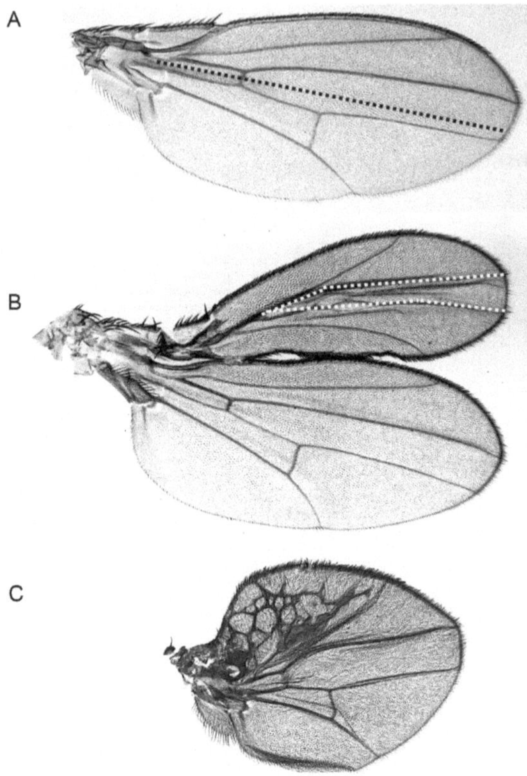

Fig. 2 . (A) Wing blade of a wild-type adult *Drosophila* showing the normal pattern of venation and the distinct rows of sensory bristles arranged along its anterior margin; the remainder of the wing margin lacks bristles but instead bears slender hairs. The boundary between the anterior and posterior compartments is indicated by a black dotted line. (B) Wing bearing a clone of *hh*-expressing cells (indicated by white dotted lines) in the anterior compartment of an otherwise genetically wild type wing. This has resulted in the wild type cells generating additional anterior compartment structures with mirror image symmetry either side of the *hh*-expressing cells. (C) wing from a ptc^{tuf}/ptc^{S2} trans heterozygous adult. The patterning of the anterior compartment is disrupted indicating a re-specification of cells to the identity of those that normally lie close to the A/P boundary at the center of the wing, where Hh activates the expression of the Col transcription factor and other targets.

promoting expression of the *decapentaplegic (dpp)* gene, which encodes another secreted protein belonging to the BMP family (Tanimoto, Itoh, ten Dijke, & Tabata, 2000).

Loss of function mutations of the *patched* (*ptc*) gene mimic the effects of *hh* overexpression both in the imaginal discs and embryo (Figs. 1 and 2), implicating it as negative regulator of the signaling system. The first mutant allele of *patched* (*ptc*) was isolated in *Drosophila* as long ago as the 1940s by Alfred Sturtevant, who gave it the name *tufted* (*tuf*) because of the ectopic tufts of sensory bristles that form on the thorax of mutant adults (Ingham, 2016; Sturtevant, 1948). These flies also showed a very slight defect in the proximal region of the wing, an effect so subtle that the involvement of the mutated gene in wing patterning went unremarked for nearly 40 years. The isolation of an embryonic lethal allele, originally designated tuf^{S2}, that in trans to *tuf* results in outgrowths and pattern duplications in the anterior half of the wing provided the first hint of a role for *ptc* in positional signaling in imaginal discs (Phillips, Roberts, Ingham, & Whittle, 1990). But prior to this, the isolation of null *ptc* alleles based on their embryonic lethal phenotype had led to its identification as a segment polarity gene (Nüsslein-Volhard et al., 1984; Nusslein-Volhard & Wieschaus, 1980). In the embryo, loss of *ptc* causes partial deletions and concomitant mirror image duplications in the cuticle of each larval segment quite distinct from those caused by *hh* and *wg* mutations (Fig. 1). Early investigations of this phenotype led to the discovery that *ptc* plays a critical role in restricting the spatial domain of *wg* expression by repressing its transcription (Martinez Arias, Baker, & Ingham, 1988) (Fig. 1), an effect that was shown to be countered by Hh: thus, in the absence of Ptc function, the expression of *wg* becomes independent of Hh (Ingham & Hidalgo, 1993; Ingham, Taylor, & Nakano, 1991). Molecular cloning of *ptc* and conceptual translation of its coding region predicted that *ptc* encodes a multipass transmembrane protein (Hooper & Scott, 1989; Nakano et al., 1989). These findings prompted the proposal that Ptc acts as the receptor for Hh, an interaction that, unusually, would lead to receptor inactivation (Ingham et al., 1991) (Fig. 3). Support for a physical interaction between the two proteins came from the finding that Hh is sequestered by Ptc in wing imaginal disc cells (Chen & Struhl, 1996); but a direct demonstration of such an interaction had to await biochemical experiments using the vertebrate orthologs of the *Drosophila* proteins described below (Marigo, Davey, Zuo, Cunningham, & Tabin, 1996; Stone et al., 1996).

The same type of epistasis analysis that placed Ptc downstream of Hh was also applied to other mutants with *hh*-like phenotypes to determine their position in the signaling pathway. In contrast to *ptc; hh* double mutants, which exhibit a *ptc* phenotype, embryos double mutant for *ptc* and *ci* display the *hh*-like phenotype typical of the *ci* single mutant, a finding that clearly places *ci* downstream of *ptc* (Forbes et al., 1993) (see Fig. 4). This is consistent

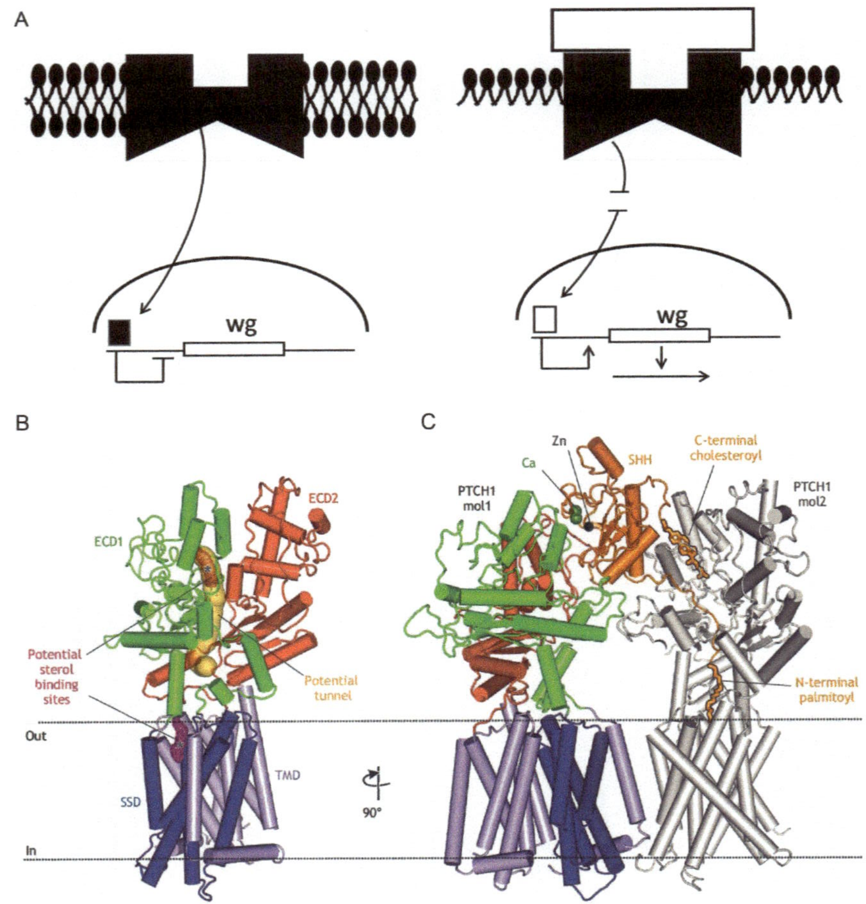

Fig. 3 See figure legend on opposite page.

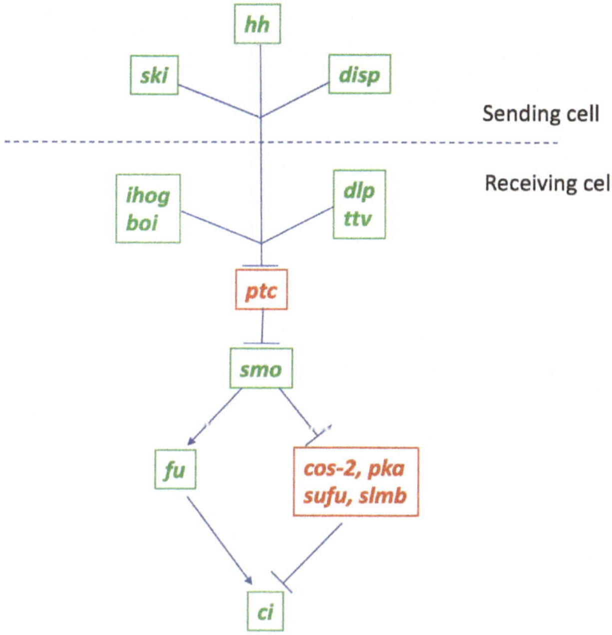

Fig. 4 The Hh signaling pathway as deduced from genetic epistasis and/or clonal analysis of mutations of the pathway components in *Drosophila*. Positive regulators of the pathway are shown in green, negative regulators in red. Arrows indicate activation, T-bars indicate repression. For details of the identification of mutant alleles of each gene, see Table 1.

Fig. 3 (A) Topology of Ptc based on hydrophobicity analysis of the conceptual translation product of the *ptc* coding region, as predicted by Nakano et al. (1989). (B) Molecular model of PTCH1 showing the sterol-sensing domain (SSD within the transmembrane domains (TMD) and the extracellular domains (ECD1 and ECD2). A possible hydrophobic tunnel (ocher surface) connects two putative sterol-binding sites (meshed surfaces, asterisks) in ECD1 and the SSD. (C) Structure of the asymmetric 1SHH:2PTCH complex reveals two distinct SHH-PTCH interfaces. The first PTCH1 molecule (mol1) binds to SHH at an interface that includes its calcium- and zinc-binding sites while the second PTCH1 molecule (mol2) engages the N-terminal palmitoyl and C-terminal cholesterol moieties of SHH that insert into the PTCH protein core. The interaction of SHH with mol1 drives PTCH endocytosis whereas the palmitate-centered interaction with mol2 inactivates the transporter function of PTCH. *Panel (A) Redrawn from Ingham, P. W., Taylor, A. M., & Nakano, Y. (1991). Role of the Drosophila patched gene in positional signalling.* Nature, 353, *184–187. Panel (B) Reproduced with permission from Kong, J. H., Siebold, C., & Rohatgi, R. (2019). Biochemical mechanisms of vertebrate hedgehog signaling.* Development, 146. *Panel (C) Reproduced with permission from Kong, J. H., Siebold, C., & Rohatgi, R. (2019). Biochemical mechanisms of vertebrate hedgehog signaling.* Development, 146.

with the finding that the *ci* gene encodes a zinc finger protein with high sequence similarity to a previously characterized mammalian transcription factor, the product of the human oncogene GLI1 (Orenic, Slusarski, Kroll, & Holmgren, 1990). Molecular genetic analysis provided strong evidence that Ci acts as the transcription factor mediating the cellular response to Hh signaling by binding to the cis-regulatory region of target genes: functional GLI binding motifs were initially found upstream of the promoter of *ptc,* the archetypal Hh-target gene in *Drosophila* (Alexandre, Jacinto, & Ingham, 1996).

The molecular cloning of *fu* had revealed that it encodes a serine threonine kinase (Preat et al., 1990), originally leading to the suggestion that it might phosphorylate and thereby potentiate the activity of the homeodomain protein encoded by another segment polarity gene, *engrailed* (Limbourg-Bouchon et al., 1991). Epistasis analysis with *ptc* mutations or ectopic *hh* expression, however, clearly established that Fu, like Ci, acts downstream of Hh to regulate target gene expression, suggesting that Fu might instead phosphorylate Ci to modulate its activity (Forbes et al., 1993; Ingham, 1993) (see Fig. 4). Molecular cloning of *smo* revealed that it belongs to the GPCR superfamily and is related to the Frizzled group of Wnt receptors (Alcedo, Ayzenzon, Von Ohlen, Noll, & Hooper, 1996; van den Heuvel & Ingham, 1996). But while this could suggest that Smo might act as a receptor Hh, genetic epistasis analysis showed that it acts downstream of Ptc to regulate Hh target gene expression (Hooper, 1994; Quirk et al., 1997; van den Heuvel & Ingham, 1996) (see Fig. 4).

Duplications of the anterior compartment of the wing similar to those induced by Hh overexpression, are also caused by mutation of the *costal-2* (*cos-2)* gene (Whittle, 1976). In line with this, loss of function of *cos-2* from both the zygotic and maternal genome results in an embryonic lethal phenotype similar—but not identical—to that of *ptc* (Forbes et al., 1993; Grau & Simpson, 1987). Notably, the *cos-2* mutation was originally discovered through its interaction with a second site mutation that enhances its phenotype: this mutation, originally named *costal-1* (Whittle, 1976) was later shown to be an allele of *pka-c1*, encoding a catalytic subunit of Protein Kinase A (PKA-C) (Collier, Suyama, Anderson, & Scott, 2004). Prior to this, mutant alleles of *pka-c1*, together with those of another gene named *supernumerary limbs* (*slmb*), had been isolated in somatic mosaic screens, in which embryonic lethal mutations can be identified through their effects in clones of imaginal disc cells, based on the same type of wing duplications caused by ectopic Hh (Jiang & Struhl, 1995, 1998; Lepage, Cohen, Diaz-Benjumea, & Parkhurst, 1995; Li, Ohlmeyer, Lane, & Kalderon, 1995; Pan & Rubin, 1995).

The phenotypes of clones of wing imaginal disc cells mutant for each of these four genes show that they are required specifically in anterior compartment cells to inhibit the expression of Hh target genes, since in each case, homozygosity for a loss of function allele results in changes in Hh target gene expression that underly the patterning defects in the wing. Thus, it could be concluded that these genes act downstream of Hh to suppress activity of its transduction pathway, rather than regulating the production or secretion of the Hh itself (see Fig. 4). Cloning of *cos-2* revealed that is encodes a member of the kinesin superfamily of microtubule motors (Sisson, Ho, Suyama, & Scott, 1997) while *slmb* encodes a subunit of the E3 ubiquitin ligase (Jiang & Struhl, 1998).

3. The developmental genetics of Hh signaling in vertebrates

The genes encoding vertebrate homologs of Hh were identified not in genetic screens, but through molecular biology (see Table 1). Complementary DNA (cDNA) clones with sequence similarity to the *Drosophila* gene were isolated using the techniques of reduced stringency DNA hybridization and degenerate polymerase chain reaction (PCR), revealing the existence of three distinct *hh* homologs, named *Desert, Indian* and *Sonic hedgehog* (Echelard et al., 1993; Krauss, Concordet, & Ingham, 1993; Riddle, Johnson, Laufer, & Tabin, 1993; Roelink et al., 1994). Initial analyses focused on *Sonic Hedgehog* (*Shh*) and highlighted three striking domains of its transcription in zebrafish, mouse and chicken embryos: along the antero-posterior axis in the notochord and the overlying floorplate of the ventral neural tube and in the posterior region of the developing fin/limb buds, corresponding to the Zone of Polarizing Activity (ZPA). Strikingly, all three of these structures were already known to act as embryonic organizing centers, making Shh an attractive candidate for the agent that mediates their activities. This hypothesis was initially supported by the effects of its over expression (Echelard et al., 1993; Krauss et al., 1993; Riddle et al., 1993) and confirmed through generation of a targeted null mutation of Shh in the mouse, which demonstrated that Shh is indeed required for the correct patterning of both the limbs and the central nervous system (Chiang et al., 1996). In homozygous mutant embryos, the proximal parts of each limb develop fairly normally; however, in the forelimb, the radius and ulna are fused and the autopod has a single digit, while in the hindlimb, the femur is relatively normal, but the tibia and fibula are severely reduced and only a single anterior-most digit is present (Chiang et al., 2001); these effects are consistent with disruption of the organizing function of the ZPA. In the neural tube, the floor plate and motor neurons are missing (Chiang et al., 1996), while the V0 and V1 interneurons are reduced in number and displaced towards the ventral midline (Pierani, Brenner-Morton, Chiang, & Jessell, 1999), consistent with an attenuation of the inducing activity of the notochord (Placzek, Jessell, & Dodd, 1993). In addition, loss of Shh activity from the prechordal plate has a dramatic effect on the development of the brain and optic vesicles resulting in cyclopia, effects characteristic of holoprosencephaly, a relatively common birth defect in human that was shown to be associated with mutations in SHH (Belloni et al., 1996).

Extensive genetic analyses as well as *in vivo* and *in vitro* manipulation of chick embryonic tissue, have since provided compelling evidence of the morphogen properties of Hh proteins. The best understood example of this is in the patterning of the neural tube, where the activation of region-specific target genes in response to the both the levels and duration of Shh signaling specifies the identity of different neuronal cell types along it dorso-ventral axis (Dessaud et al., 2010) (see Fig. 5).

Experimental manipulations in the chick embryo had also revealed a requirement for the notochord in signaling to the adjacent paraxial mesoderm to support skeletal muscle differentiation (Rong, Teillet, Ziller, & Le Douarin, 1992) and consistent with this, transcription of the myogenic factor, *Myf5*, was found to be downregulated in the medial paraxial mesoderm of *Shh* mutant mice (Chiang et al., 1996). In the zebrafish, an interaction between the notochord and paraxial mesoderm had been implicated in the specification of a distinct population of slow-twitch muscle fibers, known as muscle pioneers (MPs) (Halpern et al., 1995) that are distinguished by their expression of vertebrate orthologs of the *Drosophila* Engrailed protein (Hatta, Bremiller, Westerfield, & Kimmel, 1991). That Hh signaling might mediate their induction was supported by the finding that ectopic Hh expression is sufficient to cause an increase in number of MPs (Blagden, Currie, Ingham, & Hughes, 1997; Currie & Ingham, 1996; Hammerschmidt, Bitgood, & McMahon, 1996) and confirmed by the isolation in a systematic forward genetic screen of a number of mutations

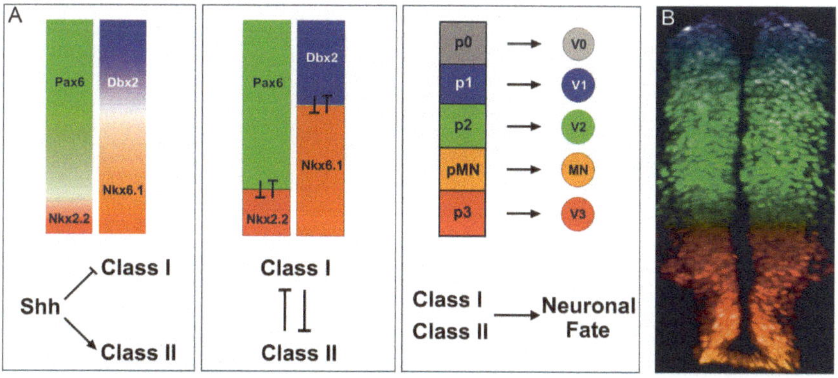

Fig. 5 Spatially restricted expression of transcription factors in neural progenitors in response to different levels of Shh activity. (A) Shh represses genes encoding class I proteins such as Pax6 and Dbx2 and activates genes encoding class II proteins such as Nkx2.2 and Nkx6.1. Discrete switches in gene expression are then generated by cross-repressive interactions between pairs of class I and II proteins. Five distinct domains of progenitors are defined by the combinatorial expression of class I and II proteins, each of which gives rise to one of the distinct neuronal subtypes characteristic of the ventral neural tube. (B) Spatially restricted Pax7, Pax6, Olig2 and Nkx6.1 expression domains exemplify the distinct progenitor domains in the neural tube. *Panels (A and B) Reproduced from Briscoe, J. (2009). Making a grade: Sonic Hedgehog signalling and the control of neural cell fate.* The EMBO Journal, 28, *457–465.*

affecting MP specification (van Eeden et al., 1996). Loss of MPs affects the organization of each somite such that they become U—rather than V-shaped (see Fig. 6). Accordingly, these mutations are known collectively as the "u-shaped somite" class and several were shown to disrupt zebrafish orthologs of the *Drosophila* Hedgehog pathway genes, namely *shh* (*syu*) itself, as well as *dispatched* (*con*) and *gli2* (*yot*), one of the vertebrate paralogs of *cubitus interruptus* (Karlstrom, Talbot, & Schier, 1999; Nakano et al., 2004; Schauerte et al., 1998) (see Table 1). Subsequent screens led to the isolation of *smoothened* mutants that display a similar u-shaped phenotype (Barresi, Stickney, & Devoto, 2000; Varga et al., 2001).

The MPs are a subset of the slow-twitch muscle fibers in the zebrafish, all of which characteristically express the homeodomain protein Prox1. The "u-shaped somite" mutants also show varying degrees of loss of the non-MP slow twitch fibers, known as superficial slow fibers (SSFs), the strongest phenotypes being a complete absence of both MPs and SSFs (Barresi et al., 2000; Lewis et al., 1999; Varga et al., 2001) (Fig. 6). These findings support a model in which different levels and/or duration of exposure to notochord derived HH signals specify distinct types of slow twitch fiber, the SSFs

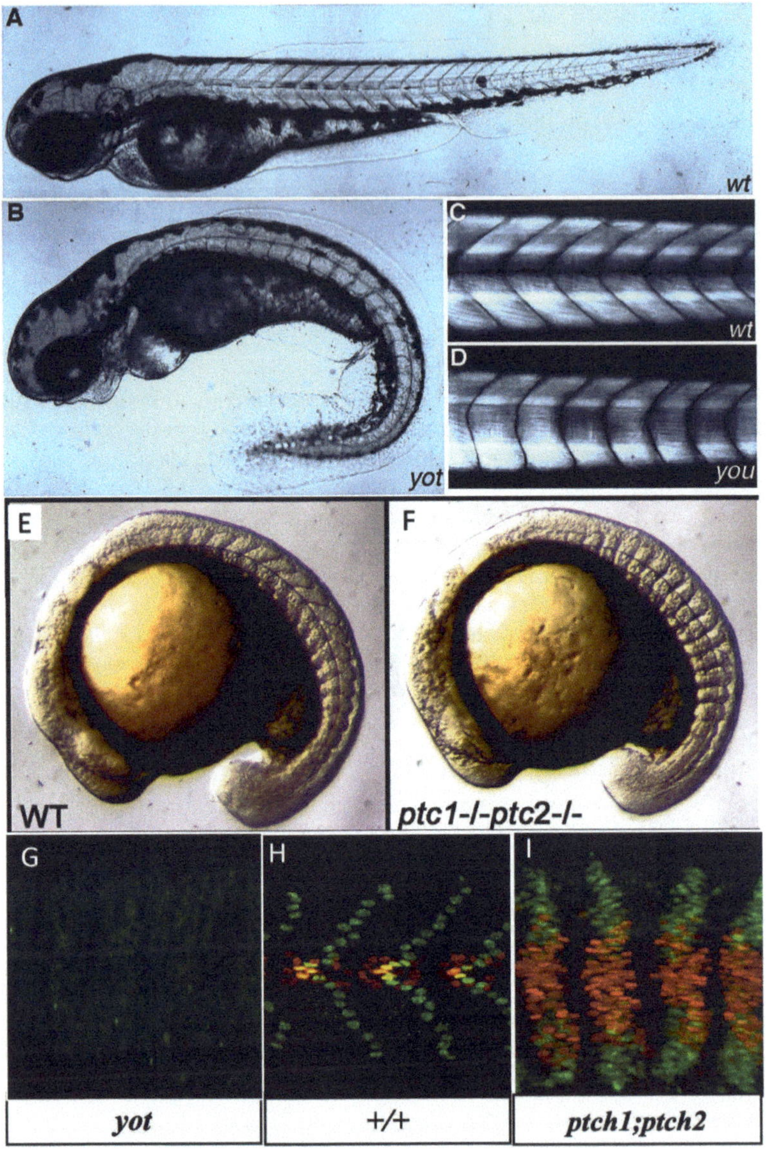

Fig. 6 Phenotypic effects of Hedgehog pathway mutants in the zebrafish. (A) Wild-type and (B) *yot* mutant zebrafish larvae at 72 hours post fertilization (hpf). The horizontal myoseptum, visible in the wild-type as a black line running through the middle of the myotome is absent from the *yot* mutant. The mutants also have a variably curved tail and their eyes lie closer together. (C and D) Higher magnification images of wild-type (C) and *you* mutant (D) larvae at 120 hpf showing the absence of the horizontal myoseptum and the resultant U-shaped myotome, in contrast to the V-shaped myotome typical of wild-type. (E) wild-type and (F) *ptch1;ptch2* double mutant zebrafish

requiring a lower level of activity than MPs (Wolff, Roy, & Ingham, 2003). In addition, intermediate levels of HH activity induce a distinct type of fast-twitch muscle fiber, known as the medial fast fibers (MFFs) (Wolff et al., 2003). The specification of different muscle fibers thus provides a sensitive assay for Hh activity and has been used extensively in the characterization of genes implicated in the signaling system.

Mutant alleles of both zebrafish paralogs of *ptc* (denoted *ptch* in all vertebrate species) were also isolated in forward genetic screens, though not through defects in the myotome: the *ptch1* mutant, named *leprechaun*, was identified through its effect on cell proliferation (Koudijs et al., 2005) while a *ptch2* mutant, known as *blowout*, was recovered through its defects in retinotectal axon projections as well as a persistent cleft at the back of the eye known as a coloboma (Karlstrom et al., 1996; Koudijs et al., 2008; Lee et al., 2008). Whereas the phenotypes of fish homozygous for either mutant alone are relatively subtle, double mutant animals display a strong Hh gain of function phenotype, which in the myotome manifests as a complete transformation of fast-twitch muscles to the slow-twitch fiber type, a large proportion of which differentiate as MPs (Fig. 6). This indicates that the two genes have overlapping and partially redundant functions in suppressing the Hh pathway, with Ptch2 making the major contribution.

Like *blowout*, the *umleitung* (*uml*) and *detour* (*dtr*) mutations were also isolated based on defects in retinal-tectal projections (Karlstrom et al., 1996). Analysis of neural tube patterning in embryos homozygous for these

---

embryos at 18hpf. The somites in the *ptch* mutant are rectangular rather than the wild-type chevron shape, reflecting the increased activity of the Hh pathway. (G–I) Hh dependent specification of skeletal muscle cell types in wild-type (+/+) and *hh* pathway mutant embryos. Prox1 protein (green) accumulates in the nuclei of superficial slow-twitch fibers (SSFs) and Muscle Pioneer (MP) fibers: Engrailed proteins (red) also accumulate in the MPs, hence they appear yellow due to the combined fluorescence emission. Medial fast-twitch fibers (MFFs) by contrast accumulate Engrailed but not Prox1 and hence appear red. In *yot* mutants (G) which constitutively express the repressor form of Gli2a, specification all of three of these muscle fiber types is blocked. While in the *ptch1;ptch2* double mutants (I) all fibers adopt the slow twitch identity (indicated by the significant increase in Prox1 +ve nuclei) and most also differentiate into MPs. *Panels (A–D) Reproduced from van Eeden, F. J., Granato, M., Schach, U., Brand, M., Furutani-Seiki, M., Haffter, P., et al. (1996). Mutations affecting somite formation and patterning in the zebrafish, Danio rerio.* Development, 123, *153–164. Panels (E and F) Reproduced from Koudijs, M. J., den Broeder, M. J., Groot, E., & van Eeden, F. J. (2008). Genetic analysis of the two zebrafish patched homologues identifies novel roles for the hedgehog signaling pathway.* BMC Developmental Biology, 8, *15. Panels (G–I) Unpublished images courtesy of Rosemary Kim and Freek van Eeden.*

mutations, however, reveals that in contrast to *blowout*, they diminish the response of cells to Shh, suggesting that both genes normally act as positive regulators of the pathway. Molecular cloning of the *dtr* locus showed that it encodes another Ci paralog, Gli1 (Karlstrom et al., 2003) while the *uml* gene encodes the Brother of Cdo (Boc) protein that acts a co-receptor for Hh ligands (Bergeron, Tyurina, Miller, Bagas, & Karlstrom, 2011) (see Table 1) as will be discussed.

Prior to the identification of *leprechaun* and *blowout* as mutant alleles of zebrafish *ptch1* and *ptch2* respectively, a null allele of mouse *Ptch1* had already been generated by homologous recombination in ES cells (Goodrich, Johnson, Milenkovic, McMahon, & Scott, 1996; Goodrich, Milenkovic, Higgins, & Scott, 1997). Strikingly, and in contrast to zebrafish, it is *Ptch1* that plays the major role in suppressing the Hh response in the mouse: the *Ptch1* null mutation is homozygous lethal and displays changes in the specification of cell fates consistent with a derepression of Shh signaling. This is clearly seen in the neural tube where expression of Foxa2 and Shh, normally restricted to the floorplate, and Nkx2.2 and Isl1/2, which define the ventrally located interneuron progenitors and motorneurons respectively, extends throughout the entire dorso-ventral axis (Goodrich et al., 1997; Motoyama et al., 2003). A similar, though less extreme neural tube phenotype is manifest by embryos homozygous for the ENU induced *Ptch1* mutant allele, *Wiggable* (Iulianella, Sakai, Kurosaka, & Trainor, 2018) while the spontaneously arising *Mesenchymal dysplasia* (*Mes*) allele causes polydactyly but has no effect on neural tube patterning (Makino, Masuya, Ishijima, Yada, & Shiroishi, 2001). By contrast, animals homozygous for a hypomorphic *Ptch2* mutation are adult viable but have skin defects (Nieuwenhuis et al., 2006).

Polydactyly is also caused by the spontaneously arising mouse mutation *extra-toes* (Johnson, 1967) which is associated with a partial deletion of the *Gli3* gene (Vortkamp, Franz, Gessler, & Grzeschik, 1992), one of three mammalian homologs of *Drosophila ci* (Table 1).

Systematic forward genetic screens carried out in the mouse by Kathryn Anderson's laboratory isolated mutant alleles of *Dispatched* (*Disp*) and *Smoothened* (*Smo*) (Caspary et al., 2002) as well as of the vertebrate ortholog of *cos-2,* known as *Kif7* (Liem Jr., He, Ocbina, & Anderson, 2009). In addition, these screens uncovered mutations with more complex phenotypes suggestive of disrupted Hh signaling. The first two of these to be described, *wimpole* (*wim*) and *flexo* (*fxo*) both display a loss of floorplate and ventral cell types from the neural tube, defects similar to the *Smo* mutant phenotype and symptomatic of a loss of Hh signaling; additionally, *fxo* mutants exhibit preaxial polydactyly, typically associated with a gain of Hh signaling (Huangfu et al., 2003). The *fxo* mutant was found to be allelic to a previously described mutant named *Polaris*, generated by insertional mutagenesis and identified based on its polycystic kidney phenotype (Moyer et al., 1994). This mutation had been shown to inactivate a gene homologous to the *Chlamydomonas ift88* gene, the product of which is essential for the assembly of cilia (Pazour et al., 2000). Remarkably, Anderson and her colleagues found that the *wim* mutation maps to the locus of another gene involved in ciliogenesis, encoding the IFT172 protein (Huangfu et al., 2003). Both IFT172 and IFT88 are components of the IFT-B complex that functions in the anterograde transport of cargo into the Primary Cilium (PC), and consistent with these functions, both *wim* and *Polaris/fxo* mutant embryos lack PC. Importantly, the construction and analysis of double mutants for *wim* and *Ptch1* or *Polaris* and *Smo* indicated that these IFT proteins are required for the proper response of cells to Hh signaling (Huangfu et al., 2003; Huangfu & Anderson, 2005). Consistent with this finding, other mutants affecting the integrity of the PC were subsequently also shown to disrupt Hh signaling. As well as various mouse mutations (Bangs & Anderson, 2017) a rare example of a developmental mutant in the chicken known as *Talpid3* based on its synpolydactyly phenotype, was found to lack PC (Yin et al., 2009). And in the zebrafish, the *iguana* (*igu*) mutant, which like the *ptch2* mutant *blowout* (*blw*), was originally isolated on the basis of retino-tectal projection defects (Karlstrom et al., 1996) was also found to have severely truncated PC (Kim, Richardson, van Eeden, & Ingham, 2010; Tay et al., 2010).

Thus, it was through these genetic analyses pioneered by Kathryn Anderson that the PC, which had previously been implicated in the sensing of extracellular stimuli but had remained poorly characterized, came to be identified as a key organelle in the reception and processing of Hh signals across the vertebrates. The significance of this discovery will be elaborated in what follows.

4. Multiple protein interactions mediate reception of the Hh signal

While the cellular response to Hh depends ultimately on the physical interaction between the protein and its principal receptor, Ptc/Ptch, several other membrane-associated proteins are also required for Hh reception.

Two of these, Cell adhesion molecule Downregulated by ONcogenes (Cdon) and the structurally related Brother of CDON (Boc), were first functionally characterized in vertebrates as cell adhesion molecules that promote muscle differentiation (Kang, Mulieri, Hu, Taliana, & Krauss, 2002; Kang, Mulieri, Miller, Sassoon, & Krauss, 1998). Although a loss of function mutation of *Cdon* had been found to cause a type of holoprosencephaly in mouse (Cole & Krauss, 2003) similar, though weaker than that caused by loss of Shh activity, it was the identification of the *Drosophila* orthologs of *Cdon* and *Boc*, named *interference hedgehog* (*ihog*) and *brother of interference hedgehog* (*boi*) respectively (Table 1), that first pointed to the involvement of these proteins in Hh signaling (Lum et al., 2003).

In contrast to the other *Drosophila* genes encoding Hh pathway components described above, *ihog* was originally identified in a cell-culture based RNAi screen for novel factors required for the cellular response to Hh, while *boi* was identified through its sequence similarity to *ihog* (Lum et al., 2003), RNAi-mediated knockdown experiments in cultured *Drosophila* cells demonstrated an additive effect of the two genes on the response to Hh, suggesting that they have partially redundant functions in mediating Hh pathway activation (Yao, Lum, & Beachy, 2006). Consistent with this, *Drosophila* mutant for one or other gene are homozygous viable and fertile, whereas double mutant animals that also lack maternal as well as zygotic expression of both genes, are embryonic lethal and manifest a phenotype identical to that of *hh* mutants, lacking the segmental expression of *wg* and displaying the typical "lawn" of denticles in the larval cuticle (Zheng, Mann, Sever, & Beachy, 2010). Similarly, clones of imaginal discs cells lacking both genes display a loss of *ptc* expression, consistent with a failure to respond to the Hh signal (Camp, Currie, Labbe, van Meyel, & Charron, 2010; Zheng et al., 2010), an effect that can be suppressed by simultaneous removal of Ptc function (Camp et al., 2014) indicating that Ihog and Boi act upstream of Ptc to mediate the Hh signal.

Like their vertebrate counterparts, *ihog* and *boi* encode Type1 transmembrane proteins belonging to the immunoglobulin superfamily and contain a variable number of Ig-like repeats as well as three sets of fibronectin type III (FnIII)-like repeats in their extracellular domains (ECDs). Biochemical analyses established that *Drosophila* Hh binds to Ihog/Boi via the first FnIII (Fn1) repeat (Yao et al., 2006; Zheng et al., 2010) while the Hh receptor, Ptc, interacts with them via the second FnIII (Fn2) repeat (Zheng et al., 2010). Reflecting the former interaction, cells in the wing imaginal disc lacking both *ihog* and *boi* fail to sequester Hh secreted by neighboring posterior compartment cells (Hsia et al., 2017) whereas interaction with Ptc is important for the cell surface localization of the Ptc receptor (Zheng et al., 2010) and for the endocytosis of Ptc once bound to Hh (Hsia et al., 2017). Notably both Ihog and Boi can also promote interactions between imaginal disc cells via homotypic interactions, a property that likely underlies the maintenance of the boundary between the anterior and posterior lineage compartments (Hsia et al., 2017).

The vertebrate Hh-family proteins similarly bind to Cdon and Boc, though notably they do so via interaction with the second and third rather than the first set of FnIII repeats (Okada et al., 2006; Tenzen et al., 2006; Yao et al., 2006). Consistent with this interaction, extensive genetic analyses have established analogous partially redundant roles for the two proteins in the response to Hh signals in vertebrates. Like its *Drosophila* ortholog, Boc is dispensable for embryonic development in mammals, since mice homozygous for a loss of function mutation are adult viable and exhibit none of the defects in neural tube patterning characteristic of loss of Hh signaling (Allen et al., 2011). By contrast, zebrafish homozygous for the *boc* mutant allele, *umleitung*, are embryonic lethal and show defects in the patterning of the ventral neural tube, indicated by a loss of *nkx2.2a* expression and a reduction in the number of cells expressing *foxa2* (Bergeron et al., 2011; Odenthal, van Eeden, Haffter, Ingham, & Nusslein-Volhard, 2000). A similar reduction in *Foxa2* expressing cells is seen in *Cdon* mutant mice (Allen et al., 2011) suggesting that, as with the Ptch proteins, the respective roles of Cdon and Boc may be reversed between mammals and fish; this could explain why forward genetic screens in zebrafish have not identified a *cdon* mutation, though confirmation must await the generation of such a mutant allele by targeted mutagenesis. Mice double mutant for *Cdon* and *Boc* display a much stronger phenotype, including the loss of floorplate gene expression as well as a marked reduction in *Olig2* and *Nkx2.2* expression in the ventral neural tube. As in *Drosophila*, the double mutant phenotype can be fully suppressed by inactivation of *Ptch1*, consistent with Boc and Cdon acting upstream of Ptch1 to facilitate reception of the signal (Camp et al., 2014).

Nevertheless, the *Cdon; Boc* neural tube phenotype is less extreme than that characteristic of complete loss of Hh signaling—and notably double mutant animals have the normal number of digits in the fore and hind-limb (Allen et al., 2011). This contrasts with the absolute dependence of Hh signaling on Ihog and Boi in *Drosophila*. Instead, complete abrogation of the response to Hh in mammals occurs upon simultaneous inactivation of a third gene, *Gas1*, encoding a GPI-linked protein first implicated in Hh signaling through the effects of its over expression in chick embryos (Lee, Buttitta, & Fan, 2001). Loss of *Gas1* alone is sufficient to disrupt neural tube patterning, causing incomplete specification of floor plate and vp3 interneuron progenitors (Allen, Tenzen, & McMahon, 2007), while *Gas1; Cdon* or *Gas1;Boc* double mutants display stronger phenotypes than the *Cdon;Boc* double mutants. Simultaneous elimination of all three genes results in a phenotype essentially identical to a complete loss of Hh activity: all Hh dependent ventral neural tube cell types are missing and the embryos display severe forebrain and cardiovascular defects, cyclopia, and holoprosencephaly (Allen et al., 2011).

5. Structure and biosynthesis of Hh proteins

Probably the best known and most unusual feature of Hh proteins is their covalent coupling to cholesterol. First characterized by Beachy and colleagues (Porter, Young, & Beachy, 1996), such a protein modification had not previously been described and remains exceptional, although a recent report has provided evidence for a similar modification of the Smo protein (Xiao et al., 2017). The primary Hh translation product possesses a signal sequence that targets it to the endoplasmic reticulum (ER), which is where the cholesterol modification occurs, as part of an autocleavage reaction promoted by the C-terminal half of the full-length protein. This contains a highly conserved domain that includes a region with sequence and structural homology to self-splicing inteins, accordingly referred to as the Hedgehog-intein or Hint fold, as well as a putative sterol recognition region (SSR) that is thought to recruit cholesterol (Hall et al., 1997). Cleavage proceeds via the formation of an internal thioester linkage between Gly-257 and Cys-258 which is then split by a nucleophilic attack using cholesterol as the electron donor, yielding the 19kDa N-terminal signaling fragment, HhN.

As well as the cholesterol moiety at its C-terminus, the HhN fragment is further lipidated through the addition of a palmitoyl adduct at its N-terminus. This modification is independent of the autocleavage and cholesterol coupling since it proceeds even when the former processing is blocked by a specific point mutation in Shh (Buglino & Resh, 2008). It was originally discovered through biochemical characterization of mammalian Shh expressed in tissue culture cells and shown to be essential for Shh activity (Pepinsky et al., 1998). Genetic analyses in *Drosophila* revealed that the palmitoylation is catalyzed by the so-called Hedgehog acyl-transferase Hhat, a member of the membrane-bound O-acyltransferase (MBOAT) family that also includes the Porcupine protein, required for palmitoylation of the Wg protein (Kadowaki, Wilder, Klingensmith, Zachary, & Perrimon, 1996). Hhat is the product of the gene identified by the *skinny hedgehog/sightless/central missing/rasp* mutations described earlier (see Table 1), all of which cause a complete loss of Hh activity, consistent with the biochemical analysis of Shh.

Whereas Hh proteins are located in the lumen of the ER, palmitoyl-CoA is synthesized in the cytosol, a separation that must be overcome for the modification of Hh proteins to occur. The Hhat protein contains membrane spanning domains and localizes to the endoplasmic reticulum membrane. Recently, cryogenic electron microscopy (CryoEM) studies have revealed that the cytosolic facing side of the protein contains two distinct cavities that can act as binding sites for the acyl chain and the CoA moiety respectively, while a V-shaped cleft accommodates HhN on its luminal side. In this way, the protein brings together its two substrates into a membrane embedded active site to promote the modification of HhN (Jiang, Benz, & Long, 2021).

6. Release and movement of Hh proteins from producing cells

One puzzling aspect of the lipid modifications is that they give Hh proteins a propensity to associate with the membranes of the cells in which they are expressed. This raises the question as to how Hh proteins can signal over several cell diameters if they remain trapped in the membranes of secreting cells. Studies in both *Drosophila* and vertebrates have suggested a variety of solutions to this conundrum. Some of these have presented evidence that removal of the palmitoyl adduct via the cleavage of the N-terminal lipidated peptide is a key step in the release of Hh from producing cells (Dierker, Dreier, Petersen, Bordych, & Grobe, 2009; Ohlig et al., 2011), while other studies have suggested that Hh forms multimeric complexes in which the lipid moieties are sequestered thereby reducing the hydrophobicity of the signal (Zeng et al., 2001). Notably, Suzanne Eaton and her colleagues presented evidence that both *Drosophila* Hh and vertebrate Shh are incorporated into lipoprotein particles, which facilitate their release and dispersal from secreting cells (Palm et al., 2013; Panakova, Sprong, Marois, Thiele, & Eaton, 2005). Studies in *Drosophila* identified lipophorin as the lipoprotein with which Hh associates, an association that can be mediated by either the cholesterol or palmitoyl moiety (Palm et al., 2013).

It is generally accepted that extraction of Hh from the plasma membrane requires the function of the *dispatched* (*disp*) gene product. This multipass transmembrane protein has significant sequence identity to Ptc and like the latter, belongs to the RND—Resistance Nodulation and Division—family of proteins found in many organisms across multiple kingdoms, including prokaryotes as well as eukaryotes (Nikaido, 2018). Indeed, RND proteins have been most extensively characterized in bacteria, where they act as efflux pumps, transporting a variety of molecules, including antibiotics, across the cell membrane. Analysis of the effects of loss of *disp* function in the wing of *Drosophila* revealed that its activity is required exclusively in Hh-expressing posterior (P) compartment cells but not in the anterior (A) compartment Hh-responding cells, pointing to a role in Hh production or secretion, the latter possibility being supported by the observed accumulation of Hh protein in mutant cells (Burke et al., 1999). Analysis of *disp* mutants in mouse and zebrafish showed that they similarly function in the release of Hh (Caspary et al., 2002; Ma et al., 2002; Nakano et al., 2004). Notably, Disp is required specifically for the release of the cholesterol modified form of HhN, since a C-terminally truncated form that is not modified (HhNu) is released from expressing cells in the absence of Disp (Burke et al., 1999; Nakano et al., 2004). In line with this, biochemical cross-linking experiments in cultured mammalian cells established that Disp interacts directly with the cholesterol moiety of Hh proteins (Tukachinsky, Kuzmickas, Jao, Liu, & Salic, 2012). CryoEM analysis of the *Drosophila* Disp protein (Cannac et al., 2020) has revealed multiple potential sterol binding sites within its Sterol Sensing Domain (SSD), a region embracing five transmembrane domains that is highly conserved between Disp, Ptc and other eukaryotic RND family proteins (Chang, Chang, Ohgami, & Yamauchi, 2006). In addition, the cryoEM structure indicates that the two large extracellular loops of Disp adopt an open conformation, forming a cleft that accommodates the Hh protein, which it contacts at several sites (Cannac et al., 2020).

While Disp is necessary for Hh release it is not sufficient, since Disp-expressing cells cultured in serum free medium retain cholesterol-coupled Shh (Tukachinsky et al., 2012). The key component of serum that facilitates the release of Shh was shown to be the Scube2 (Signal sequence CUbulin domain Epidermal growth factor-related) protein. This protein was first characterized in mice on the basis of its expression profile (Grimmond et al., 2000), but originally implicated in Hh signaling through studies in zebrafish, which showed that the *you* mutation, founding member of the u-shaped somite class described earlier (van Eeden et al., 1996) maps to and inactivates the *scube2* gene (Hollway et al., 2006; Kawakami et al., 2005; Woods & Talbot, 2005). Compared to other members of the u-shaped class, the *you* mutants display a relatively weak Hh loss of function phenotype, reflecting the duplication and partial functional redundancy of the *scube* genes (Johnson et al., 2012). Genetic mosaic analysis showed that Scube2 function can be supplied either by Hh sending or receiving cells, implying that it acts non-cell autonomously to facilitate signaling (Woods & Talbot, 2005). In line with this, medium conditioned by Scube2-expressing mammalian cells is sufficient to release cholesterol modified ShhN from cells expressing Shh (Creanga et al., 2012; Tukachinsky et al., 2012). More recent studies using purified recombinant protein have confirmed this role of Scube2 and shown that it interacts with the lipid moieties of ShhN to form a complex (Petrov, Wierbowski, Liu, & Salic, 2020). The release of ShhN mediated by Scube2 seems to be absolutely dependent upon Disp function, since it is blocked in cells lacking Disp. Moreover, the rate of Shh release was found to be potentiated by increasing levels of Disp, implying that Disp acts in a catalytic manner. Notably, reducing the Na+ concentration in the culture medium caused a dose dependent reduction in the Shh release rate, whereas increasing Na+ concentration accelerated its release. This implies that Disp acts as a Hh transporter driven by a Na+ gradient, a view supported by the fact that two amino acid residues critical for the H+ antiporter activity of bacterial RND proteins are highly conserved and required for Disp function (Petrov et al., 2020).

These findings provide an elegant solution to the conundrum of how cells can release lipidated ShhN and suggest a way in which the lipid moieties can be shielded to allow ShhN—in complex with Scube2—to pass through the aqueous extracellular environment. This complex has significantly reduced affinity for Ptch compared to the free lipidated form of ShhN (Wierbowski et al., 2020). It follows, therefore, that ShhN must be dissociated from Scube2 for optimal binding and pathway activation. Salic and colleagues have proposed a model of how this is accomplished, based on biochemical analyses of the interaction of the ShhN:Scube2 complex with cultured cells. According to this model the ShhN:Scube2 complex is first recruited to responding cells through interactions with

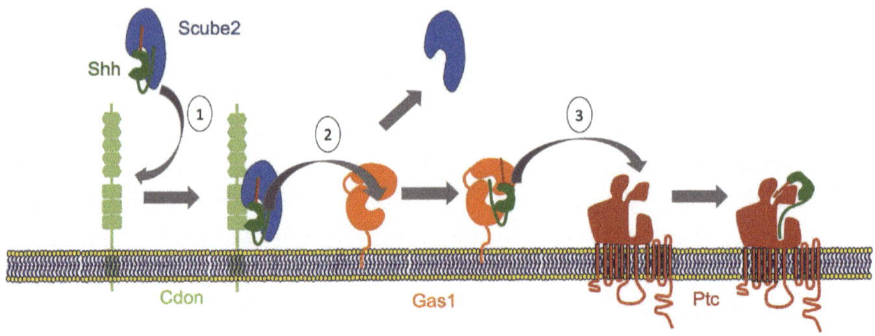

Fig. 7 A model for the transfer of Shh from the extracellular milieu to its receptor Ptch on target cells. The release of Shh from producing cells is promoted by its interaction with Scube2 (not shown); in the first step (1) the Scube2-Shh complex binds to the Cdon/Boc proteins on the cell surface of receiving cells; this facilitates the transfer of Shh form Scube2 to Gas1 (2) and finally (3) from Gas1 to Ptch. *Modified from Wierbowski, B. M., Petrov, K., Aravena, L., Gu, G., Xu, Y., & Salic, A. (2020). Hedgehog pathway activation requires coreceptor-catalyzed, lipid-dependent relay of the sonic Hedgehog ligand.* Developmental Cell, 55, *450–467 e458.*

the Cdon and Boc proteins. ShhN is then transferred from this quaternary complex to the Gas-1 protein which in turn presents and transfers ShhN to Ptch (see Fig. 7). Whereas ShhN binds all three of these co-receptors, Scube2 binds only to Cdon and Boc, which it does more strongly than Shh and via distinct binding sites. Thus, Scube2 drives the formation of a complex between Shh and its co-receptors on the surface of receiving cells, but this is not sufficient to activate the signaling response; rather it serves to recruit the ligand for presentation to its primary receptor, Ptch. Notably, when cells expressing Gas1 are incubated with the ShhN-Scube2 complex, they bind ShhN but not Scube2, indicating that ShhN can be transferred from Scube2 to Gas1, a transfer that is dependent upon presence of either the palmitoyl or cholesterol moieties, both of which bind more readily to Gas1 than to Scube2. In this view, ShhN is transferred from the Scube2-ShhN-Cdon-Boc complex to Gas1, which in turn transfers it to Ptch, driven by its higher affinity for cholesterol.

While this model provides an elegant explanation of how lipidated Hh ligands can be released from and received by vertebrate cells, it cannot fully account for the same processes in *Drosophila*, since this species lacks orthologs of both the Scube2 and Gas1 genes. Nevertheless, the requirements for Disp, Ihog/Cdon and Boi/Boc are shared with vertebrates by *Drosophila*, implying that a similar process may be deployed in flies.

Notably, studies of the *shifted* gene, first identified through a spontaneously arising mutation that narrows the spacing of veins in the middle of the wing that is patterned by Hh, have implicated it in modulating Hh signaling in *Drosophila* (Glise et al., 2005; Gorfinkiel, Sierra, Callejo, Ibanez, & Guerrero, 2005). Surprisingly, *shifted* encodes the *Drosophila* ortholog of WIF-1, a secreted protein originally characterized in vertebrates as an antagonist of WNT signaling that acts by binding to WNTs and sequestering them from their receptors (Hsieh et al., 1999). By contrast, loss of Shifted has no effect on the activity of the *Drosophila* WNT protein Wg, but was found to lead to a decrease in the levels and range of Hh protein in wing imaginal discs (Glise et al., 2005; Gorfinkiel et al., 2005). Initial studies interpreted these and other effects as evidence for a role of Shifted in stabilizing interactions between Hh and the heparan sulphate proteoglycans encoded by the *dally* and *dally-like protein* (*dlp*) genes. Originally identified through the effects of its mutation on cell division and patterning in the larval nervous system (Nakato, Futch, & Selleck, 1995), Dally was implicated in the response of cells to the Dpp and Wg signals, while Dlp was shown to be required for cells to respond to Hh (Desbordes & Sanson, 2003). Subsequent studies have implicated both proteins in the release and movement of Hh from producing cells as well as in its reception by responding cells (Callejo, Torroja, Quijada, & Guerrero, 2006; Han, Belenkaya, Wang, & Lin, 2004; Lum et al., 2003; McLellan et al., 2006; Yan et al., 2010). Further analyses of *shifted*, however, suggest that it potentiates the interaction of Hh with Ihog and Boi in a manner reminiscent of the interaction between Scube2 and Cdon and Boc (Avanesov & Blair, 2013). One possibility based on these findings is that Shifted/WIF-1 plays an analogous role to Scube2 in the reception of *Drosophila* Hh, a view supported by structural analyses suggesting it is a lipid binding protein (Liepinsh, Banyai, Patthy, & Otting, 2006; Malinauskas, Aricescu, Lu, Siebold, & Jones, 2011). In such a scenario, release of Hh from Shifted could be affected by Dlp, which is known to bind the lipid moiety in Wg (McGough et al., 2020). These parallels between Shifted/WIF1 and Scube2 could provide the basis for a common mode of Hh release and movement away from producing cells, whereby the hydrophobicity of the Hh signal imparted by its lipid moieties is overcome by the shielding effects of the Scube2 or Shifted/WIF1 proteins, allowing the signal thus chaperoned to move freely through the extracellular space.

An alternative scenario envisages Hh being transported to its target cells along actin based filipodia-like extensions. First described in sea urchin (Miller, Fraser, & McClay, 1995), such extension were subsequently discovered in *Drosophila* by Kornberg and colleagues, who named them cytonemes (Ramirez-Weber & Kornberg, 1999). Initial studies implicated cytonemes in the transmission of Wnt, FGF and BMP family signals (Kornberg & Roy, 2014), but several studies over the last decade have provided evidence that they also convey Hh signals. A key approach to visualizing Hh-associated cytonemes has been the overexpression of Ihog, which stabilizes those projecting from the basal surface of wing imaginal disc cells (Bischoff et al., 2013). Cytonemes extending from posterior compartment cells contain Hh protein as well as the Disp and Dlp proteins in puncta, which EM analysis suggests are exosomes (Gradilla et al., 2014), while those that extend from the anterior compartment cells contain the Hh receptor Ptc in distinct puncta as well as Hh itself (Chen, Huang, Hatori, & Kornberg, 2017; Gonzalez-Mendez, Seijo-Barandiaran, & Guerrero, 2017). Both types of cytoneme cross the compartment boundary, invading the adjacent territory and aligning with one another to form discrete contacts that have been described as synapses. Notably, both Ptc and Hh co-localize at these contacts, suggesting that they represent sites of signal reception (Gonzalez-Mendez et al., 2017; Gonzalez-Mendez et al., 2020).

The close juxtaposition of the two types of cytoneme has been proposed to be mediated by homophilic interactions between the Ihog molecules present in both. Such a role is consistent with the role of Ihog as a cell adhesion molecule (Wu et al., 2019) and can explain the stabilizing effect of Ihog overexpression, as well as the de-stabilizing effects of its knock-down (Yang et al., 2021). Homodimerization of Ihog requires its Fn1 domain and heparin binding domain and is promoted *in vitro* by heparin. So it is notable that overexpression of Ihog results in a significant accumulation of the HSPG Dlp, whereas clones of cells lacking Dlp and Dally—or the Ttv and Botv proteins that synthesize the glycosaminoglycans to which they are attached—are refractory to cytoneme extension (Bischoff et al., 2013; Gonzalez-Mendez et al., 2017). An attractive interpretation of these data is that these glypicans mediate Ihog homodimer formation *in vivo* (Yang et al., 2021), which would account for their requirement for Hh signal movement. The finding that the interaction between Ihog and Hh is stronger than its homophilic interaction suggests a sequence of events whereby Ihog homodimers first bring the sending and receiving cytonemes into close apposition: interaction with Hh then disrupts the homodimer, the resultant Hh-Ihog heterodimer facilitating transport of the captured ligand to the Ptc receptor (see Fig. 8). Whether or not

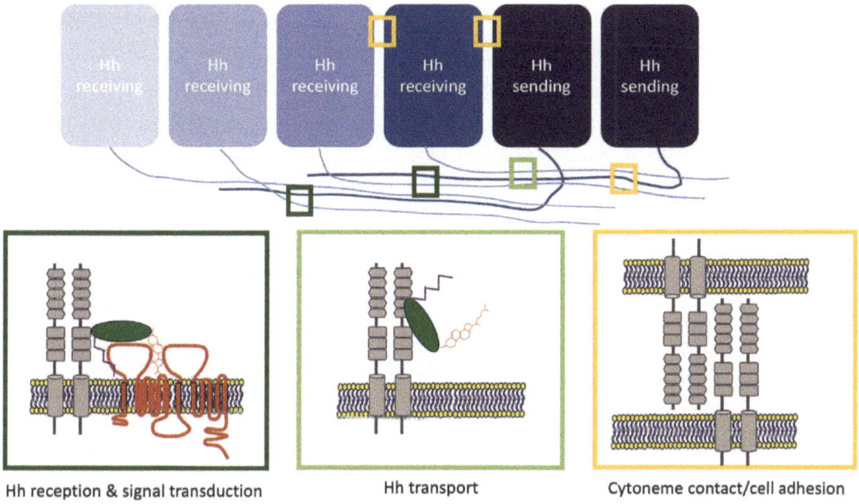

Fig. 8 A model of the competitive coordination of the dual roles of Ihog in homophilic adhesion and signal reception proposed by Yang et al. (2021). The model postulates that Ihog-Ihog trans interactions promote and stabilize cytoneme-cytoneme contacts; this in turn facilitates the "capture" by Ihog of Hh released from those cytonemes extending from Hh-expressing cells, a process driven by the higher affinity of Ihog for Hh than for itself. The Hh-Ihog complex is then transported along the cytoneme until it ultimately interacts with Ptc to activate the signaling pathway. The heparan sulfate necessary for the Ihog-Hh or trans Ihog-Ihog interactions may be supplied by Dally or Dlp, either as the membrane-associated forms of these glypicans or as the form released upon shedding (not shown). *Modified from Yang, S., Zhang, Y., Yang, C., Wu, X., El Oud, S.M., Chen, R., et al. (2021). Competitive coordination of the dual roles of the Hedgehog co-receptor in homophilic adhesion and signal reception.* eLife, 10.

Shifted/WIF1 would be required for this process is unclear, but consistent with such a role, Shifted/WIF1 is also found in cytonemes (Bilioni et al., 2013).

While these largely descriptive analyses provide a persuasive case for cytonemes mediating Hh distribution in *Drosophila*, functional evidence for their role rests largely on the disruption of their integrity through the manipulation of actin binding proteins such as Scar (Machesky & Insall, 1998) and diaphanous (Castrillon & Wasserman, 1994). These manipulations result in a reduction in cytoneme length, an effect that it accompanied by a decrease in Ptc and Hh puncta in anterior compartment cells (Bischoff et al., 2013; Chen et al., 2017). In line with this, Hh signaling activity is attenuated, as assayed by target gene expression in the anterior compartment and manifest in a narrowing of the region between the 3rd and 4th veins of the adult wing, similar to the *shifted* mutant phenotype.

Although most of the evidence implicating cytonemes in Hh signaling has come from studies in *Drosophila*, a few reports suggest that they may be similarly involved in vertebrates. Barna and colleagues first described the presence of long thin actin-based extensions emanating from mesenchymal cells in the chick limb bud (Sanders, Llagostera, & Barna, 2013). Like the *Drosophila* wing, the vertebrate limb is patterned by Hh signaling, Shh emanating from the ZPA at the posterior margin of the limb bud to control proliferation as well as antero-posterior positional identity of more anterior cells (Tickle & Towers, 2017). Cytoneme-like structures, up to 150 μM long, have been shown to extend from the ZPA and contain particles of GFP-tagged Shh, expressed from a transgene specifically in the ZPA, that move along them in both an anterograde and retrograde direction with a velocity similar to that of actin-based myosin motors. These cytonemes also contain Myosin-X, which moves towards and accumulates at their tips. Tagged forms of the Cdon and Boc proteins also localize to cytonemes, accumulating in distinct static microdomains along their length and causing them to stabilize. As in *Drosophila*, the cytonemes are not restricted to signal sending cells but also project from more anterior receiving cells. And notably, as in the imaginal disc, the receiving cytonemes interact directly with their sending counterparts extending from the ZPA (Sanders et al., 2013).

Cultured mammalian cells (NIH3T3 and mouse MEFS) have also been shown to extend cytoneme-like projections, the frequency of which can be increased by expression of several signaling proteins, including Shh (Hall et al., 2021). Overexpression of Cdon and Boc—both of which localize to these cytonemes, can also increase their occurrence. As in the chick limb, cytonemes projecting from Shh-expressing cells accumulate Shh in distinct puncta which EM analysis indicates are vesicles inside the cytonemes. These vesicles move along the cytoneme in an anterograde manner, a process that depends upon the activity of Myo10 which, as in the chick limb, accumulates at their tips (Hall et al., 2021).

7. Reception of HH at the membrane: the role of Patched and Smoothened

While the logic and simplicity of the Hh-Ptc model (Fig. 3) is seductive, it poses two important questions: how can a member of the RND superfamily of transporter proteins be involved in the reception of a secreted protein? And how can the activity of a signal be transduced into a cell by the *inactivation* of its receptor?

Early studies proposed that Ptc might form a complex with Smo, which as a member of the GPCR superfamily resembles a *bone fide* receptor, a view supported by data implying a direct physical interaction between the two proteins (Stone et al., 1996). Several subsequent findings, however, cast doubt on this model. The activity of Ptc regulates the stability, subcellular localization and posttranslational modification of Smo, which becomes phosphorylated and accumulates in the plasma membrane in response to Ptc inhibition by Hh (Denef, Neubuser, Perez, & Cohen, 2000). Notably, Ptc was found to inhibit the accumulation of Smo protein in the *Drosophila* embryo even when *smo* is overexpressed (Ingham et al., 2000) while experiments using cultured cells expressing varying levels of Ptc also indicated that Smo activity is regulated by Ptc in a non-stoichiometric manner (Taipale, Cooper, Maiti, & Beachy, 2002). This latter study also found that residues essential for the transport function of bacterial RND proteins, which are driven by proton (H^+) gradients, are conserved and required for Hh pathway inhibition by Ptc, suggesting that a similar transport activity might underlie the regulation of Smo by Ptc (Taipale et al., 2002). Ptc shares significant sequence similarity with the NPC1 protein, a eukaryotic member of the RND family originally identified through its mutation in patients suffering from Niemann Pick Syndrome (hence NPC), a disease of defective cholesterol transport (Lopez & Scott, 2013). The NPC1 protein acts as a cholesterol transporter, a function that depends critically upon its Sterol Sensing Domain (SSD), which it shares with Disp and Ptc as well as other proteins involved in cholesterol homeostasis (Michaux, Gansmuller, Hindelang, & Labouesse, 2000). This structural connection to cholesterol homeostasis/transport is reinforced by the finding that inhibitors of cholesterol biosynthesis or depletion of cholesterol can attenuate the response of cells to Hh activity (Cooper et al., 2003; Cooper, Porter, Young, & Beachy, 1998; Kinnebrew et al., 2019); moreover, experiments in cultured NIH 3T3 cells provided early evidence that Ptc can modulate cellular cholesterol levels (Bidet et al., 2011).

Support for a link between cholesterol transport and the control of Smo activity initially came with the discovery that a naturally occurring sterol alkaloid Cyclopamine (CycA) can bind to and inhibit Smo activity (Chen, Taipale, Cooper, & Beachy, 2002). Subsequently, oxysterols, derivatives of cholesterol that had been found to be sufficient to activate the Hh pathway in medulloblastoma cells (Corcoran & Scott, 2006) were shown to bind directly to Smo (Nachtergaele et al., 2012), an interaction that occurs via a shallow hydrophobic groove within the N-terminal extracellular cysteine rich domain (CRD) of the protein (Myers et al., 2013; Nachtergaele et al., 2013; Nedelcu, Liu, Xu, Jao, & Salic, 2013).

While these findings support a functional link between Smo regulation and a putative sterol transport activity of Ptc, CRISPR based screens for lipid related genes that modulate Hh signaling activity failed to uncover any of the enzymes involved in oxysterol synthesis (Kinnebrew et al., 2019). On the other hand, these screens did recover loss of function mutations in *Lss*, encoding lanosterol synthase that catalyzes an early step in cholesterol biosynthesis, as well as *Dhcr7* and *Dhcr24*, which encode the terminal pathway enzymes 7-dehydrocholesterol reductase and 24-dehydrocholesterol reductase respectively. Taken together, these mutations strongly implicated cholesterol itself as the Smo activator (Kinnebrew et al., 2019).

Indeed, further structural analyses of Smo had already pointed to cholesterol being the endogenous activating ligand. Crystals of an inactive form of Smo, containing the complete CRD linked to the transmembrane domain by the juxtamembrane linker domain but lacking the intracellular C-terminal tail, revealed a cholesterol molecule bound to the CRD (Byrne et al., 2016), the same site previously shown to bind oxysterol, an interaction confirmed by analysis of crystals of a full-length active form of Smo in complex with cholesterol (Huang et al., 2018). Notably point mutations in the CRD that disrupt the response of Smo to oxysterols remain responsive both to cholesterol and Shh in cell culture assays, implying that cholesterol rather than its derivative is the endogenous activator of Smo (Huang et al., 2016; Luchetti et al., 2016). Subsequent X-ray crystallographic analysis of Smo bound to a small molecule agonist known as SAG (Frank-Kamenetsky et al., 2002) have revealed a cholesterol molecule deep inside a hydrophobic pocket created by specific residues in the heptahelical transmembrane domain (Deshpande et al., 2019; Qi, Friedberg, De Bose-Boyd, Long, & Li, 2020). Notably, although deletion of the CRD causes constitutive activation of Smo, this can be suppressed by high levels of PTC, implying that binding of cholesterol to this second site within the membrane spanning domain may be required for the full activation of Smo (Byrne, Luchetti, Rohatgi, & Siebold, 2018).

Identifying cholesterol as the likely endogenous activator of Smo reinforced the notion that Patched proteins function as cholesterol transporters, a model strongly supported by several cryoEM analyses of the structure of Ptch, all of which have revealed its likely association with cholesterol, both in the presence or absence of its ligand, Shh. These cryoEM structures show that Ptch contains twelve transmembrane domains (TMDs) and two large extracellular domains, designated ECD1 and ECD2, containing 4 and 2 glycosylation sites respectively (Gong et al., 2018; Zhang et al., 2018), a topology and modification originally predicted by analysis of the conceptual amino acid sequence of the *Drosophila* protein (Nakano et al., 1989). The structure shows Ptch to be composed of two segments with two-fold pseudosymmetry around an axis perpendicular to the plasma membrane, each segment being composed of six TMDs, the first of which constitutes the SSD, and an ECD inserted between the first and second TMD of each repeat, an organisation typical of RND family proteins (Gong et al., 2018; Zhang et al., 2018).

The structures of monomeric Ptch have identified two sites likely to be occupied by cholesterol: the first, termed the Extracellular Steroid Biding Site (ESDS) is located in a pocket enclosed by ECD1 and the second in a membrane facing cavity adjacent to the SSD. Together, the ECDs encompass a hydrophobic tunnel that connects the ESBS and the SSD, which potentially could act as diffusion path for cholesterol when Ptch is in its active unbound state (Kowatsch, Woolley, Kinnebrew, Rohatgi, & Siebold, 2019).

Modeling based on these structures suggest that Ptch cycles between distinct conformations in which gates open and close to facilitate transfer of sterol out of plasma membrane (Kowatsch et al., 2019). And analysis of structures of the protein in complex with Shh have suggested how ligand binding could lead to its inactivation. These show that ShhN binds to two distinct sites in Ptch1: the first is composed of an extended loop in ECD1 which engages with the pseudo active site groove of the Shh globular domain, leading to the ECD1 and ECD2 moving towards each other. The second site consists of the tunnel above the SSD enclosed by the two ECDs into which the N-terminal peptide along with its palmitoyl moiety are inserted, while the C-terminal cholesterol moiety inserting into ECD1; this results in the occlusion of the tunnel which could in turn inhibit its postulated role in cholesterol transport (see Fig. 3).

While these structural studies provide compelling support for cholesterol mediating the regulation of Smo by Ptch, they raise further questions about how this process operates. Specifically, given that cholesterol is highly abundant in the plasma membrane, how can Ptch modulate its levels with the dynamic range necessary to control Smo activity? An important consideration germane to this conundrum is that cholesterol is partitioned into so-called "accessible" and "inaccessible" pools; while the latter represents the major fraction in the plasma membrane, the former has significantly higher chemical activity, enhancing its ability to interact with proteins (Das, Brown, Anderson, Goldstein, & Radhakrishnan, 2014; Lange, Ye, & Steck, 2004; Radhakrishnan, Anderson, & McConnell, 2000). Cholesterol becomes inaccessible when in complex with phospholipids, especially sphingomyelin; and notably, in the same CRISPR based screens for modulators of Smo activity that selected cholesterol biosynthetic gene as positive regulators, genes encoding enzymes involved in sphingolipid biosynthesis were identified as negative regulators (Kinnebrew et al., 2019). This finding implies that increasing the levels of accessible cholesterol by depleting sphingomyelin is sufficient to activate Smo. Consistent with this, recent studies have shown that Ptch activity can deplete accessible cholesterol from the outer leaflet of the plasma membrane in HEK293 cells by transporting it to the inner leaflet, a process that is inhibited by Shh and driven by K^+ export (Kinnebrew et al., 2021).

While these data provide a potential solution to how Ptc can achieve the necessary dynamic range of cholesterol modulation, they do not address its selectivity for Smo regulation: in other words, how could Ptch modulate plasma membrane accessible cholesterol levels without impacting on cellular cholesterol homeostasis? An appealing solution to this dilemma is based on the key role of the Primary Cilium (PC) in Hh signaling, which, as described earlier, was first revealed by the discovery that developmental processes controlled by Hh are disrupted by mutations in genes encoding IFT and other ciliary proteins. Consistent with the proposal that the PC serves as a critical hub for Hh signaling, both Smo and Ptch localize to the PC (Corbit et al., 2005; Rohatgi, Milenkovic, & Scott, 2007) a localization that is modulated by Hh itself. Thus, in the absence of ligand, Ptch accumulates in puncta along the PC as well as around its base, a localisation that depends upon sequences in its C-terminal domain (Kim et al., 2015), whereas Smo is excluded; on exposure of cells to Shh, however, Ptch moves out of the PC and Smo moves in via lateral transport from the plasma membrane to accumulate along the length of the cilia membrane (Milenkovic, Scott, & Rohatgi, 2009; Rohatgi et al., 2007). Localization of active Smo to the PC is critical to its function since, as will be discussed in the following section, it is here that is engages with the downstream components of the signal transduction pathway. The close correlation of the active states of

Ptch and Smo with their ciliary localisation implies that the PC may function as a subcellular compartment within which the levels of accessible cholesterol can be modulated by Ptch without impacting on global cellular cholesterol homeostasis. Notably, the ciliary membrane contains lower levels of accessible cholesterol than the plasma membrane, but these levels can be increased by Shh mediated inactivation of Ptch1, suggesting that the reduced levels in the PC are maintained by Ptch1 (Kinnebrew et al., 2019). In this way, Ptch1 would prevent Smo activation within the PC, while its inhibition would promote accumulation of active Smo. An interesting corollary of this model is that compartmentalization of Ptc function should be unnecessary in *Drosophila* since they have no surveillance mechanism for plasma membrane cholesterol. This is likely due to their auxotrophy for cholesterol, which represents only a minor fraction of plasma membrane sterols, the major species being ergosterol. This provides a plausible explanation of why a primary cilium is not required for Hh signaling in *Drosophila* (Kinnebrew et al., 2019).

8. The intracellular response to Hh

In most contexts, cells respond to the activation of Smo by initiating or upregulating the transcription of so-called Hh-target genes, a process that genetic epistasis analysis shows requires the activity of the Fu S/T kinase and the Ci transcription factor (Forbes et al., 1993). Despite the appealing simplicity of this pathway, elucidating the role of Fu in the regulation of Ci activity has proven far from straightforward and it is only recently that Ci has been shown to be a direct target of Fu kinase activity (Han et al., 2019). In the interim, the regulation of Ci/GLI activity by other kinases, became a major focus of investigation. A key insight was provided by the discovery that the *Drosophila* Ci protein exists in two distinct forms—the 155 kDa full-length (FL) version, containing transcriptional activation and repressor domains as well as the Zinc Finger (ZF) DNA binding domain (Alexandre et al., 1996) and a C-terminally truncated 75 kDa form that lacks the activation domain but retains the repressor and ZF DNA binding domains (Aza-Blanc, Ramirez-Weber, Laget, Schwartz, & Kornberg, 1997) Crucially, the relative abundance of these two species is modulated in response to Hh activity suggesting a potential basis for differential target gene activation in response to graded Hh activity, through varying ratios of the full length and repressor forms.

The C-terminally truncated repressor form, Ci-R, is generated through partial proteolysis of Ci-FL a process mediated by Protein Kinase A (PKA). As described earlier, PKA was first identified as a negative regulator of the pathway through the isolation of mutant alleles of the gene encoding its catalytic subunit; and clones of cells homozygous for such mutant alleles in the anterior compartment of the wing imaginal disc were shown to accumulate high levels of the Ci protein (Johnson, Grenier, & Scott, 1995), suggesting it to be a potential target of PKA activity. Compelling evidence that Ci is indeed a direct target of PKA was provided by the identification of a number of consensus PKA phosphorylation sites in its C-terminal trans-activation domain and the demonstration that their mutation abrogates the proteolytic cleavage of CiFL (Chen, Gallaher, Goodman, & Smolik, 1998). Subsequent analyses revealed that phosphorylation of these sites by PKA primes them for further phosphorylation by two other kinases, Glycogen synthase kinase 3 (GSK3) and casein kinase 1 (CK1) (Price & Kalderon, 2002). This sequential phosphorylation creates an atypical extended docking site for the Slimb protein, a substrate recognition component of an E3 SCF ubiquitin ligase complex, leading to the ubiquitination of Ci-FL and its consequent targeting to the proteasome, where it undergoes partial degradation to remove the C-terminal activation domain (Jia et al., 2005; Smelkinson & Kalderon, 2006; Smelkinson, Zhou, & Kalderon, 2007).

In contrast to *Drosophila*, vertebrates possess multiple genes encoding Gli proteins, all of which are involved in the transcriptional response to Hh ligands. Analysis of mutant alleles of all three mammalian *Gli* genes (Gli1, 2 and 3) in the mouse revealed that the activator and repressor functions are distributed between them (Park et al., 2000): Gli3 acts principally to repress Hh target gene expression, while Gli1 and Gli2 respectively function exclusively or primarily as transcriptional activators. Thus, for instance, Gli3 mutants are characterized by polydactyly and defects in the neural tube and brain (Motoyama et al., 2003; Persson et al., 2002; Tole, Ragsdale, & Grove, 2000; Wijgerde, McMahon, Rule, & McMahon, 2002), consistent with a de-repression of Shh target genes, whereas Gli2 mutants lack the floorplate, the ventral-most region of the neural tube that is induced by Shh from the underlying notochord (Matise, Epstein, Park, Platt, & Joyner, 1998). Surprisingly, Gli1 homozygous mutants are fully viable and phenotypically normal, but show a strong loss of Hh phenotype if simultaneously heterozygous for a Gli2 null allele.

In zebrafish, the situation is complicated by the presence of two *gli2* paralogs, mutant alleles of only one of which, *gli2a*, have been isolated to date. The first *gli2a* mutant alleles were isolated based on their embryonic lethality and u-shaped somite phenotype, which led to their being named *you-too* (*yot*). Both alleles were found to harbour premature stop codons C-terminal to the Zinc finger domain, such that they encode truncated forms of the protein lacking the activation domain but retaining the repressor domain, consistent with their strong loss of Hh signaling phenotype (Karlstrom et al., 1999). Notably, however, fish homozygous for a complete loss of function *gli2a* allele are viable and have no discernible mutant phenotype (Wang et al., 2013). One possible explanation for this striking difference from the mouse Gli2 mutant is the presence of the *gli2b* paralog. However, no mutant allele of the *gli2b* gene has been generated, so the potential functional redundancy remains unproven. A further divergence between mammals and fish lies in the differing requirements for Gli1 activity. Whereas the mouse *Gli1* mutant is viable and has no apparent defects, zebrafish homozygous for the *detour* (*dtr*) mutation, which inactivates *gli1,* are homozygous lethal and display defects in retinotectal axon pathfinding, a phenotype they share with the dominant negative *yot* mutant alleles of *gli2a*. Strikingly, while the *dtr* mutant phenotype is restricted to the nervous system, in combination with the *gli2a* null mutation, it results in a strong u-shaped somite phenotype (Wang et al., 2013), indicating a functional redundancy between *gli2a* and *gli1*.

Whereas Gli1 lacks a repressor domain, the Gli2 and Gli3 proteins are, like their *Drosophila* counterpart Ci, comprised of both a C-terminal activator domain and an N-terminal repressor domain; and as in *Drosophila*, both proteins can undergo partial proteolytic cleavage. In mammals, the Gli3 protein is cleaved much more efficiently than Gli2, reflecting its principal function as a repressor, whereas in zebrafish, the Gli2a protein is efficiently processed into a repressor form *in vivo*, a difference that may be attributable to the divergence in the Processing Determinant Domain of Gli2 (Pan & Wang, 2007) between mammals and zebrafish (Ben, Elworthy, Ng, van Eeden, & Ingham, 2011). As in *Drosophila*, cleavage of the vertebrate Gli proteins is promoted by their phosphorylation by PKA, GSK3β and CK1 (Pan, Bai, Joyner, & Wang, 2006; Pan, Wang, & Wang, 2009). This leads to their docking with the vertebrate ortholog of Slimb, Beta TRCP, which catalyses their ubiquitination and consequent targeting to the proteasome.

Modulating the phosphorylation of Ci/GLI proteins by PKA is thus central to the cellular response to Hh signaling both in flies and vertebrates; how this is accomplished has been a major focus of Hh research over the past 25 years. In *Drosophila*, the Cos-2 and Fu proteins play a key role in mediating Ci processing. Cos-2 is a member of the superfamily of kinesin motor (KIF) proteins and like other family members, binds to microtubules (Robbins et al., 1997; Sisson et al., 1997). In the absence of Smo activation, Cos-2 nucleates the assembly of the so-called Hedgehog Signaling Complex (HSC), acting as a scaffold for the recruitment of both Ci and Fu as well as the three kinases, PKA, GSK3β and CK1 that prime Ci for ubiquitination (Zhang et al., 2005). Accordingly, when Cos-2 is inactivated through mutation, Ci-FL fails to be targeted to the proteasome and instead accumulates in cells independently of Hh activity, leading to the ectopic activation of Hh target genes that underlies the *cos-2* mutant phenotype. Somewhat surprisingly, given that its mutant phenotype is opposite to that of *cos-2*, loss of Fu also leads to the accumulation of Ci-FL. This reflects a dual role for Fu in Ci regulation, promoting both its cleavage as well as the activation of its full-length form (Zadorozny, Little, & Kalderon, 2015). While the latter function depends upon its kinase activity, the involvement of Fu in processing Ci-FL reflects its structural role as a bridge between Cos-2, which it binds via its C-terminal domain, and PKA, which it binds via its N-terminus (Ranieri, Therond, & Ruel, 2014).

A key step in the abrogation of Ci phosphorylation and hence accumulation of the full-length form of the protein, is the interaction of the HSC with the intracellular C-terminal tail (CTT) of Smo, an interaction promoted by the stabilization of Smo and conformational changes induced when Ptc is inactivated by the Hh signal. Recruitment of the HSC by Smo brings its CTT into close proximity with PKA, resulting in a switch in substrate from Ci to Smo (Li, Ma, Wang, & Jiang, 2014; Ranieri et al., 2014). This switch has two consequences: first, it attenuates the phosphorylation of Ci and hence its cleavage; and second it promotes the phosphorylation of S/T residues within the so-called Smo Auto Inhibitory Domain (SAID) of the CTT, thereby disrupting the electrostatic interactions that maintain the CTT in a closed conformation. Several lines of evidence indicate this unfolding of the CTT leads to its dimerization, which in turn promotes the dimerization and consequent autoactivation of Fu (Shi, Li, Jia, & Jiang, 2011). The activation of Fu is critical because blocking the processing of Ci-FL into Ci-R is not in itself sufficient for pathway activation; in addition, Ci-FL must enter the nucleus, a process that is promoted by Fu kinase activity. Nuclear import of Ci-FL is inhibited by its physical interaction with Cos2 as well as with another protein, encoded by *Suppressor of fused* (*Sufu*) which, as its name implies, was identified through mutations that suppress the *fu* phenotype (Preat, 1992). It is in this context that Fu plays a positive regulatory role, its serine-threonine kinase activity being required to overcome the sequestration of Ci-FL by Cos-2 and SuFu. Although release of Ci is associate with the Fu-dependent phosphorylation of both SuFu and Cos-2, in neither case is this phosphorylation essential for Ci-FL activation (Zadorozny et al., 2015; Zhou & Kalderon, 2011). Instead, mass spectrometry analysis has recently revealed serine residues in Ci-FL that are phosphorylated by Fu, a modification that promotes the dissociation of Ci-FL from SuFu (Han et al., 2019). Dissociation from SuFu in turn potentiates binding of Ci-FL to the nuclear transport receptor Transportin-1, as well as to the transcriptional co-activator CBP, thereby facilitating Ci-FL entry into the nucleus and its activation of target gene transcription (Han et al., 2019). This dependence of Ci-FL on Fu for its activation explains why accumulation of Ci-FL in the absence of Fu does not result in Hh target gene activation, as it does in the absence of Cos-2.

As well as phosphorylating Ci, Fu also phosphorylates a cluster of sites within the Smo CTT, a positive feedback loop that further potentiates Smo activity (Sanial et al., 2017). Several other kinases have also been shown to potentiate Smo activity through phosphorylation of its CTT. Like PKA, Casein kinase1α/ε (CK1α/ε) phosphorylates three clusters of S/T residues within the SAID. These modifications prime the CTT for further phosphorylation by the G-protein coupled receptor kinase 2 (Gprk2), originally implicated in Hh signaling through studies of its vertebrate ortholog, Grk2 (Chen et al., 2004). In addition, a genetic modifier screen in *Drosophila* identified CK1α/ε as a positive regulator of Hh signaling, a role reflecting its physical interaction with and phosphorylation of Smo.

While orthologs of all three of the dedicated HSC components first identified in *Drosophila*—Fu, Cos-2 and SuFu—are present in vertebrates, there is no clear evidence that they form a complex functionally equivalent to the *Drosophila* HSC. Instead, a key process in the modulation Gli protein activity is their localization to the PC (Haycraft et al., 2005), in which as described earlier, activated Smo accumulates in response to Ptc inactivation by Hh signals. This localisation is dependent upon specific sequence motifs (Santos & Reiter, 2014) and is mediated by a non-canonical nuclear localization signal and the nuclear import protein Karyopherin-β2 (Han et al., 2017). Significantly, PKA-C is also enriched in the PC suggesting that the targeting of Glis to this organelle may facilitate their phosphorylation and subsequent ubiquitination and cleavage (Mick et al., 2015). In line with this view, targeted expression of the heat stable Protein Kinase Inhibitor (PKI) protein in the PC was found to reduce the levels of Gli3R relative to the Gli3FL form (Mick et al., 2015); and targeted expression of the dominant negative form of PKA dnPKA in the PC was shown to be sufficient to activate the pathway in myoblasts in the zebrafish embryo (Truong et al., 2021). Recent studies have now provided compelling evidence that activated Smo functions as an endogenous inhibitor of PKA-C within the PC, thereby abrogating the processing of Glis and promoting their activation. Experiments in HEK293 cells revealed that Smo can inactivate PKA-C by directly binding to it via its intracellular CTT. Remarkably, the inactivation is mediated by a motif that is highly conserved across species—including *Drosophila*—and has sequence similarities with the pseudosubstrate region of PKI (Happ et al., 2021). While vertebrate Smo lacks the PKA phosphorylation sites found in its *Drosophila* counterpart, it does contain multiple GRK2 consensus sites (Chen et al., 2004); and interaction between PKA-C and the Smo CTT is dependent upon phosphorylation of the Serines and Threonines in these sites by GRK2 (Arveseth et al., 2021). Previous studies of the interaction between GRK2 and other GPCRs have shown that it binds to activated receptors in part through an interaction with the third intracellular loop which undergoes an outward movement on ligand binding (Komolov & Benovic, 2018).

Taken together, these findings suggest a model in which the conformational changes induced by Smo activation lead to the recruitment and activation of GRK2; GRK2 then phosphorylates the CTT, promoting the accumulation of Smo in the PC as well as its recruitment of PKA-C, which it inhibits via its PKI pseudosubstrate motif. In line with this model, loss of *grk2* function in the zebrafish results in a phenotype indistinguishable from that of *smo* mutants, consistent with Grk2 being absolutely required for transduction of the Hh signal (Zhao et al., 2016), in contrast to its role in *Drosophila*. Nevertheless, there are clear parallels between the proposed mechanisms of PKA inhibition in *Drosophila* and vertebrates, the Smo CTT sequestering the PKA-C subunit in both cases; but whereas in *Drosophila* the interaction is mediated by Cos2, in vertebrates the interaction is direct.

This divergence in mechanism aligns with the finding that, Kif 7, the vertebrate ortholog of Cos2, does not interact physically with Smo in cultured cells, a finding that led initially to the erroneous conclusion that its role in Hh signaling is not conserved between flies and vertebrates (Varjosalo, Li, & Taipale, 2006), despite evidence for a conserved role being adduced in the zebrafish on the basis of the effects of transient knockdown of *kif7* (Tay, Ingham, & Roy, 2005). The proposal was refuted by the isolation and phenotypic characterization of the ENU induced Kif7 mutation, *maki* (Liem Jr. et al., 2009), as well of Kif 7 knock out alleles generated contemporaneously by homologous recombination (Cheung et al., 2009; Endoh-Yamagami et al., 2009). Mice homozygous for each of these display phenotypes indicative of a gain of Hh activity, and consistent with this, the level of the repressor form of Gli3 is reduced relative to that of the FL form. While these findings are compatible with Kif7 playing a role in regulating the PKA dependent phosphorylation of Glis, the effects of loss of Kif7 are notably less severe than might be predicted by extrapolation from the *cos2* mutant phenotype in *Drosophila*. In particular, the changes in neuronal specification in the ventral neural tube are relatively subtle and certainly much less extensive than those seen in the absence of Ptch1 (Cheung et al., 2009; Endoh-Yamagami et al., 2009; Liem Jr. et al., 2009) or PKA-C (Tuson, He, & Anderson, 2011). This difference is reinforced by findings in human where loss of function *KIF7* mutations have been found in Jeune syndrome or Acrocollosal syndrome patients. Remarkably, some of these individuals can be homozygous for the null alleles (Chen et al., 2004; Putoux et al., 2011), arguing against an essential role for KIF7 in human embryonic development. Likewise in zebrafish, animals homozygous for a similar premature termination mutation in the N-terminal domain of the protein are viable to adulthood; and even the progeny of homozygous parents, which are therefore devoid of all Kif7 activity, both zygotic and maternally derived, survive post-embryonically (Maurya et al., 2013). While such animals exhibit a partial Hh de-repression phenotype in tissues of mesodermal origin, most notably the myotome, the specification of neuronal identity appears largely unaffected. And although a reduction in levels of the repressor form of Gli2a can be detected in mutant embryos, this effect appears transient, with levels recovering at later embryonic stages (Maurya et al., 2013).

While it remains unclear how Kif7 contributes to the regulation of Gli processing, what is apparent is that Kif7 forms a complex with the Gli and SuFu proteins, a property it shares with Cos2. A notable feature of this complex is its translocation to the tip of the PC in response to pathway activation, a process thought to be essential for the dissociation of the Gli proteins from SuFu and their entry into the nucleus (Tukachinsky,

Lopez, & Salic, 2010). In contrast to its *Drosophila* counterpart, vertebrate SuFu plays a major role in regulating Gli activity, at least in mammals. Mice homozygous for a null mutation of SuFu exhibit a complete derepression of Hh pathway activity, a phenotype resembling that of Ptch1 null mutants (Cooper et al., 2005; Svard et al., 2006). Consistent with this, heterozygosity for *SUFU* mutations in human leads to the same Gorlin's syndrome caused by heterozygosity for PTCH1 mutations (Smith et al., 2014); and while zebrafish *sufu* mutants have a relatively mild zygotic phenotype (Koudijs et al., 2005) it is probable that the full effect of its loss of function is masked in mutant animals by maternally expressed gene product, since morpholino mediated knockdown of the gene product results in a stronger phenotype. This pivotal role of SuFu in regulating Gli activity contrasts with the dispensability of SuFu in *Drosophila* and indicates a divergence in the relative contributions of SuFu and Cos-2/Kif7 in restraining Ci/Gli activity between flies and vertebrates

Given that kinesins are typically microtubule motor proteins and based on its binding and co-localisation with the Gli proteins, it was originally proposed that Kif7 might be responsible for transporting the Glis along the axoneme to the tip of the PC. However, in vitro analyses established that Kif7 lacks motor activity and instead suggested that it accumulates at the PC tip by binding preferentially to the plus ends of microtubules. This led to a model whereby Kif7 establishes a binding site for the Gli-SuFu at the PC tip (He et al., 2014), a proposal supported by the finding that the accumulation of Gli proteins is disrupted in the absence of Kif7: in mouse embryonic fibroblasts lacking Kif7, Gli2 is distributed along the PC axoneme, whereas in zebrafish kif7 mutant embryos, Gli2a accumulates at the base of the PC (Maurya et al., 2013). However, in vivo studies have failed to replicate these findings, raising doubts about the proposed role in anchoring the Gli-SuFu complex. In any event, it is unclear how this role relates to the principal effect of loss of Kif7, namely the partial activation of the pathway.

In *Drosophila*, a key function of Cos-2 is to recruit the Fu kinase, facilitating its activation and bringing it into close proximity with its substrate Ci. In vertebrates, two kinases, Stk36 and Ulk3 have been identified as potential functional homologs of *Drosophila* Fu. Initial studies focused exclusively on the former, also referred to as Fu, due to its having closer amino acid sequence identity to the *Drosophila* protein. Cell based assays showed that Stk36 can potentiate activity of Gli1 and Gli2 (Murone et al., 2000), while morpholino knockdown of zebrafish *stk36* was reported to cause a loss of MP cells in the myotome, an effect indicative of a reduction rather than a complete loss of Hh signaling activity (Wilson et al., 2009; Wolff et al., 2003). By contrast, targeted mutation of the *Stk36* gene in the mouse showed no effects on Hh dependent processes, leading to the conclusion that its role is not conserved, at least in mammals (Chen, Gao, Kawakami, & Chuang, 2005; Merchant et al., 2005).

Similar to Stk36, Ulk3 was shown to potentiate Gli activity in cultured cells (Maloverjan et al., 2010) but no description of *ulk3* loss of function has been reported in any organism to date. However, mass spectrometric analysis has revealed that Shh activity promotes phosphorylation of Gli2 on Serine residues equivalent to those phosphorylated by Fu in Ci in response to Hh activity; and Crispr mediated knock out in cultured cells has shown that this phosphorylation is mediated by Ulk3 and Stk36 (Han et al., 2019). Consistent with this, Shh induced transcription of Ptch1 is attenuated by knockdown of Ulk3 and Stk36, suggesting that the two kinases have overlapping roles in mediating Gli activation, analogous to the role of Fu in activating Ci (Han et al., 2019). A strong prediction that follows from these results is that Stk36 and Ulk3 should also localize to the Primary Cilium. In this regard, it is notable that Stk36 has been shown to localize to motile cilia and to associate physically with Kif27 (Nozawa et al., 2013) the mammalian paralog of Kif7. However, there is to date no report of either protein localizing to the PC and neither Ulk3 nor Stk36 where amongst the kinases identified in the PC proteome described by Mick et al. (2015).

9. Concluding comments

It is almost exactly 30 years since we proposed that Ptc acts as the receptor for the Hh protein, a proposal that defined the first step in the now familiar signaling pathway (Ingham et al., 1991). In the intervening period, Hedgehog signaling has been transformed from an esoteric process of interest to a handful of *Drosophila* developmental biologists to one that is recognized as impacting on multiple aspects of development, tissue homeostasis and disease. Yet despite the explosion of interest triggered by the discovery of the vertebrate Hh genes, uncovering the innerworkings of the pathway has followed a long and tortuous path. In this review I have purposely focused on the mechanisms by which Hh signals are conveyed and sensed by cells to elicit changes in gene expression that underlie the specification of cell identity. In so doing, I have restricted consideration to what is regarded as the "canonical" signaling pathway; but it would be remiss to end without at least a mention of "non-canonical" Hh signaling. Perhaps the best documented example is in the context of axon guidance, in which Shh induces cytoskeletal remodeling in the growth cones of commissural axons by activating Src family kinases, a process that depends upon its interaction with Boc and Ptc and the consequent activation of Smo, but is independent of the Gli transcription factors (Yam, Langlois, Morin, & Charron, 2009). A second example is based on evidence that Smo couples to heterotrimeric G-proteins, specifically Gai, to regulate adenylate cyclase activity and thereby control PKA via the modulation of cAMP levels (Ogden et al., 2008). How this relates to the recent studies demonstrating a direct interaction between Smo and PKA remains to be seen. Indeed, many questions remain to be addressed and answered before we have a complete picture of this fascinating pathway—but the progress that has been made in recent years, not least through the structural analyses of the Patched and Smoothened proteins, is both impressive and exciting.

Acknowledgments

I would like to thank Konrad Basler, James Briscoe and Freek van Eeden for going to the trouble of locating and sending me images from many years ago—these have certainly enhanced the quality of the figures in this article. I am grateful to Ben Myers for communicating results prior to publication and to him, Arun-Kumar Kaliya-Perumal, Simon Larsson, Raj Rohatgi and Yasunori Saheki for reading the manuscript at short notice and making helpful suggestions. I am indebted to the Toh Kian Chui Foundation for its continuing support of my Distinguished Professorship.

References

Alcedo, J., Ayzenzon, M., Von Ohlen, T., Noll, M., & Hooper, J. E. (1996). The Drosophila smoothened gene encodes a seven-pass membrane protein, a putative receptor for the hedgehog signal. *Cell, 86*, 221–232.

Alexandre, C., Jacinto, A., & Ingham, P. W. (1996). Transcriptional activation of hedgehog target genes in Drosophila is mediated directly by the cubitus interruptus protein, a member of the GLI family of zinc finger DNA-binding proteins. *Genes & Development, 10*, 2003–2013.

Allen, B. L., Song, J. Y., Izzi, L., Althaus, I. W., Kang, J. S., Charron, F., et al. (2011). Overlapping roles and collective requirement for the coreceptors GAS1, CDO, and BOC in SHH pathway function. *Developmental Cell, 20*, 775–787.

Allen, B. L., Tenzen, T., & McMahon, A. P. (2007). The Hedgehog-binding proteins Gas1 and Cdo cooperate to positively regulate Shh signaling during mouse development. *Genes & Development, 21*, 1244–1257.

Amanai, K., & Jiang, J. (2001). Distinct roles of Central missing and Dispatched in sending the Hedgehog signal. *Development, 128*, 5119–5127.

Arveseth, C. D., Happ, J. T., Hedeen, D. S., Zhu, J. F., Capener, J. L., Klatt Shaw, D., et al. (2021). Smoothened transduces Hedgehog signals via activity-dependent sequestration of PKA catalytic subunits. *PLoS Biology, 19*, e3001191.

Avanesov, A., & Blair, S. S. (2013). The Drosophila WIF1 homolog Shifted maintains glypican-independent Hedgehog signaling and interacts with the Hedgehog co-receptors Ihog and Boi. *Development, 140*, 107–116.

Aza-Blanc, P., Ramirez-Weber, F. A., Laget, M. P., Schwartz, C., & Kornberg, T. B. (1997). Proteolysis that is inhibited by hedgehog targets Cubitus interruptus protein to the nucleus and converts it to a repressor. *Cell*, *89*, 1043–1053.

Bangs, F., & Anderson, K. V. (2017). Primary cilia and Mammalian Hedgehog signaling. *Cold Spring Harbor Perspectives in Biology*, *9*, a028175.

Barresi, M. J., Stickney, H. L., & Devoto, S. H. (2000). The zebrafish slow-muscle-omitted gene product is required for Hedgehog signal transduction and the development of slow muscle identity. *Development*, *127*, 2189–2199.

Basler, K., & Struhl, G. (1994). Compartment boundaries and the control of Drosophila limb pattern by hedgehog protein. *Nature*, *368*, 208–214.

Bellaiche, Y., The, I., & Perrimon, N. (1998). Tout-velu is a Drosophila homologue of the putative tumour suppressor EXT-1 and is needed for Hh diffusion. *Nature*, *394*, 85–88.

Belloni, E., Muenke, M., Roessler, E., Traverso, G., Siegel-Bartelt, J., Frumkin, A., et al. (1996). Identification of Sonic hedgehog as a candidate gene responsible for holoprosencephaly. *Nature Genetics*, *14*, 353–356.

Ben, J., Elworthy, S., Ng, A. S., van Eeden, F., & Ingham, P. W. (2011). Targeted mutation of the talpid3 gene in zebrafish reveals its conserved requirement for ciliogenesis and Hedgehog signalling across the vertebrates. *Development*, *138*, 4969–4978.

Bergeron, S. A., Tyurina, O. V., Miller, E., Bagas, A., & Karlstrom, R. O. (2011). Brother of cdo (umleitung) is cell-autonomously required for Hedgehog-mediated ventral CNS patterning in the zebrafish. *Development*, *138*, 75–85.

Bidet, M., Joubert, O., Lacombe, B., Ciantar, M., Nehme, R., Mollat, P., et al. (2011). The hedgehog receptor patched is involved in cholesterol transport. *PLoS One*, *6*, e23834.

Bilioni, A., Sanchez-Hernandez, D., Callejo, A., Gradilla, A. C., Ibanez, C., Mollica, E., et al. (2013). Balancing Hedgehog, a retention and release equilibrium given by Dally, Ihog, Boi and shifted/DmWif. *Developmental Biology*, *376*, 198–212.

Bischoff, M., Gradilla, A. C., Seijo, I., Andres, G., Rodriguez-Navas, C., Gonzalez-Mendez, L., et al. (2013). Cytonemes are required for the establishment of a normal Hedgehog morphogen gradient in Drosophila epithelia. *Nature Cell Biology*, *15*, 1269–1281.

Blagden, C. S., Currie, P. D., Ingham, P. W., & Hughes, S. M. (1997). Notochord induction of zebrafish slow muscle mediated by Sonic hedgehog. *Genes & Development*, *11*, 2163–2175.

Buglino, J. A., & Resh, M. D. (2008). Hhat is a palmitoylacyltransferase with specificity for N-palmitoylation of Sonic Hedgehog. *The Journal of Biological Chemistry*, *283*, 22076–22088.

Burke, R., Nellen, D., Bellotto, M., Hafen, E., Senti, K. A., Dickson, B. J., et al. (1999). Dispatched, a novel sterol-sensing domain protein dedicated to the release of cholesterol-modified hedgehog from signaling cells. *Cell*, *99*, 803–815.

Byrne, E. F., Luchetti, G., Rohatgi, R., & Siebold, C. (2018). Multiple ligand binding sites regulate the Hedgehog signal transducer Smoothened in vertebrates. *Current Opinion in Cell Biology*, *51*, 81–88.

Byrne, E. F. X., Sircar, R., Miller, P. S., Hedger, G., Luchetti, G., Nachtergaele, S., et al. (2016). Structural basis of Smoothened regulation by its extracellular domains. *Nature*, *535*, 517–522.

Callejo, A., Torroja, C., Quijada, L., & Guerrero, I. (2006). Hedgehog lipid modifications are required for Hedgehog stabilization in the extracellular matrix. *Development*, *133*, 471–483.

Camp, D., Currie, K., Labbe, A., van Meyel, D. J., & Charron, F. (2010). Ihog and Boi are essential for Hedgehog signaling in Drosophila. *Neural Development*, *5*, 28.

Camp, D., Haitian He, B., Li, S., Althaus, I. W., Holtz, A. M., Allen, B. L., et al. (2014). Ihog and Boi elicit Hh signaling via Ptc but do not aid Ptc in sequestering the Hh ligand. *Development*, *141*, 3879–3888.

Cannac, F., Qi, C., Falschlunger, J., Hausmann, G., Basler, K., & Korkhov, V. M. (2020). Cryo-EM structure of the Hedgehog release protein Dispatched. *Science Advances*, *6*, eaay7928.

Capdevila, J., Estrada, M. P., Sanchez-Herrero, E., & Guerrero, I. (1994). The Drosophila segment polarity gene patched interacts with decapentaplegic in wing development. *The EMBO Journal*, *13*, 71–82.

Caspary, T., Garcia-Garcia, M. J., Huangfu, D., Eggenschwiler, J. T., Wyler, M. R., Rakeman, A. S., et al. (2002). Mouse Dispatched homolog1 is required for long-range, but not juxtacrine, Hh signaling. *Current Biology*, *12*, 1628–1632.

Castrillon, D. H., & Wasserman, S. A. (1994). Diaphanous is required for cytokinesis in Drosophila and shares domains of similarity with the products of the limb deformity gene. *Development*, *120*, 3367–3377.

Chamoun, Z., Mann, R. K., Nellen, D., von Kessler, D. P., Bellotto, M., Beachy, P. A., et al. (2001). Skinny hedgehog, an acyltransferase required for palmitoylation and activity of the hedgehog signal. *Science*, *293*, 2080–2084.

Chang, T. Y., Chang, C. C., Ohgami, N., & Yamauchi, Y. (2006). Cholesterol sensing, trafficking, and esterification. *Annual Review of Cell and Developmental Biology*, *22*, 129–157.

Chen, Y., Gallaher, N., Goodman, R. H., & Smolik, S. M. (1998). Protein kinase A directly regulates the activity and proteolysis of cubitus interruptus. *Proceedings of the National Academy of Sciences of the United States of America*, *95*, 2349–2354.

Chen, M. H., Gao, N., Kawakami, T., & Chuang, P. T. (2005). Mice deficient in the fused homolog do not exhibit phenotypes indicative of perturbed hedgehog signaling during embryonic development. *Molecular and Cellular Biology*, *25*, 7042–7053.

Chen, W., Huang, H., Hatori, R., & Kornberg, T. B. (2017). Essential basal cytonemes take up Hedgehog in the Drosophila wing imaginal disc. *Development*, *144*, 3134–3144.

Chen, W., Ren, X. R., Nelson, C. D., Barak, L. S., Chen, J. K., Beachy, P. A., et al. (2004). Activity-dependent internalization of smoothened mediated by beta-arrestin 2 and GRK2. *Science*, *306*, 2257–2260.

Chen, Y., & Struhl, G. (1996). Dual roles for patched in sequestering and transducing Hedgehog. *Cell*, *87*, 553–563.

Chen, J., Taipale, J., Cooper, M., & Beachy, P. (2002). Inhibition of Hedgehog signaling by direct binding of cyclopamine to Smoothened. *Genes & Development*, *16*, 2743–2748.

Cheung, H. O., Zhang, X., Ribeiro, A., Mo, R., Makino, S., Puviindran, V., et al. (2009). The kinesin protein Kif7 is a critical regulator of Gli transcription factors in mammalian hedgehog signaling. *Science Signaling*, *2*, ra29.

Chiang, C., Litingtung, Y., Harris, M. P., Simandl, B. K., Li, Y., Beachy, P. A., et al. (2001). Manifestation of the limb prepattern: Limb development in the absence of sonic hedgehog function. *Developmental Biology*, *236*, 421–435.

Chiang, C., Litingtung, Y., Lee, E., Young, K. E., Corden, J. L., Westphal, H., et al. (1996). Cyclopia and defective axial patterning in mice lacking Sonic hedgehog gene function. *Nature*, *383*, 407–413.

Cole, F., & Krauss, R. S. (2003). Microform holoprosencephaly in mice that lack the Ig superfamily member Cdon. *Current Biology*, *13*, 411–415.

Collier, L. S., Suyama, K., Anderson, J. H., & Scott, M. P. (2004). Drosophila Costal1 mutations are alleles of protein kinase A that modulate hedgehog signaling. *Genetics*, *167*, 783–796.

Cooper, M. K., Porter, J. A., Young, K. E., & Beachy, P. A. (1998). Teratogen-mediated inhibition of target tissue response to Shh signaling. *Science*, *280*, 1603–1607.

Cooper, M. K., Wassif, C. A., Krakowiak, P. A., Taipale, J., Gong, R., Kelley, R. I., et al. (2003). A defective response to Hedgehog signaling in disorders of cholesterol biosynthesis. *Nature Genetics*, *33*, 508–513.

Cooper, A. F., Yu, K. P., Brueckner, M., Brailey, L. L., Johnson, L., McGrath, J. M., et al. (2005). Cardiac and CNS defects in a mouse with targeted disruption of suppressor of fused. *Development*, *132*, 4407–4417.

Corbit, K. C., Aanstad, P., Singla, V., Norman, A. R., Stainier, D. Y., & Reiter, J. F. (2005). Vertebrate Smoothened functions at the primary cilium. *Nature*, *437*, 1018–1021.

Corcoran, R. B., & Scott, M. P. (2006). Oxysterols stimulate Sonic hedgehog signal transduction and proliferation of medulloblastoma cells. *Proceedings of the National Academy of Sciences of the United States of America*, *103*, 8408–8413.

Creanga, A., Glenn, T. D., Mann, R. K., Saunders, A. M., Talbot, W. S., & Beachy, P. A. (2012). Scube/You activity mediates release of dually lipid-modified Hedgehog signal in soluble form. *Genes & Development*, *26*, 1312–1325.

Currie, P. D., & Ingham, P. W. (1996). Induction of a specific muscle cell type by a hedgehog-like protein in zebrafish. *Nature*, *382*, 452–455.

Das, A., Brown, M. S., Anderson, D. D., Goldstein, J. L., & Radhakrishnan, A. (2014). Three pools of plasma membrane cholesterol and their relation to cholesterol homeostasis. *eLife*, *3*, e02882.

Denef, N., Neubuser, D., Perez, L., & Cohen, S. M. (2000). Hedgehog induces opposite changes in turnover and subcellular localization of patched and smoothened. *Cell*, *102*, 521–531.

Desbordes, S. C., & Sanson, B. (2003). The glypican Dally-like is required for Hedgehog signalling in the embryonic epidermis of Drosophila. *Development*, *130*, 6245–6255.

Deshpande, I., Liang, J., Hedeen, D., Roberts, K. J., Zhang, Y., Ha, B., et al. (2019). Smoothened stimulation by membrane sterols drives Hedgehog pathway activity. *Nature*, *571*, 284–288.

Dessaud, E., Ribes, V., Balaskas, N., Yang, L. L., Pierani, A., Kicheva, A., et al. (2010). Dynamic assignment and maintenance of positional identity in the ventral neural tube by the morphogen sonic hedgehog. *PLoS Biology*, *8*, e1000382.

Dierker, T., Dreier, R., Petersen, A., Bordych, C., & Grobe, K. (2009). Heparan sulfate-modulated, metalloprotease-mediated sonic hedgehog release from producing cells. *The Journal of Biological Chemistry*, *284*, 8013–8022.

Echelard, Y., Epstein, D. J., St-Jacques, B., Shen, L., Mohler, J., McMahon, J. A., et al. (1993). Sonic hedgehog, a member of a family of putative signaling molecules, is implicated in the regulation of CNS polarity. *Cell*, *75*, 1417–1430.

Endoh-Yamagami, S., Evangelista, M., Wilson, D., Wen, X., Theunissen, J. W., Phamluong, K., et al. (2009). The mammalian Cos2 homolog Kif7 plays an essential role in modulating Hh signal transduction during development. *Current Biology*, *19*, 1320–1326.

Fietz, M. J., Concordet, J. P., Barbosa, R., Johnson, R., Krauss, S., McMahon, A. P., et al. (1994). The hedgehog gene family in Drosophila and vertebrate development. *Development. Supplement*, 43–51.

Forbes, A. J., Nakano, Y., Taylor, A. M., & Ingham, P. W. (1993). Genetic analysis of hedgehog signalling in the Drosophila embryo. *Development. Supplement*, 115–124.

Frank-Kamenetsky, M., Zhang, X. M., Bottega, S., Guicherit, O., Wichterle, H., Dudek, H., et al. (2002). Small-molecule modulators of Hedgehog signaling: Identification and characterization of Smoothened agonists and antagonists. *Journal of Biology*, *1*, 10.

Glise, B., Miller, C. A., Crozatier, M., Halbisen, M. A., Wise, S., Olson, D. J., et al. (2005). Shifted, the Drosophila ortholog of Wnt inhibitory factor-1, controls the distribution and movement of Hedgehog. *Developmental Cell, 8*, 255–266.

Gong, X., Qian, H., Cao, P., Zhao, X., Zhou, Q., Lei, J., et al. (2018). Structural basis for the recognition of Sonic Hedgehog by human Patched1. *Science, 361.*

Gonzalez-Mendez, L., Gradilla, A. C., Sanchez-Hernandez, D., Gonzalez, E., Aguirre-Tamaral, A., Jimenez-Jimenez, C., et al. (2020). Polarized sorting of Patched enables cytoneme-mediated Hedgehog reception in the Drosophila wing disc. *The EMBO Journal, 39*, e103629.

Gonzalez-Mendez, L., Seijo-Barandiaran, I., & Guerrero, I. (2017). Cytoneme-mediated cell-cell contacts for Hedgehog reception. *eLife, 6*, e24045.

Goodrich, L. V., Johnson, R. L., Milenkovic, L., McMahon, J. A., & Scott, M. P. (1996). Conservation of the hedgehog/patched signaling pathway from flies to mice: Induction of a mouse patched gene by Hedgehog. *Genes & Development, 10*, 301–312.

Goodrich, L. V., Milenkovic, L., Higgins, K. M., & Scott, M. P. (1997). Altered neural cell fates and medulloblastoma in mouse patched mutants. *Science, 277*, 1109–1113.

Gorfinkiel, N., Sierra, J., Callejo, A., Ibanez, C., & Guerrero, I. (2005). The Drosophila ortholog of the human Wnt inhibitor factor Shifted controls the diffusion of lipid-modified Hedgehog. *Developmental Cell, 8*, 241–253.

Gradilla, A. C., Gonzalez, E., Seijo, I., Andres, G., Bischoff, M., Gonzalez-Mendez, L., et al. (2014). Exosomes as Hedgehog carriers in cytoneme-mediated transport and secretion. *Nature Communications, 5*, 5649.

Grau, Y., & Simpson, P. (1987). The segment polarity gene costal-2 in Drosophila. I. The organization of both primary and secondary embryonic fields may be affected. *Developmental Biology, 122*, 186–200.

Grimmond, S., Larder, R., Van Hateren, N., Siggers, P., Hulsebos, T. J., Arkell, R., et al. (2000). Cloning, mapping, and expression analysis of a gene encoding a novel mammalian EGF-related protein (SCUBE1). *Genomics, 70*, 74–81.

Hall, E. T., Dillard, M. E., Stewart, D. P., Zhang, Y., Wagner, B., Levine, R. M., et al. (2021). Cytoneme delivery of Sonic Hedgehog from ligand-producing cells requires Myosin 10 and a Dispatched-BOC/CDON co-receptor complex. *eLife, 10.*

Hall, T. M., Porter, J. A., Young, K. E., Koonin, E. V., Beachy, P. A., & Leahy, D. J. (1997). Crystal structure of a Hedgehog autoprocessing domain: Homology between Hedgehog and self-splicing proteins. *Cell, 91*, 85–97.

Halpern, M. E., Thisse, C., Ho, R. K., Thisse, B., Riggleman, B., Trevarrow, B., et al. (1995). Cell-autonomous shift from axial to paraxial mesodermal development in zebrafish floating head mutants. *Development, 121*, 4257–4264.

Hammerschmidt, M., Bitgood, M. J., & McMahon, A. P. (1996). Protein kinase A is a common negative regulator of Hedgehog signaling in the vertebrate embryo. *Genes & Development, 10*, 647–658.

Han, C., Belenkaya, T. Y., Wang, B., & Lin, X. (2004). Drosophila glypicans control the cell-to-cell movement of Hedgehog by a dynamin-independent process. *Development, 131*, 601–611.

Han, Y., Wang, B., Cho, Y. S., Zhu, J., Wu, J., Chen, Y., et al. (2019). Phosphorylation of Ci/Gli by fused family kinases promotes Hedgehog signaling. *Developmental Cell, 50*, 610–626. e614.

Han, Y., Xiong, Y., Shi, X., Wu, J., Zhao, Y., & Jiang, J. (2017). Regulation of Gli ciliary localization and Hedgehog signaling by the PY-NLS/karyopherin-beta2 nuclear import system. *PLoS Biology, 15*, e2002063.

Happ, J. T., Arveseth, C. D., Bruystens, J., Bertinetti, D., Nelson, I. B., Olivieri, C., et al. (2021). A PKA inhibitor motif within Smoothened controls Hedgehog signal transduction. bioRxiv 2021.07.05.451193.

Hatta, K., Bremiller, R., Westerfield, M., & Kimmel, C. B. (1991). Diversity of expression of engrailed-like antigens in zebrafish. *Development*, *112*, 821–832.

Haycraft, C. J., Banizs, B., Aydin-Son, Y., Zhang, Q., Michaud, E. J., & Yoder, B. K. (2005). Gli2 and Gli3 localize to cilia and require the intraflagellar transport protein polaris for processing and function. *PLoS Genetics*, *1*, e53.

He, M., Subramanian, R., Bangs, F., Omelchenko, T., Liem, K. F., Jr., Kapoor, T. M., et al. (2014). The kinesin-4 protein Kif7 regulates mammalian Hedgehog signalling by organizing the cilium tip compartment. *Nature Cell Biology*, *16*, 663–672.

Hidalgo, A., & Ingham, P. (1990). Cell patterning in the Drosophila segment: Spatial regulation of the segment polarity gene patched. *Development*, *110*, 291–301.

Hollway, G. E., Maule, J., Gautier, P., Evans, T. M., Keenan, D. G., Lohs, C., et al. (2006). Scube2 mediates Hedgehog signalling in the zebrafish embryo. *Developmental Biology*, *294*, 104–118.

Hooper, J. (1994). Distinct pathways for autocrine and paracrine Wingless signalling in Drosophila embryos. *Nature*, *372*, 461–464.

Hooper, J. E., & Scott, M. P. (1989). The Drosophila patched gene encodes a putative membrane protein required for segmental patterning. *Cell*, *59*, 751–765.

Hsia, E. Y. C., Zhang, Y., Tran, H. S., Lim, A., Chou, Y. H., Lan, G., et al. (2017). Hedgehog mediated degradation of Ihog adhesion proteins modulates cell segregation in Drosophila wing imaginal discs. *Nature Communications*, *8*, 1275.

Hsieh, J. C., Kodjabachian, L., Rebbert, M. L., Rattner, A., Smallwood, P. M., Samos, C. H., et al. (1999). A new secreted protein that binds to Wnt proteins and inhibits their activities. *Nature*, *398*, 431–436.

Huang, P., Nedelcu, D., Watanabe, M., Jao, C., Kim, Y., Liu, J., et al. (2016). Cellular cholesterol directly activates smoothened in Hedgehog signaling. *Cell*, *166*, 1176–1187. e1114.

Huang, P., Zheng, S., Wierbowski, B. M., Kim, Y., Nedelcu, D., Aravena, L., et al. (2018). Structural basis of Smoothened activation in Hedgehog signaling. *Cell*, *175*, 295–297.

Huangfu, D., & Anderson, K. V. (2005). Cilia and Hedgehog responsiveness in the mouse. *Proceedings of the National Academy of Sciences of the United States of America*, *102*, 11325–11330.

Huangfu, D., Liu, A., Rakeman, A. S., Murcia, N. S., Niswander, L., & Anderson, K. V. (2003). Hedgehog signalling in the mouse requires intraflagellar transport proteins. *Nature*, *426*, 83–87.

Ingham, P. W. (1993). Localized hedgehog activity controls spatial limits of wingless transcription in the Drosophila embryo. *Nature*, *366*, 560–562.

Ingham, P. W. (2016). Drosophila segment polarity mutants and the rediscovery of the Hedgehog pathway genes. *Current Topics in Developmental Biology*, *116*, 477–488.

Ingham, P. W., & Fietz, M. J. (1995). Quantitative effects of hedgehog and decapentaplegic activity on the patterning of the Drosophila wing. *Current Biology*, *5*, 432–440.

Ingham, P. W., & Hidalgo, A. (1993). Regulation of wingless transcription in the Drosophila embryo. *Development*, *117*, 283–291.

Ingham, P. W., Nystedt, S., Nakano, Y., Brown, W., Stark, D., van den Heuvel, M., et al. (2000). Patched represses the Hedgehog signalling pathway by promoting modification of the Smoothened protein. *Current Biology*, *10*, 1315–1318.

Ingham, P. W., Taylor, A. M., & Nakano, Y. (1991). Role of the Drosophila patched gene in positional signalling. *Nature*, *353*, 184–187.

Iulianella, A., Sakai, D., Kurosaka, H., & Trainor, P. A. (2018). Ventral neural patterning in the absence of a Shh activity gradient from the floorplate. *Developmental Dynamics*, *247*, 170–184.

Jia, J., Zhang, L., Zhang, Q., Tong, C., Wang, B., Hou, F., et al. (2005). Phosphorylation by double-time/CKIepsilon and CKIalpha targets cubitus interruptus for Slimb/beta-TRCP-mediated proteolytic processing. *Developmental Cell*, *9*, 819–830.

Jiang, Y., Benz, T. L., & Long, S. B. (2021). Substrate and product complexes reveal mechanisms of Hedgehog acylation by HHAT. *Science, 372*, 1215–1219.
Jiang, J., & Struhl, G. (1995). Protein kinase A and hedgehog signaling in Drosophila limb development. *Cell, 80*, 563–572.
Jiang, J., & Struhl, G. (1998). Regulation of the Hedgehog and Wingless signalling pathways by the F-box/WD40-repeat protein Slimb. *Nature, 391*, 493–496.
Johnson, D. R. (1967). Extra-toes: Anew mutant gene causing multiple abnormalities in the mouse. *Journal of Embryology and Experimental Morphology, 17*, 543–581.
Johnson, R. L., Grenier, J. K., & Scott, M. P. (1995). patched overexpression alters wing disc size and pattern: Transcriptional and post-transcriptional effects on hedgehog targets. *Development, 121*, 4161–4170.
Johnson, J. L., Hall, T. E., Dyson, J. M., Sonntag, C., Ayers, K., Berger, S., et al. (2012). Scube activity is necessary for Hedgehog signal transduction in vivo. *Developmental Biology, 368*, 193–202.
Jürgens, G., Wieschaus, E., Nüsslein-Volhard, C., & Kluding, H. (1984). Mutations affecting the pattern of the larval cuticle in Drosophila melanogaster: II. Zygotic loci on the third chromosome. *Wilehm Roux Archives of Developmental Biology, 193*, 283–295.
Kadowaki, T., Wilder, E., Klingensmith, J., Zachary, K., & Perrimon, N. (1996). The segment polarity gene porcupine encodes a putative multitransmembrane protein involved in Wingless processing. *Genes & Development, 10*, 3116–3128.
Kang, J. S., Mulieri, P. J., Hu, Y., Taliana, L., & Krauss, R. S. (2002). BOC, an Ig superfamily member, associates with CDO to positively regulate myogenic differentiation. *The EMBO Journal, 21*, 114–124.
Kang, J. S., Mulieri, P. J., Miller, C., Sassoon, D. A., & Krauss, R. S. (1998). CDO, a robo-related cell surface protein that mediates myogenic differentiation. *The Journal of Cell Biology, 143*, 403–413.
Karlstrom, R. O., Talbot, W. S., & Schier, A. F. (1999). Comparative synteny cloning of zebrafish you-too: Mutations in the Hedgehog target gli2 affect ventral forebrain patterning. *Genes & Development, 13*, 388–393.
Karlstrom, R. O., Trowe, T., Klostermann, S., Baier, H., Brand, M., Crawford, A. D., et al. (1996). Zebrafish mutations affecting retinotectal axon pathfinding. *Development, 123*, 427–438.
Karlstrom, R. O., Tyurina, O. V., Kawakami, A., Nishioka, N., Talbot, W. S., Sasaki, H., et al. (2003). Genetic analysis of zebrafish gli1 and gli2 reveals divergent requirements for gli genes in vertebrate development. *Development, 130*, 1549–1564.
Kawakami, A., Nojima, Y., Toyoda, A., Takahoko, M., Satoh, M., Tanaka, H., et al. (2005). The zebrafish-secreted matrix protein you/scube2 is implicated in long-range regulation of hedgehog signaling. *Current Biology, 15*, 480–488.
Kim, J., Hsia, E. Y., Brigui, A., Plessis, A., Beachy, P. A., & Zheng, X. (2015). The role of ciliary trafficking in Hedgehog receptor signaling. *Science Signaling, 8*, ra55.
Kim, H. R., Richardson, J., van Eeden, F., & Ingham, P. W. (2010). Gli2a protein localization reveals a role for Iguana/DZIP1 in primary ciliogenesis and a dependence of Hedgehog signal transduction on primary cilia in the zebrafish. *BMC Biology, 8*, 65.
Kinnebrew, M., Iverson, E. J., Patel, B. B., Pusapati, G. V., Kong, J. H., Johnson, K. A., et al. (2019). Cholesterol accessibility at the ciliary membrane controls hedgehog signaling. *eLife, 8*, e50051.
Kinnebrew, M., Luchetti, G., Sircar, R., Frigui, S., Viti, L. V., Naito, T., et al. (2021). Patched 1 reduces the accessibility of cholesterol in the outer leaflet of membranes. *eLife, 10*, e70504.
Kojima, T., Michiue, T., Orihara, M., & Saigo, K. (1994). Induction of a mirror-image duplication of anterior wing structures by localized hedgehog expression in the anterior compartment of Drosophila melanogaster wing imaginal discs. *Gene, 148*, 211–217.

Komolov, K. E., & Benovic, J. L. (2018). G protein-coupled receptor kinases: Past, present and future. *Cellular Signalling*, *41*, 17–24.
Kornberg, T. B., & Roy, S. (2014). Cytonemes as specialized signaling filopodia. *Development*, *141*, 729–736.
Koudijs, M. J., den Broeder, M. J., Groot, E., & van Eeden, F. J. (2008). Genetic analysis of the two zebrafish patched homologues identifies novel roles for the hedgehog signaling pathway. *BMC Developmental Biology*, *8*, 15.
Koudijs, M. J., den Broeder, M. J., Keijser, A., Wienholds, E., Houwing, S., van Rooijen, E. M., et al. (2005). The zebrafish mutants dre, uki, and lep encode negative regulators of the hedgehog signaling pathway. *PLoS Genetics*, *1*, e19.
Kowatsch, C., Woolley, R. E., Kinnebrew, M., Rohatgi, R., & Siebold, C. (2019). Structures of vertebrate Patched and Smoothened reveal intimate links between cholesterol and Hedgehog signalling. *Current Opinion in Structural Biology*, *57*, 204–214.
Krauss, S., Concordet, J. P., & Ingham, P. W. (1993). A functionally conserved homolog of the Drosophila segment polarity gene hh is expressed in tissues with polarizing activity in zebrafish embryos. *Cell*, *75*, 1431–1444.
Lange, Y., Ye, J., & Steck, T. L. (2004). How cholesterol homeostasis is regulated by plasma membrane cholesterol in excess of phospholipids. *Proceedings of the National Academy of Sciences of the United States of America*, *101*, 11664–11667.
Lee, C. S., Buttitta, L., & Fan, C. M. (2001). Evidence that the WNT-inducible growth arrest-specific gene 1 encodes an antagonist of sonic hedgehog signaling in the somite. *Proceedings of the National Academy of Sciences of the United States of America*, *98*, 11347–11352.
Lee, J. D., & Treisman, J. E. (2001). Sightless has homology to transmembrane acyltransferases and is required to generate active Hedgehog protein. *Current Biology*, *11*, 1147–1152.
Lee, J. J., von Kessler, D. P., Parks, S., & Beachy, P. A. (1992). Secretion and localized transcription suggest a role in positional signaling for products of the segmentation gene hedgehog. *Cell*, *71*, 33–50.
Lee, J., Willer, J. R., Willer, G. B., Smith, K., Gregg, R. G., & Gross, J. M. (2008). Zebrafish blowout provides genetic evidence for Patched1-mediated negative regulation of Hedgehog signaling within the proximal optic vesicle of the vertebrate eye. *Developmental Biology*, *319*, 10–22.
Lepage, T., Cohen, S. M., Diaz-Benjumea, F. J., & Parkhurst, S. M. (1995). Signal transduction by cAMP-dependent protein kinase A in Drosophila limb patterning. *Nature*, *373*, 711–715.
Lewis, K. E., Currie, P. D., Roy, S., Schauerte, H., Haffter, P., & Ingham, P. W. (1999). Control of muscle cell-type specification in the zebrafish embryo by Hedgehog signalling. *Developmental Biology*, *216*, 469–480.
Li, S., Ma, G., Wang, B., & Jiang, J. (2014). Hedgehog induces formation of PKA-Smoothened complexes to promote Smoothened phosphorylation and pathway activation. *Science Signaling*, *7*, ra62.
Li, W., Ohlmeyer, J. T., Lane, M. E., & Kalderon, D. (1995). Function of protein kinase A in hedgehog signal transduction and Drosophila imaginal disc development. *Cell*, *80*, 553–562.
Liem, K. F., Jr., He, M., Ocbina, P. J., & Anderson, K. V. (2009). Mouse Kif7/Costal2 is a cilia-associated protein that regulates Sonic hedgehog signaling. *Proceedings of the National Academy of Sciences of the United States of America*, *106*, 13377–13382.
Liepinsh, E., Banyai, L., Patthy, L., & Otting, G. (2006). NMR structure of the WIF domain of the human Wnt-inhibitory factor-1. *Journal of Molecular Biology*, *357*, 942–950.
Limbourg-Bouchon, B., Busson, D., & Lamour-Isnard, C. (1991). Interactions between fused, a segment-polarity gene in Drosophila, and other segmentation genes. *Development*, *112*, 417–429.

Lopez, M. E., & Scott, M. P. (2013). Genetic dissection of a cell-autonomous neurodegenerative disorder: Lessons learned from mouse models of Niemann-Pick disease type C. *Disease Models & Mechanisms*, *6*, 1089–1100.

Luchetti, G., Sircar, R., Kong, J. H., Nachtergaele, S., Sagner, A., Byrne, E. F., et al. (2016). Cholesterol activates the G-protein coupled receptor Smoothened to promote Hedgehog signaling. *eLife*, 5.

Lum, L., Yao, S., Mozer, B., Rovescalli, A., Von Kessler, D., Nirenberg, M., et al. (2003). Identification of Hedgehog pathway components by RNAi in Drosophila cultured cells. *Science*, *299*, 2039–2045.

Lynch, C. J. (1919). An analysis of certain cases of intra-specific sterility. *Genetics*, *4*, 501–533.

Ma, Y., Erkner, A., Gong, R., Yao, S., Taipale, J., Basler, K., et al. (2002). Hedgehog-mediated patterning of the mammalian embryo requires transporter-like function of dispatched. *Cell*, *111*, 63–75.

Machesky, L. M., & Insall, R. H. (1998). Scar1 and the related Wiskott-Aldrich syndrome protein, WASP, regulate the actin cytoskeleton through the Arp2/3 complex. *Current Biology*, *8*, 1347–1356.

Makino, S., Masuya, H., Ishijima, J., Yada, Y., & Shiroishi, T. (2001). A spontaneous mouse mutation, mesenchymal dysplasia (mes), is caused by a deletion of the most C-terminal cytoplasmic domain of patched (ptc). *Developmental Biology*, *239*, 95–106.

Malinauskas, T., Aricescu, A. R., Lu, W., Siebold, C., & Jones, E. Y. (2011). Modular mechanism of Wnt signaling inhibition by Wnt inhibitory factor 1. *Nature Structural & Molecular Biology*, *18*, 886–893.

Maloverjan, A., Piirsoo, M., Kasak, L., Peil, L., Osterlund, T., & Kogerman, P. (2010). Dual function of UNC-51-like kinase 3 (Ulk3) in the Sonic hedgehog signaling pathway. *The Journal of Biological Chemistry*, *285*, 30079–30090.

Marigo, V., Davey, R. A., Zuo, Y., Cunningham, J. M., & Tabin, C. J. (1996). Biochemical evidence that patched is the Hedgehog receptor. *Nature*, *384*, 176–179.

Martinez Arias, A., Baker, N. E., & Ingham, P. W. (1988). Role of segment polarity genes in the definition and maintenance of cell states in the Drosophila embryo. *Development*, *103*, 157–170.

Matise, M. P., Epstein, D. J., Park, H. L., Platt, K. A., & Joyner, A. L. (1998). Gli2 is required for induction of floor plate and adjacent cells, but not most ventral neurons in the mouse central nervous system. *Development*, *125*, 2759–2770.

Maurya, A. K., Ben, J., Zhao, Z., Lee, R. T., Niah, W., Ng, A. S., et al. (2013). Positive and negative regulation of Gli activity by Kif7 in the zebrafish embryo. *PLoS Genetics*, *9*, e1003955.

McGough, I. J., Vecchia, L., Bishop, B., Malinauskas, T., Beckett, K., Joshi, D., et al. (2020). Glypicans shield the Wnt lipid moiety to enable signalling at a distance. *Nature*, *585*, 85–90.

McLellan, J. S., Yao, S., Zheng, X., Geisbrecht, B. V., Ghirlando, R., Beachy, P. A., et al. (2006). Structure of a heparin-dependent complex of Hedgehog and Ihog. *Proceedings of the National Academy of Sciences of the United States of America*, *103*, 17208–17213.

Merchant, M., Evangelista, M., Luoh, S. M., Frantz, G. D., Chalasani, S., Carano, R. A., et al. (2005). Loss of the serine/threonine kinase fused results in postnatal growth defects and lethality due to progressive hydrocephalus. *Molecular and Cellular Biology*, *25*, 7054–7068.

Micchelli, C. A., The, I., Selva, E., Mogila, V., & Perrimon, N. (2002). Rasp, a putative transmembrane acyltransferase, is required for Hedgehog signaling. *Development*, *129*, 843–851.

Michaux, G., Gansmuller, A., Hindelang, C., & Labouesse, M. (2000). CHE-14, a protein with a sterol-sensing domain, is required for apical sorting in C. elegans ectodermal epithelial cells. *Current Biology*, *10*, 1098–1107.

Mick, D. U., Rodrigues, R. B., Leib, R. D., Adams, C. M., Chien, A. S., Gygi, S. P., et al. (2015). Proteomics of primary cilia by proximity labeling. *Developmental Cell, 35*, 497–512.
Milenkovic, L., Scott, M. P., & Rohatgi, R. (2009). Lateral transport of Smoothened from the plasma membrane to the membrane of the cilium. *The Journal of Cell Biology, 187*, 365–374.
Miller, J., Fraser, S. E., & McClay, D. (1995). Dynamics of thin filopodia during sea urchin gastrulation. *Development, 121*, 2501–2511.
Mohler, J. (1988). Requirements for hedgehog, a segmental polarity gene, in patterning larval and adult cuticle of Drosophila. *Genetics, 120*, 1061–1072.
Motoyama, J., Milenkovic, L., Iwama, M., Shikata, Y., Scott, M. P., & Hui, C. C. (2003). Differential requirement for Gli2 and Gli3 in ventral neural cell fate specification. *Developmental Biology, 259*, 150–161.
Moyer, J. H., Lee-Tischler, M. J., Kwon, H. Y., Schrick, J. J., Avner, E. D., Sweeney, W. E., et al. (1994). Candidate gene associated with a mutation causing recessive polycystic kidney disease in mice. *Science, 264*, 1329–1333.
Murone, M., Luoh, S. M., Stone, D., Li, W., Gurney, A., Armanini, M., et al. (2000). Gli regulation by the opposing activities of fused and suppressor of fused. *Nature Cell Biology, 2*, 310–312.
Myers, B. R., Sever, N., Chong, Y. C., Kim, J., Belani, J. D., Rychnovsky, S., et al. (2013). Hedgehog pathway modulation by multiple lipid binding sites on the smoothened effector of signal response. *Developmental Cell, 26*, 346–357.
Nachtergaele, S., Mydock, L. K., Krishnan, K., Rammohan, J., Schlesinger, P. H., Covey, D. F., et al. (2012). Oxysterols are allosteric activators of the oncoprotein Smoothened. *Nature Chemical Biology, 8*, 211–220.
Nachtergaele, S., Whalen, D. M., Mydock, L. K., Zhao, Z., Malinauskas, T., Krishnan, K., et al. (2013). Structure and function of the Smoothened extracellular domain in vertebrate Hedgehog signaling. *eLife, 2*, e01340.
Nakano, Y., Guerrero, I., Hidalgo, A., Taylor, A., Whittle, J. R., & Ingham, P. W. (1989). A protein with several possible membrane-spanning domains encoded by the Drosophila segment polarity gene patched. *Nature, 341*, 508–513.
Nakano, Y., Kim, H. R., Kawakami, A., Roy, S., Schier, A. F., & Ingham, P. W. (2004). Inactivation of dispatched 1 by the chameleon mutation disrupts Hedgehog signalling in the zebrafish embryo. *Developmental Biology, 269*, 381–392.
Nakato, H., Futch, T. A., & Selleck, S. B. (1995). The division abnormally delayed (dally) gene: A putative integral membrane proteoglycan required for cell division patterning during postembryonic development of the nervous system in Drosophila. *Development, 121*, 3687–3702.
Nedelcu, D., Liu, J., Xu, Y., Jao, C., & Salic, A. (2013). Oxysterol binding to the extracellular domain of Smoothened in Hedgehog signaling. *Nature Chemical Biology, 9*, 557–564.
Nieuwenhuis, E., Motoyama, J., Barnfield, P. C., Yoshikawa, Y., Zhang, X., Mo, R., et al. (2006). Mice with a targeted mutation of patched2 are viable but develop alopecia and epidermal hyperplasia. *Molecular and Cellular Biology, 26*, 6609–6622.
Nikaido, H. (2018). RND transporters in the living world. *Research in Microbiology, 169*, 363–371.
Nozawa, Y. I., Yao, E., Lin, C., Yang, J. H., Wilson, C. W., Gacayan, R., et al. (2013). Fused (Stk36) is a ciliary protein required for central pair assembly and motile cilia orientation in the mammalian oviduct. *Developmental Dynamics, 242*, 1307–1319.
Nusslein-Volhard, C., & Wieschaus, E. (1980). Mutations affecting segment number and polarity in Drosophila. *Nature, 287*, 795–801.

Nüsslein-Volhard, C., Wieschaus, E., & Kluding, H. (1984). Mutations affecting the pattern of the larval cuticle in Drosophila melanogaster: I. Zygotic loci on the second chromosome. *Wilehm Roux Archives of Developmental Biology*, *193*, 267–282.

Odenthal, J., van Eeden, F. J., Haffter, P., Ingham, P. W., & Nusslein-Volhard, C. (2000). Two distinct cell populations in the floor plate of the zebrafish are induced by different pathways. *Developmental Biology*, *219*, 350–363.

Ogden, S. K., Fei, D. L., Schilling, N. S., Ahmed, Y. F., Hwa, J., & Robbins, D. J. (2008). G protein Galphai functions immediately downstream of Smoothened in Hedgehog signalling. *Nature*, *456*, 967–970.

Ohlig, S., Farshi, P., Pickhinke, U., van den Boom, J., Hoing, S., Jakuschev, S., et al. (2011). Sonic hedgehog shedding results in functional activation of the solubilized protein. *Developmental Cell*, *20*, 764–774.

Okada, A., Charron, F., Morin, S., Shin, D. S., Wong, K., Fabre, P. J., et al. (2006). Boc is a receptor for sonic hedgehog in the guidance of commissural axons. *Nature*, *444*, 369–373.

Orenic, T., Slusarski, D., Kroll, K., & Holmgren, R. (1990). Cloning and characterization of the segment polarity gene cubitus interruptus Dominant of Drosophila. *Genes & Development*, *4*, 1053–1067.

Palm, W., Swierczynska, M. M., Kumari, V., Ehrhart-Bornstein, M., Bornstein, S. R., & Eaton, S. (2013). Secretion and signaling activities of lipoprotein-associated hedgehog and non-sterol-modified hedgehog in flies and mammals. *PLoS Biology*, *11*, e1001505.

Pan, Y., Bai, C. B., Joyner, A. L., & Wang, B. (2006). Sonic hedgehog signaling regulates Gli2 transcriptional activity by suppressing its processing and degradation. *Molecular and Cellular Biology*, *26*, 3365–3377.

Pan, D., & Rubin, G. M. (1995). cAMP-dependent protein kinase and hedgehog act antagonistically in regulating decapentaplegic transcription in Drosophila imaginal discs. *Cell*, *80*, 543–552.

Pan, Y., & Wang, B. (2007). A novel protein-processing domain in Gli2 and Gli3 differentially blocks complete protein degradation by the proteasome. *The Journal of Biological Chemistry*, *282*, 10846–10852.

Pan, Y., Wang, C., & Wang, B. (2009). Phosphorylation of Gli2 by protein kinase A is required for Gli2 processing and degradation and the Sonic Hedgehog-regulated mouse development. *Developmental Biology*, *326*, 177–189.

Panakova, D., Sprong, H., Marois, E., Thiele, C., & Eaton, S. (2005). Lipoprotein particles are required for Hedgehog and Wingless signalling. *Nature*, *435*, 58–65.

Park, H. L., Bai, C., Platt, K. A., Matise, M. P., Beeghly, A., Hui, C. C., et al. (2000). Mouse Gli1 mutants are viable but have defects in SHH signaling in combination with a Gli2 mutation. *Development*, *127*, 1593–1605.

Pazour, G. J., Dickert, B. L., Vucica, Y., Seeley, E. S., Rosenbaum, J. L., Witman, G. B., et al. (2000). Chlamydomonas IFT88 and its mouse homologue, polycystic kidney disease gene tg737, are required for assembly of cilia and flagella. *The Journal of Cell Biology*, *151*, 709–718.

Pepinsky, R. B., Zeng, C., Wen, D., Rayhorn, P., Baker, D. P., Williams, K. P., et al. (1998). Identification of a palmitic acid-modified form of human Sonic hedgehog. *The Journal of Biological Chemistry*, *273*, 14037–14045.

Persson, M., Stamataki, D., te Welscher, P., Andersson, E., Bose, J., Ruther, U., et al. (2002). Dorsal-ventral patterning of the spinal cord requires Gli3 transcriptional repressor activity. *Genes & Development*, *16*, 2865–2878.

Petrov, K., Wierbowski, B. M., Liu, J., & Salic, A. (2020). Distinct cation gradients power cholesterol transport at different key points in the Hedgehog signaling pathway. *Developmental Cell*, *55*, 314–327. e317.

Phillips, R. G., Roberts, I. J., Ingham, P. W., & Whittle, J. R. (1990). The Drosophila segment polarity gene patched is involved in a position-signalling mechanism in imaginal discs. *Development, 110*, 105–114.

Pierani, A., Brenner-Morton, S., Chiang, C., & Jessell, T. M. (1999). A sonic hedgehog-independent, retinoid-activated pathway of neurogenesis in the ventral spinal cord. *Cell, 97*, 903–915.

Placzek, M., Jessell, T. M., & Dodd, J. (1993). Induction of floor plate differentiation by contact-dependent, homeogenetic signals. *Development, 117*, 205–218.

Porter, J. A., Young, K. E., & Beachy, P. A. (1996). Cholesterol modification of hedgehog signaling proteins in animal development. *Science, 274*, 255–259.

Preat, T. (1992). Characterization of Suppressor of fused, a complete suppressor of the fused segment polarity gene of Drosophila melanogaster. *Genetics, 132*, 725–736.

Preat, T., Therond, P., Lamour-Isnard, C., Limbourg-Bouchon, B., Tricoire, H., Erk, I., et al. (1990). A putative serine/threonine protein kinase encoded by the segment-polarity fused gene of Drosophila. *Nature, 347*, 87–89.

Price, M. A., & Kalderon, D. (2002). Proteolysis of the Hedgehog signaling effector Cubitus interruptus requires phosphorylation by glycogen synthase kinase 3 and casein kinase 1. *Cell, 108*, 823–835.

Putoux, A., Thomas, S., Coene, K. L., Davis, E. E., Alanay, Y., Ogur, G., et al. (2011). KIF7 mutations cause fetal hydrolethalus and acrocallosal syndromes. *Nature Genetics, 43*, 601–606.

Qi, X., Friedberg, L., De Bose-Boyd, R., Long, T., & Li, X. (2020). Sterols in an intramolecular channel of Smoothened mediate Hedgehog signaling. *Nature Chemical Biology, 16*, 1368–1375.

Quirk, J., van den Heuvel, M., Henrique, D., Marigo, V., Jones, T. A., Tabin, C., et al. (1997). The smoothened gene and hedgehog signal transduction in Drosophila and vertebrate development. *Cold Spring Harbor Symposia on Quantitative Biology, 62*, 217–226.

Radhakrishnan, A., Anderson, T. G., & McConnell, H. M. (2000). Condensed complexes, rafts, and the chemical activity of cholesterol in membranes. *Proceedings of the National Academy of Sciences of the United States of America, 97*, 12422–12427.

Ramirez-Weber, F. A., & Kornberg, T. B. (1999). Cytonemes: Cellular processes that project to the principal signaling center in Drosophila imaginal discs. *Cell, 97*, 599–607.

Ranieri, N., Therond, P. P., & Ruel, L. (2014). Switch of PKA substrates from Cubitus interruptus to Smoothened in the Hedgehog signalosome complex. *Nature Communications, 5*, 5034.

Riddle, R. D., Johnson, R. L., Laufer, E., & Tabin, C. (1993). Sonic hedgehog mediates the polarizing activity of the ZPA. *Cell, 75*, 1401–1416.

Rijsewijk, F., Schuermann, M., Wagenaar, E., Parren, P., Weigel, D., & Nusse, R. (1987). The Drosophila homolog of the mouse mammary oncogene int-1 is identical to the segment polarity gene wingless. *Cell, 50*, 649–657.

Robbins, D. J., Nybakken, K. E., Kobayashi, R., Sisson, J. C., Bishop, J. M., & Therond, P. P. (1997). Hedgehog elicits signal transduction by means of a large complex containing the kinesin-related protein costal2. *Cell, 90*, 225–234.

Roelink, H., Augsburger, A., Heemskerk, J., Korzh, V., Norlin, S., Ruiz i Altaba, A., et al. (1994). Floor plate and motor neuron induction by vhh-1, a vertebrate homolog of hedgehog expressed by the notochord. *Cell, 76*, 761–775.

Rohatgi, R., Milenkovic, L., & Scott, M. P. (2007). Patched1 regulates hedgehog signaling at the primary cilium. *Science, 317*, 372–376.

Rong, P. M., Teillet, M. A., Ziller, C., & Le Douarin, N. M. (1992). The neural tube/notochord complex is necessary for vertebral but not limb and body wall striated muscle differentiation. *Development, 115*, 657–672.

Sanders, T. A., Llagostera, E., & Barna, M. (2013). Specialized filopodia direct long-range transport of SHH during vertebrate tissue patterning. *Nature, 497*, 628–632.

Sanial, M., Becam, I., Hofmann, L., Behague, J., Arguelles, C., Gourhand, V., et al. (2017). Dose-dependent transduction of Hedgehog relies on phosphorylation-based feedback between the G-protein-coupled receptor Smoothened and the kinase Fused. *Development, 144*, 1841–1850.

Santos, N., & Reiter, J. F. (2014). A central region of Gli2 regulates its localization to the primary cilium and transcriptional activity. *Journal of Cell Science, 127*, 1500–1510.

Schauerte, H. E., van Eeden, F. J., Fricke, C., Odenthal, J., Strahle, U., & Haffter, P. (1998). Sonic hedgehog is not required for the induction of medial floor plate cells in the zebrafish. *Development, 125*, 2983–2993.

Shi, Q., Li, S., Jia, J., & Jiang, J. (2011). The Hedgehog-induced Smoothened conformational switch assembles a signaling complex that activates Fused by promoting its dimerization and phosphorylation. *Development, 138*, 4219–4231.

Sisson, J. C., Ho, K. S., Suyama, K., & Scott, M. P. (1997). Costal2, a novel kinesin-related protein in the Hedgehog signaling pathway. *Cell, 90*, 235–245.

Smelkinson, M. G., & Kalderon, D. (2006). Processing of the Drosophila hedgehog signaling effector Ci-155 to the repressor Ci-75 is mediated by direct binding to the SCF component Slimb. *Current Biology, 16*, 110–116.

Smelkinson, M. G., Zhou, Q., & Kalderon, D. (2007). Regulation of Ci-SCFSlimb binding, Ci proteolysis, and hedgehog pathway activity by Ci phosphorylation. *Developmental Cell, 13*, 481–495.

Smith, M. J., Beetz, C., Williams, S. G., Bhaskar, S. S., O'Sullivan, J., Anderson, B., et al. (2014). Germline mutations in SUFU cause Gorlin syndrome-associated childhood medulloblastoma and redefine the risk associated with PTCH1 mutations. *Journal of Clinical Oncology, 32*, 4155–4161.

Stone, D. M., Hynes, M., Armanini, M., Swanson, T. A., Gu, Q., Johnson, R. L., et al. (1996). The tumour-suppressor gene patched encodes a candidate receptor for Sonic hedgehog. *Nature, 384*, 129–134.

Sturtevant, A. (1948). Sequence of loci near the centromere of chromosome 2. *Drosophila Information Service, 22*, 56.

Svard, J., Heby-Henricson, K., Persson-Lek, M., Rozell, B., Lauth, M., Bergstrom, A., et al. (2006). Genetic elimination of Suppressor of fused reveals an essential repressor function in the mammalian Hedgehog signaling pathway. *Developmental Cell, 10*, 187–197.

Tabata, T., Eaton, S., & Kornberg, T. B. (1992). The Drosophila hedgehog gene is expressed specifically in posterior compartment cells and is a target of engrailed regulation. *Genes & Development, 6*, 2635–2645.

Tabata, T., & Kornberg, T. B. (1994). Hedgehog is a signaling protein with a key role in patterning Drosophila imaginal discs. *Cell, 76*, 89–102.

Taipale, J., Cooper, M. K., Maiti, T., & Beachy, P. A. (2002). Patched acts catalytically to suppress the activity of Smoothened. *Nature, 418*, 892–897.

Takei, Y., Ozawa, Y., Sato, M., Watanabe, A., & Tabata, T. (2004). Three Drosophila EXT genes shape morphogen gradients through synthesis of heparan sulfate proteoglycans. *Development, 131*, 73–82.

Tanimoto, H., Itoh, S., ten Dijke, P., & Tabata, T. (2000). Hedgehog creates a gradient of DPP activity in Drosophila wing imaginal discs. *Molecular Cell, 5*, 59–71.

Tashiro, S., Michiue, T., Higashijima, S., Zenno, S., Ishimaru, S., Takahashi, F., et al. (1993). Structure and expression of hedgehog, a Drosophila segment-polarity gene required for cell-cell communication. *Gene, 124*, 183–189.

Tay, S. Y., Ingham, P. W., & Roy, S. (2005). A homologue of the Drosophila kinesin-like protein Costal2 regulates Hedgehog signal transduction in the vertebrate embryo. *Development, 132*, 625–634.

Tay, S. Y., Yu, X., Wong, K. N., Panse, P., Ng, C. P., & Roy, S. (2010). The iguana/DZIP1 protein is a novel component of the ciliogenic pathway essential for axonemal biogenesis. *Developmental Dynamics*, *239*, 527–534.

Taylor, A. M., Nakano, Y., Mohler, J., & Ingham, P. W. (1993). Contrasting distributions of patched and hedgehog proteins in the Drosophila embryo. *Mechanisms of Development*, *42*, 89–96.

Tenzen, T., Allen, B. L., Cole, F., Kang, J. S., Krauss, R. S., & McMahon, A. P. (2006). The cell surface membrane proteins Cdo and Boc are components and targets of the Hedgehog signaling pathway and feedback network in mice. *Developmental Cell*, *10*, 647–656.

Tickle, C., & Towers, M. (2017). Sonic Hedgehog signaling in limb development. *Frontiers in Cell and Development Biology*, *5*, 14.

Tiniakow, G. G., & Terentieva, E. L. (1933). Cubitus Interruptus, a New Genovariation of the Fourth Chromosome of DROSOPHILA MELANOGASTER. *Genetics*, *18*, 117–120.

Tole, S., Ragsdale, C. W., & Grove, E. A. (2000). Dorsoventral patterning of the telencephalon is disrupted in the mouse mutant extra-toes(J). *Developmental Biology*, *217*, 254–265.

Truong, M. E., Bilekova, S., Choksi, S. P., Li, W., Bugaj, L. J., Xu, K., et al. (2021). Vertebrate cells differentially interpret ciliary and extraciliary cAMP. *Cell*, *184*, 2911–2926. e2918.

Tukachinsky, H., Kuzmickas, R. P., Jao, C. Y., Liu, J., & Salic, A. (2012). Dispatched and scube mediate the efficient secretion of the cholesterol-modified hedgehog ligand. *Cell Reports*, *2*, 308–320.

Tukachinsky, H., Lopez, L. V., & Salic, A. (2010). A mechanism for vertebrate Hedgehog signaling: Recruitment to cilia and dissociation of SuFu-Gli protein complexes. *The Journal of Cell Biology*, *191*, 415–428.

Tuson, M., He, M., & Anderson, K. V. (2011). Protein kinase A acts at the basal body of the primary cilium to prevent Gli2 activation and ventralization of the mouse neural tube. *Development*, *138*, 4921–4930.

van den Heuvel, M., & Ingham, P. W. (1996). smoothened encodes a receptor-like serpentine protein required for hedgehog signalling. *Nature*, *382*, 547–551.

van Eeden, F. J., Granato, M., Schach, U., Brand, M., Furutani-Seiki, M., Haffter, P., et al. (1996). Mutations affecting somite formation and patterning in the zebrafish, Danio rerio. *Development*, *123*, 153–164.

Varga, Z. M., Amores, A., Lewis, K. E., Yan, Y. L., Postlethwait, J. H., Eisen, J. S., et al. (2001). Zebrafish smoothened functions in ventral neural tube specification and axon tract formation. *Development*, *128*, 3497–3509.

Varjosalo, M., Li, S. P., & Taipale, J. (2006). Divergence of hedgehog signal transduction mechanism between Drosophila and mammals. *Developmental Cell*, *10*, 177–186.

Vortkamp, A., Franz, T., Gessler, M., & Grzeschik, K. H. (1992). Deletion of GLI3 supports the homology of the human Greig cephalopolysyndactyly syndrome (GCPS) and the mouse mutant extra toes (Xt). *Mammalian Genome*, *3*, 461–463.

Wang, X., Zhao, Z., Muller, J., Iyu, A., Khng, A. J., Guccione, E., et al. (2013). Targeted inactivation and identification of targets of the Gli2a transcription factor in the Zebrafish. *Biology Open*, *2*, 1203–1213.

Whittle, J. R. (1976). Clonal analysis of a genetically caused duplication of the anterior wing in Drosophila melanogaster. *Developmental Biology*, *51*, 257–268.

Wierbowski, B. M., Petrov, K., Aravena, L., Gu, G., Xu, Y., & Salic, A. (2020). Hedgehog pathway activation requires coreceptor-catalyzed, lipid-dependent relay of the sonic Hedgehog ligand. *Developmental Cell*, *55*, 450–467. e458.

Wieschaus, E., Nusslein-Volhard, C., & Jurgens, G. (1984). Mutations affecting the pattern of the larval cuticle in Drosophila melanogaster: III. Zygotic loci on the X-chromosome and fourth chromosome. *Wilehm Roux Archives of Developmental Biology*, *193*, 296–307.

Wijgerde, M., McMahon, J. A., Rule, M., & McMahon, A. P. (2002). A direct requirement for Hedgehog signaling for normal specification of all ventral progenitor domains in the presumptive mammalian spinal cord. *Genes & Development, 16*, 2849–2864.

Wilson, C. W., Nguyen, C. T., Chen, M. H., Yang, J. H., Gacayan, R., Huang, J., et al. (2009). Fused has evolved divergent roles in vertebrate Hedgehog signalling and motile ciliogenesis. *Nature, 459*, 98–102.

Wolff, C., Roy, S., & Ingham, P. W. (2003). Multiple muscle cell identities induced by distinct levels and timing of hedgehog activity in the zebrafish embryo. *Current Biology, 13*, 1169–1181.

Woods, I. G., & Talbot, W. S. (2005). The you gene encodes an EGF-CUB protein essential for Hedgehog signaling in zebrafish. *PLoS Biology, 3*, e66.

Wu, X., Zhang, Y., Chuang, K. H., Cai, X., Ajaz, H., & Zheng, X. (2019). The Drosophila Hedgehog receptor component Interference hedgehog (Ihog) mediates cell-cell interactions through trans-homophilic binding. *The Journal of Biological Chemistry, 294*, 12339–12348.

Xiao, X., Tang, J. J., Peng, C., Wang, Y., Fu, L., Qiu, Z. P., et al. (2017). Cholesterol modification of Smoothened is required for Hedgehog signaling. *Molecular Cell, 66*, 154–162. e110.

Yam, P. T., Langlois, S. D., Morin, S., & Charron, F. (2009). Sonic hedgehog guides axons through a noncanonical, Src-family-kinase-dependent signaling pathway. *Neuron, 62*, 349–362.

Yan, D., Wu, Y., Yang, Y., Belenkaya, T. Y., Tang, X., & Lin, X. (2010). The cell-surface proteins Dally-like and Ihog differentially regulate Hedgehog signaling strength and range during development. *Development, 137*, 2033–2044.

Yang, S., Zhang, Y., Yang, C., Wu, X., El Oud, S. M., Chen, R., et al. (2021). Competitive coordination of the dual roles of the Hedgehog co-receptor in homophilic adhesion and signal reception. *eLife, 10*, e65770.

Yao, S., Lum, L., & Beachy, P. (2006). The ihog cell-surface proteins bind Hedgehog and mediate pathway activation. *Cell, 125*, 343–357.

Yin, Y., Bangs, F., Paton, I. R., Prescott, A., James, J., Davey, M. G., et al. (2009). The Talpid3 gene (KIAA0586) encodes a centrosomal protein that is essential for primary cilia formation. *Development, 136*, 655–664.

Zadorozny, E. V., Little, J. C., & Kalderon, D. (2015). Contributions of costal 2-Fused interactions to Hedgehog signaling in Drosophila. *Development, 142*, 931–942.

Zeng, X., Goetz, J. A., Suber, L. M., Scott, W. J., Jr., Schreiner, C. M., & Robbins, D. J. (2001). A freely diffusible form of Sonic hedgehog mediates long-range signalling. *Nature, 411*, 716–720.

Zhang, Y., Bulkley, D. P., Xin, Y., Roberts, K. J., Asarnow, D. E., Sharma, A., et al. (2018). Structural basis for cholesterol transport-like activity of the Hedgehog receptor Patched. *Cell, 175*, 1352–1364. e1314.

Zhang, W., Zhao, Y., Tong, C., Wang, G., Wang, B., Jia, J., et al. (2005). Hedgehog-regulated Costal2-kinase complexes control phosphorylation and proteolytic processing of Cubitus interruptus. *Developmental Cell, 8*, 267–278.

Zhao, Z., Lee, R. T., Pusapati, G. V., Iyu, A., Rohatgi, R., & Ingham, P. W. (2016). An essential role for Grk2 in Hedgehog signalling downstream of Smoothened. *EMBO Reports, 17*, 739–752.

Zheng, X., Mann, R. K., Sever, N., & Beachy, P. A. (2010). Genetic and biochemical definition of the Hedgehog receptor. *Genes & Development, 24*, 57–71.

Zhou, Q., & Kalderon, D. (2011). Hedgehog activates fused through phosphorylation to elicit a full spectrum of pathway responses. *Developmental Cell, 20*, 802–814.

CHAPTER TWO

Different strategies by distinct Wnt-signaling pathways in activating a nuclear transcriptional response

Linh T. Vuong and Marek Mlodzik*
Department of Cell, Developmental, & Regenerative Biology, Graduate School of Biomedical Sciences, Icahn School of Medicine at Mount Sinai, New York, NY, United States
*Corresponding author: e-mail address: marek.mlodzik@mssm.edu

Contents

Abstract

The Wnt family of secreted glycolipo-proteins signals through multiple signal transduction pathways and is essential for embryonic development and organ development and homeostasis. The Wnt-pathways are conserved and critical in all metazoans. Wnt signaling pathways comprise the canonical Wnt/β-catenin pathway and several non-canonical signaling branches, of which Wnt-Planar Cell Polarity (PCP) signaling and the Wnt/Calcium pathway have received the most attention and are best understood. nterestingly, all Wnt-pathways have a nuclear signaling branch and also can affect many cellular processes independent of its nuclear transcriptional regulation. Canonical Wnt/β-catenin signaling is the most critical for a nuclear transcriptional response, in both development and disease, yet the mechanism(s) on how the "business end" of the pathway, β-catenin, translocates to the nucleus to act as co-activator to the TCF/Lef transcription factor family still remains obscure. Here we discuss and compare the very different

Current Topics in Developmental Biology, Volume 149
ISSN 0070-2153
https://doi.org/10.1016/bs.ctdb.2022.02.008

strategies on how the respective Wnt signaling pathways activate a nuclear transcriptional response. We also highlight some recent new insights into how β-catenin is translocated to the nucleus via an IFT-A, Kinesin-2, and microtubule dependent mechanism and how this aspect of canonical Wnt-signaling uses ciliary proteins in a cilium independent manner, conserved between *Drosophila* and mammalian cells.

1. Overview of Wnt-signaling pathways

The founding member of the Wnt gene family was first discovered in *Drosophila*, with its original mutant allele lacking wings, and hence it was named *wingless* (*wg*) (Sharma & Chopra, 1976). Subsequently, it was shown to be an essential gene and required throughout development including embryogenesis (Nusslein-Volhard & Wieschaus, 1980). Conserved sequences were identified in mice via gene disruption strategies using viral insertions as mutagens in a cancer screening context looking for malignant mammary tissue transformations (Nusse, van Ooyen, Cox, Fung, & Varmus, 1984). As this original insertion sequence in the mouse genome was named *int1*, and the associated mouse gene was found to be highly homologous to the *Drosophila* gene, the associated gene family was named *Wnt* as a combination of *Drosophila* Wingless and mouse "int." The family of Wnt/Wg genes is conserved in all metazoans, with multiple members in most species, and encodes proteins of a secreted glycolipoprotein class. Wnt genes are not only essential and critical for animal development, organogenesis, and homeostasis of all metazoans, but are also critically linked to regeneration and stem cell maintenance, as well as being dysregulated in many diseases (Cadigan & Waterman, 2012; Clevers, 2006; Logan & Nusse, 2004; MacDonald, Tamai, & He, 2009).

Strikingly, over the years, it was shown that Wnt protein family associated pathways comprise a fairly large set of distinct signaling branches, which share the Wnt ligands and Frizzled (Fz, or Fzd, for short) receptor family and the downstream effector Dishevelled (Dsh, a.k.a. Dvl in mammals) (see below for review references). Among these, the canonical Wnt/β-catenin signaling pathway stands out as the most studied and overall best understood (Cadigan & Waterman, 2012; Clevers, 2006; Eisenmann, 2005; Logan & Nusse, 2004; MacDonald et al., 2009). For an updated comprehensive pathway overview see also the Wnt homepage (www.stanford.edu/~rnusse/wntwindow.html). In addition to the canonical pathway, there exist several so-called non-canonical Wnt signaling pathways that have emerged and received significant attention. Among these non-canonical pathways, the following received the most attention: Wnt-Planar Cell Polarity (PCP) signaling (reviewed in (Adler, 2002; Goodrich & Strutt, 2011; Harrison, Shao, Strutt, & Strutt, 2020; Humphries & Mlodzik, 2018; Peng & Axelrod, 2012) and the Wnt/Calcium pathway (see reviews Kohn & Moon, 2005; Kuhl, Sheldahl, Park, Miller, & Moon, 2000). These are the most studied and (at least in part) understood and will be discussed further here in the context of nuclear signaling and transcriptional activation. For a complete listing of additional non-canonical Wnt-pathways see (Semenov, Habas, Macdonald, & He, 2007).

Each of the above mentioned three Wnt-pathways have both a nuclear signaling branch and can also affect several cellular processes independently of their effects on nuclear transcriptional regulation. Canonical Wnt/β-catenin signaling is the most critical for a nuclear transcriptional response, in both development and disease (reviewed in (Clevers, 2006; Clevers & Nusse, 2012; MacDonald et al., 2009)). Although quite a bit is understood about the nuclear complexes that activate or repress transcription of target genes in the nucleus with co-repressors and co-activators of the TCF/Lef transcription factor family, the key mechanism(s) of how nuclear translocation of β-catenin is achieved still remains obscure. As β-catenin is the absolutely essential co-activator, around which a transcription activation complex is assembled, this missing link is critical not just for better understanding of the pathway cell biology, but also for potentially better focused therapeutic approaches. In contrast, the two non-canonical Wnt-signaling branches (see below) are better understood at the nuclear signaling level, although there are significant gaps in their understanding mostly at the level of membrane proximal pathway relay mechanisms.

In this chapter, we discuss and compare the very different strategies on how the respective Wnt signaling pathways generate an active nuclear response.

1.1 The canonical Wnt/β-catenin signaling pathway

Canonical Wnt/Wingless signaling is a highly conserved pathway with critical roles in the regulation of a vast catalogue of developmental processes, including but not restricted to tissue patterning as a morphogen across the animal kingdom, cell fate specification, cell proliferation, cell survival, and migration (Cadigan & Waterman, 2012; Clevers, 2006; Logan & Nusse, 2004; MacDonald et al., 2009). Dysregulated pathway components and or misexpression of pathway proteins are in most cases associated with

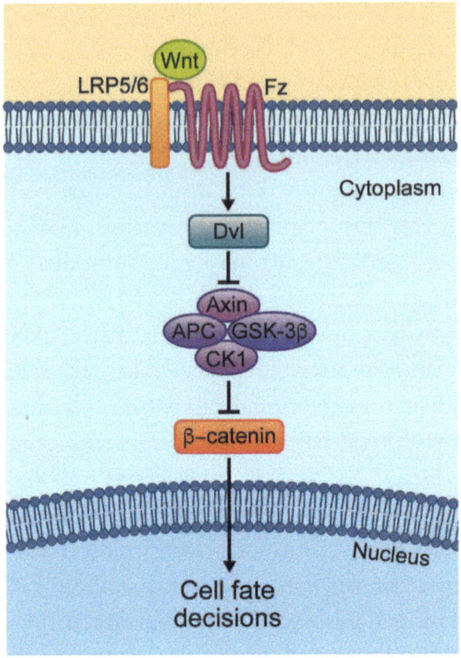

Fig. 1 Simplified overview of the canonical Wnt (β-catenin) pathway. Upon pathway activation, Wnt binding causes the dimerization of the Fz (Fzd in vertebrates) receptor and the LRP5/6 co-receptor. The receptor complex then recruit a Dsh/Dvl-Axin multimer, which leads to the disassembly of the "destruction complex," which—in the inactive state—includes Axin, APC, GSK-3β and CK1 among other proteins, and β-catenin. Thus upon the disassembly of the DC, β-catenin is freed and relieved from proteasome-dependent degradation and accumulates in the cytoplasm. Subsequently, β-catenin translocates to the nucleus, where its associates with the TCF/Lef transcription factors and several co-activators to turn on Wnt target gene transcription, which regulates cell fate decisions, proliferation, and cell survival. See main text for details.

diseases, including several cancer types (e.g. Clevers, 2006; Clevers & Nusse, 2012; Coombs, Covey, & Virshup, 2008; Prestwich & Macdougald, 2007).

At the center of active canonical Wnt signaling is the stabilization of β-catenin (Armadillo/Arm in *Drosophila*) (Cadigan & Waterman, 2012; MacDonald et al., 2009; Valenta, Hausmann, & Basler, 2012) (see Fig. 1 for simplified schematic). β-catenin/Arm is a multi-functional protein that is critical as both a component of adherens junctions (AJs) and their link to cytoskeletal elements (reviewed in (Bienz, 2005; McEwen, Escobar, & Gottardi, 2012; Tian et al., 2011) and as the key component in Wnt/Wg signaling. In the canonical Wnt pathway, β-catenin/Arm is the essential transcriptional co-activator for Wnt-target gene expression (reviewed in (Cadigan & Waterman, 2012; MacDonald et al., 2009; Valenta et al., 2012). Wnt/Wg proteins bind to Frizzled (Fz/Fzd in mammals) receptors and LRP5/6 (Arrow in *Drosophila*) co-receptors, resulting in the disassembly of the so-called "destruction complex," composed of Axin, APC (Adenomatous Polyposis Coli), and the kinases GSK3β and Casein kinase I (CKI), which phosphorylate the cytoplasmic pool of β-catenin/Arm and thus target it for degradation (reviewed in (Cadigan & Waterman, 2012; MacDonald et al., 2009; Valenta et al., 2012)). Note that there are two pools of β-catenin/Arm, one being stably associated with Cadherins in AJ complexes and the other, the cytoplasmic, free pool that is targeted for degradation in the absence of Wnt-signaling. Wnt pathway activation results in the break-up of the destruction complex and re-localization of Axin to the plasma membrane, where it associates with Dsh/Dvl and the Fz/Fzd receptors and LRP5/6 co-receptors, forming large complexes of Axin-Dsh-Fz-LRP5/6 aggregates in endocytosed vesicles, generally referred to as signalosomes (Cadigan & Waterman, 2012; Cliffe, Hamada, & Bienz, 2003; MacDonald et al., 2009). Generally, the Dsh/Dvl and Axin proteins behave like protein scaffolds that hold together either the destruction complex (Axin) or the signalosome (DshDvl) (Cadigan & Waterman, 2012; MacDonald et al., 2009; Valenta et al., 2012). Axin removal from the destruction complex leads not only to complex disassembly, but importantly frees up and thus stabilizes cytoplasmic β-catenin/Arm, which can then translocate to the nucleus to act as a co-activator of the TCF/Lef transcription factors (Cadigan & Waterman, 2012; MacDonald et al., 2009). Again, in the absence of Wnt/Wg presence, β-catenin/Arm is phosphorylated by the "destruction complex" and targeted to the proteasome (Clevers & Nusse, 2012; Niehrs, 2012). While these membrane proximal and cytoplasmic aspects of the pathway are quite well defined, a big mystery remains in how β-catenin/Arm gets into the nucleus. Its primary sequence does not contain a bona fide nuclear localization signal (NLS), and thus while once in the nucleus its molecular function and interaction partners are again well known, the steps in between remain largely unclear. New insight into the potential mechanism(s) of how β-catenin/Arm is translocated in the nucleus have recently emerged (Balmer et al., 2015; Vuong et al., 2018) and these are discussed in a dedicated section below.

In the context of Wnt/Wg-signaling, *Drosophila* and in particular wing development in the fly serves as an excellent model system and paradigm for pathway dissection and functional analyses. Importantly, once the wing primordia are established and a larval wing discs has formed, Wg is expressed as a two-cell stripe directly at the dorso-ventral (D/V) boundary of the future wing pouch. Emanating from there it forms an extracellular concentration gradient and acts as a morphogen, activating targets in a concentration dependent manner (e.g. (Baena-Lopez, Nojima, & Vincent, 2012; Danielson et al., 1995; Strigini & Cohen, 2000) and thus patterning the wing and specifying future wing structures, including the blade and margin associated cell fates (Neumann & Cohen, 1997; Zecca, Basler, & Struhl, 1996). In this context, high threshold transcriptional targets of Wg, include *senseless (sens)*, a gene responsible for the formation of margin sensory bristles of adult wings, in cells directly adjacent to D/V boundary, while lower threshold target genes expressed in the wing blade include *Distalless* (*Dll*), required for wing growth (Neumann & Cohen, 1997; Zecca et al., 1996). *Dll* is expressed in a graded fashion, decreasing towards the edges of the wing pouch. Similar to most of the genetic dissection of Wg/Wnt signaling having relied on the use of the *Drosophila* wing, the discovery of a novel and unexpected mechanism for the nuclear translocation of β-catenin/Arm also resulted from genetic screens in this model, using the above mentioned target genes as functional read-out for β-catenin/Arm nuclear function (Balmer et al., 2015; Vuong et al., 2018; Vuong, Mukhopadhyay, & Choi, 2014).

1.2 Wnt/Fz-PCP signaling

Like the canonical Wnt/Wg-signaling pathway, Wnt/Planar Cell Polarity (PCP) signaling was first discovered in insects and is best studied and characterized in *Drosophila* (e.g. (Adler, 2002; Seifert & Mlodzik, 2007). In flies, PCP is evident in all adult cuticular structures and the compound eye and thus the simple easy of observation has been instrumental in identifying PCP pathway components and assembling these into a Wnt/PCP signaling pathway (e.g. Axelrod, Miller, Shulman, Moon, & Perrimon, 1998; Boutros, Paricio, Strutt, & Mlodzik, 1998; Das, Jenny, Klein, Eaton, & Mlodzik, 2004; Jenny, Reynolds-Kenneally, Das, Burnett, & Mlodzik, 2005; Tree et al., 2002; Tree, Ma, & Axelrod, 2002; Usui et al., 1999; Vinson & Adler, 1987; Vinson, Conover, & Adler, 1989; Wolff & Rubin, 1998) (see Fig. 2 for simplified schematic overview). Cellular patterning in tissues often requires directional information, either at the level of distinct cell fate specification or cellular orientation among equivalent cells. Both, mesenchymal cells and most epithelia are polarized within the planar axis, which is referred to as PCP. Such polarity-type arrangements

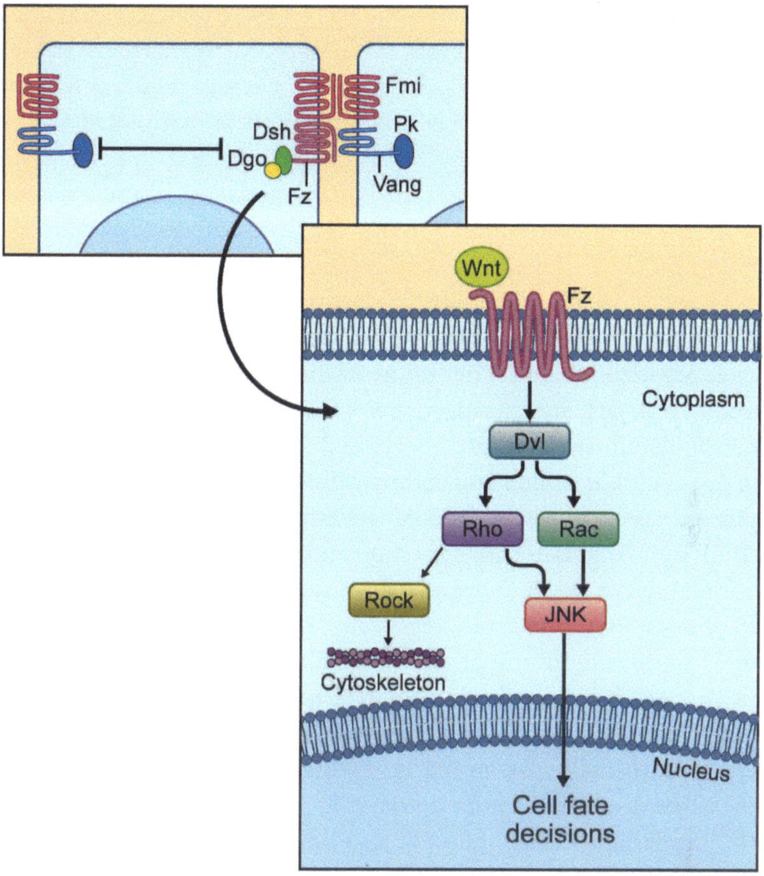

Fig. 2 Overview of the non-canonical Wnt-Fz/PCP pathway. Upper part: a simplified diagram of the interactions among the PCP core components, which results in the stable resolution and establishment of two complexes, the Fz-Fmi-Dsh-Dgo complex (presumed to be on the side of the Wnt ligand) and the Vang-Fmi-Pk complex. Lower part: Simplified schematic view of the Wnt/PCP signaling pathway downstream of the Fz-Dsh complex. The PCP pathway is activated when a Wnt ligand binds to a Frizzled (Fz/Fzd) receptor and induces a clustering of the core PCP protein to either the Vang/Vangl-Pk complex or Fz-Dsh-Dgo complex (upper part). Downstream activation of the PCP pathway is associated with Dsh/Dvl phosphorylation and its interactions with downstream effectors Rho and Rac, possibly also cdc42 (not shown), with for example RhoA then acting through either Rock or the JNK-type MAPK cascade, with the respective kinases regulating either the cytoskeleton assembly (Rock) or triggering the expression of the target genes in the nucleus (JNK) regulating cell fate decisions. See main text for details.

provide cells and tissues with positional information allowing them to (i) generate polarized structures that are oriented with respect to tissue axes, (ii) embed specialized cells (e.g. sensory cells) with a specific orientation in any given organ, (iii) regulate cellular movement in a directed fashion, or (iv) instruct directed asymmetric cellular differentiation to generate asymmetric clusters of cells with specialized functions. All these PCP aspects are conserved across species and critical for development and homeostasis, and often linked to human disease (see recent reviews (Aw & Devenport, 2017; Butler & Wallingford, 2017; Davey & Moens, 2017; Goodrich & Strutt, 2011; Humphries & Mlodzik, 2018; Peng & Axelrod, 2012; Simons & Mlodzik, 2008)).The feature of directed, asymmetrically oriented differential cell fate specification, involves the nuclear branch of Wnt/Fz-PCP signaling (see dedicated Section 2.2).

For general PCP studies and pathway dissection, two tissues in *Drosophila* have served as exceptional model systems to unravel key features in Wnt/PCP signaling: these are the wings and eyes. *Drosophila* wing development, arguably the best understood PCP system, serves as an excellent system to dissect PCP mediated regulation of cytoskeletal orientation and assembly and the establishment and maintenance of the asymmetric distribution of core PCP factors from late larval to pupal stages of development. Here, all wing cells form a single actin-based cellular hair that is oriented in the proximo-distal axis, pointing distally, and this uniform orientation pattern, when disrupted in mutants of PCP components is easily appreciated as such. Similarly, in eyes, the very regular arrangement of ommatidia (or unit eyes, a.k.a. facets) with respect to the anterior-posterior (AP) and dorsal-ventral (DV) axes is altered in PCP mutants (Adler, 2002; Goodrich & Strutt, 2011; Peng & Axelrod, 2012; Seifert & Mlodzik, 2007). Based on the associated phenotypes, a core group of evolutionarily conserved Wnt-PCP genes have been identified, which are generally referred to as the "core Frizzled (Fz)/PCP factors." The core Fz/PCP molecules include three transmembrane proteins: Fz, Flamingo (Fmi, a.k.a. Starry night/Stan; Celsr in vertebrates), and Van Gogh (Vang; a.k.a. Strabismus/Stbm; Vangl1/2 in vertebrates), and three cytoplasmic factors Dishevelled (Dsh, Dvl in mammals), Prickle (Pk), and Diego (Dgo, Inversin/Diversin in vertebrates) (reviewed in (Aw & Devenport, 2017; Butler & Wallingford, 2017; Davey & Moens, 2017; Goodrich & Strutt, 2011; Humphries & Mlodzik, 2018; Peng & Axelrod, 2012). These core PCP factors interact with each other, resolving into two stable complexes: one composed of Fmi-Fz-Dsh-Dgo and the other of Fmi-Vang-Pk.

These complexes are stabilizing each other between cells across cell membranes, with a Fmi-Fz and Fmi-Vang intercellular adhesive interaction, and they antagonize each other within a given cell, via negative interactions between Dsh-Dgo and Pk. As a result these complexes become stably and asymmetrically localized along an axis (Aw & Devenport, 2017; Aw, Heck, Joyce, & Devenport, 2016; Butler & Wallingford, 2017; Davey & Moens, 2017; Goodrich & Strutt, 2011; Humphries & Mlodzik, 2018; Peng & Axelrod, 2012; Yang & Mlodzik, 2015). Generally, it has been demonstrated that the polarization axes are generated by a localized expression of Wnt, with cells orienting towards the Wnt source (for example Carvajal-Gonzalez, Mulero-Navarro, & Mlodzik, 2016; Gao et al., 2011; Minegishi et al., 2017; Wu, Roman, Carvajal-Gonzalez, & Mlodzik, 2013).

Once a polarization axis is established, largely by the interactions between the core PCP components, the asymmetric localization of the core complexes then regulates downstream effectors to elicit tissue-specific responses affecting both cytoskeletal orientation (including the alignment of the mitotic spindle and localization centrioles and cilia) and nuclear signaling (to induce cell fate) (reviewed in Aw et al., 2016; Butler & Wallingford, 2017; Carvajal-Gonzalez et al., 2016; Goodrich & Strutt, 2011; Humphries & Mlodzik, 2018; Peng & Axelrod, 2012; Segalen & Bellaiche, 2009; Wallingford, 2010). Importantly, not only the core Wnt-Fz/PCP factors are conserved across the animal kingdom, but also the set of downstream cellular responses and read-outs that, like the molecular PCP (see references above). Two of the most obvious and common read-outs (i) polarization of acto-myosin cytoskeletal elements and (ii) positioning of the mitotic spindle/centriole/cilia are discussed in several reviews (see references above). Here, we focus on the nuclear signaling mechanisms of the Wnt/Fz-PCP pathway and compare it to other nuclear Wnt-signaling mechanisms. This read-out, although it shares the membrane proximal regulatory input, can largely be separated from the others downstream of the action of the Dsh/Dvl protein family.

1.3 Wnt/calcium signaling

There are several proposed additional non-canonical Wnt-signaling pathways beyond PCP signaling (see Snapshot: Semenov et al., 2007), which include specific functions of Wnt-signaling in neuronal pathfinding (also linked to PCP) or synaptic functions and bouton formation or stability (reviewed for example in McLeod et al., 2020; McLeod & Salinas, 2018; Teo & Salinas, 2021; Zou, 2020). Of these additional pathways, the

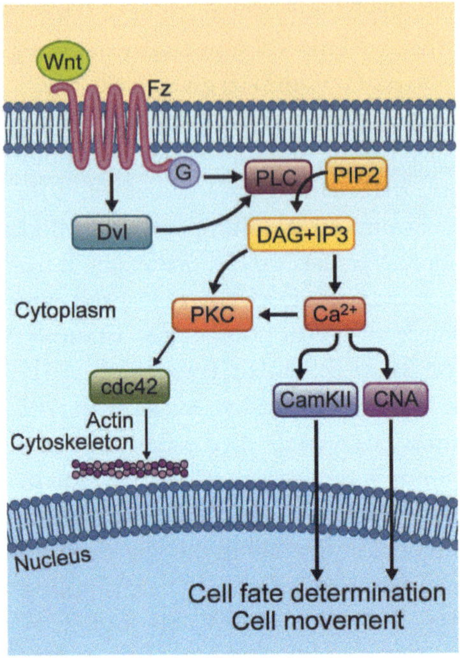

Fig. 3 Overview of the Wnt-Ca2+ pathway. Wnt ligand-Fz binding is mediating Dvl activation with possibly trimeric G proteins-coupled to the process. Dvl also can activate PLC to decompose PIP2 into DAG and IP3. DAG is activated by released calcium from ER, which activates PKC, which then activates Cdc42 to regulate actomyosin cytoskeletal elements. In parallel, IP3 leads to intracellular calcium release via ER, which in turn results in phosphorylation and activation of CamKII and CNA. These proteins activate NFkB and NFAT which translocate to the nucleus to regulate cell movement and cell fate decisions.

Wnt/calcium signaling pathway has had significant attention and is molecularly quite well defined (De, 2011; Kohn & Moon, 2005; Kuhl, Sheldahl, Park, et al., 2000).

The Wnt/calcium pathway leads to the activation of calcium/calmodulin-dependent kinase II (CamKII) and PKC (see Fig. 3 for schematic overview). In this pathway, PKC activates actomyosin cytoskeletal elements via cdc42 and its downstream effectors. The nuclear branch of the Wnt/calcium pathway acts via CamKII and Calcineurin (CNA) on rather general nuclear factors like ATF, NF-AT or NFκB. Wnt/calcium signaling has been suggested to be activated by Wnt5a, in particular, and its analyses have mostly been performed in non-mammalian vertebrate model systems, namely zebrafish and frog/*Xenopus* (Kohn & Moon, 2005; Kuhl, Sheldahl,

Park, et al., 2000). For example, Wnt5 has been linked to intracellular calcium release. Such intracellular calcium signaling is required for body plan specification and shown to affect dorsalization of embryos, and accordingly, zebrafish embryos deficient for Wnt5 display reduced fluxes of calcium and dorsalization defects. Interestingly, such Wnt5 mutant zebrafish embryos (loss-of-function), which display a hyperdorsalization, can in part be rescued by the expression of a constitutively active CamKII, a key proposed effector of the Wnt/calcium pathway (Westfall et al., 2003). Consistent with the studies in zebrafish, expression of dominant negative isoforms of either Wnt11 or CamKII lead to embryonic dorsalization in *Xenopus* (Kuhl, Sheldahl, Malbon, & Moon, 2000). Other developmental contexts where the Wnt/calcium pathway has been suggested to function include mammalian T-cell differentiation. The NF-AT transcription factor is a nuclear effector regulated by the calcium/calmodulin-dependent phosphatase called Calcineurin/CNA. Co-expression of NF-AT with Wnt5a in Xenopus induces nuclear translocation of NF-AT during embryonic dorsalization (reviewed in (Rao, Luo, & Hogan, 1997). Similarly, treatment of T-cells with Wnt5a results in nuclear translocation and accumulation of NF-AT (Murphy & Hughes, 2002). Taken together, these data are consistent with the notion that a Wnt5/calcium pathway uses CNA to regulate nuclear accumulation of NF-AT as a pathway read out.

It has also been suggested that trimeric G-proteins are part of the Wnt/calcium pathway (summarized in (Semenov et al., 2007). Potential evidence for that comes from a study in *Drosophila*: here a trimeric G-protein complex, Gαo, has been linked to Wnt/Wg signaling in the context of wing development (Katanaev, Ponzielli, Semeriva, & Tomlinson, 2005). However, a link to calcium signaling has not been made in the *Drosophila* study, and so the proposed Wnt/calcium pathway based on experiments in fish and frogs has not been reconstructed in *Drosophila* or mice. Whether the potentially associated trimeric G-proteins, as suggested from a study in *Drosophila* (Katanaev et al., 2005), contribute to these effects or act independently of or parallel to these kinases is not resolved, nor is a formal link established of this trimeric G-protein unit to the Wnt/calcium-pathway.

The direct molecular information flow is less well defined in this Wnt pathway, as compared to canonical Wnt or Wnt/PCP signaling. Interestingly, some of the proposed components of this pathway have been linked to it through derivative analyses in heterologous systems and basically the pathway takes advantage of other existing signaling modules, which were defined in signaling contexts outside to Wnt signaling (reviewed in Kohn & Moon, 2005;

Kuhl, Sheldahl, Malbon, & Moon, 2000). Thus, in general, information flow here has been largely derived from the common knowledge base, as again it uses components of other signaling mechanisms (see reviews Kohn & Moon, 2005; Kuhl, Sheldahl, Malbon, & Moon, 2000; Semenov et al., 2007). As such, downstream of Dsh/Dvl, the PLC enzyme gets activated which converts PIP2 into the second messengers di-acylglycerol (DAG) and inositol-3-phosphate (IP3). These either activate PKC directly (DAG) or lead to an increase in intracellular calcium/Ca^{2+}, which in turn activates CamKII and CNA, for example. In summary, Wnt/calcium signaling redeploys these existing, more general, signaling modules, for example a PLC-DAG-PKC activation sequence or PLC-IP3-Ca^{2+}-CAMKII signaling flow, as downstream effector modules of a Wnt-Frizzled-Dsh/Dvl interaction. Similarly, the resulting effectors that would act in the context of cytoskeletal regulation, e.g. PKC-cdc42 etc, or in nuclear contexts, resulting in the activation of NF-AT, ATF2 or NFκB for example, are rather generic cellular or nuclear effectors used downstream of several other pathways.

2. Induction of nuclear responses downstream of Wnt-signaling

2.1 Mechanisms of canonical Wnt/β-catenin signaling in the nucleus

2.1.1 Wnt/β-catenin mediated transcriptional activation

Upon activation of Wnt signaling, the inhibition of β-catenin destruction leads to increased levels of β-catenin in the cytoplasm, and subsequent β-catenin translocation into the nucleus (Cadigan & Waterman, 2012; MacDonald et al., 2009). Once in the nucleus, β-catenin associates with TCF/Lef through its central region containing the Arm-repeats to activate transcription of Wnt/β-catenin specific target genes (Fig. 4) (Behrens et al., 1996; Cadigan & Waterman, 2012; MacDonald et al., 2009). In the absence of β-catenin, TCF recruits a transcriptional repressor complex with TLE (*Groucho* in *Drosophila*), which promotes histone deacetylation and chromatin compaction (Cavallo et al., 1998; Daniels & Weis, 2005). In *Drosophila* wing imaginal discs, the high threshold Wg transcriptional target gene *senseless* (*sens*) directly controls the formation of adult wing margin sensory bristles, while *Distalless* (*Dll*) and *vestigial* (*vg*) act as medium and low threshold Wg target genes, respectively, and are required for adult wing growth and cell survival (Neumann & Cohen, 1997; Zecca et al., 1996). β-catenin/Arm acts as a transcriptional co-activator and, besides the TCF/Lef transcription

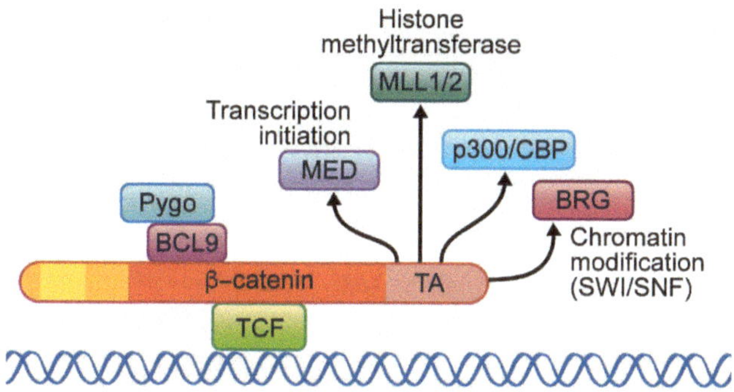

Fig. 4 TCF and β-catenin and its co-activator complexes in the nucleus. Upon Wnt signaling activation, β-catenin translocates to the nucleus and binds to TCF. The TCF/β-catenin complex recruits many co-activators required for the initiation and maintenance of active gene expression. These include for example, but are not restricted to, BCL9 (Legless/Lgs) and Pygopus (Pygo), interacting within the Arm repeat region of β-catenin (red region of β-catenin in schematic), and Mediator (for transcription initiation), p300/CBP, MLL1/2 histone methyltransferase and BRG (for chromatin modification (SWI/SNF) in the C-terminal transactivation (TA) region of β-catenin. Note that while the central Arm repeats region of β-catenin (red) associates with TCF, and the more N-terminal Arm-repeats bind to BCL9, most of other co-activator complexes interact with β-catenin in its C-terminal TA domain, creating a dazzling interplay between β-catenin and the transcriptional apparatus and the chromatin. Note that no nuclear interactors have thus far been identified for the most N-terminal region of β-catenin (N-terminal to the Arm repeats, orange and yellow in schematic).

factor, also has other nuclear binding partners that play critical roles in the activation of Wnt target genes. For example, through its C-terminal transactivation (TA) domain it interacts with CREB binding protein (CBP)/p300 (Sustmann, Flach, Ebert, Eastman, & Grosschedl, 2008) and SET domain-containing protein 1 (SET-1) (Sierra, Yoshida, Joazeiro, & Jones, 2006) among others (Fig. 4). β-catenin/Arm also interacts directly with BCL9, *legless* (*lgs*) in *Drosophila* (Hoffmans & Basler, 2007). Unlike most other co-activators binding β-catenin/Arm, BCL9/Lgs interacts with the first (most N-terminal) Arm repeat of β-catenin, which is specifically required for β-catenin-dependent transcription and it acts to bridge the complex of TCF-β-catenin/Arm-coactivators to activate transcription of Wnt target genes (Fig. 4) (Mosimann, Hausmann, & Basler, 2009). In *Drosophila*, the BCL9/Lgs protein is essential for Wg signaling and it acts by recruiting an additional protein, Pygopus, which is believed to provide a link to transcriptional initiation of Wnt signaling target genes. Lack of either BCL9/Lgs or Pygopus phenocopies Arm/β-catenin null mutants (Kramps et al., 2002; Thompson, Townsley, Rosin-Arbesfeld, Musisi, & Bienz, 2002).

Without question β-catenin is the main regulator of Wnt signaling in the nucleus. However, it is poorly understood how β-catenin is retained inside the nucleus. It has been proposed that the maintenance of β-catenin level in the nucleus is regulated by TCF and BCL9/Lgs via multiple mechanisms. TCF/Lef can enrich β-catenin in the nucleus (Behrens et al., 1996; Huber et al., 1996), and BCL9 and its nuclear interactor Pygopus can also strongly bind to β-catenin and maintain its levels high in that cell compartment (Kramps et al., 2002; Townsley, Cliffe, & Bienz, 2004). This function is tightly regulated through phosphorylation of β-catenin at Tyrosine 142, which not only diminishes β-catenin affinity to α-catenin in the E-cadherin junctional complex but also enhances β-catenin binding to the transcriptional coactivator BCL9/Lgs (Brembeck et al., 2004). This leads, in conjunction with TCF/Lef, to increased transcriptional activity. Another protein that has been proposed to affect β-catenin nuclear localization is the protein kinase AKT, which can phosphorylate β-catenin at Serine 552 and thus promote its nuclear localization (Fang et al., 2007). However, the mechanism by which these factors regulate and keep β-catenin localized in the nucleus, when Wnt is activated, has not been elucidated.

2.1.2 Wnt/β-catenin mediated transcriptional repression

It has been shown that activation of the Wnt/β-catenin pathway not only stimulates transcription but can also repress certain genes. In *Drosophila*, nearly 20% of Wnt/Wg responsive genes are predicted to be repressed by Wnt/Wg signaling (Franz, Shlyueva, Brunner, Stark, & Basler, 2017). Wnt-repressed genes have TCT/Lef binding sites, but they are distinct from the classical Wnt response Elements (WRE). For these genes, TCF promotes transcription via recruitment of coactivators in the absence of β-catenin; and, in the presence of β-catenin, the formation of a TCF/β-catenin complex represses transcription via recruitment of corepressors (Archbold, Broussard, Chang, & Cadigan, 2014). For example, the *Drosophila stripe* gene, whose TCF binding site overlaps with a binding site for Cubitus interruptus (Ci), a Hedgehog pathway dependent transcription factor that is essential for *stripe* transcription. Binding of the TCF/β-catenin complex most likely displaces Ci and thus represses transcription (Piepenburg, Vorbruggen, & Jackle, 2000). It will be interesting to further dissect how the different TCF motifs and β-catenin/Arm regulatory factors bring about activation or repression in a context dependent manner.

2.2 Nuclear branch of Wnt/planar cell polarity (PCP) signaling

During *Drosophila* eye development, Fz/PCP signaling regulates two patterning process. It establishes specific cell fates in the 5-cell photoreceptor precluster, which governs the general ommatidial patterning via nuclear signaling activating transcription of specific target genes (reviewed in (Jenny, 2010; Mlodzik, 1999), the mechanism of which will be discussed here, and subsequently it regulates a directed movement, ommatidial rotation, of the photoreceptor cluster, or ommatidium (reviewed in for example (Davey & Moens, 2017; Seifert & Mlodzik, 2007)). Wnt-PCP nuclear signaling has been primarily dissected in the *Drosophila* eye, and hence we focus on this tissue for its description. However, importantly, it is conserved in other species and contexts (Veeman, Slusarski, Kaykas, Louie, & Moon, 2003).

The *Drosophila* eye consists of ~800 regularly arranged ommatidia, each consisting of 8 photoreceptor neurons arranged into an invariant trapezoidal pattern and 12 accessory (cone, pigment, and bristle) cells. During larval stages, the eye develops from an epithelial imaginal disc, which is initially composed of identical pluripotent precursor cells. Development and differentiation is initiated when a wave of cell proliferation and differentiation (called morphogenetic furrow, MF) moves across the eye imaginal disc from posterior to anterior; the furrow leaves in its wake evenly spaced preclusters, which comprise initially a 5-cell precluster consisting of the R8, R2/R5, and R3/R4 photoreceptor neurons, of differentiating groups of cells that will mature into ommatidia, consisting of all 8 R-cell neurons and accessory cells (Heberlein, Wolff, & Rubin, 1993; Roignant & Treisman, 2009; Tomlinson & Ready, 1987; Wolff & Ready, 1991). The critical PCP signaling aspect involves, posterior to the furrow, an inductive signaling event specifying the R3 fate within the R3/R4 cell pair, and the associated transcriptional upregulation of R3 specific genes (including upregulation of the Notch ligand *Dl* to induce the neighboring cell as R4; reviewed in (Jenny, 2010; Mlodzik, 1999)). In addition, and unrelated to the nuclear branch, Wnt/PCP signaling also coordinates the direction and degree of ommatidial rotation via input to cytoskeletal and cell adhesion regulation. The coordinated induction of R3 in every ommatidial precluster, and the resulting asymmetric, directed signaling to R4, leads to a very regular ommatidial arrangement, with identical chiral forms in each ommatidium in the respective half of the eye, and opposite chiral forms in the ventral and dorsal halves. This type of mirror-symmetric arrangement of ommatidia is thought to be coordinated by Wnt/PCP signaling with Wg and Wnt4, both being expressed on the dorsal and ventral pole of the eye field, providing positional information (reviewed in Jenny, 2010; Mlodzik, 1999; Roignant & Treisman, 2009; Wu et al., 2013).

As discussed above, due to the coordinated inductive nuclear signaling requirement from R3 to R4, loss-of-function (LOF) mutations in core PCP genes generally randomize cell fate specification within the R3/R4 pair (and thus result in random chiral ommatidial arrangements) or even complete loss of chirality, giving rise to R3-R3 or R4-R4 symmetrical ommatidia (reviewed in Jenny, 2010; Mlodzik, 1999), see also (Boutros et al., 1998; Das et al., 1998; Jenny et al., 2005; Rawls & Wolff, 2003; Wolff & Rubin, 1998; Zheng, Zhang, & Carthew, 1995). Similarly, interference with PCP signaling via overexpression of the core components causes phenotypes similar to LOF defects, affecting again R3/R4 cell fate specification and subsequently rotation of ommatidia (Das, Reynolds-Kenneally, & Mlodzik, 2002; Fanto & Mlodzik, 1999; Jenny et al., 2005; Rawls & Wolff, 2003).

The *Drosophila* eye imaginal disc giving rise to the precisely patterned retina has provided the first and still the most precise insight on how Wnt-Fz/PCP signaling employs a nuclear response. Following an initial bias (presumably provided from the Wnt source at the poles of the eye field) and due to an upregulation of core PCP factors in precursor cells of the R3/R4 pair (Djiane, Yogev, & Mlodzik, 2005), the Fz-Dsh-Dgo complex is enriched on the R3 side of the R3/R4 border, whereas the Vang-Pk complex (inhibiting Fz-Dsh) is enriched on the R4 side (Fig. 2). As such Fz-Dsh signaling is enhanced in R3 and repressed in R4 leading to obvious differences in Fz-Dsh/PCP signaling activity between the two initially equivalent R3/R4 precursors (Djiane et al., 2005). Thus, the cell with the higher Fz-Dsh/PCP activity always becomes specified as R3 via a nuclear signaling branch and then induces its neighbor within the R3/R4 pair to adopt the R4 fate (via Notch signaling activation in R4, (Cooper & Bray, 1999; del Alamo & Mlodzik, 2006; Fanto & Mlodzik, 1999; Tomlinson & Struhl, 1999)). Importantly, specification of R3 and R4 depend on transcriptional responses in the R3 precursor generated by Fz-Dsh/PCP signaling, directly (R3) and indirectly (R4), respectively.

How is a Fz-Dsh/PCP signal relayed to the nucleus? First indications that such a pathway branch exists came from the realization that Fz-Dsh/PCP signaling acts through Rho family GTPases and can activate the JNK-type MAPK cascade (Boutros et al., 1998; Strutt, Weber, & Mlodzik, 1997). Strikingly, these studies revealed that the Dsh isoforms/truncations containing domains that are required in PCP signaling also activate the JNK MAPK-cascade in both cell culture assays and *in vivo* in Xenopus (Boutros et al., 1998; Shulman, Perrimon, & Axelrod, 1998). These insights came initially from gain of function assays, but they were quickly confirmed in genetic LOF interactions in *Drosophila* (Boutros et al., 1998; Paricio, Feiguin, Boutros, Eaton, & Mlodzik, 1999). Subsequently, a key requirement for Rho-family GTPases, as downstream effectors of Fz and Dsh in nuclear PCP signaling, was established, which confirmed the existence of such a pathway and provided new insights into it (Boutros et al., 1998; Fanto & Mlodzik, 1999; Paricio et al., 1999; Strutt et al., 1997), see below for more detail. The combination of these studies and related experiments in vertebrates established not only that the canonical Wnt-pathway and Wnt-PCP signaling diverge at the level of Dsh, downstream of the Wnt-Frizzled interactions, but also firmly linked Dsh to the activation of a JNK-type MAPK cascade via Rho-family GTPases (reviewed in Boutros & Mlodzik, 1999; Wallingford & Habas, 2005).

The above insights led the way to further dissecting the requirements of the Rho GTPases as Dsh effectors. The initial identification of RhoA (a.k.a. Rho1 in *Drosophila*) as a mediator of PCP signaling and characterization of its mutant alleles (Strutt et al., 1997) already suggested that RhoA is likely to act on both the cytoskeletal branch and a nuclear branch of the Fz-Dsh-RhoA PCP pathway. This expectation of a pathway split at the level of RhoA was subsequently confirmed and refined by demonstrating that one of the RhoA effectors, Rho-kinase (Rock; dROK in *Drosophila*), only affects cytoskeletal signaling in both the *Drosophila* eye and wing (Winter et al., 2001), while RhoA itself also affects nuclear signaling (Boutros et al., 1998; Strutt et al., 1997). A detailed dissection of RhoA signaling mutants, which allow to distinguish between cytoplasmic and nuclear signaling branches, further confirmed this notion, and defined RhoA mutant isoforms that acted predominantly via nuclear effectors and these affected the transcription dependent cell fate outcomes within the R3/R4 pair (Fanto, Weber, Strutt, & Mlodzik, 2000). Subsequent work analyzing the potential function of the Rac GPTases confirmed this and also pointed to a nuclear effect of this in R3/R4 cell fate induction (Munoz-Descalzo, Gomez-Cabrero, Mlodzik, & Paricio, 2007). In both cases, the GTPases activate the JNK-cascade, which translocates into the nucleus to act as an activator of the AP-1 transcription factor, consisting of Jun and Fos (reviewed for example in (Kockel, Homsy, & Bohmann, 2001)).

Jun and Fos, the dimerization partners in the AP-1 transcription factor (reviewed in (Kockel et al., 2001; Shaulian & Karin, 2001), are required to both specify R3 (Weber, Paricio, & Mlodzik, 2000) and also to upregulate transcription of *Dl* and *neuralized* (*neur*) in R3 to induce the R4 fate in the neighboring cell (Cooper & Bray, 1999; del Alamo & Mlodzik, 2006; Fanto & Mlodzik, 1999; Weber et al., 2000). In this nuclear signaling branch, the Wnt-PCP pathway takes advantage of an existing nuclear signaling module, activating a MAPK (JNK)-kinase cascade with ultimately JNK, the MAPK, entering the nucleus to phosphorylate and activate the AP-1 members Jun and Fos (*kayak/kay* in *Drosophila*). This in turn allows the Jun-Fos homo- or hetero-dimers to recruit coactivators and turn on transcription of tissue and cell type specific target genes in the R3 cell (Weber et al., 2000; Weber, Pataki, Mihaly, & Mlodzik, 2008) and other contexts (Veeman et al., 2003). Thus, the Wnt/PCP pathway uses a rather generic module/set of factors for its nuclear response, one that is also used downstream of several other signaling pathways. There is slight a variation of the theme though, as the common link of the MAPK cascade to AP-1 (Jun/Fos) requires factors from both families, Jun and Fos (for example in RTK/Ras/MAPK signaling), downstream of nuclear Wnt/PCP signaling Jun appears at least in part dispensable, as displayed by redundancy with Fos (Weber et al., 2000). As in Drosophila there is only one family member for each factor, Jun and Fos (*kay* in Drosophila), this is real redundancy and cannot be explained by another Jun family member taking over. Moreover, there is no such redundancy in other signaling context in *Drosophila*, notably dorsal closure and thorax closure for example, where both Jun and Fos are essential (Kockel, Zeitlinger, Staszewski, Mlodzik, & Bohmann, 1997; Zeitlinger et al., 1997). As such this variation of the JNK-module signaling to Jun/Fos is unique in the Wnt/PCP context.

Similar to the MAPK-Jun/Fos signaling module downstream of other pathways, the Wnt-PCP-JNK branch signaling through Jun/Fos cooperates with other nuclear signaling pathways in its response. For example, in the context of the R3 and R4 cell fate specification, nuclear Wnt/PCP signaling acts in concert of Egfr (RTK)/Ras/MAPK signaling and its nuclear effectors of the ETS transcription factor family (Weber et al., 2008), and its effects on Notch signaling in R4 are amplified through feedback loops that also affect R3. It is worth noting that a similar link between Wnt/PCP signaling and asymmetric Notch pathway activation has been documented in the fish lateral line neuromasts—the hearing and balance organ of fish—which is also comprised of a multicellular sensory unit system comparable to the ommatidia. Here also the Wnt/PCP pathway sets up the direction of Notch signaling (Dale, Sisson, & Topczewski, 2009; Mirkovic, Pylawka, & Hudspeth, 2012; Navajas Acedo et al., 2019). Possibly, Wnt/PCP signaling and the Notch pathway also link up to govern the asymmetric formation of the feather buds in birds (Crowe, Henrique, Ish-Horowicz, & Niswander, 1998; Widelitz et al., 1999).

In summary, the core Fz/PCP nuclear signaling branch uses a well-known (and not pathway specific) set of effectors, both cytoplasmic kinases and nuclear factors, and it also cooperates with several other signaling pathways during R3/R4 specification, and likely in general. Thus the analysis of nuclear Wnt/PCP signaling also underscores how multiple signaling pathways and local cellular interactions are linked in achieving precision in a highly regulated cell fate induction process.

3. How does β-catenin get transported into the nucleus

3.1 New insights into β-catenin nuclear translocation

β-catenin is a multifunctional protein associated with adherens junctions, linking these to the actomyosin cytoskeleton, and canonical Wnt signaling, where it acts as a nuclear co-factor (see above). As such its subcellular localization needs to be very tightly regulated. The entry of β-catenin into the nucleus is a key signaling step in the Wnt/β-catenin pathway. Although the high concentration of free cytoplasmic β-catenin appears to be a prerequisite for its subsequent nuclear translocation, it does not explain how β-catenin gets into the nucleus to promote transcription (Henderson & Fagotto, 2002; Stadeli, Hoffmans, & Basler, 2006). In the classic nuclear transport model, nuclear localization sequence (NLS) containing proteins bind to Importin-α, which is brought to the nuclear pore complex (NPC) and transported through the nuclear pore via the interaction with Importin-β. With the assistance of Ran GTPase, the cargo is then moved through the nuclear envelope and released inside the nucleus (Fagotto, Gluck, & Gumbiner, 1998; Yokoya, Imamoto, Tachibana, & Yoneda, 1999). However, β-catenin has no identifiable NLS (Suh & Gumbiner, 2003). It has been shown that β-catenin can enter the nucleus in a Ran-independent manner (Yokoya et al., 1999). Since β-catenin and Importin share similar Armadillo repeat domains (Kutay, Bischoff, Kostka, Kraft, & Gorlich, 1997), β-catenin might directly interact with nuclear core components, bypassing the β-importin/karyopherin proteins (Fagotto et al., 1998; Yokoya et al., 1999). Another proposed model for the nuclear translocation of β-catenin is its transport into the nucleus in a complex with other proteins, such as FOXM1 or BCL9 (Townsley et al., 2004). Once in the nucleus, interaction with TCF, BCL9 and Pygopus may function as a bridge to β-catenin (Krieghoff, Behrens, & Mayr, 2006; Tolwinski & Wieschaus, 2001). Furthermore, a role for Rac1 GTPase and Jun N-terminal kinase 2 (JNK2) has been suggested in the nuclear localization of β-catenin upon Wnt pathway activation, although it remains unclear how Rac1, normally linked to Wnt/Fz-PCP signaling (see above), is involved in canonical Wnt signaling (Wu et al., 2008).

Recently, a genetic screen to test for potential roles of ciliary proteins in non-ciliated epithelial cells in *Drosophila* identified Intraflagellar Transport complex A (IFT-A) proteins to modulate canonical Wnt/Wg signaling, independently of the ciliary role of IFTs (Balmer et al., 2015). Among the five conserved core IFT-A proteins in *Drosophila*, mutants for four of the corresponding genes, IFT122, IFT121, IFT140 and IFT143, displayed growth defects and wing notches or missing wing margin bristles in adult wings, indicating that they play a role in *Drosophila* wing development and canonical Wg signaling. It remained unclear whether the cytoplasmic IFT-A proteins associate with microtubular structures and whether such association is required for their function in Wnt/Wg signaling. Interestingly, we have shown separately that Klp64D, the *Drosophila* motor protein Kinesin 2 subunit, also plays a role in Wg/Wnt signaling and Arm/β-catenin trafficking (Vuong et al., 2014). Klp64D recruits Arm/β-catenin and its interacting partner Dishevelled (Dsh) for Wingless signal transduction. In the absence of Kinesin 2 function, Arm/β-catenin abnormally accumulated in the cytoplasm, indicating that Klp64D might be involved in trafficking of Arm/β-catenin for Wg/Wnt signaling pathway.

Subsequent studies demonstrated that IFT-A proteins, and IFT140 in particular, associate with Kinesin 2 and promote nuclear translocation of β-catenin upon Wnt/Wg pathway activation (Vuong et al., 2018). This requirement, acting downstream of β-catenin stabilization, has been also shown to be conserved from *Drosophila* to mammalian cells, e.g. mouse embryonic fibroblasts (Vuong et al., 2018). Loss of function mutants of either Kinesin 2 or IFT-A proteins in *Drosophila* wing tissues displayed indistinguishable effects on Wnt target gene expression and wing development. Kinesin 2 and IFT140 interact genetically and physically with each other.

Kap 3, a non-motor component of Kinesin 2, directly interacts with IFT140, and acts as the cargo domain to transport IFT140 along microtubules. Both single mutant clones and double mutant clones for *kinesin 2* and *ift140* or *kap 3* fail to activate Wg/Wnt signaling targets in *Drosophila*. Moreover, double mutant clones for either *kinesin 2* or *IFT140* with *axin* in *Drosophila* displayed high levels of stabilized cytoplasmic Arm/β-catenin in both wing imaginal disc cells and salivary glands, but its nuclear translocation was markedly reduced as nicely visualized in the large salivary gland cells (Vuong et al., 2018). Similarly, *ift140*$^{-/-}$ mutant mouse embryonic fibroblast (MEFs) showed a striking reduction of nuclear β-catenin upon Wnt3a mediated pathway activation, which importantly was not affected in control *ift88* mutant MEFs (IFT88 is a component of IFT-B), confirming the non-ciliary role of IFT-A in Wnt-signaling in mammalian contexts. Taken together, these data suggest a conserved non-ciliary role for the Kinesin 2/IFT-A complex in transporting Arm/β-catenin to the nucleus (Vuong et al., 2018).

In this context, IFT140 and Arm/β-catenin directly interact through a small domain (53 amino acids) at N-terminal region of Arm/β-catenin, Arm$^{34-87}$ (Vuong et al., 2018). The deletion of this small region of Arm is a known mutant stable isoform, ArmS10, which is commonly used in *Drosophila* as a "constitutively active" isoform of Arm/β-catenin (Blauwkamp, Chang, & Cadigan, 2008). Strikingly, ArmS10 can enter the nucleus independently of the Kinesin 2/IFT140 function, which suggests that this small region is critical for the differential behavior relative to wild-type stable Arm/β-catenin in terms of Kinesin 2/IFT140 effects.

Taken together, these findings allowed us to propose a new molecular mechanism for the entry of Arm/β-catenin to the nucleus. When Wg/Wnt signaling is activated, the complex of Kinesin 2/Kap3/IFT-A binds cytoplasmic Arm/β-catenin via IFT140 and this complex might then be transported along the microtubules to the nucleus (Fig. 5) (Vuong et al., 2018). Once near (or in) the nucleus, Arm/β-catenin may be released from the IFT-A complex to bind to TCF/Lef to mediate transcriptional activation of Wnt target genes. Moreover, as binding of Arm/β-catenin to the Kinesin2/IFT-A complex promotes nuclear localization, it is likely that IFT140/IFT-A with other proteins for Arm/β-catenin binding to the specific domain in the N-terminal region of Arm/β-catenin. Such cytoplasmic proteins could trap/sequester Arm/β-catenin in the cytoplasm, and therefore limit its capacity to enter the nucleus.

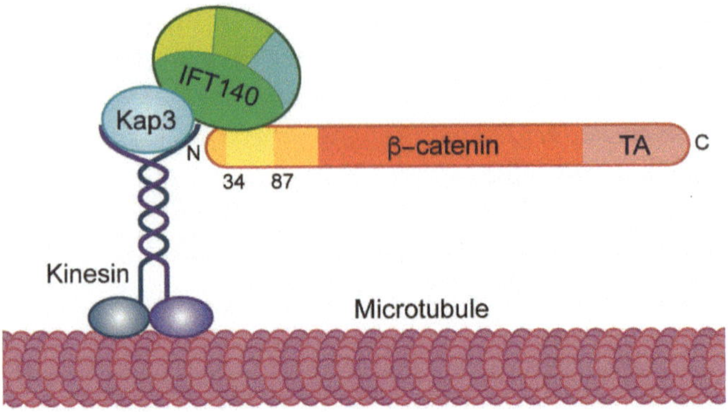

Fig. 5 Proposed model for β-catenin associated complex that mediates its translocation to the nucleus. Upon Wnt/β-catenin signaling activation, Kinesin 2 interacts with IFT-A via a direct Kap3-IFT140 interaction. Note that this complex forms on cytoplasmic microtubules. IFT140 also binds free cytoplasmic β-catenin in the N-terminal region of β-catenin (yellow in schematic), thus recruiting β-catenin to the Kinesin 2/IFT-A complex, which could allow its movement along microtubule towards the nucleus. See main text for details.

3.2 Missing information in the context β-catenin nuclear translocation

Our study (Vuong et al., 2018) has identified a novel mechanistic function of the kinesin2/IFT-A complex in Wnt signaling, independent of their role in the cilium. The Kinesin 2/IFT-A complex might promote nuclear translocation of Arm/β-catenin by two unrelated mechanisms: (i) protecting it from binding by a cytoplasmic inhibitor/retention factor, and (ii) moving it along microtubules to the nucleus. As these mechanistic features of the complex are conserved in mammalian cells, these exciting observations are potentially amenable to therapeutic treatment.

Nonetheless, several questions remain and should be the focus of future studies in Drosophila and in mammalian contexts: (i) How does the Kinesin-2/IFT-A complex promote nuclear localization of Arm/β-catenin? While the existing data suggest that the complex moves along microtubules to transport β-catenin to the nucleus (Vuong et al., 2018), this remains a hypothesis and needs to be confirmed by live imaging in *Drosophila* and mammalian cells. (ii) What is the identity of the antagonistic factors that compete with IFT140 for binding to the β-catenin/Arm N-terminal region, and do they really serve an inhibitory function for β-catenin/Arm nuclear localization as cytoplasmic retention factors? The APC/Axin complex was suggested as a cytoplasmic anchor, keeping β-catenin/Arm out of the nucleus (Tolwinski & Wieschaus, 2001) and our data in the salivary glands assay are consistent with this function (Vuong et al., 2018). However, Axin can be excluded as the inhibitory factor competing with IFT140, because in the absence of Axin, β-catenin/Arm fails to translocate to the nucleus in *Klp64D* or *IFT140* mutant backgrounds. Efforts to identify IFT140 competing factors await further investigation. And (iii), it remains to be tested whether the small N-terminal domain of β-catenin/Arm, required for binding to IFT140 (Arm^{34-87}), plays a critical role in Wg/Wnt signaling. If so, it might serve as an entry point for potential new drug development and/or therapeutic application to inhibit an overactive Wnt/β-catenin signaling scenario.

4. Concluding remarks

β-catenin (Armadillo/Arm in *Drosophila*) is a multitasking and evolutionary conserved molecule that exerts a crucial role in a multitude of developmental and homeostatic processes in metazoans. β-catenin has been established to be a critical component of the activation of the Wnt signaling pathway. In this canonical Wnt pathway, β-catenin is the key effector responsible for transduction of the signal to the nucleus, as it triggers transcriptional responses of Wnt-specific target genes and is thus responsible for the control of cell fate decisions and growth in many cells and tissues. The second well defined type of Wnt signaling, the Wnt/PCP pathway, is independent of β-catenin function. β-catenin is however not just a component of the Wnt signaling cascade. In the late 1980s, β-catenin was identified together with two other molecules (∝-catenin and γ-catenin/plakoglobin), as critical proteins associated with E-cadherin, the key molecule of Ca^{2+}-dependent cell adhesion (Ozawa, Baribault, & Kemler, 1989). How can β-catenin as a single molecule affect so many cellular phenomena in animals, and regulate and mediate so many different outputs and read-outs? It is a complex question that will require many more studies to be fully answered.

In this review chapter, we have mainly focused on the role of β-catenin in the canonical Wnt/β-catenin signaling pathway and how it can translocate into the nucleus to activate the target genes of the pathway. As we proposed, upon Wnt signaling activation, the complex of Kinesin-2/Kap3/IFT-A binds to β-catenin and transports it to the nucleus by moving along cytoplasmic microtubules. The binding of β-catenin to microtubules has been first reported during neuronal morphogenesis in mice. In this context, β-catenin is transported to microtubule ends in a complex with APC and

N-cadherin. Kap3 is required for localization of β-catenin to the plasma membrane of neuroepithelial cells (Teng et al., 2005). Binding of APC to Kap3 is required for localization of APC to the tip of epithelial extensions (Jimbo et al., 2002). However, it remains unclear whether these proteins are transported in the same kinesin complex and, importantly, how APC might regulate the transport of β-catenin and N-cadherin. Furthermore, APC has also been reported to play a role in the transport of β-catenin to the nucleus upon activation of Wnt signaling (Tolwinski & Wieschaus, 2001). Therefore, future studies should tackle these questions to investigate the mechanism of how the Kinesin2/IFT-A complex might promote β-catenin translocation to the nucleus and whether there are other proteins involved in this complex. Moreover, since the N-terminal region of β-catenin is required for direct binding to IFT140 (within the IFT-A), it is necessary to establish the role of this domain of β-catenin in the nuclear translocation and the Wnt signaling pathway in general. A better understanding of the β-catenin structure, as this N-terminal region is thus far reported as unstructured, as well as the identification and characterization of new factors involved in nuclear Wnt signaling will not only provide significant new insights into the mechanism(s) of how β-catenin is translocated and retained inside the nucleus, but might also facilitate therapeutic approaches and development of drugs to target this pathway in the context Wnt/β-catenin-driven cancers and other disease in humans.

In contrast to canonical Wnt/β-catenin signaling, the nuclear branches of the non-canonical Wnt pathways discussed here seem rather derived, with the redeployment of fairly standard kinase cascade-transcription factor interactions that were defined in other signaling contexts initially. As such, β-catenin is the only Wnt signaling dedicated nuclear effector. Nonetheless, novel insights will be exciting for either Wnt signaling pathway, and importantly the complexity of interactions, positive and negative, between the distinct Wnt signaling branches promise lots of unexpected discoveries to follow.

Acknowledgments

We are grateful to all Mlodzik lab members for their helpful discussions and support, and Philippe Soriano for thoughtful comments on the manuscript. We wish to thank Jill Gregory at the Academic Medical Illustration Department of the Icahn School of Medicine at Mount Sinai for excellent assistance with figure preparation. Related research in the Mlodzik laboratory is supported by a National Institutes of Health grant from the NIGMS R35 GM127103.

References

Adler, P. N. (2002). Planar signaling and morphogenesis in Drosophila. *Developmental Cell, 2*, 525–535.

Archbold, H. C., Broussard, C., Chang, M. V., & Cadigan, K. M. (2014). Bipartite recognition of DNA by TCF/Pangolin is remarkably flexible and contributes to transcriptional responsiveness and tissue specificity of wingless signaling. *PLoS Genetics, 10*, e1004591.

Aw, W. Y., & Devenport, D. (2017). Planar cell polarity: Global inputs establishing cellular asymmetry. *Current Opinion in Cell Biology, 44*, 110–116.

Aw, W. Y., Heck, B. W., Joyce, B., & Devenport, D. (2016). Transient tissue-scale deformation coordinates alignment of planar cell polarity junctions in the mammalian skin. *Current Biology, 26*, 2090–2100.

Axelrod, J. D., Miller, J. R., Shulman, J. M., Moon, R. T., & Perrimon, N. (1998). Differential recruitment of dishevelled provides signaling specificity in the planar cell polarity and wingless signaling pathways. *Genes & Development, 12*, 2610–2622.

Baena-Lopez, L. A., Nojima, H., & Vincent, J. P. (2012). Integration of morphogen signalling within the growth regulatory network. *Current Opinion in Cell Biology, 24*, 166–172.

Balmer, S., Dussert, A., Collu, G. M., Benitez, E., Iomini, C., & Mlodzik, M. (2015). Components of intraflagellar transport complex A function independently of the cilium to regulate canonical Wnt signaling in Drosophila. *Developmental Cell, 34*, 705–718.

Behrens, J., von Kries, J. P., Kuhl, M., Bruhn, L., Wedlich, D., Grosschedl, R., et al. (1996). Functional interaction of beta-catenin with the transcription factor LEF-1. *Nature, 382*, 638–642.

Bienz, M. (2005). beta-Catenin: A pivot between cell adhesion and Wnt signalling. *Current Biology, 15*, R64–R67.

Blauwkamp, T. A., Chang, M. V., & Cadigan, K. M. (2008). Novel TCF-binding sites specify transcriptional repression by Wnt signalling. *The EMBO Journal, 27*, 1436–1446.

Boutros, M., & Mlodzik, M. (1999). Dishevelled: At the crossroads of divergent intracellular signaling pathways. *Mechanisms of Development, 83*, 27–37.

Boutros, M., Paricio, N., Strutt, D. I., & Mlodzik, M. (1998). Dishevelled activates JNK and discriminates between JNK pathways in planar polarity and wingless signaling. *Cell, 94*, 109–118.

Brembeck, F. H., Schwarz-Romond, T., Bakkers, J., Wilhelm, S., Hammerschmidt, M., & Birchmeier, W. (2004). Essential role of BCL9-2 in the switch between beta-catenin's adhesive and transcriptional functions. *Genes & Development, 18*, 2225–2230.

Butler, M. T., & Wallingford, J. B. (2017). Planar cell polarity in development and disease. *Nature Reviews. Molecular Cell Biology, 18*, 375–388.

Cadigan, K. M., & Waterman, M. L. (2012). TCF/LEFs and Wnt signaling in the nucleus. *Cold Spring Harbor Perspectives in Biology, 4*.

Carvajal-Gonzalez, J. M., Mulero-Navarro, S., & Mlodzik, M. (2016). Centriole positioning in epithelial cells and its intimate relationship with planar cell polarity. *BioEssays, 38*, 1234–1245.

Cavallo, R. A., Cox, R. T., Moline, M. M., Roose, J., Polevoy, G. A., Clevers, H., et al. (1998). Drosophila Tcf and Groucho interact to repress wingless signalling activity. *Nature, 395*, 604–608.

Clevers, H. (2006). Wnt/beta-catenin signaling in development and disease. *Cell, 127*, 469–480.

Clevers, H., & Nusse, R. (2012). Wnt/beta-catenin signaling and disease. *Cell, 149*, 1192–1205.

Cliffe, A., Hamada, F., & Bienz, M. (2003). A role of dishevelled in relocating axin to the plasma membrane during wingless signaling. *Current Biology, 13*, 960–966.

Coombs, G. S., Covey, T. M., & Virshup, D. M. (2008). Wnt signaling in development, disease and translational medicine. *Current Drug Targets, 9*, 513–531.
Cooper, M. T., & Bray, S. J. (1999). Frizzled regulation of notch signalling polarizes cell fate in the Drosophila eye. *Nature, 397*, 526–530.
Crowe, R., Henrique, D., Ish-Horowicz, D., & Niswander, L. (1998). A new role for Notch and Delta in cell fate decisions: Patterning the feather array. *Development, 125*, 767–775.
Dale, R. M., Sisson, B. E., & Topczewski, J. (2009). The emerging role of Wnt/PCP signaling in organ formation. *Zebrafish, 6*, 9–14.
Daniels, D. L., & Weis, W. I. (2005). Beta-catenin directly displaces Groucho/TLE repressors from Tcf/Lef in Wnt-mediated transcription activation. *Nature Structural & Molecular Biology, 12*, 364–371.
Danielson, K. G., Pillarisetti, J., Cohen, I. R., Sholehvar, B., Huebner, K., Ng, L. J., et al. (1995). Characterization of the complete genomic structure of the human WNT-5A gene, functional analysis of its promoter, chromosomal mapping, and expression in early human embryogenesis. *The Journal of Biological Chemistry, 270*, 31225–31234.
Das, G., Jenny, A., Klein, T. J., Eaton, S., & Mlodzik, M. (2004). Diego interacts with prickle and Strabismus/Van Gogh to localize planar cell polarity complexes. *Development, 131*, 4467–4476.
Das, P., Maduzia, L. L., Wang, H., Finelli, A. L., Cho, S. H., Smith, M. M., et al. (1998). The Drosophila gene Medea demonstrates the requirement for different classes of Smads in dpp signaling. *Development, 125*, 1519–1528.
Das, G., Reynolds-Kenneally, J., & Mlodzik, M. (2002). The atypical cadherin Flamingo links frizzled and notch signaling in planar polarity establishment in the Drosophila eye. *Developmental Cell, 2*, 655–666.
Davey, C. F., & Moens, C. B. (2017). Planar cell polarity in moving cells: Think globally, act locally. *Development, 144*, 187–200.
De, A. (2011). Wnt/Ca2+ signaling pathway: A brief overview. *Acta Biochimica et Biophysica Sinica Shanghai, 43*, 745–756.
del Alamo, D., & Mlodzik, M. (2006). Frizzled/PCP-dependent asymmetric neuralized expression determines R3/R4 fates in the Drosophila eye. *Developmental Cell, 11*, 887–894.
Djiane, A., Yogev, S., & Mlodzik, M. (2005). The apical determinants aPKC and dPatj regulate frizzled-dependent planar cell polarity in the Drosophila eye. *Cell, 121*, 621–631.
Eisenmann, D. M. (2005). *Wnt signaling* (pp. 1–17). WormBook.
Fagotto, F., Gluck, U., & Gumbiner, B. M. (1998). Nuclear localization signal-independent and importin/karyopherin-independent nuclear import of beta-catenin. *Current Biology, 8*, 181–190.
Fang, D., Hawke, D., Zheng, Y., Xia, Y., Meisenhelder, J., Nika, H., et al. (2007). Phosphorylation of beta-catenin by AKT promotes beta-catenin transcriptional activity. *The Journal of Biological Chemistry, 282*, 11221–11229.
Fanto, M., & Mlodzik, M. (1999). Asymmetric notch activation specifies photoreceptors R3 and R4 and planar polarity in the Drosophila eye. *Nature, 397*, 523–526.
Fanto, M., Weber, U., Strutt, D. I., & Mlodzik, M. (2000). Nuclear signaling by Rac and Rho GTPases is required in the establishment of epithelial planar polarity in the Drosophila eye. *Current Biology, 10*, 979–988.
Franz, A., Shlyueva, D., Brunner, E., Stark, A., & Basler, K. (2017). Probing the canonicity of the Wnt/wingless signaling pathway. *PLoS Genetics, 13*, e1006700.
Gao, B., Song, H., Bishop, K., Elliot, G., Garrett, L., English, M. A., et al. (2011). Wnt signaling gradients establish planar cell polarity by inducing Vangl2 phosphorylation through Ror2. *Developmental Cell, 20*, 163–176.
Goodrich, L. V., & Strutt, D. (2011). Principles of planar polarity in animal development. *Development, 138*, 1877–1892.

Harrison, C., Shao, H., Strutt, H., & Strutt, D. (2020). Molecular mechanisms mediating asymmetric subcellular localisation of the core planar polarity pathway proteins. *Biochemical Society Transactions*, *48*, 1297–1308.

Heberlein, U., Wolff, T., & Rubin, G. M. (1993). The TGF beta homolog dpp and the segment polarity gene hedgehog are required for propagation of a morphogenetic wave in the Drosophila retina. *Cell*, *75*, 913–926.

Henderson, B. R., & Fagotto, F. (2002). The ins and outs of APC and beta-catenin nuclear transport. *EMBO Reports*, *3*, 834–839.

Hoffmans, R., & Basler, K. (2007). BCL9-2 binds Arm/beta-catenin in a Tyr142-independent manner and requires Pygopus for its function in Wg/Wnt signaling. *Mechanisms of Development*, *124*, 59–67.

Huber, O., Korn, R., McLaughlin, J., Ohsugi, M., Herrmann, B. G., & Kemler, R. (1996). Nuclear localization of beta-catenin by interaction with transcription factor LEF-1. *Mechanisms of Development*, *59*, 3–10.

Humphries, A. C., & Mlodzik, M. (2018). From instruction to output: Wnt/PCP signaling in development and cancer. *Current Opinion in Cell Biology*, *51*, 110–116.

Jenny, A. (2010). Planar cell polarity signaling in the Drosophila eye. *Current Topics in Developmental Biology*, *93*, 189–227.

Jenny, A., Reynolds-Kenneally, J., Das, G., Burnett, M., & Mlodzik, M. (2005). Diego and Prickle regulate Frizzled planar cell polarity signalling by competing for dishevelled binding. *Nature Cell Biology*, 7, 691–697.

Jimbo, T., Kawasaki, Y., Koyama, R., Sato, R., Takada, S., Haraguchi, K., et al. (2002). Identification of a link between the tumour suppressor APC and the kinesin superfamily. *Nature Cell Biology*, *4*, 323–327.

Katanaev, V. L., Ponzielli, R., Semeriva, M., & Tomlinson, A. (2005). Trimeric G protein-dependent frizzled signaling in Drosophila. *Cell*, *120*, 111–122.

Kockel, L., Homsy, J. G., & Bohmann, D. (2001). Drosophila AP-1: Lessons from an invertebrate. *Oncogene*, *20*, 2347–2364.

Kockel, L., Zeitlinger, J., Staszewski, L. M., Mlodzik, M., & Bohmann, D. (1997). Jun in Drosophila development: Redundant and nonredundant functions and regulation by two MAPK signal transduction pathways. *Genes & Development*, *11*, 1748–1758.

Kohn, A. D., & Moon, R. T. (2005). Wnt and calcium signaling: Beta-catenin-independent pathways. *Cell Calcium*, *38*, 439–446.

Kramps, T., Peter, O., Brunner, E., Nellen, D., Froesch, B., Chatterjee, S., et al. (2002). Wnt/wingless signaling requires BCL9/legless-mediated recruitment of pygopus to the nuclear beta-catenin-TCF complex. *Cell*, *109*, 47–60.

Krieghoff, E., Behrens, J., & Mayr, B. (2006). Nucleo-cytoplasmic distribution of beta-catenin is regulated by retention. *Journal of Cell Science*, *119*, 1453–1463.

Kuhl, M., Sheldahl, L. C., Malbon, C. C., & Moon, R. T. (2000). Ca(2+)/calmodulin-dependent protein kinase II is stimulated by Wnt and Frizzled homologs and promotes ventral cell fates in Xenopus. *The Journal of Biological Chemistry*, *275*, 12701–12711.

Kuhl, M., Sheldahl, L. C., Park, M., Miller, J. R., & Moon, R. T. (2000). The Wnt/Ca2+ pathway: A new vertebrate Wnt signaling pathway takes shape. *Trends in Genetics*, *16*, 279–283.

Kutay, U., Bischoff, F. R., Kostka, S., Kraft, R., & Gorlich, D. (1997). Export of importin alpha from the nucleus is mediated by a specific nuclear transport factor. *Cell*, *9*, 1061–1071.

Logan, C. Y., & Nusse, R. (2004). The Wnt signaling pathway in development and disease. *Annual Review of Cell and Developmental Biology*, *20*, 781–810.

MacDonald, B. T., Tamai, K., & He, X. (2009). Wnt/beta-catenin signaling: Components, mechanisms, and diseases. *Developmental Cell*, *17*, 9–26.

McEwen, A. E., Escobar, D. E., & Gottardi, C. J. (2012). Signaling from the adherens junction. *Sub-Cellular Biochemistry, 60*, 171–196.

McLeod, F., Boyle, K., Marzo, A., Martin-Flores, N., Moe, T. Z., Palomer, E., et al. (2020). Wnt signaling through nitric oxide synthase promotes the formation of multi-innervated spines. *Frontiers in Synaptic Neuroscience, 12*, 575863.

McLeod, F., & Salinas, P. C. (2018). Wnt proteins as modulators of synaptic plasticity. *Current Opinion in Neurobiology, 53*, 90–95.

Minegishi, K., Hashimoto, M., Ajima, R., Takaoka, K., Shinohara, K., Ikawa, Y., et al. (2017). A Wnt5 activity asymmetry and intercellular signaling via PCP proteins polarize node cells for left-right symmetry breaking. *Developmental Cell, 40*(439-452), e434.

Mirkovic, I., Pylawka, S., & Hudspeth, A. J. (2012). Rearrangements between differentiating hair cells coordinate planar polarity and the establishment of mirror symmetry in lateral-line neuromasts. *Biology Open, 1*, 498–505.

Mlodzik, M. (1999). Planar polarity in the Drosophila eye: A multifaceted view of signaling specificity and cross-talk. *The EMBO Journal, 18*, 6873–6879.

Mosimann, C., Hausmann, G., & Basler, K. (2009). Beta-catenin hits chromatin: Regulation of Wnt target gene activation. *Nature Reviews. Molecular Cell Biology, 10*, 276–286.

Munoz-Descalzo, S., Gomez-Cabrero, A., Mlodzik, M., & Paricio, N. (2007). Analysis of the role of the Rac/Cdc42 GTPases during planar cell polarity generation in Drosophila. *The International Journal of Developmental Biology, 51*, 379–387.

Murphy, L. L., & Hughes, C. C. (2002). Endothelial cells stimulate T cell NFAT nuclear translocation in the presence of cyclosporin A: Involvement of the wnt/glycogen synthase kinase-3 beta pathway. *Journal of Immunology, 169*, 3717–3725.

Navajas Acedo, J., Voas, M. G., Alexander, R., Woolley, T., Unruh, J. R., Li, H., et al. (2019). PCP and Wnt pathway components act in parallel during zebrafish mechanosensory hair cell orientation. *Nature Communications, 10*, 3993.

Neumann, C. J., & Cohen, S. M. (1997). Long-range action of wingless organizes the dorsal-ventral axis of the Drosophila wing. *Development, 124*, 871–880.

Niehrs, C. (2012). The complex world of WNT receptor signalling. *Nature Reviews. Molecular Cell Biology, 13*, 767–779.

Nusse, R., van Ooyen, A., Cox, D., Fung, Y. K., & Varmus, H. (1984). Mode of proviral activation of a putative mammary oncogene (int-1) on mouse chromosome 15. *Nature, 307*, 131–136.

Nusslein-Volhard, C., & Wieschaus, E. (1980). Mutations affecting segment number and polarity in Drosophila. *Nature, 287*, 795–801.

Ozawa, M., Baribault, H., & Kemler, R. (1989). The cytoplasmic domain of the cell adhesion molecule uvomorulin associates with three independent proteins structurally related in different species. *The EMBO Journal, 8*, 1711–1717.

Paricio, N., Feiguin, F., Boutros, M., Eaton, S., & Mlodzik, M. (1999). The Drosophila STE20-like kinase misshapen is required downstream of the Frizzled receptor in planar polarity signaling. *The EMBO Journal, 18*, 4669–4678.

Peng, Y., & Axelrod, J. D. (2012). Asymmetric protein localization in planar cell polarity: Mechanisms, puzzles, and challenges. *Current Topics in Developmental Biology, 101*, 33–53.

Piepenburg, O., Vorbruggen, G., & Jackle, H. (2000). Drosophila segment borders result from unilateral repression of hedgehog activity by wingless signaling. *Molecular Cell, 6*, 203–209.

Prestwich, T. C., & Macdougald, O. A. (2007). Wnt/beta-catenin signaling in adipogenesis and metabolism. *Current Opinion in Cell Biology, 19*, 612–617.

Rao, A., Luo, C., & Hogan, P. G. (1997). Transcription factors of the NFAT family: Regulation and function. *Annual Review of Immunology, 15*, 707–747.

Rawls, A. S., & Wolff, T. (2003). Strabismus requires Flamingo and Prickle function to regulate tissue polarity in the Drosophila eye. *Development, 130*, 1877–1887.

Roignant, J. Y., & Treisman, J. E. (2009). Pattern formation in the Drosophila eye disc. *The International Journal of Developmental Biology, 53*, 795–804.
Segalen, M., & Bellaiche, Y. (2009). Cell division orientation and planar cell polarity pathways. *Seminars in Cell & Developmental Biology, 20*, 972–977.
Seifert, J. R., & Mlodzik, M. (2007). Frizzled/PCP signalling: A conserved mechanism regulating cell polarity and directed motility. *Nature Reviews, 8*, 126–138.
Semenov, M. V., Habas, R., Macdonald, B. T., & He, X. (2007). SnapShot: Noncanonical Wnt signaling pathways. *Cell, 131*, 1378.
Sharma, R. P., & Chopra, V. L. (1976). Effect of the wingless (wg1) mutation on wing and haltere development in Drosophila melanogaster. *Developmental Biology, 48*, 461–465.
Shaulian, E., & Karin, M. (2001). AP-1 in cell proliferation and survival. *Oncogene, 20*, 2390–2400.
Shulman, J. M., Perrimon, N., & Axelrod, J. D. (1998). Frizzled signaling and the developmental control of cell polarity. *Trends in Genetics, 14*, 452–458.
Sierra, J., Yoshida, T., Joazeiro, C. A., & Jones, K. A. (2006). The APC tumor suppressor counteracts beta-catenin activation and H3K4 methylation at Wnt target genes. *Genes & Development, 20*, 586–600.
Simons, M., & Mlodzik, M. (2008). Planar cell polarity signaling: From fly development to human disease. *Annual Review of Genetics, 42*, 517–540.
Stadeli, R., Hoffmans, R., & Basler, K. (2006). Transcription under the control of nuclear Arm/beta-catenin. *Current Biology, 16*, R378–R385.
Strigini, M., & Cohen, S. M. (2000). Wingless gradient formation in the Drosophila wing. *Current Biology, 10*, 293–300.
Strutt, D. I., Weber, U., & Mlodzik, M. (1997). The role of RhoA in tissue polarity and frizzled signalling. *Nature, 387*, 292–295.
Suh, E. K., & Gumbiner, B. M. (2003). Translocation ofbeta-catenin into the nucleus independent of interactions with FG-rich nucleoporins. *Experimental Cell Research, 290*, 447–456.
Sustmann, C., Flach, H., Ebert, H., Eastman, Q., & Grosschedl, R. (2008). Cell-type-specific function of BCL9 involves a transcriptional activation domain that synergizes with beta-catenin. *Molecular and Cellular Biology, 28*, 3526–3537.
Teng, J., Rai, T., Tanaka, Y., Takei, Y., Nakata, T., Hirasawa, M., et al. (2005). The KIF3 motor transports N-cadherin and organizes the developing neuroepithelium. *Nature Cell Biology, 7*, 474–482.
Teo, S., & Salinas, P. C. (2021). Wnt-frizzled signaling regulates activity-mediated synapse formation. *Frontiers in Molecular Neuroscience, 14*, 683035.
Thompson, B., Townsley, F., Rosin-Arbesfeld, R., Musisi, H., & Bienz, M. (2002). A new nuclear component of the Wnt signalling pathway. *Nature Cell Biology, 4*, 367–373.
Tian, X., Liu, Z., Niu, B., Zhang, J., Tan, T. K., Lee, S. R., et al. (2011). E-cadherin/beta-catenin complex and the epithelial barrier. *Journal of Biomedicine & Biotechnology, 2011*, 567305.
Tolwinski, N. S., & Wieschaus, E. (2001). Armadillo nuclear import is regulated by cytoplasmic anchor Axin and nuclear anchor dTCF/Pan. *Development, 128*, 2107–2117.
Tomlinson, A., & Ready, D. F. (1987). Neuronal differentiation in Drosophila ommatidium. *Developmental Biology, 120*, 366–376.
Tomlinson, A., & Struhl, G. (1999). Decoding vectorial information from a gradient: Sequential roles of the receptors frizzled and notch in establishing planar polarity in the Drosophila eye. *Development, 126*, 5725–5738.
Townsley, F. M., Cliffe, A., & Bienz, M. (2004). Pygopus and Legless target Armadillo/beta-catenin to the nucleus to enable its transcriptional co-activator function. *Nature Cell Biology, 6*, 626–633.
Tree, D. R., Ma, D., & Axelrod, J. D. (2002). A three-tiered mechanism for regulation of planar cell polarity. *Seminars in Cell & Developmental Biology, 13*, 217–224.

Tree, D. R., Shulman, J. M., Rousset, R., Scott, M. P., Gubb, D., & Axelrod, J. D. (2002). Prickle mediates feedback amplification to generate asymmetric planar cell polarity signaling. *Cell*, *109*, 371–381.

Usui, T., Shima, Y., Shimada, Y., Hirano, S., Burgess, R. W., Schwarz, T. L., et al. (1999). Flamingo, a seven-pass transmembrane cadherin, regulates planar cell polarity under the control of frizzled. *Cell*, *98*, 585–595.

Valenta, T., Hausmann, G., & Basler, K. (2012). The many faces and functions of beta-catenin. *The EMBO Journal*, *31*, 2714–2736.

Veeman, M. T., Slusarski, D. C., Kaykas, A., Louie, S. H., & Moon, R. T. (2003). Zebrafish prickle, a modulator of noncanonical Wnt/Fz signaling, regulates gastrulation movements. *Current Biology*, *13*, 680–685.

Vinson, C. R., & Adler, P. N. (1987). Directional non-cell autonomy and the transmission of polarity information by the frizzled gene of Drosophila. *Nature*, *329*, 549–551.

Vinson, C. R., Conover, S., & Adler, P. N. (1989). A Drosophila tissue polarity locus encodes a protein containing seven potential transmembrane domains. *Nature*, *338*, 263–264.

Vuong, L. T., Iomini, C., Balmer, S., Esposito, D., Aaronson, S. A., & Mlodzik, M. (2018). Kinesin-2 and IFT-A act as a complex promoting nuclear localization of beta-catenin during Wnt signalling. *Nature Communications*, *9*, 5304.

Vuong, L. T., Mukhopadhyay, B., & Choi, K. W. (2014). Kinesin-II recruits Armadillo and dishevelled for wingless signaling in Drosophila. *Development*, *141*, 3222–3232.

Wallingford, J. B. (2010). Planar cell polarity signaling, cilia and polarized ciliary beating. *Current Opinion in Cell Biology*, *22*, 597–604.

Wallingford, J. B., & Habas, R. (2005). The developmental biology of dishevelled: An enigmatic protein governing cell fate and cell polarity. *Development*, *132*, 4421–4436.

Weber, U., Paricio, N., & Mlodzik, M. (2000). Jun mediates frizzled-induced R3/R4 cell fate distinction and planar polarity determination in the Drosophila eye. *Development*, *127*, 3619–3629.

Weber, U., Pataki, C., Mihaly, J., & Mlodzik, M. (2008). Combinatorial signaling by the Frizzled/PCP and Egfr pathways during planar cell polarity establishment in the Drosophila eye. *Developmental Biology*, *316*, 110–123.

Westfall, T. A., Brimeyer, R., Twedt, J., Gladon, J., Olberding, A., Furutani-Seiki, M., et al. (2003). Wnt-5/pipetail functions in vertebrate axis formation as a negative regulator of Wnt/beta-catenin activity. *The Journal of Cell Biology*, *162*, 889–898.

Widelitz, R. B., Jiang, T. X., Chen, C. W., Stott, N. S., Jung, H. S., & Chuong, C. M. (1999). Wnt-7a in feather morphogenesis: Involvement of anterior-posterior asymmetry and proximal-distal elongation demonstrated with an in vitro reconstitution model. *Development*, *126*, 2577–2587.

Winter, C. G., Wang, B., Ballew, A., Royou, A., Karess, R., Axelrod, J. D., et al. (2001). Drosophila Rho-associated kinase (Drok) links frizzled-mediated planar cell polarity signaling to the actin cytoskeleton. *Cell*, *105*, 81–91.

Wolff, T., & Ready, D. F. (1991). Cell death in normal and rough eye mutants of Drosophila. *Development*, *113*, 825–839.

Wolff, T., & Rubin, G. M. (1998). Strabismus, a novel gene that regulates tissue polarity and cell fate decisions in Drosophila. *Development*, *125*, 1149–1159.

Wu, J., Roman, A. C., Carvajal-Gonzalez, J. M., & Mlodzik, M. (2013). Wg and Wnt4 provide long-range directional input to planar cell polarity orientation in Drosophila. *Nature Cell Biology*, *15*, 1045–1055.

Wu, X., Tu, X., Joeng, K. S., Hilton, M. J., Williams, D. A., & Long, F. (2008). Rac1 activation controls nuclear localization of beta-catenin during canonical Wnt signaling. *Cell*, *133*, 340–353.

Yang, Y., & Mlodzik, M. (2015). Wnt-Frizzled/planar cell polarity signaling: Cellular orientation by facing the wind (Wnt). *Annual Review of Cell and Developmental Biology*, *31*, 623–646.
Yokoya, F., Imamoto, N., Tachibana, T., & Yoneda, Y. (1999). beta-Catenin can be transported into the nucleus in a Ran-unassisted manner. *Molecular Biology of the Cell*, *10*, 1119–1131.
Zecca, M., Basler, K., & Struhl, G. (1996). Direct and long-range action of a wingless morphogen gradient. *Cell*, *87*, 833–844.
Zeitlinger, J., Kockel, L., Peverali, F. A., Jackson, D. B., Mlodzik, M., & Bohmann, D. (1997). Defective dorsal closure and loss of epidermal decapentaplegic expression in Drosophila fos mutants. *The EMBO Journal*, *16*, 7393–7401.
Zheng, L., Zhang, J., & Carthew, R. W. (1995). Frizzled regulates mirror-symmetric pattern formation in the Drosophila eye. *Development*, *121*, 3045–3055.
Zou, Y. (2020). Breaking symmetry—Cell polarity signaling pathways in growth cone guidance and synapse formation. *Current Opinion in Neurobiology*, *63*, 77–86.

CHAPTER THREE

Hippo signaling in cardiac fibroblasts during development, tissue repair, and fibrosis

Chang-Ru Tsai[a] and James F. Martin[a,b,*]
[a]Department of Molecular Physiology and Biophysics, Baylor College of Medicine, Houston, TX, United States
[b]Cardiomyocyte Renewal Laboratory, Texas Heart Institute, Houston, TX, United States
*Corresponding author: e-mail address: jfmartin@bcm.edu

Contents

Abstract

The evolutionarily conserved Hippo signaling pathway plays key roles in regulating the balance between cell proliferation and apoptosis, cell differentiation, organ size control, tissue repair, and regeneration. Recently, the Hippo pathway has been shown to regulate heart fibrosis, defined as excess extracellular matrix (ECM) deposition and increased tissue stiffness. Cardiac fibroblasts (CFs) are the primary cell type that produces, degrades, and remodels the ECM during homeostasis, aging, inflammation, and tissue repair and regeneration. Here, we review the available evidence from the current literature regarding how the Hippo pathway regulates the formation and function of CFs during heart development and tissue repair.

Current Topics in Developmental Biology, Volume 149
ISSN 0070-2153
https://doi.org/10.1016/bs.ctdb.2022.02.010

1. Introduction

The leading cause of death in the developed world is heart disease, primarily coronary artery diseases, including myocardial infarction (MI) and heart failure (Aparicio et al., 2021). Within minutes or hours after MI, millions of cardiomyocytes (CMs) in the infarcted region die due to prolonged ischemia. The injured ventricular wall is replaced with reparative fibrotic scar, called replacement fibrosis, which is essential to preventing cardiac rupture. However, cardiac fibrosis increases tissue stiffness and spontaneous arrhythmias, impairing cardiac function and eventually resulting in heart failure (Burke, Burgos Villar, & Small, 2021; Frangogiannis, 2019b). Cardiac fibrosis, defined as the excess deposition of ECM, is also associated with other myocardial diseases, such as hypertension, heart failure, and arrhythmia. In addition, metabolic dysfunction, systemic inflammation, and aging can also induce heart fibrosis in the absence of tissue injury (Frangogiannis, 2019b).

Aside from replacement fibrosis, interstitial fibrosis, defined as ECM accumulation in the interstitial space, can be initiated by chronic hypertension (Frangogiannis, 2019a). High blood pressure increases resistance against the pumping heart, leading to CM hypertrophic remodeling and eventual CM apoptosis. Hypertension also increases stress to the coronary artery wall, inducing interstitial fibrosis around perivascular regions to support the vessels and replace the dying CMs in the myocardium. Ultimately, the fibrosis and thickening of the myocardium decreases cardiac function and results in heart failure (Frey & Olson, 2003; Heineke & Molkentin, 2006). Various mouse models have been developed to study cardiac fibrosis (Bacmeister et al., 2019; Pakshir et al., 2020).

Resident cardiac fibroblasts (CFs) are the primary cell type that contributes to cardiac fibrosis. Like other fibroblasts, CFs also show high plasticity (Bretherton, Bugg, Olszewski, & Davis, 2020; Daseke, Tenkorang, Chalise, Konfrst, & Lindsey, 2020; Tallquist, 2020a). In response to injury, CFs differentiate into multiple cell states with different and sometimes overlapping markers (Tallquist & Molkentin, 2017). After MI in mice, CFs become activated (activated fibroblasts, AFs) and proliferate on days 2 to 4 after MI. Subsequently, AFs differentiate into myofibroblasts with increased alpha smooth muscle actin (αSMA, encoded by *Acta2*) expression, which peaks around 7 days after MI. Around 10 days after MI, myofibroblasts further transition into matrifibrocytes, which do not express αSMA and only express low levels of type 1 collagen and type 3 collagen (Fu et al., 2018). Interestingly, matrifibrocytes express some genes that are related to cartilage and tendon, such as cartilage oligomeric matrix protein (*Comp*), chondroadherin (*Chad*), and cartilage intermediate layer protein (*Cilp2*). These cells also express anti-apoptosis genes and have low turnover rates. Matrifibrocytes are presumably required to maintain the cardiac scar, given that removing these cells by cryo-injury reduces cardiac function and promotes dilation (Fu et al., 2018).

The Hippo pathway is highly conserved from *Drosophila* to mammals. In mice, the genetic manipulation of Hippo pathway components affects organ size. Furthermore, Hippo signaling is dysregulated in cancer, similar to the overgrowth phenotypes induced by loss of Hippo signaling components in *Drosophila.* In addition to regulating the balance between cell proliferation and apoptosis, the Hippo pathway also regulates cell differentiation, cell growth, the cell fate decision during development (Wu & Guan, 2021; Zheng & Pan, 2019), cancer (Dey, Varelas, & Guan, 2020), and tissue repair and regeneration (Moya & Halder, 2019). Recently, the Hippo pathway has been shown to play essential roles in organ fibrosis (Dey et al., 2020; Kim, Choi, & Mo, 2019), including cardiac fibrosis (Francisco et al., 2020, 2021; Mia et al., 2021; Xiao et al., 2019). Furthermore, several studies have shown that the Hippo pathway plays critical roles in CFs during development (Singh et al., 2016), adult homeostasis, and tissue repair (Francisco et al., 2020, 2021; Mia et al., 2021; Xiao et al., 2019), which are the focus of this chapter.

2. Components of the Hippo pathway and their regulation

2.1 Hippo core kinase cascade

The Hippo core complex regulates the balance between cell proliferation and apoptosis in flies, mice, and humans (Fig. 1). The first gene in the Hippo kinase cascade identified from genetic screens in *Drosophila* was *warts* (*wts*), also named large tumor suppressor kinase (*lats, LATS1/2* in mammals) (Justice, Woods, Noll, & Bryant, 1995; Xu, Wang, Zhang, Stewart, & Wan, 1995). The loss of *wts* led to the cell-autonomous overproliferation of diverse developing epithelial tissues and organs and the superficial wart-like phenotype in adult fly legs (Justice et al., 1995). Several years after the discovery of *wts*, a gene encoding a WW domain–containing protein named *salvador* (*sav*, also known as *shar-pei, SAV* in mammals) was identified.

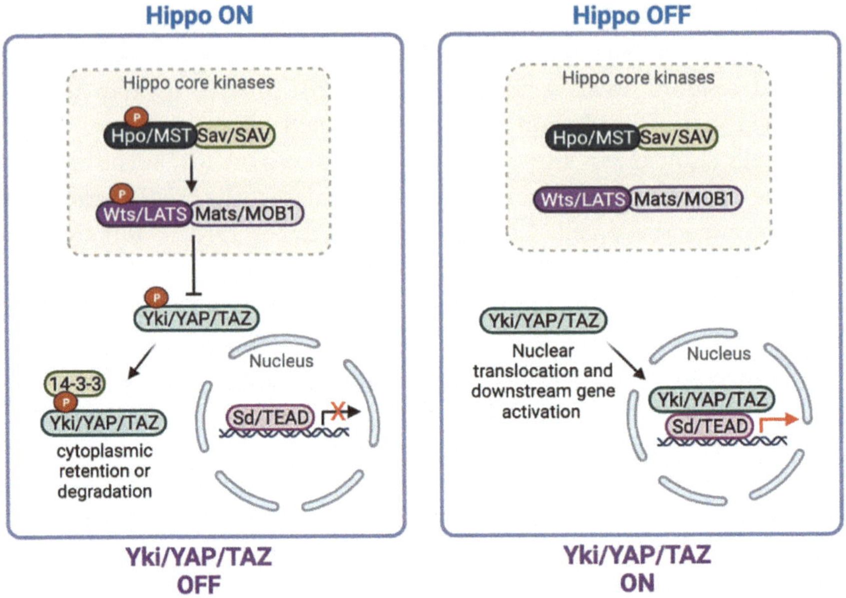

Fig. 1 Hippo pathway core components. When Hippo signaling is activated, Hpo/MST kinase binds to its adaptor protein Sav/SAV to promote the phosphorylation and activation of Wts/LATS kinase. Next, Wts/LATS, with its adaptor protein Mats/MOB1, phosphorylates Yki/YAP/TAZ and inhibits their function through cytoplasmic retention or degradation. When Hippo signaling is inactive, Yki/YAP/TAZ translocate into the nucleus and bind to transcription factors, such as TEAD, to promote downstream gene expression that promotes cell proliferation, survival, or cell differentiation.

The loss of *sav* caused hyperproliferation phenotypes similar to those caused by the loss of *wts* (Kango-Singh et al., 2002; Tapon et al., 2002). Furthermore, both loss-of-*sav* and loss-of-*wts* enhanced cell proliferation and decreased cell death, indicating the importance of these genes in regulating the balance between cell proliferation and cell death. Tapon and colleagues (Tapon et al., 2002) went on to show that Wts formed a complex with Sav in vitro, revealing part of the Hippo core complex. Next, the tumor suppressor *hippo* (*hpo, MST1/2* in mammals) was also identified by phenotypes of overgrowth and reduced apoptosis (Harvey, Pfleger, & Hariharan, 2003; Jia, Zhang, Wang, Trinko, & Jiang, 2003; Pantalacci, Tapon, & Léopold, 2003; Udan, Kango-Singh, Nolo, Tao, & Halder, 2003; Wu, Huang, Dong, & Pan, 2003). Five separate studies showed that Hpo interacts with Sav in vitro, and Jia et al. (2003) and Wu et al. (2003) further demonstrated the Hpo-Wts interaction. Wu et al. showed that the interaction between Hpo and Sav promoted the phosphorylation of

Wts by Hpo. In addition to Hpo/Sav, Mob as tumor suppressor (Mats in flies, MOB1A/B in mammals) was also shown to interact with Wts and stimulate its kinase activity. The loss of *mats* led to tissue overgrowth and reduced cell death, similar to the loss of *wts* (Lai et al., 2005). Importantly, introducing the human homolog of *mats1*, *MOB1A*, rescued the overgrowth phenotype in *Drosophila* caused by the loss of *mats1* (Lai et al., 2005), indicating the conservation of this Hippo pathway component between flies and humans. Similarly, mutations in Hippo pathway gene homologs have been shown to lead to dramatic tissue overgrowth in mice. Loss of the *hpo* murine homolog *Mst1/2* (Lee et al., 2010; Lu et al., 2010; Song et al., 2010; Zhou et al., 2009) in the mouse liver led to liver overgrowth.

2.2 Transcriptional cofactors downstream of the Hippo signaling pathway

A key finding that revealed how the Hippo core kinases regulate gene expression was the discovery of the transcriptional cofactor Yki (YAP/TAZ in mammals), which functions downstream of the Hippo signaling pathway (Fig. 1), in a yeast two-hybrid screen for Wts binding partners (Huang, Wu, Barrera, Matthews, & Pan, 2005). Different from the loss of core kinases, the loss of *yki* led to reduced cell proliferation, whereas the gain of *yki* promoted cell proliferation and inhibited apoptosis, in part by regulating the expression of genes encoding the cell cycle and apoptosis regulators Cyclin E (CycE) and Death-associated inhibitor of apoptosis 1 (Diap1), respectively. Using both genetic and biochemical approaches, Huang and colleagues (Huang et al., 2005) showed that Hpo/Sav promoted the phosphorylation of Yki by Wts, which inhibited the transcriptional activator function of Yki. In determining how Yki phosphorylation inactivates Yki function, the mammalian Yki homolog YAP was found to contain a conserved 14-3-3 binding motif that promoted its cytoplasmic retention upon AKT phosphorylation (Basu, Totty, Irwin, Sudol, & Downward, 2003). Three separate groups identified this serine residue in Yki/YAP as a critical Hippo-responsive phosphorylation site in both flies and mammals (Dong et al., 2007; Oh & Irvine, 2008; Zhao et al., 2007). Thus, the Hippo kinase cascade phosphorylates and inhibits the transcriptional activity of Yki or YAP/TAZ by nuclear exclusion. Because Yki and YAP/TAZ cannot directly bind to DNA, they must couple with other transcription factors, such as ERBB4 (Komuro, Nagai, Navin, & Sudol, 2003),

RUNX2 (Yagi, Chen, Shigesada, Murakami, & Ito, 1999), TEAD (Vassilev, Kaneko, Shu, Zhao, & DePamphilis, 2001), or TP73 (Basu et al., 2003). To identify transcription factors that YAP can activate, Zhao and colleagues (Zhao et al., 2008) screened a human transcription factor library and found that TEAD1-4 transcriptional activities were strongly activated by YAP (Zhao et al., 2008) and TAZ (Heng Zhang et al., 2009). Furthermore, Zhao and colleagues showed that TEAD is required for YAP-induced cell proliferation and epithelial-to-mesenchymal transition (EMT) in vitro and that the TEAD homolog in *Drosophila*, called Scalloped (Sd), is necessary for Yki-induced tissue overgrowth (Zhao et al., 2008). Consistent with this finding, another study showed that Yki physically interacts with Sd and that Sd is critical for Yki activity (Goulev et al., 2008). Thus, when Hippo core kinases are activated, Wts (LATS) phosphorylates Yki (YAP/TAZ) and promotes its cytoplasmic localization. In contrast, when Hippo kinases are inactive, Yki enters the nucleus and binds to the transcription factor Sd (TEAD1–4 in mammals) to promote downstream target gene expression (Fig. 1). Here, the discovery of the canonical Hippo pathway components are highlighted. Other mechanisms of core component regulation have been reported. For example, The Thousand and one kinases (TAOKs) (Boggiano, Vanderzalm, & Fehon, 2011; Poon, Lin, Zhang, & Harvey, 2011) phosphorylate and activate Hpo/MST, while striatin-interacting phosphatase and kinase and protein phosphatase 2A complex (STRIPAK/PP2A) (Bae et al., 2017; Ribeiro et al., 2010) can dephosphorylate and inactivate Hpo/MST. Detailed reviews of the Hippo pathway (Ma, Meng, Chen, & Guan, 2019; Misra & Irvine, 2018; Moya & Halder, 2019; Zheng & Pan, 2019) and the posttranscriptional regulation of Hippo components (Meng, Xie, & Martin, 2021) can be found elsewhere.

2.3 Upstream regulators of the Hippo signaling pathway

Many developmental signaling pathways, such as Wnt, Notch, Hedgehog, and TGF-β pathways, are regulated by designated ligand-receptor pairs. In contrast, the Hippo pathway is regulated by a wide variety of upstream signals (Fig. 2) related to cell junctions and polarity (Karaman & Halder, 2018), the cytoskeleton, mechanical forces (Herum, Lunde, McCulloch, & Christensen, 2017; Panciera, Azzolin, Cordenonsi, & Piccolo, 2017; Pocaterra, Romani, & Dupont, 2020), G protein coupled receptor (GPCR) (Luo & Yu, 2019), and cellular stress (Fig. 2). The importance of upstream

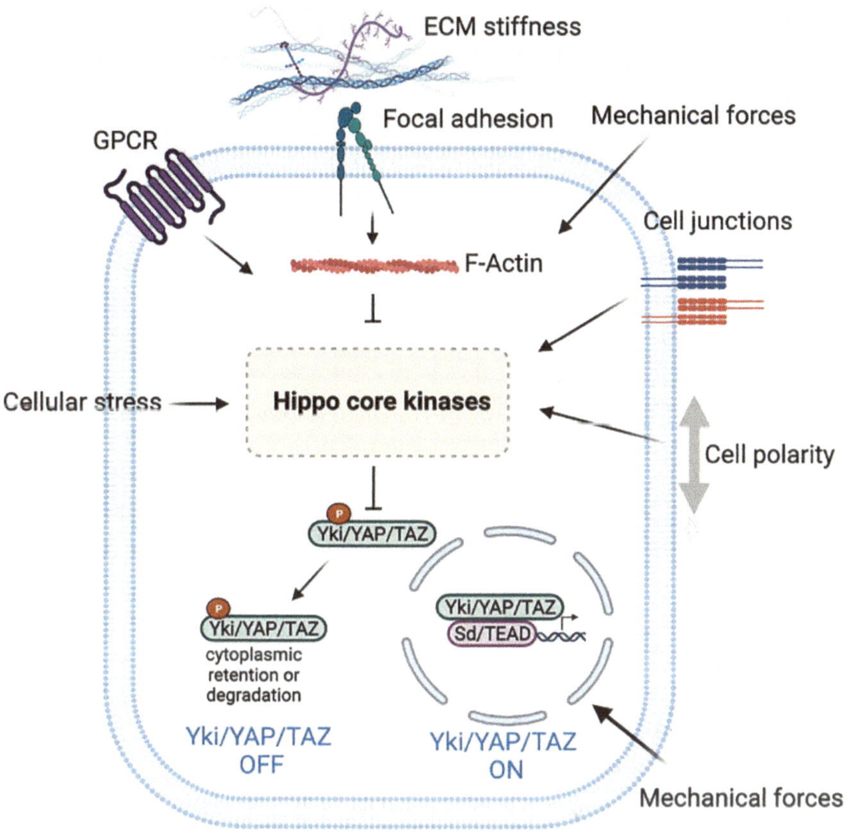

Fig. 2 Upstream signals of the Hippo pathway core components. The Hippo pathway is regulated by diverse upstream signals related to cell junctions, cell polarity, G protein-coupled receptor signaling (GPCR), mechanical forces, extracellular matrix (ECM) stiffness, focal adhesions, the cytoskeleton, and cellular stress.

signals is likely to be dependent on the cell type and context. For example, cell polarity and cell junctions may play a more important role in epithelial tissue than in mesenchymal cells, such as fibroblasts. Below, we focus on focal adhesions and the cytoskeleton.

2.3.1 Focal adhesions

Cells sense and interact with the ECM through focal adhesions (Burridge & Chrzanowska-Wodnicka, 1996). Integrins are key transmembrane focal adhesion receptors that are essential for the adhesion of cells to the ECM (Hynes, 2002). Upon binding to extracellular proteins, integrins are activated and change conformation, promoting their cytoplasmic tails to interact with intracellular signaling molecules, such as integrin-linked kinase (ILK), SRC, FAK kinase, and Vinculin, to regulate the actin cytoskeleton and Hippo signaling pathway. Different integrins have been shown to promote YAP/TAZ nuclear localization and activation (Chang et al., 2015; Elbediwy et al., 2016; Kaneko, Ito, Naoe, Lacy-Hulbert, & Ikeda, 2014; Tang et al., 2013). In addition, ILK phosphorylates and inhibits myosin phosphatase targeting subunit 1 (MYPT1), leading to Merlin (Jin, Sperka, Herrlich, & Morrison, 2006) and Hippo core kinase inactivation and thus YAP activation (Serrano, McDonald, Lock, Muller, & Dedhar, 2013). Moreover, integrin-associated kinases FAK and SRC also act upstream of Hippo signaling. Inducing cell adhesion by plating starved epithelial cells on a fibronectin-coated surface promotes YAP dephosphorylation and nuclear localization in an FAK-dependent manner. Furthermore, SRC, the critical downstream kinase of FAK, inhibits LATS1 activation, leading to YAP activation in this cell culture model (Kim & Gumbiner, 2015). Similarly, SRC also acts downstream of integrin signaling to promote YAP nuclear localization in primary cultured osteoblasts (Kaneko et al., 2014). Thus, focal adhesions regulate Hippo/YAP signaling through integrins, ILK, FAK, and SRC.

2.3.2 Cytoskeleton

The cytoskeleton has been shown to regulate Yki/YAP localization and activity. Growing cells on stiff surfaces promotes actin polymerization (increased stress fibers) and YAP nuclear localization and activation (Aragona et al., 2013; Dupont et al., 2011; Wada, Itoga, Okano, Yonemura, & Sasaki, 2011). Blocking actin polymerization prevents YAP activation, indicating that filamentous actin (F-actin) promotes YAP activation (Dupont et al., 2011; Sansores-Garcia et al., 2011; Wada et al., 2011; Zhao et al., 2012). In contrast, increasing F-actin induces YAP/TAZ nuclear localization and activation (Aragona et al., 2013; Sansores-Garcia et al., 2011). Importantly, in *Drosophila*, the actin cytoskeleton has also been shown to regulate Yki activity. Increased F-actin induced by the loss of the actin capping protein (encoded by *capulet*) or gain of the actin formin diaphanous promotes Yki activity and tissue overgrowth (Fernández et al., 2011; Sansores-Garcia et al., 2011). On the other hand, Hippo signaling also regulates actin polymerization in vivo. Loss of *hpo*, *sav*, *mats*, or *wts* increased F-actin (Fernández et al., 2011). Furthermore, overexpression of the Hippo core components partially rescued the overgrowth and F-actin accumulation phenotypes induced by the gain of *diaphanous* (Sansores-Garcia et al., 2011). Thus, these results suggests that the Hippo core components and F-actin negatively regulate each other. Yki also regulates actin polymerization. Silencing Yki in the larval epidermal cells blocked wound-induced actin polymerization around the wound edges (Tsai, Anderson, Burra, Jo, & Galko, 2017). Furthermore, Yki localized at the apical junctional region (AJR) has been shown to play a non-transcriptional role in promoting tissue growth (Xu et al., 2018) by activating myosin through the Yki N-terminal domain. Because myosin activation promotes Yki/YAP activities, the authors proposed a positive feed-forward loop in which cortical Yki promotes myosin activation, which in turn further enhances Yki activity (Xu et al., 2018).

Another cytoskeleton protein that has been linked to Hippo signaling is spectrin, which forms a supportive network under the cell membrane (MacHnicka, Grochowalska, Bogusławska, Sikorski, & Lecomte, 2012). However, spectrin has a different effect on the Hippo signaling pathway than F-actin. Loss of spectrin promotes Yki/YAP activation in mammalian cells and tissue overgrowth in *Drosophila* (Deng et al., 2015; Fletcher et al., 2015; Wong et al., 2015). Fletcher and colleagues (Fletcher et al., 2015) showed that apical spectrin promotes Hippo core kinases through the clustering of components upstream of the Hippo pathway, including Expended, Crumb, Merlin, and Kibra, leading to Yki inactivation. In addition, in *Drosophila* the multi-PDZ scaffold big bang has been shown to bind and inhibit spectrin (Forest et al., 2018), leading to Yki activation and increased actomyosin activity at the cell cortex (Deng et al., 2015).

3. The contribution of resident CFs to cardiac fibrosis

Myofibroblasts are the major cell type that deposits collagens and other types of extracellular matrices in fibrotic tissues (Fig. 3) (Darby, Zakuan, Billet, & Desmoulière, 2016; Pakshir et al., 2020; Wang, Zohar, & McCulloch, 2006). The mammalian heart is composed of diverse cell types, such as CMs, fibroblasts, endothelial cells, endocardial cells, epicardial cells, smooth muscle cells, and pericytes (Pinto et al., 2016). Among these cell types, resident fibroblasts, endothelial cells, epicardial cells, and smooth muscle cells have been proposed to differentiate into myofibroblasts during cardiac injury and fibrosis. To investigate these possibilities, several groups have conducted genetic lineage-tracing experiments in different injury models in mice using cell type–specific Cre recombinase. Two studies in which the tamoxifen-inducible Cre that specifically labels resident fibroblasts ($Tcf21^{iCre}$) was used showed that most of the myofibroblasts,

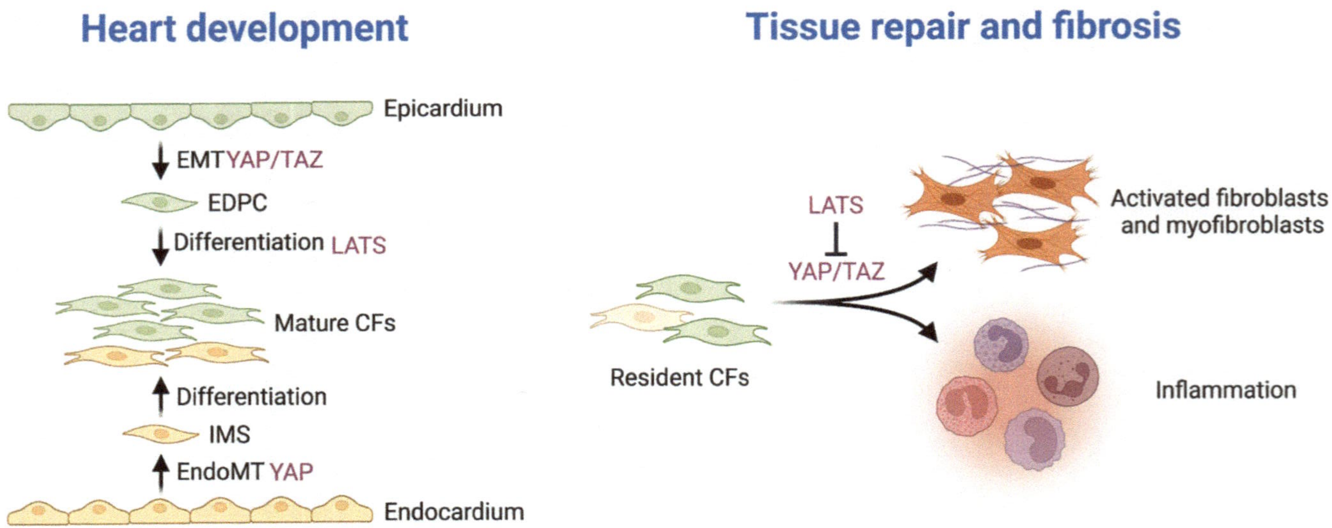

Fig. 3 Hippo signaling in cardiac fibroblast formation and function. During heart development, resident cardiac fibroblasts (CFs) are primarily derived from the epicardium and endocardium. YAP and TAZ are required for epicardial cells to proliferate and undergo epithelial-to--mesenchymal transition (EMT) to differentiate into epicardial derived progenitor cells (EDPC). LATS1/2 are required for EDPCs to differentiate into CFs. YAP is required for endocardial cells to undergo endocardial-to-mesenchymal transition (EndoMT) to an intermediate mesenchymal stage (IMS) before CF differentiation. Upon injury, resident CFs become active and differentiate into myofibroblasts to promote fibrosis. YAP activation upon the loss of LATS1/2 promotes CF activation and inflammation.

labeled by αSMA, were derived from the Tcf21 lineage cells in different injury models, including MI and pressure overload (PO) induced by transverse aortic constriction (TAC) or chronic treatment with angiotensin II (AngII) or isoproterenol (Ivey et al., 2018; Kanisicak et al., 2016). In contrast, lineages of endothelial cells, smooth muscle cells, and monocytes and macrophages did not contribute to (myo)fibroblasts after injury, as indicated by transcriptional reporters of CF markers, type 1 collagen and Periostin (encoded by *Postn*) (Kanisicak et al., 2016; Moore-Morris et al., 2014). In line with this finding, heat shock protein 47 (HSP47), a critical chaperone for collagen maturation (Ishida et al., 2006; Nagata, 1988; Satoh, Hirayoshi, Yokota, Hosokawa, & Nagata, 1996) during organ fibrosis (Abdelnasir et al., 2014; Masuda, Fukumoto, Hirayoshi, & Nagata, 1994; Razzaque, Hossain, Kohno, & Taguchi, 1998; Xiao et al., 2012), is essential in the CF lineage, but not in CM or endothelial cell lineages, for generating scar tissue after MI (Khalil et al., 2019). Furthermore, although most CFs are derived from the epicardium during heart development (Fig. 3) (Acharya, Baek, Banfi, Eskiocak, & Tallquist, 2011; Smith, Baek, Sung, & Tallquist, 2011), adult-stage epicardial cell lineage–tracing experiments with $Wt1^{CreERT2}$ revealed that the contribution of the epicardium to the fibroblast pool was limited in both a PO model (Moore-Morris et al., 2014) and an MI model (Zhou et al., 2011). Together, these results indicate that myofibroblasts are primarily derived from resident CFs during cardiac fibrosis.

It is worth noting that CFs are essential for generating protective scars that prevent cardiac rupture after MI. Eliminating myofibroblasts genetically by using another myofibroblast-specific inducible Cre (*Postn-iCre*) and *Rosa26-DTA* (diphtheria toxin A) significantly increased lethality after MI (Kanisicak et al., 2016). Along the same lines, the loss of the myofibroblast-secreted extracellular proteins, periostin, (Norris et al., 2007; Shimazaki et al., 2008) or fibroblast cytokine follistatin-like 1 (Maruyama et al., 2016) led to increased cardiac rupture after MI. Additional studies of CF-specific gene deletion or overexpression can be found elsewhere (Fu, Liu, Li, Li, & Wang, 2020; Umbarkar, Ejantkar, Tousif, & Lal, 2021).

3.1 Hippo signaling pathway in CFs during development and tissue repair

In mice, there are two major developmental sources of CFs (Tallquist, 2020b) (Fig. 3). The first is the mesothelial covering of the heart, known as the epicardium. During heart development, epicardial cells undergo EMT and differentiate into epicardial-derived progenitor cells (EDPCs).

EDPCs further differentiate into CFs and other cell types, such as smooth muscle cells and endothelial cells (Quijada, Trembley, & Small, 2020). The loss of *Yap/Taz* in the mouse epicardium impaired epicardial cell proliferation, EMT, and reduced epicardial-derived CFs, smooth muscle cells and endothelial cells and led to coronary vasculature defects and caused embryonic lethality between mouse embryonic day (E) 11.5 and E12.5 (Singh et al., 2016) (Table 1). It is likely that loss of epicardial *Yap/Taz* reduced the number of EDPCs and their derivatives leading to the coronary defects. On the other hand, knocking out *Lats1/2* in mouse epicardial cells promoted their proliferation (Xiao et al., 2018) (Fig. 3) (Table 1). Interestingly, despite the expansion of the epicardial cells, CFs number were actually reduced in the *Lats1/2* conditional knockout hearts (Xiao et al., 2018). Xiao and colleagues identified a cell state intermediate between epicardial cells and fibroblasts which was expanded in the *Lats1/2* conditional knockout hearts, indicating that LATS1/2 are required for fibroblast differentiation and maturation. Along the same line, they identified DHRS3, an inhibitor of retinoic acid synthesis (Billings et al., 2013), as a YAP target that inhibited CF differentiation (Xiao et al., 2018). In addition, Xiao and colleagues also identify DPP4, a cell surface serine protease, as a YAP target in the epicardium that disrupted coronary vessel patterning in the *Lats1/2* conditional knockout hearts (Xiao et al., 2018). Taken together, the loss of either *Yap/Taz* or loss of *Lats1/2* abolishes CF formation, indicating that the level of Hippo signaling in the epicardium is critical for CF formation.

Another source of CFs is the endothelial/endocardial population during development (Fig. 3) (Ali et al., 2014; Moore-Morris et al., 2014). Genetic lineage-tracing experiments with *Tie2-Cre* revealed that endothelial/endocardial-derived cells contributed to 64% of the fibroblasts in the interventricular septum and 12% of the fibroblasts in the left ventricle (Moore-Morris et al., 2014). Another study in which the same Cre line was used also showed evidence for endocardial/endothelial-derived fibroblasts (Ali et al., 2014). In that study, epicardial-derived CFs were similarly more abundant than were endocardial/endothelial-derived CFs across different chambers and the septum. A recent single-cell transcriptomic study has also shown the presence of minor endocardial/endothelial-derived CFs in the ventricle (Forte et al., 2020).

How do endothelial/endocardial cells differentiate into fibroblasts? During early heart development, some endocardial cells undergo endothelial-to-mesenchymal transition (EndoMT) to become valve mesenchyme (De Lange et al., 2004) and then mature into valve interstitial fibroblasts

Table 1 Genetic studies in mice showing Hippo pathway function in cardiac fibroblasts during development and after injury.

Gene(s)	Genetic tools	Stage and disease models	Phenotypes and major findings	Reference
Yap, *Taz*	$Wt1^{CreERT2/+}$; $Yap^{f/f}$; $Taz^{f/f}$	Development	Impaired epicardial proliferation, epithelial-to-mesenchymal transition, and coronary vasculature formation	Singh et al. (2016)
Lats1, *Lats2*	$Wt1^{CreERT2/+}$; $Lats1^{f/f}$; $Lats2^{f/f}$	Development	Expansion of epicardium, defective cardiac fibroblast differentiation/maturation, and coronary vasculature structure	Xiao et al. (2018)
Yap	$Tie2^{Cre/+}$; $Yap^{f/f}$	Development	Impaired proliferation and defective endothelial-to-mesenchymal transition	Zhang et al. (2014)
Lats1, *Lats2*, *Yap*, *Taz*	$Tcf21^{iCre/+}$; $Lats1^{f/f}$; $Lats2^{f/f}$ $Tcf21^{iCre/+}$; $Lats1^{f/f}$; $Lats2^{f/f}$; $Yap^{f/+}$; $Taz^{f/+}$	Adult, sham and MI	Myofibroblast proliferation and differentiation, fibrosis, inflammation Removing one copy of *Yap* and *Taz* reduced *Lats* conditional knockout-induced fibrosis after MI	Xiao et al. (2019)
Yap	$Tcf21^{iCre/+}$; $Yap^{f/f}$ $Col1a1^{CreERT2/+}$; $Yap^{f/f}$	Adult, MI, chronic AngII/PE	Myofibroblast differentiation, fibrosis	Francisco et al. (2020)
Yap	*AAV-hTCF21-YAP*	Adult, homeostasis	Myofibroblast differentiation, inflammation	Francisco et al. (2021)
Yap, *Taz*	$Col1a2^{CreERT/+}$; $Yap^{f/f}$; $Taz^{f/f}$	Adult, MI	Myofibroblast differentiation, inflammation	Mia et al. (2021)

AngII, angiotensin II; PE, phenylephrine; MI, myocardial infarction; *f*, *flox*.

(Horne et al., 2015). If endothelial/endocardial cells differentiate into CFs through EndoMT, there should be an intermediate mesenchymal stage, and lineage-tracing these intermediate cells should label fibroblasts. Indeed, dual-lineage reporter experiments in which cells were labeled with the endocardial marker Nfatc1 and the mesenchymal marker Sox9 revealed an intermediate mesenchymal stage of endocardial-derived CFs (Huang, Feng, et al., 2019) (Fig. 3). Similar to the epicardial loss of *Yap/Taz*, the loss of *Yap* in endocardial cells induced by *Tie2-Cre* impaired endocardial cell proliferation and the EndoMT that forms valve mesenchyme (Zhang et al., 2014) (Table 1 and Fig. 3). Zhang and colleagues (Zhang et al., 2014) showed that YAP interacts with SMAD2/3/4 complexes in vivo and promotes the expression of EndoMT/EMT genes, such as *SNAIL*, *TWIST1*, and *SLUG*, upon TGF-β treatment in cultured human endothelial cells. Whether the loss of *YAP* in the endocardium blocked endocardial-derived CF formation could not be directly examined because of early embryonic lethality (Zhang et al., 2014).

Do endocardial/endothelial-derived CFs respond to injuries similarly to epicardial-derived CFs? In the PO model, the ratios of epicardial-derived and endothelial/endocardial-derived CFs remained the same (Ali et al., 2014; Moore-Morris et al., 2014), suggesting that, despite the ontology, epicardial-derived and endocardial-derived fibroblasts proliferate at similar rates in response to PO. Moreover, transcriptomic analyses showed similar gene expression profiles between epicardial-derived and endothelial/endocardial-derived CFs after PO. These findings suggest that epicardial-derived and endothelial/endocardial-derived CFs may be functionally similar after PO. Interestingly, in a recent study, single-cell transcriptomic analysis of hearts after MI showed that most myofibroblasts were derived from epicardial lineages, whereas the endothelial/endocardial-derived CF population remained stable, not contributing to myofibroblast clusters after MI (Forte et al., 2020). These studies suggest that CFs from different origins can respond differently to various types of injuries.

In humans, altered Hippo pathway signaling has been observed in different fibrotic heart diseases, including hypertrophic cardiomyopathy (Wang et al., 2014), ischemic and nonischemic heart failure (Leach et al., 2017; Sharifi-sanjani et al., 2021), and arrhythmogenic cardiomyopathy (Chen et al., 2014). Because most of the studies described above involved western blot analyses of human samples, it remains unclear whether Hippo pathway signaling is altered in CFs.

In a pioneering study by Xiao and colleagues (Xiao et al., 2019) (Table 1), the function of the Hippo core kinases LATS1/2 were examined in adult resident CFs in both uninjured mouse hearts and hearts after MI. When *Lats1/2* was deleted by using inducible fibroblast-specific Cre ($Tcf21^{iCre}$), CFs spontaneously promoted heart fibrosis that was exacerbated after MI. To determine whether *Lats1/2* deletion promoted fibroblast differentiation and other injury responses, Xiao and colleagues performed single-cell RNA sequencing of control and *Lats1/2* conditional knockout ventricles. Strikingly, the cell composition was significantly altered in the *Lats1/2* conditional knockout hearts, with the significant expansion of myofibroblast-like cells and monocytes/macrophages.

Myofibroblast-like cells express classical myofibroblast markers, such as *Acta2* and *Fn1*. Gene regulation network analyses revealed the co-expression of TEAD and its downstream targets, specifically in myofibroblast-like cells, indicating that these cells are differentiated from *Lats1/2*-deficient CFs (Fig. 3). Cardiac fibrosis induced by the conditional knockout of *Lats1/2* that was exacerbated by MI was significantly reduced by deleting one copy of *Yap* and *Taz* in CFs (Xiao et al., 2019). Thus, these results support the idea that Hippo signaling normally maintains the resting CF cell state and that loss of LATS1/2 activates downstream YAP/TAZ/TEAD to promote fibroblast proliferation, myofibroblast differentiation, and heart fibrosis. Consistent with this, the loss of *Yap* in CFs reduced myofibroblast differentiation, as indicated by αSMA expression, and thus decreased MI-induced and PO-induced heart fibrosis (Francisco et al., 2020) (Table 1).

TAZ, the paralog of YAP downstream of Hippo, also promotes profibrotic gene expression and lung fibrosis (Liu et al., 2015; Noguchi et al., 2017). Recently, the Singh group (Mia et al., 2021) (Table 1) showed that the nuclear localization of YAP and TAZ was increased after MI and that the CF-specific deletion of both *Yap* and *Taz* abolished the expression of multiple profibrotic genes and cardiac fibrosis after MI. On the other hand, gain of YAP function increased profibrotic gene expression and cardiac fibrosis after MI (Francisco et al., 2020; Mia et al., 2021). Francisco and colleagues (Francisco et al., 2020) proposed that myocardin-related transcription factor A (MRTF-A) is a critical target of YAP that promotes myofibroblast differentiation. Although MRTF-A induces myofibroblast gene programs after MI (Small et al., 2010), follow-up studies are needed to compare the roles of the Hippo pathway and MRTF-A in CF proliferation and activation (Small & Brooks, 2020). In addition, the Singh group found that the MI-induced cytokine interleukin (IL)-33 is regulated by YAP. The CF-specific overexpression of YAP is sufficient to induce IL-33 expression, and the fibroblast-specific knockout of *Yap/Taz* or the inhibition of YAP with the pharmaceutical inhibitor verteporfin blocked IL-33 expression after MI (Mia et al., 2021). Results of a luciferase assay and chromatin immunoprecipitation experiments further indicated that YAP directly binds and activates the *IL-33* promoter. Moreover, in cultured CFs, IL-33 promoted αSMA and type 1 collagen expression (Mia et al., 2021).

In addition to its role in myofibroblast differentiation, the Hippo pathway also functions non-autonomously in CFs to regulate inflammation (Fig. 3). Intercellular communication network analysis showed that *Lats1/2*-deficient CFs had a larger number of significant ligand-receptor pairs (~5 fold increase) than did control fibroblasts (Xiao et al., 2019). Many of these ligands are involved in wound responses, including THBS1, SERPINE1, HBEGF, EREG, and TIMP1. Several top ligand-receptor pairs, such as CCL2, CCL7, CXCL12, and CSF1, were shown to be secreted from *Lats1/2*-deficient fibroblasts to myeloid cells. Chemokine/cytokine protein array data from *Lats* conditional knockout hearts further confirmed the upregulation of TIMP1, CSF1, CCL2, CCL12, CXCL1, CXCL10, and IL-1RA levels. Consistently, YAP overexpression in CFs has been shown to promote the expression of inflammatory-associated genes in vitro and in mouse hearts (Francisco et al., 2021) (Table 1), as well as after MI (Mia et al., 2021). The Del Re group (Francisco et al., 2021) performed chromatin immunoprecipitation using YAP antibody and found that YAP binds to the *Ccl2* promoter in cultured CFs, indicating that *Ccl2* is a direct target of YAP. This is consistent with a previous study showing that YAP and TEAD1 bind to the *Ccl2* promoter (Guo et al., 2017). Together, these studies revealed that the Hippo pathway in CFs regulates fibrosis and inflammation. Since resident and recruited macrophages play distinct roles in regulating tissue repair and fibrosis, understanding how these cardiac cells interact at various time points after wounding remains critical.

4. Outstanding questions and future directions

4.1 What are the signals upstream of the Hippo signaling pathway upon different types of cardiac injury?

The upregulation of YAP/TAZ in fibroblasts after injury has been reported in different fibrotic diseases and organs, such as kidney (Liang et al., 2017b), liver (Mannaerts et al., 2015), and lung (Noguchi et al., 2017). Five days after MI, YAP and TAZ were found to be upregulated in CFs at both the RNA and protein level (Mia et al., 2021). As mentioned previously, at the protein level, YAP/TAZ are inhibited by the Hippo kinase cascade through phosphorylation. At the transcriptional level, *Yap* has been reported to be upregulated by the Ets transcription factor GA binding protein (GABP) (H. Wu et al., 2013). Interestingly, the same study also showed that LATS1/2 inhibited GABP. Therefore, LATS1/2 inhibits Yap function both at the protein and RNA levels. LATS1/2 may be inactivated after MI, leading to the upregulation of GABP and its target YAP. *Lats1/2* deletion in CFs was shown to induce myofibroblast differentiation and inflammation without injury (Xiao et al., 2019), indicating the importance of LATS1/2s function in preventing CF activation and differentiation.

GPCR signaling has been shown to activate YAP/TAZ by inhibiting LATS1/2 (Yu et al., 2012). A different study on lung fibrosis showed that the GPCR ligand sphingosine-1-phosphate could activate YAP in lung fibroblasts to promote lung fibrosis (Huang et al., 2020). Another GPCR, angiotensin type 1 receptor, has been shown to activate YAP in cultured CFs. YAP has been shown to be required for AngII-induced type 1 *collagen* expression (Harikrishnan, Titus, Cowling, & Kailasam, 2019) and *Acta2* expression, as well as myofibroblast differentiation in cultured rat CFs (Francisco et al., 2020; Jin, Zhu, Shi, Wen, & Wu, 2019). Interestingly, in rat CFs, αSMA is also required for AngII-induced Yap activation via calcium influx through TRCP6 (Harikrishnan et al., 2019), suggesting a positive feedback loop in AngII-treated CFs. Further studies are required to understand how GPCRs regulate YAP/TAZ after injury in vivo.

Recently, the Singh group (Mia et al., 2021) showed in cultured CFs that YAP and TAZ are required for TGF-β–induced profibrotic gene expression and myofibroblast differentiation. Similarly, YAP and TAZ have been shown to be critical for the TGF-β–induced differentiation of myofibroblasts from kidney fibroblasts (Liang et al., 2017a) and vocal fold fibroblasts (Nakamura, Hiwatashi, Bing, Doyle, & Branski, 2021). These results suggest that YAP and TAZ can act downstream of TGF-β. In addition, in human primary vocal fold fibroblasts, TGF-β stimulated the localization of YAP/TAZ in close proximity to SMAD2/3, the downstream transcription factor of TGF-β signaling (Nakamura et al., 2021), suggesting that YAP/TAZ can form a complex with SMAD2/3. Thus, it will be interesting to determine whether interactions between YAP/TAZ and SMAD2/SMAD3 are essential in CFs during cardiac fibrosis.

In addition to biochemical cues, mechanical inputs can also activate YAP and TAZ (Panciera et al., 2017). Thus, PO-induced stress to the ventricular and vessel walls may activate YAP and TAZ in nearby CFs and promote their activation and differentiation into myofibroblasts. Moreover, as determined by using atomic force microscopy, ECM stiffness increases during cardiac fibrosis. For example, the ECM stiffness of the healthy myocardium is around 18 kilopascal (kPa), whereas the fibrotic myocardium after MI is around 55 kPa (Berry et al., 2006). ECM of high stiffness is known to promote YAP/TAZ nuclear localization and downstream gene expression (Dupont et al., 2011). Although the detailed mechanisms of YAP/TAZ in mechanical sensing remain unclear, cytoskeletal dynamics are central in this regulation (Pocaterra et al., 2020). The Rho-associated coiled-coil containing kinase (ROCK) acts downstream of small GTP-binding protein RHOA to regulate the actin cytoskeleton (Riento & Ridley, 2003). The CF-specific knockout of *Rock2* attenuates AngII-induced fibroblast activation, cardiac fibrosis, and diastolic dysfunction (Shimizu et al., 2017). Along the same lines, the Del Re group showed that AngII-induced YAP activation is RHO A dependent in cultured rat CFs (Francisco et al., 2020). *Ctgf* expression in CFs was significantly reduced in mouse hearts with the CF-specific knockout of *Rock2* after PO. The transcription factor that upregulates *Ctgf* in this context remains unknown. Because *Ctgf* is a direct target of YAP/TAZ, and actin cytoskeletal contractility is critical for YAP/TAZ activation, the loss of *Rock2*-dependent actin contractility may impair AngII-induced YAP/TAZ activation and *Ctgf* expression. Further study is warranted to test this hypothesis.

Recently, compared with randomly oriented mimetics, aligned collagen organization was shown to promote the YAP-induced expression of ECM genes (Bugg et al., 2020). Thus, it is likely that during cardiac fibrosis, increased tissue stiffness further activates YAP/TAZ. In turn, activated YAP and TAZ further drive profibrotic gene expression to enhance ECM stiffness in a positive feed-forward loop as the scar matures. However, 10 days after MI, myofibroblasts were shown to further transition into matrifibrocytes with diminished αSMA and collagen expression (Fu et al., 2018), suggesting that the YAP/TAZ-induced myofibroblast program is turned off through an unknown mechanism. Exploring the on and off switch of the Hippo signaling pathway after wounding will provide important insight into cardiac fibrosis.

Finally, because the Hippo pathway integrates diverse upstream biochemical and mechanical inputs, it is conceivable that multiple inputs activate YAP and TAZ during different disease states. These inputs could differ according to the type of cardiac injury. For example, PO-induced mechanical forces on the chamber and vessel walls are not seen in MI. In addition, MI-induced CM death and inflammation did not occur, at least in the early stages of PO. Thus, the upstream inputs of Hippo signaling in CFs remain to be rigorously tested in different injury models and at various time points.

4.2 What are the downstream targets of YAP/TAZ in CFs upon different types of cardiac injury and at different disease stages?

The loss of *Yap* reduced MI-induced or AngII/phenylephrine (PE)-induced myofibroblast expansion and fibrosis, but it did not improve cardiac function (Francisco et al., 2020). The loss of both *Yap* and *Taz* in CFs decreased fibrosis after MI and cardiac function (Mia et al., 2021), suggesting that YAP and TAZ have their own unique roles. Although YAP and TAZ both act downstream of Hippo and interact with TEAD to regulate gene expression, downstream targets can have distinct differences (Plouffe et al., 2018; Reggiani, Gobbi, Ciarrocchi, & Sancisi, 2021). In the MI model, CFs become proliferative and activated around days 2–4 after MI. The activated fibroblasts differentiate into myofibroblasts around day 4–7 after MI (Fu et al., 2018). The regulation of CF proliferation and differentiation is not fully understood. A recent study suggested that CF proliferation and activation are genetically separate (Burke et al., 2018). In the MI model, proliferation declines as activated fibroblasts differentiate into myofibroblasts. Because the Hippo signaling pathway regulates both proliferation and myofibroblast differentiation, YAP/TAZ may activate different targets over time. In addition, varying cardiac injuries may activate different YAP/TAZ targets at unique time points. Thus, it will be interesting to investigate which downstream targets of YAP/TAZ are activated upon different types of injury and at various disease stages. To this end, using a temporally regulated genetic tool, such as the combination of Cre-mediated recombination and a tetracycline/doxycycline-inducible system (Belteki et al., 2005), will be helpful for further pinpointing the functions of YAP/TAZ in CFs.

4.3 What are the non-cell autonomous HSP functions in CFs during homeostasis and tissue repair?

Using CF-specific Cre mouse lines (Fu et al., 2020; Plikus et al., 2021; Swonger, Liu, Ivey, & Tallquist, 2016) and the adeno-associated virus (AAV) vector with a fibroblast-specific promoter (Francisco et al., 2020), researchers are now able to understand the functions of Hippo pathway components specifically in CFs. In addition to CF proliferation and myofibroblast differentiation, Hippo signaling pathway in CFs also promote inflammation (Francisco et al., 2021; Mia et al., 2021; Xiao et al., 2019). The drastic expansion of macrophages in hearts with *Lats1/2*-deficient CFs demonstrates the non–cell-autonomous effects of the CF Hippo signaling pathway (Xiao et al., 2019). Nonetheless, the importance of these YAP/TAZ-induced ligands and whether they have beneficial or deleterious effects on tissue repair remain to be tested. Moreover, CF-secreted ligands may signal to every cardiac cell type, including themselves. Complicated cell-cell communications can now be investigated in an efficient way by using the single-cell genetic methodologies and bioinformatic approaches available (Svensson, Vento-Tormo, & Teichmann, 2018). In addition, the advanced spatial transcriptomic technologies (Dries et al., 2021; Eng et al., 2019; Grisanti Canozo, Zuo, Martin, & Samee, 2021; Rodriques et al., 2020; Vickovic et al., 2019) that permit in situ tissue profiling will also facilitate the identification of critical signaling between neighboring cells.

5. Conclusion

Despite the tremendous medical burden of cardiac fibrosis, no treatment option is available. However, recent studies in animals have revealed that cardiac fibrosis can be reduced and even reversed. First, similar to the loss of YAP/TAZ, blocking TGF-β signaling components, including TGF-β receptors 1/2 (Bhandary et al., 2018; Khalil et al., 2017; Meng et al., 2018) and the downstream transcription factors SMAD2/3 (Huang, Chen, et al., 2019; Khalil et al., 2017; Kong et al., 2018), was reported to reduce cardiac fibrosis and improve cardiac function after injury. Second, the overexpression of the matricellular protein CCN5 reverses TAC-induced fibrosis (Jeong et al., 2016). Third, eliminating myofibroblasts with engineered T-cells reverses AngII/PE-induced cardiac fibrosis and cardiac dysfunction (Aghajanian et al., 2019). Finally, the manipulation of Hippo signaling in CMs promotes heart regeneration and—importantly—reverses fibrosis (Leach et al., 2017). These studies of different injury models demonstrate that cardiac fibrosis is highly dynamic and reversible. Although the detailed molecular and cellular mechanisms that regulate cardiac fibrosis remain to be explored, more effective treatments will likely be developed in the near future.

Acknowledgments

We thank Nicole Stancel, Ph.D., ELS(D), of Scientific Publications at Texas Heart Institute, for scientific editing. We apologize to colleagues whose work could not be cited in this chapter due to space limitations. All figures were drawn in BioRender.

References

Abdelnasir, A., Sun, J. R., Cheng, Y. F., Chen, H. B., Tang, S., Kemper, N., et al. (2014). Evaluation of Hsp47 expression in heat-stressed rat myocardial cells in vitro and in vivo. *Genetics and Molecular Research*, *13*(4), 10787–10802. https://doi.org/10.4238/2014.December.18.20.

Acharya, A., Baek, S. T., Banfi, S., Eskiocak, B., & Tallquist, M. D. (2011). Efficient inducible Cre-mediated recombination in Tcf21cell lineages in the heart and kidney. *Genesis*, *49*(11), 870–877. https://doi.org/10.1002/dvg.20750.

Aghajanian, H., Kimura, T., Rurik, J. G., Hancock, A. S., Leibowitz, M. S., Li, L., et al. (2019). Targeting cardiac fibrosis with engineered T cells. *Nature*. https://doi.org/10.1038/s41586-019-1546-z.

Ali, S. R., Ranjbarvaziri, S., Talkhabi, M., Zhao, P., Subat, A., Hojjat, A., et al. (2014). Developmental heterogeneity of cardiac fibroblasts does not predict pathological proliferation and activation. *Circulation Research*, *115*(7), 625–635. https://doi.org/10.1161/CIRCRESAHA.115.303794.

Aparicio, H. J., Benjamin, E. J., Callaway, C. W., Carson, A. P., Cheng, S., Elkind, M. S. V., et al. (2021). Heart disease and stroke statistics-2021 update a report from the American Heart Association. *Circulation*. https://doi.org/10.1161/CIR.0000000000000950.

Aragona, M., Panciera, T., Manfrin, A., Giulitti, S., Michielin, F., Elvassore, N., et al. (2013). A mechanical checkpoint controls multicellular growth through YAP/TAZ regulation by actin-processing factors. *Cell*, *154*(5), 1047–1059. https://doi.org/10.1016/j.cell.2013.07.042.

Bacmeister, L., Schwarzl, M., Warnke, S., Stoffers, B., Blankenberg, S., Westermann, D., et al. (2019). Inflammation and fibrosis in murine models of heart failure. In *Vol. 114, Issue 3. Basic research in cardiology* Dr. Dietrich Steinkopff Verlag GmbH and Co. KG. https://doi.org/10.1007/s00395-019-0722-5.

Bae, S. J., Ni, L., Osinski, A., Tomchick, D. R., Brautigam, C. A., & Luo, X. (2017). SAV1 promotes Hippo kinase activation through antagonizing the PP2A phosphatase STRIPAK. *eLife*, *6*, 1–28. https://doi.org/10.7554/eLife.30278.

Basu, S., Totty, N. F., Irwin, M. S., Sudol, M., & Downward, J. (2003). Akt phosphorylates the Yes-associated protein, YAP, to induce interaction with 14-3-3 and attenuation of p73-mediated apoptosis some of the metabolic effects of Akt, whereas BAD can account for some of the survival-promoting effects. The most convincingly. *Molecular Cell*, *11*, 11–23.

Belteki, G., Haigh, J., Kabacs, N., Haigh, K., Sison, K., Costantini, F., et al. (2005). Conditional and inducible transgene expression in mice through the combinatorial use of Cre-mediated recombination and tetracycline induction. *Nucleic Acids Research*, *33*(5), 1–10. https://doi.org/10.1093/nar/gni051.

Berry, M. F., Engler, A. J., Woo, Y. J., Pirolli, T. J., Bish, L. T., Jayasankar, V., et al. (2006). Mesenchymal stem cell injection after myocardial infarction improves myocardial compliance. *American Journal of Physiology - Heart and Circulatory Physiology*, *290*(6), 2196–2203. https://doi.org/10.1152/ajpheart.01017.2005.

Bhandary, B., Meng, Q., James, J., Osinska, H., Gulick, J., Valiente-Alandi, I., et al. (2018). Cardiac fibrosis in proteotoxic cardiac disease is dependent upon myofibroblast TGF-β signaling. *Journal of the American Heart Association*, 7(20), 1–18. https://doi.org/10.1161/JAHA.118.010013.

Billings, S. E., Pierzchalski, K., Tjaden, N. E. B., Pang, X. Y., Trainor, P. A., Kane, M. A., et al. (2013). The retinaldehyde reductase DHRS3 is essential for preventing the formation of excess retinoic acid during embryonic development. *FASEB Journal, 27*(12), 4877–4889. https://doi.org/10.1096/fj.13-227967.

Boggiano, J. C., Vanderzalm, P. J., & Fehon, R. G. (2011). Tao-1 phosphorylates Hippo/MST kinases to regulate the Hippo-Salvador-Warts tumor suppressor pathway. *Developmental Cell, 21*(5), 888–895. https://doi.org/10.1016/j.devcel.2011.08.028.

Bretherton, R., Bugg, D., Olszewski, E., & Davis, J. (2020). Regulators of cardiac fibroblast cell state. *Matrix Biology, 91–92*, 117–135. Elsevier B.V. https://doi.org/10.1016/j.matbio.2020.04.002.

Bugg, D., Bretherton, R., Kim, P., Olszewski, E., Nagle, A., Schumacher, A. E., et al. (2020). Infarct collagen topography regulates fibroblast fate via p38-yes-associated protein transcriptional enhanced associate domain signals. *Circulation Research, 127*, 1306–1322. https://doi.org/10.1161/CIRCRESAHA.119.316162.

Burke, R. M., Burgos Villar, K. N., & Small, E. M. (2021). Fibroblast contributions to ischemic cardiac remodeling. *Cellular Signalling*, 77. https://doi.org/10.1016/j.cellsig.2020.109824. Elsevier Inc.

Burke, R. M., Lighthouse, J. K., Quijada, P., Dirkx, R. A., Rosenberg, A., Moravec, C. S., et al. (2018). Small proline-rich protein 2B drives stress-dependent p53 degradation and fibroblast proliferation in heart failure. *Proceedings of the National Academy of Sciences of the United States of America, 115*(15), E3436–E3445. https://doi.org/10.1073/pnas.1717423115.

Burridge, K., & Chrzanowska-Wodnicka, M. (1996). Focal adhesions, contractility, and signaling. *Annual Review of Cell and Developmental Biology, 12*, 463–519. https://doi.org/10.1146/annurev.cellbio.12.1.463.

Chang, C., Goel, H. L., Gao, H., Pursell, B., Shultz, L. D., Greiner, D. L., et al. (2015). A laminin 511 matrix is regulated by Taz and functions as the ligand for the α6Bβ1 integrin to sustain breast cancer stem cells. *Genes and Development, 29*(1), 1–6. https://doi.org/10.1101/gad.253682.114.

Chen, S. N., Gurha, P., Lombardi, R., Ruggiero, A., Willerson, J. T., & Marian, A. J. (2014). The hippo pathway is activated and is a causal mechanism for adipogenesis in arrhythmogenic cardiomyopathy. *Circulation Research, 114*(3), 454–468. https://doi.org/10.1161/CIRCRESAHA.114.302810.

Darby, I. A., Zakuan, N., Billet, F., & Desmoulière, A. (2016). The myofibroblast, a key cell in normal and pathological tissue repair. *Cellular and Molecular Life Sciences, 73*(6), 1145–1157. https://doi.org/10.1007/s00018-015-2110-0.

Daseke, M. J., Tenkorang, M. A. A., Chalise, U., Konfrst, S. R., & Lindsey, M. L. (2020). Cardiac fibroblast activation during myocardial infarction wound healing: Fibroblast polarization after MI. *Matrix Biology, 91–92*, 109–116. https://doi.org/10.1016/j.matbio.2020.03.010.

De Lange, F. J., Moorman, A. F. M., Anderson, R. H., Männer, J., Soufan, A. T., De Gier-De Vries, C., et al. (2004). Lineage and morphogenetic analysis of the cardiac valves. *Circulation Research, 95*(6), 645–654. https://doi.org/10.1161/01.RES.0000141429.13560.cb.

Deng, H., Wang, W., Yu, J., Zheng, Y., Qing, Y., & Pan, D. (2015). Spectrin regulates Hippo signaling by modulating cortical actomyosin activity. *eLife, 2015*(4), 1–17. https://doi.org/10.7554/eLife.06567.

Dey, A., Varelas, X., & Guan, K. L. (2020). Targeting the Hippo pathway in cancer, fibrosis, wound healing and regenerative medicine. *Nature Reviews Drug Discovery, 19*(7), 480–494. Nature Research https://doi.org/10.1038/s41573-020-0070-z.

Dong, J., Feldmann, G., Huang, J., Wu, S., Zhang, N., Comerford, S. A., et al. (2007). Elucidation of a universal size-control mechanism in Drosophila and mammals. *Cell, 130*(6), 1120–1133. https://doi.org/10.1016/j.cell.2007.07.019.

Dries, R., Chen, J., Del Rossi, N., Khan, M. M., Sistig, A., & Yuan, G. C. (2021). Advances in spatial transcriptomic data analysis. *Genome Research*, *31*(10), 1706–1718. https://doi.org/10.1101/gr.275224.121.

Dupont, S., Morsut, L., Aragona, M., Enzo, E., Giulitti, S., Cordenonsi, M., et al. (2011). Role of YAP/TAZ in mechanotransduction. *Nature*, *474*(7350), 179–183. https://doi.org/10.1038/nature10137.

Elbediwy, A., Vincent-Mistiaen, Z. I., Spencer-Dene, B., Stone, R. K., Boeing, S., Wculek, S. K., et al. (2016). Integrin signalling regulates YAP and TAZ to control skin homeostasis. *Development (Cambridge)*, *143*(10), 1674–1687. https://doi.org/10.1242/dev.133728.

Eng, C. H. L., Lawson, M., Zhu, Q., Dries, R., Koulena, N., Takei, Y., et al. (2019). Transcriptome-scale super-resolved imaging in tissues by RNA seqFISH+. *Nature*, *568*(7751), 235–239. https://doi.org/10.1038/s41586-019-1049-y.

Fernández, B. G., Gaspar, P., Brás-Pereira, C., Jezowska, B., Rebelo, S. R., & Janody, F. (2011). Actin-capping protein and the Hippo pathway regulate F-actin and tissue growth in Drosophila. *Development*, *138*(11), 2337–2346. https://doi.org/10.1242/dev.063545.

Fletcher, G. C., Elbediwy, A., Khanal, I., Ribeiro, P. S., Tapon, N., & Thompson, B. J. (2015). The Spectrin cytoskeleton regulates the Hippo signalling pathway. *The EMBO Journal*, *34*(7), 940–954. https://doi.org/10.15252/embj.201489642.

Forest, E., Logeay, R., Géminard, C., Kantar, D., Frayssinoux, F., Heron-Milhavet, L., et al. (2018). The apical scaffold big bang binds to spectrins and regulates the growth of Drosophila melanogaster wing discs. *Journal of Cell Biology*, *217*(3), 1047–1062. https://doi.org/10.1083/jcb.201705107.

Forte, E., Skelly, D. A., Chen, M., Daigle, S., Morelli, K. A., Hon, O., et al. (2020). Dynamic interstitial cell response during myocardial infarction predicts resilience to rupture in genetically diverse mice. *Cell Reports*, *30*(9), 3149–3163.e6. https://doi.org/10.1016/j.celrep.2020.02.008.

Francisco, J., Zhang, Y., Jeong, J. I., Mizushima, W., Ikeda, S., Ivessa, A., et al. (2020). Blockade of fibroblast YAP attenuates cardiac fibrosis and dysfunction through MRTF-A inhibition. *JACC: Basic to Translational Science*, *5*(9), 931–945. https://doi.org/10.1016/j.jacbts.2020.07.009.

Francisco, J., Zhang, Y., Nakada, Y., Jeong, J. I., Huang, C. Y., Ivessa, A., et al. (2021). AAV-mediated YAP expression in cardiac fibroblasts promotes inflammation and increases fibrosis. *Scientific Reports*, *11*(1). https://doi.org/10.1038/s41598-021-89989-5.

Frangogiannis, N. G. (2019a). The extracellular matrix in ischemic and nonischemic heart failure. *Circulation Research*, *125*(1), 117–146. https://doi.org/10.1161/CIRCRESAHA.119.311148.

Frangogiannis, N. G. (2019b). Cardiac fibrosis: Cell biological mechanisms, molecular pathways and therapeutic opportunities. *Molecular Aspects of Medicine*, *65*, 70–99. Elsevier Ltd. https://doi.org/10.1016/j.mam.2018.07.001.

Frey, N., & Olson, E. N. (2003). Cardiac hypertrophy: The good, the bad, and the ugly. *Annual Review of Physiology*, *65*, 45–79. https://doi.org/10.1146/annurev.physiol.65.092101.142243.

Fu, X., Khalil, H., Kanisicak, O., Boyer, J. G., Vagnozzi, R. J., Maliken, B. D., et al. (2018). Specialized fibroblast differentiated states underlie scar formation in the infarcted mouse heart. *The Journal of Clinical Investigation*, *128*(5), 2127–2143. https://doi.org/10.1172/JCI98215.

Fu, X., Liu, Q., Li, C., Li, Y., & Wang, L. (2020). Cardiac fibrosis and cardiac fibroblast lineage-tracing: Recent advances. *Frontiers in Physiology*, *11*(May). https://doi.org/10.3389/fphys.2020.00416.

Goulev, Y., Fauny, J. D., Gonzalez-Marti, B., Flagiello, D., Silber, J., & Zider, A. (2008). SCALLOPED interacts with YORKIE, the nuclear effector of the hippo tumor-suppressor pathway in Drosophila. *Current Biology, 18*(6). https://doi.org/10.1016/j.cub.2008.02.034.

Grisanti Canozo, F. J., Zuo, Z., Martin, J. F., & Samee, M. A. H. (2021). Cell-type modeling in spatial transcriptomics data elucidates spatially variable colocalization and communication between cell-types in mouse brain. *Cell Systems*. https://doi.org/10.1016/j.cels.2021.09.004.

Guo, X., Zhao, Y., Yan, H., Yang, Y., Shen, S., Dai, X., et al. (2017). Single tumor-initiating cells evade immune clearance by recruiting type II macrophages. *Genes and Development, 31*(3), 247–259. https://doi.org/10.1101/gad.294348.116.

Harikrishnan, V., Titus, A. S., Cowling, R. T., & Kailasam, S. (2019). Collagen receptor cross-talk determines α-smooth muscle actin-dependent collagen gene expression in angiotensin II-stimulated cardiac fibroblasts. *Journal of Biological Chemistry, 294*(51), 19723–19739. https://doi.org/10.1074/jbc.RA119.009744.

Harvey, K. F., Pfleger, C. M., & Hariharan, I. K. (2003). The Drosophila Mst ortholog, hippo, restricts growth and cell proliferation and promotes apoptosis. *Cell, 114*(4), 457–467. https://doi.org/10.1016/S0092-8674(03)00557-9.

Heineke, J., & Molkentin, J. D. (2006). Regulation of cardiac hypertrophy by intracellular signalling pathways. *Nature Reviews Molecular Cell Biology*, 7(8), 589–600. https://doi.org/10.1038/nrm1983.

Herum, K. M., Lunde, I. G., McCulloch, A. D., & Christensen, G. (2017). The soft- and hard-heartedness of cardiac fibroblasts: Mechanotransduction signaling pathways in fibrosis of the heart. *Journal of Clinical Medicine, 6*(5). https://doi.org/10.3390/jcm6050053.

Horne, T. E., Vandekopple, M., Sauls, K., Koenig, S. N., Anstine, L. J., Garg, V., et al. (2015). Dynamic heterogeneity of the heart valve interstitial cell population in mitral valve health and disease. *Journal of Cardiovascular Development and Disease, 2*(3), 214–232. https://doi.org/10.3390/jcdd2030214.

Huang, S., Chen, B., Su, Y., Alex, L., Humeres, C., Shinde, A. V., et al. (2019). Distinct roles of myofibroblast-specific Smad2 and Smad3 signaling in repair and remodeling of the infarcted heart. *Journal of Molecular and Cellular Cardiology, 132*(December 2018), 84–97. https://doi.org/10.1016/j.yjmcc.2019.05.006.

Huang, X., Feng, T., Jiang, Z., Meng, J., Kou, S., Lu, Z., et al. (2019). Dual lineage tracing identifies intermediate mesenchymal stage for endocardial contribution to fibroblasts, coronary mural cells, and adipocytes. *Journal of Biological Chemistry, 294*(22), 8894–8906. https://doi.org/10.1074/jbc.RA118.006994.

Huang, L. S., Sudhadevi, T., Fu, P., Punathil-Kannan, P. K., Ebenezer, D. L., Ramchandran, R., et al. (2020). Sphingosine kinase 1/s1p signaling contributes to pulmonary fibrosis by activating hippo/yap pathway and mitochondrial reactive oxygen species in lung fibroblasts. *International Journal of Molecular Sciences, 21*(6). https://doi.org/10.3390/ijms21062064.

Huang, J., Wu, S., Barrera, J., Matthews, K., & Pan, D. (2005). The Hippo signaling pathway coordinately regulates cell proliferation and apoptosis by inactivating Yorkie, the Drosophila homolog of YAP. *Cell, 122*(3), 421–434. https://doi.org/10.1016/j.cell.2005.06.007.

Hynes, R. O. (2002). Integrins. *Cell, 110*, 673–687. https://doi.org/10.1016/B978-012439510-7/50078-4.

Ishida, Y., Kubota, H., Yamamoto, A., Kitamura, A., Bächinger, H. P., & Nagata, K. (2006). Type I collagen in Hsp47-null cells is aggregated in endoplasmic reticulum and deficient in N-propeptide processing and fibrillogenesis. *Molecular Biology of the Cell*, 17(May), 2346–2355. https://doi.org/10.1091/mbc.E05.

Ivey, M. J., Kuwabara, J. T., Pai, J. T., Moore, R. E., Sun, Z., & Tallquist, M. D. (2018). Resident fibroblast expansion during cardiac growth and remodeling. *Journal of Molecular and Cellular Cardiology*, *114*, 161–174. https://doi.org/10.1016/j.yjmcc.2017.11.012.

Jeong, D., Yang, D. K., Kho, C., Oh, J. G., Lee, A., Chen, J., et al. (2016). Matricellular protein CCN5 reverses established cardiac fibrosis. *Journal of the American College of Cardiology*, *67*(13), 1556–1568. https://doi.org/10.1016/j.jacc.2016.01.030.

Jia, J., Zhang, W., Wang, B., Trinko, R., & Jiang, J. (2003). The Drosophila Ste20 family kinase dMST functions as a tumor suppressor by restricting cell proliferation and promoting apoptosis. *Genes and Development*, *17*(20), 2514–2519. https://doi.org/10.1101/gad.1134003.

Jin, H., Sperka, T., Herrlich, P., & Morrison, H. (2006). Tumorigenic transformation by CPI-17 through inhibition of a merlin phosphatase. *Nature*, *442*(7102), 576–579. https://doi.org/10.1038/nature04856.

Jin, B., Zhu, J., Shi, H. M., Wen, Z. C., & Wu, B. W. (2019). YAP activation promotes the transdifferentiation of cardiac fibroblasts to myofibroblasts in matrix remodeling of dilated cardiomyopathy. *Brazilian Journal of Medical and Biological Research*, *52*(1), 1–9. https://doi.org/10.1590/1414-431X20187914.

Justice, R. W., Woods, D. F., Noll, M., & Bryant, P. J. (1995). The Drosophila tumor suppressor gene warts encodes a homolog o-f human myotonic dystrophy kinase and is required for the control of cell shape and proliferation. *Genes & Development*, *9*, 534–546.

Kaneko, K., Ito, M., Naoe, Y., Lacy-Hulbert, A., & Ikeda, K. (2014). Integrin αv in the mechanical response of osteoblast lineage cells. *Biochemical and Biophysical Research Communications*, *447*(2), 352–357. https://doi.org/10.1016/j.bbrc.2014.04.006.

Kango-Singh, M., Nolo, R., Tao, C., Verstreken, P., Hiesinger, P. R., Bellen, H. J., et al. (2002). Shar-pei mediates cell proliferation arrest during imaginal disc growth in Drosophila. *Development*, *129*(24), 5719–5730. https://doi.org/10.1242/dev.00168.

Kanisicak, O., Khalil, H., Ivey, M. J., Karch, J., Maliken, B. D., Correll, R. N., et al. (2016). Genetic lineage tracing defines myofibroblast origin and function in the injured heart. *Nature Communications*, 7. https://doi.org/10.1038/ncomms12260.

Karaman, R., & Halder, G. (2018). Cell junctions in Hippo signaling. *Cold Spring Harbor Perspectives in Biology*, *10*(5). https://doi.org/10.1101/cshperspect.a028753.

Khalil, H., Kanisicak, O., Prasad, V., Correll, R. N., Fu, X., Schips, T., et al. (2017). Fibroblast-specific TGF-β-Smad2/3 signaling underlies cardiac fibrosis. *Journal of Clinical Investigation*, *127*(10), 3770–3783. https://doi.org/10.1172/JCI94753.

Khalil, H., Kanisicak, O., Vagnozzi, R. J., Johansen, A. K., Maliken, B. D., Prasad, V., et al. (2019). Cell-specific ablation of Hsp47 defines the collagen-producing cells in the injured heart. *JCI Insight*, *4*(15). https://doi.org/10.1172/jci.insight.128722.

Kim, C.-L., Choi, S.-H., & Mo, J.-S. (2019). Role of the Hippo pathway in fibrosis and cancer. *Cell*, *8*(5), 468. https://doi.org/10.3390/cells8050468.

Kim, N. G., & Gumbiner, B. M. (2015). Adhesion to fibronectin regulates Hippo signaling via the FAK-Src-PI3K pathway. *Journal of Cell Biology*, *210*(3), 503–515. https://doi.org/10.1083/jcb.201501025.

Komuro, A., Nagai, M., Navin, N. E., & Sudol, M. (2003). WW domain-containing protein YAP associates with ErbB-4 and acts as a co-transcriptional activator for the carboxyl-terminal fragment of ErbB-4 that translocates to the nucleus. *Journal of Biological Chemistry*, *278*(35), 33334–33341. https://doi.org/10.1074/jbc.M305597200.

Kong, P., Shinde, A. V., Su, Y., Russo, I., Chen, B., Saxena, A., et al. (2018). Opposing actions of fibroblast and cardiomyocyte smad3 signaling in the infarcted myocardium. *Circulation*, *137*(7), 707–724. https://doi.org/10.1161/CIRCULATIONAHA.117.029622.

Lai, Z. C., Wei, X., Shimizu, T., Ramos, E., Rohrbaugh, M., Nikolaidis, N., et al. (2005). Control of cell proliferation and apoptosis by mob as tumor suppressor, mats. *Cell, 120*(5), 675–685. https://doi.org/10.1016/j.cell.2004.12.036.

Leach, J. P., Heallen, T., Zhang, M., Rahmani, M., Morikawa, Y., Hill, M. C., et al. (2017). Hippo pathway deficiency reverses systolic heart failure after infarction. *Nature, 550*(7675), 260–264. https://doi.org/10.1038/nature24045.

Lee, K. P., Lee, J. H., Kim, T. S., Kim, T. H., Park, H. D., Byun, J. S., et al. (2010). The Hippo-Salvador pathway restrains hepatic oval cell proliferation, liver size, and liver tumorigenesis. *Proceedings of the National Academy of Sciences of the United States of America, 107*(18), 8248–8253. https://doi.org/10.1073/pnas.0912203107.

Liang, M., Yu, M., Xia, R., Song, K., Wang, J., Luo, J., et al. (2017a). Yap/Taz deletion in gli+ cell-derived myo fibroblasts attenuates fibrosis. *Journal of the American Society of Nephrology, 28*(11), 3278–3290. https://doi.org/10.1681/ASN.2015121354.

Liang, M., Yu, M., Xia, R., Song, K., Wang, J., Luo, J., et al. (2017b). Yap/Taz deletion in gli+ cell-derived myofibroblasts attenuates fibrosis. *Journal of the American Society of Nephrology, 28*(11), 3278–3290. https://doi.org/10.1681/ASN.2015121354.

Liu, F., Lagares, D., Choi, K. M., Stopfer, L., Marinković, A., Vrbanac, V., et al. (2015). Mechanosignaling through YAP and TAZ drives fibroblast activation and fibrosis. *American Journal of Physiology - Lung Cellular and Molecular Physiology, 308*(4), L344–L357. https://doi.org/10.1152/ajplung.00300.2014.

Lu, L., Li, Y., Kim, S. M., Bossuyt, W., Liu, P., Qiu, Q., et al. (2010). Hippo signaling is a potent in vivo growth and tumor suppressor pathway in the mammalian liver. *Proceedings of the National Academy of Sciences of the United States of America, 107*(4), 1437–1442. https://doi.org/10.1073/pnas.0911427107.

Luo, J., & Yu, F. (2019). GPCR-Hippo signaling in cancer. *2*.

Ma, S., Meng, Z., Chen, R., & Guan, K. L. (2019). The hippo pathway: Biology and pathophysiology. *Annual Review of Biochemistry, 88*(1), 577–604. Annual Reviews Inc https://doi.org/10.1146/annurev-biochem-013118-111829.

MacHnicka, B., Grochowalska, R., Bogusławska, D. M., Sikorski, A. F., & Lecomte, M. C. (2012). Spectrin-based skeleton as an actor in cell signaling. *Cellular and Molecular Life Sciences, 69*(2), 191–201. https://doi.org/10.1007/s00018-011-0804-5.

Mannaerts, I., Leite, S. B., Verhulst, S., Claerhout, S., Eysackers, N., Thoen, L. F. R., et al. (2015). The Hippo pathway effector YAP controls mouse hepatic stellate cell activation. *Journal of Hepatology, 63*(3), 679–688. https://doi.org/10.1016/j.jhep.2015.04.011.

Maruyama, S., Nakamura, K., Papanicolaou, K. N., Sano, S., Shimizu, I., Asaumi, Y., et al. (2016). Follistatin-like 1 promotes cardiac fibroblast activation and protects the heart from rupture. *EMBO Molecular Medicine, 8*(8), 949–966. https://doi.org/10.15252/emmm.201506151.

Masuda, H., Fukumoto, M., Hirayoshi, K., & Nagata, K. (1994). Coexpression of the collagen-binding stress protein HSP47 gene and the α1 (I) and α1 (III) collagen genes in carbon tetrachloride-induced rat liver fibrosis. *Journal of Clinical Investigation, 94*(6), 2481–2488. https://doi.org/10.1172/JCI117617.

Meng, Q., Bhandary, B., Bhuiyan, M. S., James, J., Osinska, H., Valiente-Alandi, I., et al. (2018). Myofibroblast-specific TGFβ receptor II signaling in the fibrotic response to cardiac myosin binding protein C-induced cardiomyopathy. *Circulation Research, 123*(12), 1285–1297. https://doi.org/10.1161/CIRCRESAHA.118.313089.

Meng, F., Xie, B., & Martin, J. F. (2021). Targeting the Hippo pathway in heart repair. *Cardiovascular Research*, 1–13. https://doi.org/10.1093/cvr/cvab291.

Mia, M. M., Cibi, D. M., Ghani, S. A. B. A., Singh, A., Tee, N., Sivakumar, V., et al. (2021). Loss of Yap/Taz in cardiac fibroblasts attenuates adverse remodelling and improves cardiac function. *Cardiovascular Research*, 1–22. https://doi.org/10.1093/cvr/cvab205.

Misra, J. R., & Irvine, K. D. (2018). The hippo signaling network and its biological functions. *Annual Review of Genetics*, *52*, 65–87. https://doi.org/10.1146/annurev-genet-120417-031621.

Moore-Morris, T., Guimarães-Camboa, N., Banerjee, I., Zambon, A. C., Kisseleva, T., Velayoudon, A., et al. (2014). Resident fibroblast lineages mediate pressure overload-induced cardiac fibrosis. *Journal of Clinical Investigation*, *124*(7), 2921–2934. https://doi.org/10.1172/JCI74783.

Moya, I. M., & Halder, G. (2019). Hippo–YAP/TAZ signalling in organ regeneration and regenerative medicine. *Nature Reviews Molecular Cell Biology*, *20*(4), 211–226. Nature Publishing Group https://doi.org/10.1038/s41580-018-0086-y.

Nagata, K. (1988). Laboratory of Molecular Biology, National Cancer Institute, National Institutes of Health, Bethesda, MD 20892 3The Second Department of Pathology, Nagoya University School of Medicine, Nagoya 466, Japan. *Biochemical and Biophysical Research Communications*, *153*(1), 428–434.

Nakamura, R., Hiwatashi, N., Bing, R., Doyle, C. P., & Branski, R. C. (2021). Concurrent YAP/TAZ and SMAD signaling mediate vocal fold fibrosis. *Scientific Reports*, *11*(1), 1–11. https://doi.org/10.1038/s41598-021-92871-z.

Noguchi, S., Saito, A., Mikami, Y., Urushiyama, H., Horie, M., Matsuzaki, H., et al. (2017). TAZ contributes to pulmonary fibrosis by activating profibrotic functions of lung fibroblasts. *Scientific Reports*, 7. https://doi.org/10.1038/srep42595.

Norris, R. A., Damon, B., Mironov, V., Kasyanov, V., Ramamurthi, A., Moreno-Rodriguez, R., et al. (2007). Periostin regulates collagen fibrillogenesis and the biomechanical properties of connective tissues. *Journal of Cellular Biochemistry*, *101*(3), 695–711. https://doi.org/10.1002/jcb.21224.

Oh, H., & Irvine, K. D. (2008). In vivo regulation of Yorkie phosphorylation and localization. *Development*, *135*(6), 1081–1088. https://doi.org/10.1242/dev.015255.

Pakshir, P., Noskovicova, N., Lodyga, M., Son, D. O., Schuster, R., Goodwin, A., et al. (2020). The myofibroblast at a glance. *Journal of Cell Science*, *133*(13). https://doi.org/10.1242/jcs.227900.

Panciera, T., Azzolin, L., Cordenonsi, M., & Piccolo, S. (2017). Mechanobiology of YAP and TAZ in physiology and disease. *Nature Reviews Molecular Cell Biology*, *18*(12), 758–770. Nature Publishing Group https://doi.org/10.1038/nrm.2017.87.

Pantalacci, S., Tapon, N., & Léopold, P. (2003). The salvador partner Hippo promotes apoptosis and cell-cycle exit in Drosophila. *Nature Cell Biology*, *5*(10), 921–927. https://doi.org/10.1038/ncb1051.

Pinto, A. R., Ilinykh, A., Ivey, M. J., Kuwabara, J. T., D'antoni, M. L., Debuque, R., et al. (2016). Revisiting cardiac cellular composition. *Circulation Research*, *118*(3), 400–409. https://doi.org/10.1161/CIRCRESAHA.115.307778.

Plikus, M. V., Wang, X., Sinha, S., Forte, E., Thompson, S. M., Herzog, E. L., et al. (2021). Fibroblasts: Origins, definitions, and functions in health and disease. *Cell*, *184*(15), 3852–3872. https://doi.org/10.1016/j.cell.2021.06.024.

Plouffe, S. W., Lin, K. C., Moore, J. L., Tan, F. E., Ma, S., Ye, Z., et al. (2018). The Hippo pathway effector proteins YAP and TAZ have both distinct and overlapping functions in the cell. *Journal of Biological Chemistry*, *293*(28), 11230–11240. https://doi.org/10.1074/jbc.RA118.002715.

Pocaterra, A., Romani, P., & Dupont, S. (2020). YAP/TAZ functions and their regulation at a glance. *Journal of Cell Science*, *133*(2). https://doi.org/10.1242/jcs.230425.

Poon, C. L. C., Lin, J. I., Zhang, X., & Harvey, K. F. (2011). The sterile 20-like kinase Tao-1 controls tissue growth by regulating the Salvador-Warts-Hippo pathway. *Developmental Cell*, *21*(5), 896–906. https://doi.org/10.1016/j.devcel.2011.09.012.

Quijada, P., Trembley, M. A., & Small, E. M. (2020). The role of the epicardium during heart development and repair. *Circulation Research*, 377–394. Lippincott Williams and Wilkins https://doi.org/10.1161/CIRCRESAHA.119.315857.

Razzaque, M. S., Hossain, M. A., Kohno, S., & Taguchi, T. (1998). Bleomycin-induced pulmonary fibrosis in rat is associated with increased expression of collagen-binding heat shock protein (HSP) 47. *Virchows Archiv*, *432*(5), 455–460. https://doi.org/10.1007/s004280050191.

Reggiani, F., Gobbi, G., Ciarrocchi, A., & Sancisi, V. (2021). YAP and TAZ are not identical twins. *Trends in Biochemical Sciences*, *46*(2), 154–168. https://doi.org/10.1016/j.tibs.2020.08.012.

Ribeiro, P. S., Josué, F., Wepf, A., Wehr, M. C., Rinner, O., Kelly, G., et al. (2010). Combined functional genomic and proteomic approaches identify a PP2A complex as a negative regulator of Hippo signaling. *Molecular Cell*, *39*(4), 521–534. https://doi.org/10.1016/j.molcel.2010.08.002.

Riento, K., & Ridley, A. J. (2003). Rocks: Multifunctional kinases in cell behaviour. *Nature Reviews Molecular Cell Biology*, *4*(6), 446–456. https://doi.org/10.1038/nrm1128.

Rodriques, S. G., Stickels, R. R., Goeva, A., Martin, C. A., Murray, E., Vanderburg, C. R., et al. (2020). *HHS Public Access*, *363*(6434), 1463–1467. https://doi.org/10.5281/zenodo.2571615.

Sansores-Garcia, L., Bossuyt, W., Wada, K. I., Yonemura, S., Tao, C., Sasaki, H., et al. (2011). Modulating F-actin organization induces organ growth by affecting the Hippo pathway. *EMBO Journal*, *30*(12), 2325–2335. https://doi.org/10.1038/emboj.2011.157.

Satoh, M., Hirayoshi, K., Yokota, S. I., Hosokawa, N., & Nagata, K. (1996). Intracellular interaction of collagen-specific stress protein HSP47 with newly synthesized procollagen. *Journal of Cell Biology*, *133*(2), 469–483. https://doi.org/10.1083/jcb.133.2.469.

Serrano, I., McDonald, P. C., Lock, F., Muller, W. J., & Dedhar, S. (2013). Inactivation of the Hippo tumour suppressor pathway by integrin-linked kinase. *Nature Communications*, *4*, 1–12. https://doi.org/10.1038/ncomms3976.

Sharifi-sanjani, M., Berman, M., Goncharov, D., Alhamaydeh, M., Avolio, T. G., Baust, J., et al. (2021). Yes-associated protein (Yap) is up-regulated in heart failure and promotes cardiac fibroblast proliferation. *International Journal of Molecular Sciences*, *22*(11), 1–15. https://doi.org/10.3390/ijms22116164.

Shimazaki, M., Nakamura, K., Kii, I., Kashima, T., Amizuka, N., Li, M., et al. (2008). Periostin is essential for cardiac healing after acute myocardial infarction. *Journal of Experimental Medicine*, *205*(2), 295–303. https://doi.org/10.1084/jem.20071297.

Shimizu, T., Narang, N., Chen, P., Yu, B., Knapp, M., Janardanan, J., et al. (2017). Fibroblast deletion of ROCK2 attenuates cardiac hypertrophy, fibrosis, and diastolic dysfunction. *JCI Insight*, *2*(13). https://doi.org/10.1172/jci.insight.93187.

Singh, A., Ramesh, S., Dasan, M., Cibi, E. N., Olson, J. A., Epstein, M. K., et al. (2016). Hippo signaling mediators Yap and Taz are required in the epicardium for coronary vasculature development. *Cell Reports*, *15*, 1384–1393. https://doi.org/10.1016/j.celrep.2016.04.027.

Small, E. M., & Brooks, A. C. (2020). Cut the YAP: Limiting fibrosis in pathologic cardiac remodeling. *JACC: Basic to Translational Science*, *5*(9), 946–948. https://doi.org/10.1016/j.jacbts.2020.08.004.

Small, E. M., Thatcher, J. E., Sutherland, L. B., Kinoshita, H., Gerard, R. D., Richardson, J. A., et al. (2010). Myocardin-related transcription factor-a controls myofibroblast activation and fibrosis in response to myocardial infarction. *Circulation Research*, *107*(2), 294–304. https://doi.org/10.1161/CIRCRESAHA.110.223172.

Smith, C. L., Baek, S. T., Sung, C. Y., & Tallquist, M. D. (2011). Epicardial-derived cell epithelial-to-mesenchymal transition and fate specification require PDGF receptor signaling. *Circulation Research*, *108*(12). https://doi.org/10.1161/CIRCRESAHA.110.235531.

Song, H., Mak, K. K., Topol, L., Yun, K., Hu, J., Garrett, L., et al. (2010). Mammalian Mst1 and Mst2 kinases play essential roles in organ size control and tumor suppression. *Proceedings of the National Academy of Sciences of the United States of America*, *107*(4), 1431–1436. https://doi.org/10.1073/pnas.0911409107.

Svensson, V., Vento-Tormo, R., & Teichmann, S. A. (2018). Exponential scaling of single-cell RNA-seq in the past decade. *Nature Protocols*, *13*(4), 599–604. https://doi.org/10.1038/nprot.2017.149.

Swonger, J. M., Liu, J. S., Ivey, M. J., & Tallquist, M. D. (2016). Genetic tools for identifying and manipulating fibroblasts in the mouse. *Differentiation*, *92*(3), 66–83. Elsevier Ltd https://doi.org/10.1016/j.diff.2016.05.009.

Tallquist, M. D. (2020a). Cardiac fibroblast diversity. *Annual Review of Physiology*. https://doi.org/10.1146/annurev-physiol-021119.

Tallquist, M. D. (2020b). Developmental pathways of cardiac fibroblasts. *Cold Spring Harbor Perspectives in Biology*, *12*(4). https://doi.org/10.1101/cshperspect.a037184.

Tallquist, M. D., & Molkentin, J. D. (2017). Redefining the identity of cardiac fibroblasts. *Nature Reviews Cardiology*, *14*(8), 484–491. Nature Publishing Group https://doi.org/10.1038/nrcardio.2017.57.

Tang, Y., Rowe, R. G., Botvinick, E. L., Kurup, A., Putnam, A. J., Seiki, M., et al. (2013). MT1-MMP-dependent control of skeletal stem cell commitment via a β1-integrin/YAP/TAZ signaling axis. *Developmental Cell*, *25*(4), 402–416. https://doi.org/10.1016/j.devcel.2013.04.011.

Tapon, N., Harvey, K. F., Bell, D. W., Wahrer, D. C. R., Schiripo, T. A., Haber, D. A., et al. (2002). Salvador promotes both cell cycle exit and apoptosis in Drosophila and is mutated in human cancer cell lines. *Cell*, *110*(4), 467–478. https://doi.org/10.1016/S0092-8674(02)00824-3.

Tsai, C.-R., Anderson, A. E., Burra, S., Jo, J., & Galko, M. J. (2017). Yorkie regulates epidermal wound healing in Drosophila larvae independently of cell proliferation and apoptosis. *Developmental Biology*, *427*(1), 61–71. https://doi.org/10.1016/J.YDBIO.2017.05.006.

Udan, R. S., Kango-Singh, M., Nolo, R., Tao, C., & Halder, G. (2003). Hippo promotes proliferation arrest and apoptosis in the Salvador/Warts pathway. *Nature Cell Biology*, *5*(10), 914–920. https://doi.org/10.1038/ncb1050.

Umbarkar, P., Ejantkar, S., Tousif, S., & Lal, H. (2021). Mechanisms of fibroblast activation and myocardial fibrosis: Lessons learned from FB-specific conditional mouse models. *Cell*, *10*(9). https://doi.org/10.3390/cells10092412.

Vassilev, A., Kaneko, K. J., Shu, H., Zhao, Y., & DePamphilis, M. L. (2001). TEAD/TEF transcription factors utilize the activation domain of YAP65, a Src/Yes-associated protein localized in the cytoplasm. *Genes and Development*, *15*(10), 1229–1241. https://doi.org/10.1101/gad.888601.

Vickovic, S., Eraslan, G., Salmén, F., Klughammer, J., Stenbeck, L., Schapiro, D., et al. (2019). High-definition spatial transcriptomics for in situ tissue profiling. *Nature Methods*, *16*(10), 987–990. https://doi.org/10.1038/s41592-019-0548-y.

Wada, K.-I., Itoga, K., Okano, T., Yonemura, S., & Sasaki, H. (2011). Hippo pathway regulation by cell morphology and stress fibers. *Development (Cambridge, England)*, *138*(18), 3907–3914. https://doi.org/10.1242/dev.070987.

Wang, P., Mao, B., Luo, W., Wei, B., Jiang, W., Liu, D., et al. (2014). The alteration of Hippo/YAP signaling in the development of hypertrophic cardiomyopathy. *Basic Research in Cardiology*, *109*(5), 1–11. https://doi.org/10.1007/s00395-014-0435-8.

Wang, J., Zohar, R., & McCulloch, C. A. (2006). Multiple roles of α-smooth muscle actin in mechanotransduction. *Experimental Cell Research, 312*(3), 205–214. https://doi.org/10.1016/j.yexcr.2005.11.004.

Wong, K. K. L., Li, W., An, Y., Duan, Y., Li, Z., Kang, Y., et al. (2015). B-spectrin regulates the Hippo signaling pathway and modulates the basal actin network. *Journal of Biological Chemistry, 290*(10), 6397–6407. https://doi.org/10.1074/jbc.M114.629493.

Wu, Z., & Guan, K. L. (2021). Hippo signaling in embryogenesis and development. *Trends in Biochemical Sciences, 46*(1), 51–63. https://doi.org/10.1016/j.tibs.2020.08.008.

Wu, S., Huang, J., Dong, J., & Pan, D. (2003). hippo encodes a Ste-20 family protein kinase that restricts cell proliferation and promotes apoptosis in conjunction with salvador and warts. *Cell, 114*(4), 445–456. https://doi.org/10.1016/S0092-8674(03)00549-X.

Wu, H., Xiao, Y., Zhang, S., Ji, S., Wei, L., Fan, F., et al. (2013). The Ets transcription factor GABP Is a component of the Hippo pathway essential for growth and antioxidant defense. *Cell Reports, 3*(5), 1663–1677. https://doi.org/10.1016/j.celrep.2013.04.020.

Xiao, Y., Hill, M. C., Li, L., Deshmukh, V., Martin, T. J., Wang, J., et al. (2019). Hippo pathway deletion in adult resting cardiac fibroblasts initiates a cell state transition with spontaneous and self-sustaining fibrosis. *Genes & Development, 33*(21–22), 1491–1505. https://doi.org/10.1101/gad.329763.119.

Xiao, Y., Hill, M. C., Zhang, M., Martin, T. J., Morikawa, Y., Wang, S., et al. (2018). Hippo signaling plays an essential role in cell state transitions during cardiac fibroblast development. *Developmental Cell, 45*(2), 153–169.e6. https://doi.org/10.1016/J.DEVCEL.2018.03.019.

Xiao, H. B., Liu, R. H., Ling, G. H., Xiao, L., Xia, Y. C., Liu, F. Y., et al. (2012). HSP47 regulates ECM accumulation in renal proximal tubular cells induced by TGF-β1 through ERK1/2 and JNK MAPK pathways. *American Journal of Physiology - Renal Physiology, 303*(5), 757–765. https://doi.org/10.1152/ajprenal.00470.2011.

Xu, J., Vanderzalm, P. J., Ludwig, M., Su, T., Tokamov, S. A., & Fehon, R. G. (2018). Yorkie functions at the cell cortex to promote myosin activation in a non-transcriptional manner. *Developmental Cell, 46*(3), 271–284.e5. https://doi.org/10.1016/j.devcel.2018.06.017.

Xu, T., Wang, W., Zhang, S., Stewart, R. A., & Wan, Y. (1995). Identifying tumor suppressors in genetic mosaics: The Drosophila latsgeneencodes a putative protein kinase. *Development, 121*, 1053–1063.

Yagi, R., Chen, L. F., Shigesada, K., Murakami, Y., & Ito, Y. (1999). A WW domain-containing Yes-associated protein (YAP) is a novel transcriptional co-activator. *EMBO Journal, 18*(9), 2551–2562. https://doi.org/10.1093/emboj/18.9.2551.

Yu, F. X., Zhao, B., Panupinthu, N., Jewell, J. L., Lian, I., Wang, L. H., et al. (2012). Regulation of the Hippo-YAP pathway by G-protein-coupled receptor signaling. *Cell, 150*(4), 780–791. https://doi.org/10.1016/j.cell.2012.06.037.

Zhang, H., Liu, C. Y., Zha, Z. Y., Zhao, B., Yao, J., Zhao, S., et al. (2009). TEAD transcription factors mediate the function of TAZ in cell growth and epithelial-mesenchymal transition. *Journal of Biological Chemistry, 284*(20), 13355–13362. https://doi.org/10.1074/jbc.M900843200.

Zhang, H., Von Gise, A., Liu, Q., Hu, T., Tian, X., He, L., et al. (2014). Yap1 Is required for endothelial to mesenchymal transition of the atrioventricular cushion. *Journal of Biological Chemistry, 289*(27), 18681–18692. https://doi.org/10.1074/jbc.M114.554584.

Zhao, B., Li, L., Wang, L., Wang, C. Y., Yu, J., & Guan, K. L. (2012). Cell detachment activates the Hippo pathway via cytoskeleton reorganization to induce anoikis. *Genes and Development, 26*(1), 54–68. https://doi.org/10.1101/gad.173435.111.

Zhao, B., Wei, X., Li, W., Udan, R. S., Yang, Q., Kim, J., et al. (2007). Inactivation of YAP oncoprotein by the Hippo pathway is involved in cell contact inhibition and tissue growth control. *Genes and Development*, *21*(21), 2747–2761. https://doi.org/10.1101/gad.1602907.

Zhao, B., Ye, X., Yu, J., Li, L., Li, W., Li, S., et al. (2008). TEAD mediates YAP-dependent gene induction and growth control. *Genes and Development*, *22*(14), 1962–1971. https://doi.org/10.1101/gad.1664408.

Zheng, Y., & Pan, D. (2019). The Hippo signaling pathway in development and disease. *Developmental Cell*, *50*(3), 264–282. https://doi.org/10.1016/j.devcel.2019.06.003.

Zhou, D., Conrad, C., Xia, F., Park, J. S., Payer, B., Yin, Y., et al. (2009). Mst1 and Mst2 maintain hepatocyte quiescence and suppress hepatocellular carcinoma development through inactivation of the Yap1 oncogene. *Cancer Cell*, *16*(5), 425–438. https://doi.org/10.1016/j.ccr.2009.09.026.

Zhou, B., Honor, L. B., He, H., Qing, M., Oh, J. H., Butterfield, C., et al. (2011). Adult mouse epicardium modulates myocardial injury by secreting paracrine factors. *Journal of Clinical Investigation*, *121*(5), 1894–1904. https://doi.org/10.1172/JCI45529.

CHAPTER FOUR

Pulling back the curtain: The hidden functions of receptor tyrosine kinases in development

James F. Clark and Philippe M. Soriano*
Department of Cell, Developmental, and Regenerative Biology, Icahn School of Medicine at Mount Sinai, New York, NY, United States
*Corresponding author: e-mail address: philippe.soriano@mssm.edu

Contents

Abstract

Receptor tyrosine kinases (RTKs) are a conserved superfamily of transmembrane growth factor receptors that drive numerous cellular processes during development and in the adult. Upon activation, multiple adaptors and signaling effector proteins are recruited to binding site motifs located within the intracellular domain of the RTK. These RTK-effector interactions drive subsequent intracellular signaling cascades involved in canonical RTK signaling. Genetic dissection has revealed that alleles of Fibroblast Growth Factor receptors (FGFRs) that lack all canonical RTK signaling still retain some kinase-dependent biological activity. Here we examine how genetic analysis can be used to understand the mechanism by which RTKs drive multiple developmental processes via canonical signaling while revealing noncanonical activities. Recent data from both FGFRs and other RTKs highlight potential noncanonical roles in cell adhesion and nuclear signaling. The data supporting such functions are discussed as are recent technologies that have the potential to provide valuable insight into the developmental significance of these noncanonical activities.

Current Topics in Developmental Biology, Volume 149
ISSN 0070-2153
https://doi.org/10.1016/bs.ctdb.2021.12.001

1. Introduction

Receptor tyrosine kinases (RTKs) are a superfamily of signaling receptors expressed throughout development that initiate a cascade of intracellular signals upon ligand binding in order to regulate a multitude of biological activities. Loss of individual RTKs can lead to a variety of defects from early development through organogenesis and postnatal life. The study of RTK function has predominantly focused on their role in activating intracellular signaling pathways. Early examinations probed the question of how a protein localized to the cell surface could induce a transcriptional change within the nucleus. During the early 1990s, multiple groups were independently uncovering the pieces needed to answer this question: biochemical analysis identified interactions between RTKs and other proteins; subsequent structural analysis identified the importance of key interaction domains such as SH2 and SH3; and genetic analysis demonstrated conservation across species and uncovered novel interactions (Schlessinger, 2014). A picture was beginning to develop that could answer the question of how RTKs transmitted information to the nucleus.

A breakthrough was made using genetic screens revolving around the Drosophila RTK, Sevenless, that finally aligned the major pieces of the puzzle. These screens were able to functionally link Sev-Drk-Sos activation to the RAF-MEK-MAPK signaling pathway identified in mammals (Fortini, Simon, & Rubin, 1992; Simon, Dodson, & Rubin, 1993). This link provided a direct path by which an activated RTK could induce transcriptional changes in the nucleus, as MAPKs drive the activation of nuclear transcription factors (Marshall, 1994; Treisman, 1994). It was soon found that most of these components were conserved from *C. elegans* through to mammals. Continued work led to the discovery of a handful of intracellular pathways that act downstream of RTKs, including ERK/MAPK, PI3K/AKT, JAK-STAT, PLCγ, JNK, and p38 (Raabe, 2000). This body of work has led to the view that RTKs solely function as signaling hubs, whereby activated receptor dimers nucleate the formation of multifaceted protein complexes that can engage a variety of intracellular signaling cascades.

However, growing evidence suggests that these canonical signal transduction pathways are not the only output of RTK activation. Genetic dissection, through various mutations of the receptors, has led to uncovering additional functions beyond canonical signal transduction. For certain RTKs such as FGFRs, disruption of effector binding sites that block further signal activation do not ablate all ligand-dependent activity. In some instances, deletion or replacement of the kinase domain does not phenocopy a complete null allele, indicating partial functionality remains in the kinase-dead receptor. Considering these findings, we must ask what further functionality do these receptors contain beyond canonical signaling and how are these functions interpreted to give rise to a cellular response?

This chapter explores the facets of both canonical and noncanonical signaling in RTKs. We first begin with an examination of canonical RTK signaling, using the FGFR family of RTKs as an example. We next turn to genetic dissection as a means of probing canonical signaling and uncovering novel noncanonical activity. Multiple groups have used knock-in mutations to examine these functions *in vivo* across multiple RTK families. We then discuss possible noncanonical activities relating to functions in cell adhesion and nuclear signaling, again with a focus on FGFR activity, that have been observed or documented in other RTKs. We conclude with a discussion of directions that can be taken to further explore and elucidate noncanonical RTK activity *in vivo*.

2. Receptor tyrosine kinase signaling

We will first examine the canonical view of RTK signaling, using the activity of Fibroblast Growth Factor Receptors (FGFRs) as a model, to detail the mechanism of signal transduction and then describe independent novel activities. The Fibroblast Growth Factor (FGF) family is a highly conserved signaling pathway that drives numerous biological outputs. The mammalian FGF signaling pathway consists of 4 tyrosine kinase receptors (FGFRs) and 15 canonical FGF ligands (FGFs). In addition to the canonical FGFs, there are endocrine and intracellular FGFs that play roles outside of paracrine signal transduction, but they will not be discussed in this chapter (Ornitz & Itoh, 2015).

The basic structure of RTKs, like the FGFRs, consist of a single-pass transmembrane (TM), an extracellular domain (ECD), and an intracellular domain (ICD) (Fig. 1A). The ECD of FGFRs consists of three immunoglobulin-like (Ig) domains: IgI acts in an autoinhibitory manner to prevent receptor dimerization in the absence of ligand, while IgII and IgIII mediate the binding of FGFs. The binding of a ligand to the ECD of its cognate RTK results in dimerization of the ligand-receptor complex. This dimerization drives the activation of the TK domain on each receptor, resulting in downstream biological activity (Lemmon & Schlessinger, 2010; Schlessinger, 2014). An acid

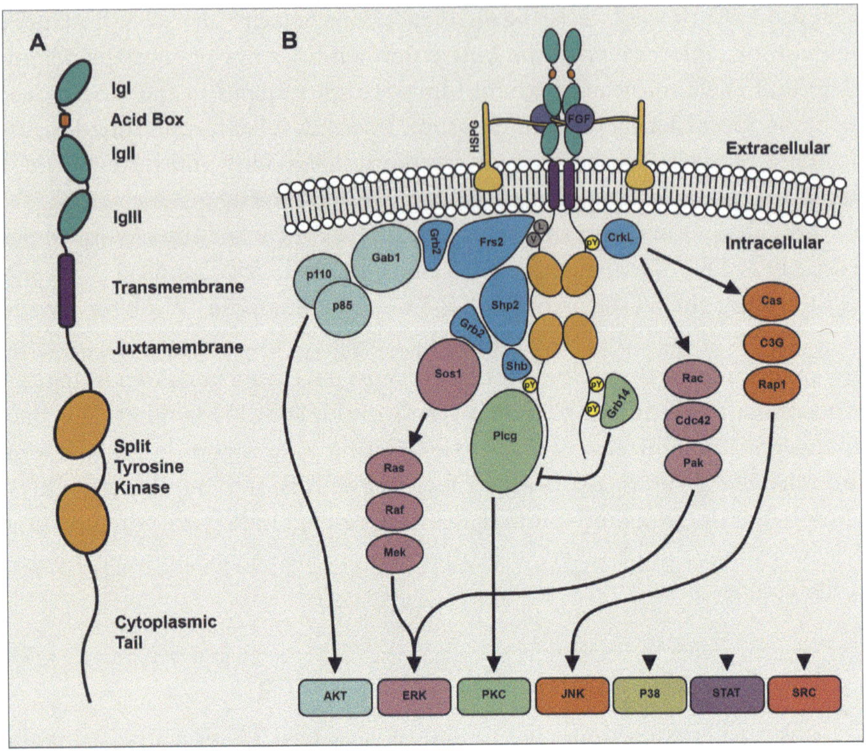

Fig. 1 FGF-RTK structure and signaling. (A) Structure of an FGF receptor consisting of an extracellular domain (immunoglobulin-like (Ig) domain I–III and an acid box motif), a single-pass transmembrane domain, and an intracellular domain (juxtamembrane region, split tyrosine kinase domain, and a cytoplasmic tail region). (B) Active FGFRs engage multiple intracellular signaling pathways through the recruitment of multiple effector proteins to the intracellular domain of the receptor. FGF signaling is a strong activator of ERK/MAPK and can engage additional signaling pathways such as AKT/PI3K, PKC, JNK, P38, STAT, and SRC. Key residues involved in recruiting adaptors are labeled as L (leucine), V (valine), or pY (phospho-tyrosine). Known effector proteins are labeled in the image.

box motif is also located between IgI and IgII that may play a role in binding to Cadherins, which will be discussed in Section 4.

FGF signaling also involves the presence of heparan sulfate proteoglycans (HSPGs), which consist of a protein core bound to multiple heparan sulfate chains, that can interact with numerous growth factors to aid the formation of stable ligand-receptor complexes. The HSPG protein cores can associate with either the cell membrane, as in Syndecan and Glypican, or the extracellular matrix, as in Perlecan and Agrin (Li & Kusche-Gullberg, 2016). HSPGs bind to FGF ligands and are required to facilitate FGF–FGFR complexes and the dimerization of FGFRs (Rapraeger, Krufka, & Olwin, 1991; Yayon, Klagsbrun, Esko, Leder, & Ornitz, 1991).

The ICD of FGFRs is composed of multiple domains: a juxtamembrane region, a split TK domain, and a cytoplasmic tail. Upon ligand binding, the ICD of the receptor undergoes a conformational change, bringing the TK domain of each receptor closer together. This proximity between the TK domains induces transphosphorylation of several tyrosine residues within the intracellular domain of the receptor. Many of these phospho-tyrosine residues are located within conserved binding motifs, such as PTB, SH2, and SH3, that can recruit various effector proteins that facilitate activation of FGFR signaling. This sequential activation via tyrosine phosphorylation is common across all RTK family members (Fig. 1B) (Brewer, Mazot, & Soriano, 2016; Lemmon & Schlessinger, 2010).

Canonical FGF signaling involves the recruitment of various intracellular effector proteins to the receptor. The activated signaling complex typically consists of the adaptor proteins FRS2/3, CRKL/CRKII, SHB, and GRB14 as well as PLCγ. Additional effectors include STAT1/3/5, SRC, GRB2, and p85. Genetic dissection has revealed that the effectors function in a combinatorial manner to induce intracellular signals. FGF signaling is a strong activator of ERK/MAPK activity; however additional pathways are also activated, which include PI3K/AKT, STAT, PLCγ, p38, and JNK (Fig. 1B) (Brewer et al., 2016).

FGFR1 and FGFR2 play important roles during early mammalian development (Kang, Garg, & Hadjantonakis, 2017; Molotkov, Mazot, Brewer, Cinalli, & Soriano, 2017), while FGFR3 and FGFR4 regulate later aspects of development (Colvin, Bohne, Harding, McEwen, & Ornitz, 1996; Weinstein, Xu, Ohyama, & Deng, 1998). The FGF ligands have varied and overlapping expression patterns throughout development. Additionally, each receptor has varied affinities for different ligands. This combination of binding affinity and spatiotemporal expression, alongside the number of ligands and receptors, provides for a complex signaling family that can provide a robust variety of specific cellular outputs in numerous developmental contexts (Ornitz & Itoh, 2015).

3. Dismantling RTK signaling one pathway at a time

For more than three decades, RTKs have been studied for their potential to drive numerous biological outputs. As discussed previously, biochemical assays were used to identify how the various RTK adaptors and signal transducers are recruited to the activated receptor. Alongside these experiments, *in vivo* analyses provided insights into the roles of signaling pathways downstream of each RTK family. Beyond generating "knock-out" null alleles, point mutations can be a valuable strategy for understanding the signaling potential of RTKs. As mentioned above, upon activation RTKs recruit numerous effector proteins to their intracellular domain. These effectors bind to specific motifs, typically initiated by the phosphorylation of tyrosine residues within those sequences. This mechanism has provided a valuable opportunity to understand signal transduction; conversion of tyrosine to phenylalanine can ablate the binding of a specific effector without altering the structure of the receptor. Therefore, these point mutations provide the ability to genetically analyze the contribution of individual RTK-effector interactions.

A classic example of biochemical dissection of RTK pathways was performed for PDGFRs (Fambrough, McClure, Kazlauskas, & Lander, 1999; Joly, Kazlauskas, Fay, & Corvera, 1994; Kundra et al., 1994; Rosenkranz, DeMali, Gelderloos, Bazenet, & Kazlauskas, 1999; Valius & Kazlauskas, 1993). Upon activation, PDGFRs recruit multiple intracellular signal transducers, such as PI3K, SRC, and PLCγ. To understand the pathways necessary for driving specific cellular behaviors, multiple tyrosine residues were converted to phenylalanine to abrogate the recruitment of specific effectors. These mutations were analyzed either individually or combinatorially to assess the contribution of each binding partner to cellular outputs. Analyses in cultured cells revealed that certain effectors were required to drive distinct cellular behaviors, as binding of PI3K to PDGFRα was found to be both necessary and sufficient to drive DNA synthesis while PI3K, SRC, and PLCγ contributed additively to drive chemotaxis (Rosenkranz et al., 1999).

Studies in cultured cells provide an ideal environment for biochemical analysis of protein interaction and phosphorylation, however, understanding the biological significance of RTKs and their various downstream activities may be limited. Early methods also utilized the transfection of plasmids to drive expression, resulting in overexpression, which could facilitate interactions not seen at endogenous expression levels. Therefore, examining RTK function in the embryo or in the animal provides a more comprehensive approach to deducing multiple outputs across varied contexts. To this end, multiple knock-in mouse models have been generated to study the effects at an organismal level, providing an environment to identify possible tissue-specific effects.

In vivo analysis is typically done by comparing embryos homozygous for the wild-type allele to embryos homozygous for the specific signaling mutant allele to examine embryonic phenotypes. Alternatively, the mutant allele could be paired with a floxed null allele and a tissue-specific Cre deleter, resulting in a hemizygous signaling mutation in the tissue of interest. Similar to the cell-based analysis discussed above, combinatorial alleles contain multiple signaling mutations to study the effects of simultaneous loss of targeted signaling pathways, thereby producing a series of signaling mutant alleles that can isolate the activity, or loss, of specific RTKs effectors *in vivo*. A discussion of example allelic series constructed in multiple RTKs will be examined below. A comprehensive list of alleles made in mice, including some that are not discussed in the text, can be found in Table 1.

The generation of an allelic series of signaling mutations in both PDGFRs has suggested different models in which signaling pathways can be utilized *in vivo*. PDGFRα signaling is integral to neural crest development, somite organization, and neural tube closure (Soriano, 1997), and is known to recruit Src family kinases (SFKs), SHP-2, PI3K, and PLCγ (Heldin & Lennartsson, 2013). The most severe allele that disrupted the binding of all effectors phenocopied an allele that only lacked PI3K recruitment, arguing that PI3K was the main driver of PDGFRα activity during development and functioned as a binary switch. Consistent with this result, mice containing mutations that abrogated PI3K binding to both PDGFRα and PDGFRβ, phenotypically copied *Pdgfra* null mice. Not only did this further argue that PI3K was the predominant output, but it also hinted at the possibility that PDGFRα and PDGFRβ could form heterodimers *in vivo* (Klinghoffer, Hamilton, Hoch, & Soriano, 2002), as had previously been seen *in vitro* (Ekman, Thuresson, Heldin, & Ronnstrand, 1999) and was subsequently verified in a later study (Fantauzzo & Soriano, 2016). Interestingly, PDGFRβ exhibits a different use of signals compared to PDGFRα. PDGFRβ plays a role in kidney and vascular development (Soriano, 1994). Like PDGFRα, PDGFRβ recruits SFKs, SHP-2, PI3K, and PLCγ in addition to GRB2 and RasGAP, however, PDGFRβ appears to use these pathways in a quantitative manner. Ablation of each individual effector binding site modulates the severity of the allele, indicating an additive nature of each signal to produce the observed phenotype. Of note, the most severe PDGFRβ allele did not phenocopy the null, leaving open the question of what functionality remained in the mutant receptor (Tallquist, French, & Soriano, 2003).

Another example of an RTK that has been studied through genetic dissection *in vivo* is RET. RET activity plays a key role in the development of

Table 1 RTK knock-in signaling mutations *in vivo* mouse models.

Family	**Receptor**	**Gene**	**Allele**	**Effectors**	**Citation**
(A) Signaling mutation alleles					
Eph	EPHA4	*Epha4*	2F	SH2-binding	Kullander et al. (2001)
EGFR	ERBB2	*Erbb2*	Y1028F	Involved in receptor recycling	Chan, Hardy, Dankort, Laing, and Muller (2004)
			Y1144F	Grb2	Chan et al. (2004)
			Y1227F	Shc	Chan et al. (2004)
FGFR	FGFR1	*Fgfr1*	ΔFRS	Frs2/3	Hoch and Soriano (2006)
			F	Frs2/3	Brewer et al. (2015)
			C	CrkL/CrkII	Brewer et al. (2015)
			CPG	CrkL/CrkII, Plcγ, Grb14	Brewer et al. (2015)
			FCPG	Frs2/3, CrkL/CrkII, Plcγ, Grb14	Brewer et al. (2015)
	FGFR2	*Fgfr2*	F	Frs2/3	Ray et al. (2020)
			C	CrkL/CrkII	Ray et al. (2020)
			PG	Plcγ, Grb14	Ray et al. (2020)
			CPG	CrkL/CrkII, Plcγ, Grb14	Ray et al. (2020)
			FCPG	Frs2/3, CrkL/CrkII, Plcγ, Grb14	Ray et al. (2020)
Met	MET	*Met*	D	Multifunctional docking site	Maina et al. (1996)
			Grb2	Grb2	Maina et al. (1996)
			2S	SFKs	Maina et al. (2001)
			2G	Grb2	Maina et al. (2001)
			2P	Plcγ	Maina et al. (2001)
PDGFR	PDGFRa	*Pdgfra*	PI3K	PI3K	Klinghoffer et al. (2002)
			Src	SFKs	Klinghoffer et al. (2002)

Table 1 RTK knock-in signaling mutations *in vivo* mouse models.—cont'd

Family	Receptor	Gene	Allele	Effectors	Citation
			F7	PI3K, SFKs, Shp-2, Plcγ	Klinghoffer et al. (2002)
	PDGFRb	*Pdgfrb*	F1	RasGAP	Tallquist et al. (2003)
			F2	PI3K	Heuchel et al. (1999)
			F3	PI3K, Plcγ	Tallquist et al. (2000)
			F5	PI3K, RasGAP, Shp-2, Plcγ	Tallquist et al. (2003)
			F7	PI3K, RasGAP, Shp-2, Plcγ, SFKs, Grb2	Tallquist et al. (2003)
	KIT	*Kit*	Y567F	Src	Agosti et al. (2004); Kimura et al. (2004)
			Y567F, Y569F	Src, Shp-1/2	Kimura et al. (2004)
			Y719F	PI3K	Blume-Jensen et al. (2000); Kissel et al. (2000)
Ret	RET	*Ret*	S697A	PKA	Asai et al. (2006)
			Y981F	Src	Jain et al. (2006)
			Y1015F	Plcγ	Jain et al. (2006)
			Y1062F	Shc	Jijiwa et al. (2004); Wong et al. (2005); Jain et al. (2006)
			F1072A	PDZ-binding	Wong et al. (2005)
Tie	TIE2	*Tek*	Y1100F	p85/PI3K, SH2-binding	Tachibana, Jones, Dumont, Puri, and Bernstein (2005)
Trk	TRKB	*Ntrk2*	Shc	Shc	Minichiello et al. (1998)
			Plc	Plcγ	Minichiello et al. (2002)
			D	Shc, Plcγ	Medina et al. (2004)
	TRKC	*Ntrk3*	Shc	Shc	Postigo et al. (2002)
VEGFR	VEGFR2	*Kdr*	Y949F	T cell specific adaptor (TSAd)	Li et al. (2016)

Continued

Table 1 RTK knock-in signaling mutations *in vivo* mouse models.—cont'd

Family	Receptor	Gene	Allele	Effectors	Citation
			Y1173F	Plcγ, PKC, MAPK	Sakurai, Ohgimoto, Kataoka, Yoshida, and Shibuya (2005)
			Y1212F	Plcγ, PKC, MAPK	Sakurai et al. (2005)

Family	Receptor	Gene	Mutation	Citation
(B) Kinase domain mutation alleles				
Axl	MER	*Mertk*	Kinase domain truncation	Camenisch, Koller, Earp, and Matsushima (1999)
Eph	EPHA4	*Epha4*	Conserved lysine mutation; K653M	Kullander et al. (2001)
	EPHA5	*Epha5*	ICD replacement with β-galactosidase	Feldheim et al. (2004)
	EPHB2	*Ephb2*	ICD replacement with β-galactosidase	Henkemeyer et al. (1996)
EGFR	ERBB2	*Erbb2*	Conserved lysine mutation in cDNA knock-in; K757M	Chan et al. (2002)
FGFR	FGFR2	*Fgfr2*	Conserved lysine mutation; K517A	Ray et al. (2020)
Met	MET	*Met*	Exon deletion in kinase domain	Bladt, Riethmacher, Isenmann, Aguzzi, and Birchmeier (1995)
	RON	*Mst1r*	Exon deletion in kinase domain	Waltz et al. (2001)
Trk	TRKC	*Ntrk3*	Exon deletion in kinase domain	Klein et al. (1994)
VEGFR	VEGFR1	*Flt1*	ICD deletion	Hiratsuka et al. (1998)
	VEGFR3	*Flt4*	Kinase domain point mutation; I1053F	Zhang et al. (2010)

(A) Knock-in point mutations in RTKs that disrupt the binding of various effector proteins. (B) Knock-in mutations that disrupt kinase domain activity, either through complete deletion of the intracellular domain (ICD), deletion of the exon containing the catalytic residues in the kinase domain, or via point mutations in the kinase domain.

the kidneys and the enteric nervous system (Schuchardt, D'Agati, Larsson-Blomberg, Costantini, & Pachnis, 1994). Activated RET receptor recruits multiple effectors to engage MAPK, PI3K, PKC, and SRC signaling pathways. Of note, RET encodes two isoforms, RET9 and RET51, that differ in the length of their cytoplasmic tails. Analysis of multiple *Ret* alleles identified PLCγ, MAPK, and PI3K as necessary signals for proper renal development, while only MAPK and PI3K were required for the formation of the enteric ganglia (Jain, Encinas, Johnson, & Milbrandt, 2006; Jijiwa et al., 2004; Wong et al., 2005). SRC is thought to play a minor role as loss of recruitment exhibited a lowly penetrant ureter defect (Jain et al., 2006; Wong et al., 2005). As observed in PDGFRs, these studies have identified specific requirements for signaling downstream of RET in various developmental contexts. However, the knock-in mutations were studied individually, leaving open several questions relating to how multiple signaling pathways are used in tandem to drive a biological response. For instance, neither disruption of PLCγ nor MAPK/PI3K binding completely phenocopied the null allele, as null mice die right after birth, compared to mice deficient for PLCγ or MAPK/PI3K binding that die at around 1 month after birth. Would an allele carrying both the PLCγ and MAPK/PI3K binding site mutations in tandem produce the phenotype of the null? Combinatorial alleles in *Ret* could provide the opportunity to uncover the role that SRC recruitment might play in RET function.

In a similar vein, mutations were introduced into the adaptor docking site of MET. MET activity is necessary during development, as null mutants display embryonic lethality with gross defects in placental and liver development as well as a reduction in migratory muscle precursors (Schmidt et al., 1995; Uehara et al., 1995). However rather than introducing mutations to completely ablate binding, the multifunctional binding motif was altered to recruit only specific adaptors while excluding others (Maina et al., 2001). This approach, while arguably less physiological, combined the previously detailed loss-of-function analysis with novel gain-of-function mutations. The study provided the valuable observation that PI3K recruitment to MET could promote proper axon outgrowth while SRC recruitment was sufficient to rescue fetal myoblast proliferation. These data provide further credence to idea that, while multiple RTK effectors may function additively to promote a specific cellular output, there may be tissue-specific requirements for a subset of effectors to engage normal development.

A second approach to analyzing RTK function involves disrupting the catalytic activity of the kinase domain. These mutations may provide insight into the existence of RTK function beyond the phosphorylation of downstream targets. If all RTK function is driven through kinase-dependent phosphorylation, then alleles that lack kinase activity should phenotypically mimic a null allele. However, there have been several instances where knock-in disruption of the kinase domain does not fully inhibit all function of the RTK. For instance, complete loss of TRKC, a neurotrophin receptor, exhibits severe agenesis of neuromuscular junctions, however, disruption of the kinase domain exhibits a lesser phenotype, indicating that the kinase-deficient TRKC still maintains some functionality (Klein et al., 1994; Liebl, Tessarollo, Palko, & Parada, 1997). Similarly, *Vegfr-1* null mutants are embryonic lethal, however, homozygous mice containing a truncated allele of *Vegfr-1*, lacking the kinase domain, are recoverable at birth but exhibited defects in macrophage migration (Hiratsuka, Minowa, Kuno, Noda, & Shibuya, 1998). Of note, both studies utilized large deletions of the intracellular domain to disrupt kinase activity which may have unintended effects on the function of the protein. Nonautophosphorylated motifs may be present in various regions of the ICD such that deletion of the entire ICD may inadvertently mask kinase-independent functions. To alleviate this uncertainty, knock-in point mutations can be used once again to impede the catalytic activity of the kinase domain with as little disruption to the surrounding sequences as possible.

The catalytic domain of protein kinases has long been characterized through a combination of structural and biochemical studies. One ideal mutation site lies within the ATP binding site of the kinase domain. An invariant lysine residue is conserved across multiple tyrosine kinase families, including RTKs, and is thought to be integral in the phosphate transfer reaction. Mutating this specific lysine *in vitro* eliminates kinase activity in multiple RTKs, including EGFR, INSR, PDGFR, and FGFR (Hanks, Quinn, & Hunter, 1988). Despite this, the method has not been utilized frequently *in vivo* to determine if kinase-independent RTK functions are present and/or relevant during development. A kinase dead (KD) allele of *EphA4* utilized this strategy to study kinase-dependent effects of EPHA4 during neuronal development. It was found that EPHA4 kinase activity is required for corticospinal tract axon guidance, while kinase activity is dispensable for the formation of the anterior commissure. Although the authors attribute this result to reverse signaling through an ephrin, with $EPHA4^{KD}$ acting noncell autonomously as a ligand, rather than kinase-independent function in EPHA4 (Kullander et al., 2001). However, the EPH/EPHRIN interactions also pose a unique situation for RTK signaling, as the combination of forward and reverse signaling adds a layer of complexity to deciphering the importance of the receptor vs the ligand (Davy & Soriano, 2005).

Similar analysis was performed using a KD allele of *ErbB2*. It was found that kinase activity was required for development as KD mutant embryos phenocopied null mutants, exhibiting similar cardiac and peripheral nervous system developmental defects. However, the KD allele was constructed by knocking-in the complete cDNA of *ErbB2*, with the kinase domain mutation, at the endogenous locus, as opposed to a point mutation knock-in. The authors did observe a mild developmental delay in embryos expressing a wild-type cDNA knock-in allele and postulated that splicing may play a role in protein expression regulation, as protein levels of both knock-in alleles were reduced compared to wild type, indicating a potential problem with a cDNA knock-in approach (Chan, Hardy, Laing, Hardy, & Muller, 2002). Key regulatory mechanisms may occur during mRNA processing or key motifs may be present in intronic regions, both of which are lost when using cDNA. While the gross defects of the KD knock-in allele are similar to the null, the changes in protein expression levels may mask the full effect of kinase domain mutation. This study highlights the need for altering as little genetic information as necessary to prevent unintended effects on expression and function.

In the analyses of *Ret* and *Met*, it was observed that the most severe knock-in mutations were able to phenocopy their respective null alleles, arguing that the loss of recruitment of known adaptors encompassed the full function of each receptor. However, as was observed in *Pdgfrb*, abrogating the binding of known adaptors is not guaranteed to mimic the null allele. Recent genetic analysis of *Fgfr1* and *Fgfr2* exhibit an example of residual activity. Mutations were made at known binding sites to disrupt the recruitment of FRS2, CRKL, PLCγ, and GRB14, alone or in combination, to either FGFR1 or FGFR2. The most severe combinatorial alleles resulted in the loss of all known intracellular canonical signaling activity (Brewer, Molotkov, Mazot, Hoch, & Soriano, 2015; Ray et al., 2020). Intriguingly, embryos containing the severe point mutation alleles of either *Fgfr1* or *Fgfr2* survived well beyond the embryonic lethality induced by their respective null alleles. Embryos containing congruent point mutations in both *Fgfr1* and *Fgfr2* that eliminate canonical signaling for each receptor survive until gastrulation while double null embryos die at implantation (Ray et al., 2020), ruling out the possibility of compensation from the other receptor or through heterodimer formation. These data therefore argue that the signaling deficient alleles of *Fgfr1* and *Fgfr2* provide a degree of activity sufficient to continue early development, surviving past the implantation lethality seen in *Fgfr1* null animals (Brewer et al., 2015; Kurowski, Molotkov, & Soriano, 2019) or the mid-gestation lethality seen in *Fgfr2* null animals (Molotkov et al., 2017). This is especially surprising in the case of *Fgfr2* as loss of all know signaling activity from FGFR2 still produces viable, fertile mice. The results of the *Fgfr1* and *Fgfr2* allelic series argue that the most severe point mutation alleles are still able to provide partial functionality for continued development while exhibiting a complete lack of canonical signaling. However, a kinase dead *Fgfr2* allele appears to broadly phenocopy the null allele, suggesting that the remaining functionality in the signaling mutation allele is kinase dependent (Ray et al., 2020). This then raises the questions: what functionality is left and how is that functionality translated into a biological output?

4. Without canonical signaling, what novel functions are revealed?

As evidenced above, genetic dissection of RTKs provides a means to determine the function and necessity of specific intracellular pathways driven by RTK activation during development. However, there exist multiple instances where removing canonical signal transduction of RTKs does not exhibit the phenotype of a null mutant. Similarly, examining kinase-dead alleles provides the ability to determine whether the catalytic activity of the RTK kinase domain is necessary for all functions. Current evidence from *EphA4* (Kullander et al., 2001), *ErbB2* (Chan et al., 2002), and *Fgfr2* (Ray et al., 2020) suggests that kinase activity is indeed required for biological activity of RTKs *in vivo*. However, this may not hold true for all RTK families and highlights the need for continued analysis within the field.

This section will focus on potential activities of FGFRs, as *in vivo* analysis indicates that loss of all known canonical signal transduction does not recapitulate the null, whereas a kinase-dead allele appears to recapitulate the null in the case of *Fgfr2*. We first examine how FGFRs may play a role in cell adhesion by regulating the formation, stability, or function of adhesion complexes. We then discuss the intracellular role of FGFRs, focusing on functions in the nucleus, by examining the means of trafficking, nuclear activity, and biological significance. Together these and possibly other noncanonical activities may account for the discrepancies observed *in vivo* between the signaling mutant and null alleles.

4.1 Adhesion

Cell adhesion molecules (CAMs) are a group of proteins involved in the formation of cell-cell and cell-extracellular matrix connections. CAMs generally consist of four groups: Cadherins, the immunoglobulin superfamily (IgSF), Selectins, and Integrins. Cell-cell contacts are formed using Cadherins, Ig SFs, and Selectins, while Integrins are involved in cell-extracellular matrix junctions. Traditionally, FGF signaling was known to regulate the expression of various CAMs during development, however a growing body of evidence suggests that FGFRs directly interact with CAMs to modulate FGF signaling and facilitate cell migration and adhesion.

Cadherins are a large family of adhesion molecules involved in the formation of cell-cell adherens junctions. Cadherins consist of a large extracellular domain, comprised mainly of Cadherin domain repeats, a single-pass transmembrane domain, and a smaller cytoplasmic domain. The Cadherin domain repeats exhibit homophilic adhesion properties, resulting in the formation of trans-interacting homodimers. Neighboring cells can form cell-cell junctions via the homodimerization of the extracellular domains of Cadherins. Thus, cells expressing the same species of Cadherin will adhere to one another, while those expressing different species of Cadherins will not. Meanwhile, the cytoplasmic domain can bind to p120 and β-catenin which then recruit α-catenin and other adaptors, such as Vinculin, that anchor the Cadherin to the actin network of the cell. Clustering of these Cadherin complexes strengthens the adhesion force of a junction (Garcia, Nelson, & Chavez, 2018; Leckband & de Rooij, 2014).

The current paradigm implicates canonical signaling in the interaction of FGFRs and Cadherins during epithelial-mesenchymal transition (EMT), a key process in multiple developmental contexts such as gastrulation or neural crest delamination. Cells undergoing EMT lose apicobasal polarity and break cell-cell contacts to adopt a more migratory state. To accomplish this, cells must switch from expressing one predominant Cadherin to another. FGF signaling through FGFR1 is involved in Cadherin switching during EMT, via the transcriptional regulation of *Snai1* (Ciruna & Rossant, 2001; Gonzalez & Medici, 2014; Theveneau & Mayor, 2012). However, there are instances where FGFR1 may play a more direct role in the function of Cadherins.

Several studies have suggested that Cadherins and Ig SFs, such as NCAM and L1CAM, can bind to the acid box domain located in the extracellular domain of FGFRs to regulate adhesion-dependent processes, including neurite outgrowth (Francavilla et al., 2007; Sanchez-Heras, Howell, Williams, & Doherty, 2006; Williams, Furness, Walsh, & Doherty, 1994). Evidence also suggests that FGFRs and Cadherins can be colocalized in complexes at the cell membrane (Boscher & Mege, 2008; Suyama, Shapiro, Guttman, & Hazan, 2002; Utton, Eickholt, Howell, Wallis, & Doherty, 2001). The presence of FGFR may help to stabilize the Cadherin along cell-cell junctions, mediated through a complex involving Src (Nguyen et al., 2019). Lines of evidence from multiple species indicate that loss of FGFR activity leads to mislocalization of Cadherins and cell polarity defects. In Drosophila, FGF signaling was found to affect polarized localization of multiple CAMs involved in EMT during gastrulation (Sun & Stathopoulos, 2018). Similarly, mislocalization of N-Cadherin was observed in Zebrafish cardiomyocytes in response to chemical FGFR inhibition (Rasouli et al., 2018). Defects in trophectoderm polarity and E-Cadherin localization were also observed during blastocyst development in mouse embryos lacking *Fgfr1* (Kurowski et al., 2019). Together, these data point toward a function for FGFRs in regulating cell-cell adhesion activity during development (Fig. 2A).

Integrins are transmembrane receptors typically involved in the formation of cell-ECM adhesions (see chapter "Integrins, anchors and signal transducers of hematopoietic stem cells during development and in adulthood" by Krenn et al., in this volume for more extensive coverage of this topic). Integrins form heterodimers consisting of an alpha and a beta subunit; the specific combination of the subunits dictates the substrate specificity of the heterodimer. ECM proteins, such as laminin, collagen, and fibronectin, function as ligands for Integrins. Upon binding a ligand, the Integrin heterodimer recruits numerous intracellular adaptors, including Talin, Vinculin, and Paxillin, and signaling proteins, including Src, Focal Adhesion Kinase (FAK), and Integrin-linked Kinase (ILK), that serve to anchor the complexes to the actin network of the cell. Clustering of these integrin complexes forms the basis for focal adhesions (Berrier & Yamada, 2007; Mishra & Manavathi, 2021).

Interactions between Integrins and various RTKs have been documented in multiple instances, however a detailed understanding of the role that Integrin-RTK interactions play in development has yet to be elucidated. These interactions can produce outputs from either signaling axis; Integrin-RTK association can induce low levels of RTK activation in the absence of ligand and potentially prime the RTK for ligand-binding, enhancing the RTK signaling output (Bill et al., 2004; Veevers-Lowe,

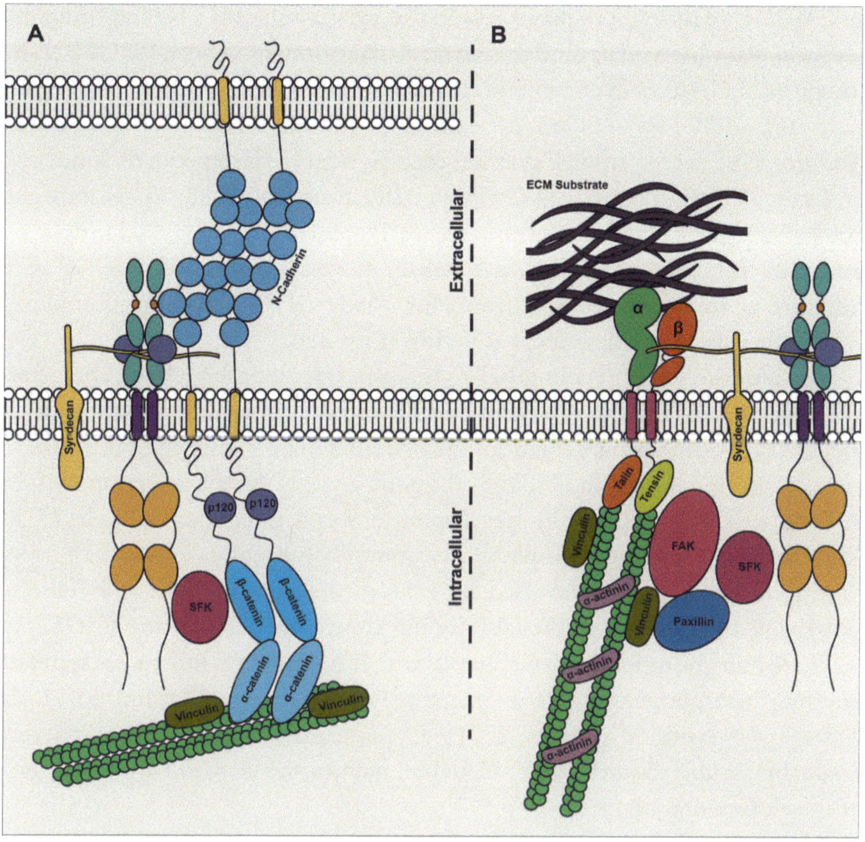

Fig. 2 Noncanonical FGFR interactions with CAMs. (A) Recruitment of FGFR to an adherens junction via direct binding of N-Cadherin to the acid box motif located in the ECD of FGFR. The active junction recruits multiple adaptors, scaffolding, and signaling proteins to link the adherens junction to the actin cytoskeleton. (B) Recruitment of FGFR to a focal adhesion complex through interaction with Syndecan bound to an Integrin dimer. The active focal adhesion complex is anchored to actin stress fibers through the recruitment of key proteins such as FAK and Paxillin, as well as many others.

Ball, Shuttleworth, & Kielty, 2011). Alternatively, the association can result in enhanced Integrin signaling in response to growth factors, promoting the stabilization of focal adhesions (Cascone, Napione, Maniero, Serini, & Bussolino, 2005; Masson-Gadais, Houle, Laferriere, & Huot, 2003). The formation of these Integrin-RTK complexes appears to be dependent on the extracellular domains of both the Integrin and the RTK rather than via intracellular interaction (Borges, Jan, & Ruoslahti, 2000).

HSPGs are thought to play a role in the association of RTKs and Integrins. Heparin was observed to bind the extracellular domains of both PDGFRβ and Integrin (ITG) α5, and enzymatic cleavage of heparan was sufficient to dissociate the PDGFRβ-ITGα5/β1 complex (Veevers-Lowe et al., 2011). Additionally, earlier studies documented Syndecan-1-dependent induction of both ITGαV/β3 and ITGαV/β5 (Beauvais, Burbach, & Rapraeger, 2004; McQuade, Beauvais, Burbach, & Rapraeger, 2006). As HSPGs are integrally involved in FGFR activation, the association of HSPGs with Integrins is an attractive explanation for how FGFRs may form complexes with Integrins to regulate focal adhesion (Fig. 2B).

Most of the work detailing RTK-Integrin interaction has been characterized using *in vitro* cell culture. This has allowed the controlled investigation of specific RTK-Integrin interactions on a defined matrix, such as fibronectin or vitronectin. However, the biological significance of each interaction may be difficult to extrapolate as the developing embryo integrates multiple ECM components, creating a complex environment. Recently, *ex vivo* culture of murine salivary glands was used to analyze the role of both cell-cell and cell-ECM adhesion in epithelial budding morphogenesis. Using CRISPR/Cas9, the study modulated Cadherin and Integrin function to recapitulate budding morphogenesis in a tractable way (Wang, Matsumoto, Lish, Cartagena-Rivera, & Yamada, 2021). *Ex vivo* organoid cultures may provide a valuable system to study RTK-CAM interaction while maintaining a degree of developmental context.

Currently, most data that implicates interaction between FGFRs and CAMs still implies that canonical RTK signaling is required for biological function. It has been difficult to tease apart the signaling-dependent vs signaling-independent functions. However, recent data suggests that there may be functions independent of canonical RTK signaling. As mentioned before, embryos carrying mutant *Fgfr1* or *Fgfr2* alleles, lacking the ability to engage known downstream signaling, can survive further into development compared to embryos with null alleles (Ray et al., 2020). Later in development, embryos containing a single copy of this signaling mutant *Fgfr1* allele exhibit proper formation of cell-cell contacts in neural-crest derived cranial mesenchyme, while embryos containing null alleles do not. Additionally, analysis of primary cultured mutant cells indicates that the most severe signaling *Fgfr1* and *Fgfr2* alleles can provide sufficient function to allow for adhesion, migration, and the formation of cell-cell contacts in response to FGF stimulation, unlike cells harboring null alleles for both *Fgfr1* and *Fgfr2*. Interestingly, these functions appear to be kinase dependent, as both a kinase dead allele of *Fgfr2* or chemical inhibition of FGFR kinase activity mimic the phenotype of *Fgfr2* null cells without engaging canonical signals such as ERK. While FGFRs have been associated with CAMs for years, these observations argue that FGFRs contain functionality beyond that of the known canonical downstream signals, but that functionality is still dependent upon the kinase activity of the receptor. There are two possible explanations for this: either there are still unknown downstream signaling pathways activated by the FGFR kinase domain that then trigger a cascade to activate cell adhesion, or the kinase domain could interact with CAMs, such as Cadherin or Integrin, directly or within adhesion complexes to stabilize and regulate adhesion and migration dynamics.

4.2 RTK activity in subcellular compartments and in the nucleus

Recent evidence has continued to elucidate the internalization pathways of RTKs and their respective functions with greater accuracy. This has led to the concept that intracellular trafficking of RTKs can modulate their biological output. RTKs have been shown to engage various intracellular signals depending on the subcellular localization of the RTK. Endocytic trafficking can regulate the internalization of RTKs either through recycling the receptor back to the membrane to prolong signaling or transporting the receptor to lysosomes for degradation and signal attenuation (Fig. 3A). Therefore, the endocytic path that RTKs take through the cell modulates how the RTK signals. However, studies analyzing signal engagement still revolve around key canonical signaling pathways (Sadowski, Pilecka, & Miaczynska, 2009).

Recent data implicate PI3K activity as a major regulator of RTK recycling. PI3K has been shown to regulate a switch between recycling an internalized receptor back to the plasma membrane or trafficking it to late endosomes for lysosomal degradation. However, rather than engage the AKT phosphorylation arm of PI3K signaling, the PI3K complex interacts with both RAB proteins and microtubule binding proteins to regulate RTK trafficking (Thapa et al., 2020). Thus, PI3K activation downstream of RTK activation can aid in prolonging RTK signaling through recycling of the receptor back to the plasma membrane (Bruggemann, Karajannis, Stanoev, Stallaert, & Bastiaens, 2021; Francavilla et al., 2013; Stallaert et al., 2018). While this mechanism describes an interesting aspect to PI3K engagement downstream of RTKs, it does not implicate novel noncanonical RTK activity.

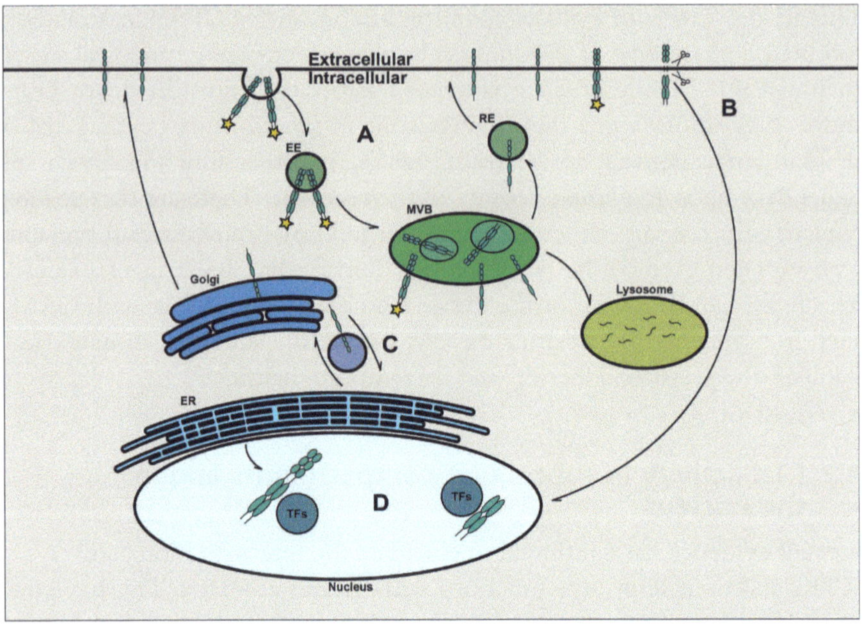

Fig. 3 Intracellular trafficking of FGFR. (A) Endocytic trafficking regulates the internalization of active FGFR (stars) to process the receptor for recycling to the cell surface or degradation in the lysosome (EE, early endosome; MVB, multivesicular body; RE, recycling endosome). (B) Enzymatic cleavage may release the intracellular domain of FGFR into the cytoplasm, which is subsequently translocated to the nucleus. The detailed mechanism of this translocation is unknown. (C) Active full-length FGFR may be imported into the nucleus through retrograde trafficking from the Golgi apparatus and endoplasmic reticulum (ER). (D) Once inside the nucleus, full-length or truncated FGFRs are proposed to interact with transcription factors (TFs), or other DNA interacting proteins, to directly regulate transcriptional targets.

One the other hand, the observation of nuclear RTKs implicates non-canonical function. Rather than being directed toward lysosomal degradation or recycling to the cell surface, activated RTKs may be redirected toward the nucleus, as either ICD fragments or intact receptors, to facilitate transcriptional regulation of target genes. Nuclear RTKs have been documented across multiple family members in cell culture and disease contexts (Carpenter & Liao, 2013; Chen, Hsu, & Hung, 2020; Hancock et al., 2019; Tuzon, Rigueur, & Merrill, 2019); however, uncovering the presence of nuclear RTKs in development and their biological significance has been more challenging.

There are thought to be two mechanisms by which RTKs are localized to the nucleus. The first is through enzymatic cleavage. Upon ligand binding, the extracellular domain of the RTK can be proteolytically cleaved, followed by a second cleavage event in the juxtamembrane domain. This mechanism of ICD cleavage is thought to be mediated by a γ-secretase in an equivalent manner as Notch signaling. The resulting cytoplasmic ICD is shuttled into the nucleus where it is thought to bind to DNA at specific promoter regions to aid in the expression of transcriptional targets (Fig. 3B). Alternatively, there have been multiple instances where full length RTKs were observed in the nucleus. This process is thought to involve retrograde transport from the endoplasmic reticulum to the nucleus via the Sec61 complex (Fig. 3C). Molecular chaperones may facilitate this translocation, with Importin-β being a requirement for nuclear translocation of members of the EGFR family (Carpenter & Liao, 2013; Chen & Hung, 2015).

Currently, multiple studies have shown evidence of nuclear FGFR activity in diseases such as Bent Bone Dysplasia and multiple forms of cancer (Neben et al., 2014; Tuzon et al., 2019). However, the extent of nuclear FGFR activity during development is still currently unknown. One set of studies found unique effects of membrane localized FGFR2 compared to nuclear localization in murine gonadal development. FGFR2 appeared to be directed to the nucleus in response to FGF9, in cells that were undergoing Sertoli cell specification (Kim et al., 2007; Schmahl, Kim, Colvin, Ornitz, & Capel, 2004). A second example of nuclear FGFR2 was observed in both developing terminal end buds of the pubertal mammary gland and in the nuclei of the cells located on the periphery of salivary gland buds in response to FGF10 treatment. In both instances, nuclear FGFR2 appeared to be associated with groups of cells actively undergoing proliferation (Lu, Ewald, Martin, & Werb, 2008; Steinberg et al., 2005). However, the biological outcomes in the two examples are inconsistent: during Sertoli cell development, nuclear FGFR2 is required for specification and commitment to the Sertoli fate, while in branching morphogenesis nuclear FGFR2 is observed in proliferating cells. This could indicate that FGFR2 interacts with cell-specific partners within the nucleus to engage different cellular outputs.

As noted previously (Schmahl et al., 2004), the presence of nuclear FGFR2 is specific to pre-Sertoli cells in the male gonad, as female gonads did not exhibit any nuclear localization. This raises the question of whether there are cell-specific cofactors that facilitate the nuclear localization of FGFR2, as opposed to a generalized epithelial phenomenon. The presence of male specific HSPGs was postulated as an explanation for why male cells responded to FGF9, but female cells did not. However, these studies do not address the mechanism of nuclear accumulation. No cryptic nuclear localization sequences (NLS) have been reported in FGFRs, however certain FGF isoforms are reported to contain an NLS, including FGF1, FGF2, and FGF10. It is postulated that the NLS of the FGF ligand may facilitate the translocation of the FGF-FGFR complex into the nucleus, as FGFs have also been observed in the nucleus *in vitro* (Tuzon et al., 2019).

Once inside the nucleus, FGFR2 is thought to potentially regulate gene expression through interaction with transcription factors (Fig. 3D). The transcription factor, SOX9, was localized within the nucleus of pre-Sertoli cells alongside FGFR2; however, a direct functional interaction was not shown (Schmahl et al., 2004). The specifics of how nuclear FGFRs exhibit control over transcriptional regulation has yet to be detailed. Alternatively, FGFRs may phosphorylate targets within the nucleus, however there is no definitive evidence for this mechanism.

Interestingly, the examples of nuclear FGFR in development seem to be specific to FGFR2. This leaves open the question as to whether other FGFRs, or other RTKS, are active in the nucleus during development, or if nuclear FGFRs are mainly seen in cell culture or disease-related contexts. Determining the actual biological significance of nuclear localization would be quite difficult to tease apart from the membrane-localized signal transduction. Two *Fgfr2* mutations commonly found in Bent Bone Dysplasia are located in the TM domain of the receptor and are thought to result in decreased hydrophobicity of the TM domain. These mutations may destabilize the domain, resulting in increased nuclear localization, however this would not account for nuclear accumulation during normal development (Neben, Tuzon, Mao, Lay, & Merrill, 2017). While genetic dissection could be used to attempt to disrupt nuclear localization, the implications of those changes would need to be characterized in detail, as there are still questions left unanswered revolving around the endocytic processing of FGFRs.

5. Conclusion

Canonical RTK function has been defined by the signal transduction pathways that result in the activation of intracellular signaling nodes, such as PI3K and ERK/MAPK, although mounting evidence suggests that multiple RTKs may engage in activity beyond canonical signaling. Studies surrounding FGFR activity have uncovered potential novel activities, as the loss of canonical intracellular signaling does not recapitulate the null phenotype *in vivo*. Conversely, the loss of catalytic activity in FGFR2, via a kinase-dead mutation, appears to phenocopy the null, indicating the presence of noncanonical, kinase-dependent activity *in vivo*. Two areas of study have provided possible mechanisms that account for this discrepancy: direct interaction with CAMs and direct nuclear activity. These interactions have been documented across multiple RTK family members, using *in vitro* assays. However, the complexity of signaling families such as PDGF and FGF imposes difficulties on dissecting the *in vivo* effects of specific receptor-effector interactions. Therefore, identifying a system that can alleviate the complexity of possible ligand-receptor redundancies could be a boon for identifying noncanonical RTK signal transduction.

Kit and *Colony Stimulating Factor-1 Receptor* (*Csfr1*) provide opportunities to study single receptor-ligand systems, as they each bind only a single ligand, *Kit Ligand* (*Kitl*) and *Csf1*, respectively. A subset of tyrosine residues was mutated in *Kit*; however, they were never examined using combinatorial alleles (Agosti et al., 2004; Blume-Jensen et al., 2000; Kimura et al., 2004; Kissel et al., 2000). Additional tyrosine residues thought to recruit other effectors were also never examined *in vivo*. Expanding upon these experiments could provide novel insights into how *Kit*, and RTKs, engage intracellular activity. Like PDGFRα/β and FGFR1/2, combinatorial alleles may hint at novel mechanisms that fall beyond canonical RTK signaling. The reduced upstream simplicity of a single receptor-ligand pair provides a more discernable *in vivo* model by which to study these noncanonical effects. Therefore, revisiting this classical RTK could prove to be quite valuable for the field. Both *Kit* and *Csf1r* are members of the PDGF subfamily of RTKs, so insights gleaned through genetic manipulation of these receptors may not be applicable to RTKs from other subfamilies. Therefore, continued analysis of various RTK subfamilies will prove valuable in uncovering noncanonical RTK activities.

Advances in genetic manipulation, such as the fast-evolving CRISPR technology, provide a simplified means of creating specific knock-in mutations to disrupt either effector recruitment or kinase activity (Clark, Dinsmore, & Soriano, 2020). In tandem, advances in multiple imaging technologies may also prove useful in deciphering RTK function. While signaling is typically engaged at the cell membrane; RTK-effector interactions may dictate the fate of internalized RTK, which may in turn govern interactions with specific proteins in various organelles. However, inferred interactions via microscopy colocalization cannot be verified via traditional microscopy techniques, as resolution is dictated by the diffraction limit. Advances in super-resolution microscopy provide the ability to discern individual proteins at subdiffraction limits, therefore providing a means by which to observe direct interactions (Huang, Bates, & Zhuang, 2009). As discussed above, nuclear localization of FGFRs has been posited as a mechanism of direct transcriptional regulation alongside cofactors. Super-resolution microscopy could provide a means to observe FGFR interaction with nuclear cofactors. Similarly, localized interactions with CAMs at cell-cell or cell-ECM junctions could be observed to provide insight into how these interactions are formed and maintained (Sieben, Douglass, Guichard, & Manley, 2018). Additionally, live imaging of both cell culture and whole-mount organisms can provide valuable insight into the role of RTK localization. Live whole-mount imaging solutions, such as light sheet microscopy, provide the ability to reduce phototoxicity while quickly imaging larger samples (Wan, McDole, & Keller, 2019). Real-time tracking of FGFR in response to developmental stimuli could provide insight into the varied routes that FGFRs may be internalized, either through recycling, degradation, or nuclear translocation. Additionally, optogenetics can be adapted to interrogate RTK signaling *in vivo* to drive specific receptor-adaptor interactions in a tractable manner (Krishnamurthy et al., 2020). These technologies, coupled with advances in genetic manipulation, can provide new perspectives to the historical view of RTK signaling.

Acknowledgments

We thank Stu Aaronson, Sergei Sokol, and our laboratory colleagues for critical comments on the manuscript. We apologize to investigators whose work we could not cite due to space limitations. J.F.C. is supported by F32 DE029387 from the National Institute of Dental and Craniofacial Research (NIDCR). Work in the authors' laboratory is supported by grants DE022363 and DE022778 from the NIDCR to P.S.

References

Agosti, V., Corbacioglu, S., Ehlers, I., Waskow, C., Sommer, G., Berrozpe, G., et al. (2004). Critical role for Kit-mediated Src kinase but not PI 3-kinase signaling in pro T and pro B cell development. *The Journal of Experimental Medicine*, *199*, 867–878.

Asai, N., Fukuda, T., Wu, Z., Enomoto, A., Pachnis, V., Takahashi, M., et al. (2006). Targeted mutation of serine 697 in the Ret tyrosine kinase causes migration defect of enteric neural crest cells. *Development*, *133*, 4507–4516.

Beauvais, D. M., Burbach, B. J., & Rapraeger, A. C. (2004). The syndecan-1 ectodomain regulates alphavbeta3 integrin activity in human mammary carcinoma cells. *The Journal of Cell Biology*, *167*, 171–181.

Berrier, A. L., & Yamada, K. M. (2007). Cell-matrix adhesion. *Journal of Cellular Physiology*, *213*, 565–573.

Bill, H. M., Knudsen, B., Moores, S. L., Muthuswamy, S. K., Rao, V. R., Brugge, J. S., et al. (2004). Epidermal growth factor receptor-dependent regulation of integrin-mediated signaling and cell cycle entry in epithelial cells. *Molecular and Cellular Biology*, *24*, 8586–8599.

Bladt, F., Riethmacher, D., Isenmann, S., Aguzzi, A., & Birchmeier, C. (1995). Essential role for the c-met receptor in the migration of myogenic precursor cells into the limb bud. *Nature*, *376*, 768–771.

Blume-Jensen, P., Jiang, G., Hyman, R., Lee, K. F., O'Gorman, S., & Hunter, T. (2000). Kit/stem cell factor receptor-induced activation of phosphatidylinositol 3′-kinase is essential for male fertility. *Nature Genetics*, *24*, 157–162.

Borges, E., Jan, Y., & Ruoslahti, E. (2000). Platelet-derived growth factor receptor beta and vascular endothelial growth factor receptor 2 bind to the beta 3 integrin through its extracellular domain. *The Journal of Biological Chemistry*, *275*, 39867–39873.

Boscher, C., & Mege, R. M. (2008). Cadherin-11 interacts with the FGF receptor and induces neurite outgrowth through associated downstream signalling. *Cellular Signalling*, *20*, 1061–1072.

Brewer, J. R., Mazot, P., & Soriano, P. (2016). Genetic insights into the mechanisms of Fgf signaling. *Genes & Development*, *30*, 751–771.

Brewer, J. R., Molotkov, A., Mazot, P., Hoch, R. V., & Soriano, P. (2015). Fgfr1 regulates development through the combinatorial use of signaling proteins. *Genes & Development*, *29*, 1863–1874.

Bruggemann, Y., Karajannis, L. S., Stanoev, A., Stallaert, W., & Bastiaens, P. I. H. (2021). Growth factor-dependent ErbB vesicular dynamics couple receptor signaling to spatially and functionally distinct Erk pools. *Science Signaling*, *14*, eabd9943.

Camenisch, T. D., Koller, B. H., Earp, H. S., & Matsushima, G. K. (1999). A novel receptor tyrosine kinase, Mer, inhibits TNF-alpha production and lipopolysaccharide-induced endotoxic shock. *Journal of Immunology*, *162*, 3498–3503.

Carpenter, G., & Liao, H. J. (2013). Receptor tyrosine kinases in the nucleus. *Cold Spring Harbor Perspectives in Biology*, *5*, a008979.

Cascone, I., Napione, L., Maniero, F., Serini, G., & Bussolino, F. (2005). Stable interaction between alpha5beta1 integrin and Tie2 tyrosine kinase receptor regulates endothelial cell response to Ang-1. *The Journal of Cell Biology*, *170*, 993–1004.

Chan, R., Hardy, W. R., Dankort, D., Laing, M. A., & Muller, W. J. (2004). Modulation of Erbb2 signaling during development: A threshold level of Erbb2 signaling is required for development. *Development*, *131*, 5551–5560.

Chan, R., Hardy, W. R., Laing, M. A., Hardy, S. E., & Muller, W. J. (2002). The catalytic activity of the ErbB-2 receptor tyrosine kinase is essential for embryonic development. *Molecular and Cellular Biology*, *22*, 1073–1078.

Chen, M. K., Hsu, J. L., & Hung, M. C. (2020). Nuclear receptor tyrosine kinase transport and functions in cancer. *Advances in Cancer Research*, *147*, 59–107.

Chen, M. K., & Hung, M. C. (2015). Proteolytic cleavage, trafficking, and functions of nuclear receptor tyrosine kinases. *The FEBS Journal*, *282*, 3693–3721.

Ciruna, B., & Rossant, J. (2001). FGF signaling regulates mesoderm cell fate specification and morphogenetic movement at the primitive streak. *Developmental Cell*, *1*, 37–49.

Clark, J. F., Dinsmore, C. J., & Soriano, P. (2020). A most formidable arsenal: Genetic technologies for building a better mouse. *Genes & Development*, *34*, 1256–1286.

Colvin, J. S., Bohne, B. A., Harding, G. W., McEwen, D. G., & Ornitz, D. M. (1996). Skeletal overgrowth and deafness in mice lacking fibroblast growth factor receptor 3. *Nature Genetics*, *12*, 390–397.

Davy, A., & Soriano, P. (2005). Ephrin signaling in vivo: Look both ways. *Developmental Dynamics*, *232*, 1–10.

Ekman, S., Thuresson, E. R., Heldin, C. H., & Ronnstrand, L. (1999). Increased mitogenicity of an alphabeta heterodimeric PDGF receptor complex correlates with lack of RasGAP binding. *Oncogene*, *18*, 2481–2488.

Fambrough, D., McClure, K., Kazlauskas, A., & Lander, E. S. (1999). Diverse signaling pathways activated by growth factor receptors induce broadly overlapping, rather than independent, sets of genes. *Cell, 97*, 727–741.

Fantauzzo, K. A., & Soriano, P. (2016). PDGFRbeta regulates craniofacial development through homodimers and functional heterodimers with PDGFRalpha. *Genes & Development, 30*, 2443–2458.

Feldheim, D. A., Nakamoto, M., Osterfield, M., Gale, N. W., DeChiara, T. M., Rohatgi, R., et al. (2004). Loss-of-function analysis of EphA receptors in retinotectal mapping. *The Journal of Neuroscience, 24*, 2542–2550.

Fortini, M. E., Simon, M. A., & Rubin, G. M. (1992). Signalling by the sevenless protein tyrosine kinase is mimicked by Ras1 activation. *Nature, 355*, 559–561.

Francavilla, C., Loeffler, S., Piccini, D., Kren, A., Christofori, G., & Cavallaro, U. (2007). Neural cell adhesion molecule regulates the cellular response to fibroblast growth factor. *Journal of Cell Science, 120*, 4388–4394.

Francavilla, C., Rigbolt, K. T., Emdal, K. B., Carraro, G., Vernet, E., Bekker-Jensen, D. B., et al. (2013). Functional proteomics defines the molecular switch underlying FGF receptor trafficking and cellular outputs. *Molecular Cell, 51*, 707–722.

Garcia, M. A., Nelson, W. J., & Chavez, N. (2018). Cell-cell junctions organize structural and signaling networks. *Cold Spring Harbor Perspectives in Biology, 10*, a029181.

Gonzalez, D. M., & Medici, D. (2014). Signaling mechanisms of the epithelial-mesenchymal transition. *Science Signaling, 7*, re8.

Hancock, M. L., Meyer, R. C., Mistry, M., Khetani, R. S., Wagschal, A., Shin, T., & Flanagan, J. G. (2019). Insulin receptor associates with promoters genome-wide and regulates gene expression. *Cell, 177*, 722–736.

Hanks, S. K., Quinn, A. M., & Hunter, T. (1988). The protein kinase family: Conserved features and deduced phylogeny of the catalytic domains. *Science, 241*, 42–52.

Heldin, C. H., & Lennartsson, J. (2013). Structural and functional properties of platelet-derived growth factor and stem cell factor receptors. *Cold Spring Harbor Perspectives in Biology, 5*, a009100.

Henkemeyer, M., Orioli, D., Henderson, J. T., Saxton, T. M., Roder, J., Pawson, T., & Klein, R. (1996). Nuk controls pathfinding of commissural axons in the mammalian central nervous system. *Cell, 86*, 35–46.

Heuchel, R., Berg, A., Tallquist, M., Ahlen, K., Reed, R. K., Rubin, K., et al. (1999). Platelet-derived growth factor beta receptor regulates interstitial fluid homeostasis through phosphatidylinositol-3′ kinase signaling. *Proceedings of the National Academy of Sciences of the United States of America, 96*, 11410–11415.

Hiratsuka, S., Minowa, O., Kuno, J., Noda, T., & Shibuya, M. (1998). Flt-1 lacking the tyrosine kinase domain is sufficient for normal development and angiogenesis in mice. *Proceedings of the National Academy of Sciences of the United States of America, 95*, 9349–9354.

Hoch, R. V., & Soriano, P. (2006). Context-specific requirements for Fgfr1 signaling through Frs2 and Frs3 during mouse development. *Development, 133*, 663–673.

Huang, B., Bates, M., & Zhuang, X. (2009). Super-resolution fluorescence microscopy. *Annual Review of Biochemistry, 78*, 993–1016.

Jain, S., Encinas, M., Johnson, E. M., Jr., & Milbrandt, J. (2006). Critical and distinct roles for key RET tyrosine docking sites in renal development. *Genes & Development, 20*, 321–333.

Jijiwa, M., Fukuda, T., Kawai, K., Nakamura, A., Kurokawa, K., Murakumo, Y., et al. (2004). A targeting mutation of tyrosine 1062 in Ret causes a marked decrease of enteric neurons and renal hypoplasia. *Molecular and Cellular Biology, 24*, 8026–8036.

Joly, M., Kazlauskas, A., Fay, F. S., & Corvera, S. (1994). Disruption of PDGF receptor trafficking by mutation of its PI-3 kinase binding sites. *Science, 263*, 684–687.

Kang, M., Garg, V., & Hadjantonakis, A. K. (2017). Lineage establishment and progression within the inner cell mass of the mouse blastocyst requires FGFR1 and FGFR2. *Developmental Cell, 41*, 496–510.

Kim, Y., Bingham, N., Sekido, R., Parker, K. L., Lovell-Badge, R., & Capel, B. (2007). Fibroblast growth factor receptor 2 regulates proliferation and Sertoli differentiation during male sex determination. *Proceedings of the National Academy of Sciences of the United States of America*, *104*, 16558–16563.

Kimura, Y., Jones, N., Kluppel, M., Hirashima, M., Tachibana, K., Cohn, J. B., et al. (2004). Targeted mutations of the juxtamembrane tyrosines in the Kit receptor tyrosine kinase selectively affect multiple cell lineages. *Proceedings of the National Academy of Sciences of the United States of America*, *101*, 6015–6020.

Kissel, H., Timokhina, I., Hardy, M. P., Rothschild, G., Tajima, Y., Soares, V., et al. (2000). Point mutation in kit receptor tyrosine kinase reveals essential roles for kit signaling in spermatogenesis and oogenesis without affecting other kit responses. *The EMBO Journal*, *19*, 1312–1326.

Klein, R., Silos-Santiago, I., Smeyne, R. J., Lira, S. A., Brambilla, R., Bryant, S., et al. (1994). Disruption of the neurotrophin-3 receptor gene trkC eliminates la muscle afferents and results in abnormal movements. *Nature*, *368*, 249–251.

Klinghoffer, R. A., Hamilton, T. G., Hoch, R., & Soriano, P. (2002). An allelic series at the PDGFalphaR locus indicates unequal contributions of distinct signaling pathways during development. *Developmental Cell*, *2*, 103–113.

Krishnamurthy, V. V., Fu, J., Oh, T. J., Khamo, J., Yang, J., & Zhang, K. (2020). A generalizable optogenetic strategy to regulate receptor tyrosine kinases during vertebrate embryonic development. *Journal of Molecular Biology*, *432*, 3149–3158.

Kullander, K., Mather, N. K., Diella, F., Dottori, M., Boyd, A. W., & Klein, R. (2001). Kinase-dependent and kinase-independent functions of EphA4 receptors in major axon tract formation in vivo. *Neuron*, *29*, 73–84.

Kundra, V., Escobedo, J. A., Kazlauskas, A., Kim, H. K., Rhee, S. G., Williams, L. T., et al. (1994). Regulation of chemotaxis by the platelet-derived growth factor receptor-beta. *Nature*, *367*, 474–476.

Kurowski, A., Molotkov, A., & Soriano, P. (2019). FGFR1 regulates trophectoderm development and facilitates blastocyst implantation. *Developmental Biology*, *446*, 94–101.

Leckband, D. E., & de Rooij, J. (2014). Cadherin adhesion and mechanotransduction. *Annual Review of Cell and Developmental Biology*, *30*, 291–315.

Lemmon, M. A., & Schlessinger, J. (2010). Cell signaling by receptor tyrosine kinases. *Cell*, *141*, 1117–1134.

Li, J. P., & Kusche-Gullberg, M. (2016). Heparan sulfate: Biosynthesis, structure, and function. *International Review of Cell and Molecular Biology*, *325*, 215–273.

Li, X., Padhan, N., Sjostrom, E. O., Roche, F. P., Testini, C., Honkura, N., et al. (2016). VEGFR2 pY949 signalling regulates adherens junction integrity and metastatic spread. *Nature Communications*, *7*, 11017.

Liebl, D. J., Tessarollo, L., Palko, M. E., & Parada, L. F. (1997). Absence of sensory neurons before target innervation in brain-derived neurotrophic factor-, neurotrophin 3-, and TrkC-deficient embryonic mice. *The Journal of Neuroscience*, *17*, 9113–9121.

Lu, P., Ewald, A. J., Martin, G. R., & Werb, Z. (2008). Genetic mosaic analysis reveals FGF receptor 2 function in terminal end buds during mammary gland branching morphogenesis. *Developmental Biology*, *321*, 77–87.

Maina, F., Casagranda, F., Audero, E., Simeone, A., Comoglio, P. M., Klein, R., et al. (1996). Uncoupling of Grb2 from the Met receptor in vivo reveals complex roles in muscle development. *Cell*, *87*, 531–542.

Maina, F., Pante, G., Helmbacher, F., Andres, R., Porthin, A., Davies, A. M., et al. (2001). Coupling Met to specific pathways results in distinct developmental outcomes. *Molecular Cell*, *7*, 1293–1306.

Marshall, C. J. (1994). MAP kinase kinase kinase, MAP kinase kinase and MAP kinase. *Current Opinion in Genetics & Development*, *4*, 82–89.

Masson-Gadais, B., Houle, F., Laferriere, J., & Huot, J. (2003). Integrin alphavbeta3, requirement for VEGFR2-mediated activation of SAPK2/p38 and for Hsp90-dependent phosphorylation of focal adhesion kinase in endothelial cells activated by VEGF. *Cell Stress & Chaperones*, *8*, 37–52.

McQuade, K. J., Beauvais, D. M., Burbach, B. J., & Rapraeger, A. C. (2006). Syndecan-1 regulates alphavbeta5 integrin activity in B82L fibroblasts. *Journal of Cell Science*, *119*, 2445–2456.

Medina, D. L., Sciarretta, C., Calella, A. M., Von Bohlen Und Halbach, O., Unsicker, K., & Minichiello, L. (2004). TrkB regulates neocortex formation through the Shc/PLCgamma-mediated control of neuronal migration. *The EMBO Journal*, *23*, 3803–3814.

Minichiello, L., Calella, A. M., Medina, D. L., Bonhoeffer, T., Klein, R., & Korte, M. (2002). Mechanism of TrkB-mediated hippocampal long-term potentiation. *Neuron*, *36*, 121–137.

Minichiello, L., Casagranda, F., Tatche, R. S., Stucky, C. L., Postigo, A., Lewin, G. R., et al. (1998). Point mutation in trkB causes loss of NT4-dependent neurons without major effects on diverse BDNF responses. *Neuron*, *21*, 335–345.

Mishra, Y. G., & Manavathi, B. (2021). Focal adhesion dynamics in cellular function and disease. *Cellular Signalling*, *85*, 110046.

Molotkov, A., Mazot, P., Brewer, J. R., Cinalli, R. M., & Soriano, P. (2017). Distinct requirements for FGFR1 and FGFR2 in primitive endoderm development and exit from pluripotency. *Developmental Cell*, *41*(5), 511–526.

Neben, C. L., Idoni, B., Salva, J. E., Tuzon, C. T., Rice, J. C., Krakow, D., et al. (2014). Bent bone dysplasia syndrome reveals nucleolar activity for FGFR2 in ribosomal DNA transcription. *Human Molecular Genetics*, *23*, 5659–5671.

Neben, C. L., Tuzon, C. T., Mao, X., Lay, F. D., & Merrill, A. E. (2017). FGFR2 mutations in bent bone dysplasia syndrome activate nucleolar stress and perturb cell fate determination. *Human Molecular Genetics*, *26*, 3253–3270.

Nguyen, T., Duchesne, L., Sankara Narayana, G. H. N., Boggetto, N., Fernig, D. D., Uttamrao Murade, C., et al. (2019). Enhanced cell-cell contact stability and decreased N-cadherin-mediated migration upon fibroblast growth factor receptor-N-cadherin cross talk. *Oncogene*, *38*, 6283–6300.

Ornitz, D. M., & Itoh, N. (2015). The fibroblast growth factor signaling pathway. *Wiley Interdisciplinary Reviews: Developmental Biology*, *4*, 215–266.

Postigo, A., Calella, A. M., Fritzsch, B., Knipper, M., Katz, D., Eilers, A., et al. (2002). Distinct requirements for TrkB and TrkC signaling in target innervation by sensory neurons. *Genes & Development*, *16*, 633–645.

Raabe, T. (2000). The sevenless signaling pathway: Variations of a common theme. *Biochimica et Biophysica Acta*, *1496*, 151–163.

Rapraeger, A. C., Krufka, A., & Olwin, B. B. (1991). Requirement of heparan sulfate for bFGF-mediated fibroblast growth and myoblast differentiation. *Science*, *252*, 1705–1708.

Rasouli, S. J., El-Brolosy, M., Tsedeke, A. T., Bensimon-Brito, A., Ghanbari, P., Maischein, H. M., et al. (2018). The flow responsive transcription factor Klf2 is required for myocardial wall integrity by modulating Fgf signaling. *eLife*, 7, e38889.

Ray, A. T., Mazot, P., Brewer, J. R., Catela, C., Dinsmore, C. J., & Soriano, P. (2020). FGF signaling regulates development by processes beyond canonical pathways. *Genes & Development*, *34*, 1735–1752.

Rosenkranz, S., DeMali, K. A., Gelderloos, J. A., Bazenet, C., & Kazlauskas, A. (1999). Identification of the receptor-associated signaling enzymes that are required for platelet-derived growth factor-AA-dependent chemotaxis and DNA synthesis. *The Journal of Biological Chemistry*, *274*, 28335–28343.

Sadowski, L., Pilecka, I., & Miaczynska, M. (2009). Signaling from endosomes: Location makes a difference. *Experimental Cell Research*, *315*, 1601–1609.

Sakurai, Y., Ohgimoto, K., Kataoka, Y., Yoshida, N., & Shibuya, M. (2005). Essential role of Flk-1 (VEGF receptor 2) tyrosine residue 1173 in vasculogenesis in mice. *Proceedings of the National Academy of Sciences of the United States of America*, *102*, 1076–1081.

Sanchez-Heras, E., Howell, F. V., Williams, G., & Doherty, P. (2006). The fibroblast growth factor receptor acid box is essential for interactions with N-cadherin and all of the major isoforms of neural cell adhesion molecule. *The Journal of Biological Chemistry*, *281*, 35208–35216.

Schlessinger, J. (2014). Receptor tyrosine kinases: Legacy of the first two decades. *Cold Spring Harbor Perspectives in Biology*, *6*, a008912.

Schmahl, J., Kim, Y., Colvin, J. S., Ornitz, D. M., & Capel, B. (2004). Fgf9 induces proliferation and nuclear localization of FGFR2 in Sertoli precursors during male sex determination. *Development*, *131*, 3627–3636.

Schmidt, C., Bladt, F., Goedecke, S., Brinkmann, V., Zschiesche, W., Sharpe, M., et al. (1995). Scatter factor/hepatocyte growth factor is essential for liver development. *Nature*, *373*, 699–702.

Schuchardt, A., D'Agati, V., Larsson-Blomberg, L., Costantini, F., & Pachnis, V. (1994). Defects in the kidney and enteric nervous system of mice lacking the tyrosine kinase receptor Ret. *Nature*, *367*, 380–383.

Sieben, C., Douglass, K. M., Guichard, P., & Manley, S. (2018). Super-resolution microscopy to decipher multi-molecular assemblies. *Current Opinion in Structural Biology*, *49*, 169–176.

Simon, M. A., Dodson, G. S., & Rubin, G. M. (1993). An SH3-SH2-SH3 protein is required for p21Ras1 activation and binds to sevenless and Sos proteins in vitro. *Cell*, *73*, 169–177.

Soriano, P. (1994). Abnormal kidney development and hematological disorders in PDGF beta-receptor mutant mice. *Genes & Development*, *8*, 1888–1896.

Soriano, P. (1997). The PDGF alpha receptor is required for neural crest cell development and for normal patterning of the somites. *Development*, *124*, 2691–2700.

Stallaert, W., Bruggemann, Y., Sabet, O., Baak, L., Gattiglio, M., & Bastiaens, P. I. H. (2018). Contact inhibitory Eph signaling suppresses EGF-promoted cell migration by decoupling EGFR activity from vesicular recycling. *Science Signaling*, *11*, eaat0114.

Steinberg, Z., Myers, C., Heim, V. M., Lathrop, C. A., Rebustini, I. T., Stewart, J. S., et al. (2005). FGFR2b signaling regulates ex vivo submandibular gland epithelial cell proliferation and branching morphogenesis. *Development*, *132*, 1223–1234.

Sun, J., & Stathopoulos, A. (2018). FGF controls epithelial-mesenchymal transitions during gastrulation by regulating cell division and apicobasal polarity. *Development*, *145*, dev161927.

Suyama, K., Shapiro, I., Guttman, M., & Hazan, R. B. (2002). A signaling pathway leading to metastasis is controlled by N-cadherin and the FGF receptor. *Cancer Cell*, *2*, 301–314.

Tachibana, K., Jones, N., Dumont, D. J., Puri, M. C., & Bernstein, A. (2005). Selective role of a distinct tyrosine residue on Tie2 in heart development and early hematopoiesis. *Molecular and Cellular Biology*, *25*, 4693–4702.

Tallquist, M. D., French, W. J., & Soriano, P. (2003). Additive effects of PDGF receptor beta signaling pathways in vascular smooth muscle cell development. *PLoS Biology*, *1*, 288–299.

Tallquist, M. D., Klinghoffer, R. A., Heuchel, R., Mueting-Nelsen, P. F., Corrin, P. D., Heldin, C. H., et al. (2000). Retention of PDGFR-beta function in mice in the absence of phosphatidylinositol 3′-kinase and phospholipase Cgamma signaling pathways. *Genes & Development*, *14*, 3179–3190.

Thapa, N., Chen, M., Horn, H. T., Choi, S., Wen, T., & Anderson, R. A. (2020). Phosphatidylinositol-3-OH kinase signalling is spatially organized at endosomal compartments by microtubule-associated protein 4. *Nature Cell Biology*, *22*, 1357–1370.

Theveneau, E., & Mayor, R. (2012). Neural crest delamination and migration: From epithelium-to-mesenchyme transition to collective cell migration. *Developmental Biology*, *366*, 34–54.

Treisman, R. (1994). Ternary complex factors: Growth factor regulated transcriptional activators. *Current Opinion in Genetics & Development*, *4*, 96–101.

Tuzon, C. T., Rigueur, D., & Merrill, A. E. (2019). Nuclear fibroblast growth factor receptor signaling in skeletal development and disease. *Current Osteoporosis Reports*, *17*, 138–146.

Uehara, Y., Minowa, O., Mori, C., Shiota, K., Kuno, J., Noda, T., et al. (1995). Placental defect and embryonic lethality in mice lacking hepatocyte growth factor/scatter factor. *Nature*, *373*, 702–705.

Utton, M. A., Eickholt, B., Howell, F. V., Wallis, J., & Doherty, P. (2001). Soluble N-cadherin stimulates fibroblast growth factor receptor dependent neurite outgrowth and N-cadherin and the fibroblast growth factor receptor co-cluster in cells. *Journal of Neurochemistry*, *76*, 1421–1430.

Valius, M., & Kazlauskas, A. (1993). Phospholipase C-gamma 1 and phosphatidylinositol 3 kinase are the downstream mediators of the PDGF receptor's mitogenic signal. *Cell*, *73*, 321–334.

Veevers-Lowe, J., Ball, S. G., Shuttleworth, A., & Kielty, C. M. (2011). Mesenchymal stem cell migration is regulated by fibronectin through alpha5beta1-integrin-mediated activation of PDGFR-beta and potentiation of growth factor signals. *Journal of Cell Science*, *124*, 1288–1300.

Waltz, S. E., Eaton, L., Toney-Earley, K., Hess, K. A., Peace, B. E., Ihlendorf, J. R., & Degen, S. J. (2001). Ron-mediated cytoplasmic signaling is dispensable for viability but is required to limit inflammatory responses. *The Journal of Clinical Investigation*, *108*, 567–576.

Wan, Y., McDole, K., & Keller, P. J. (2019). Light-sheet microscopy and its potential for understanding developmental processes. *Annual Review of Cell and Developmental Biology*, *35*, 655–681.

Wang, S., Matsumoto, K., Lish, S. R., Cartagena-Rivera, A. X., & Yamada, K. M. (2021). Budding epithelial morphogenesis driven by cell-matrix versus cell-cell adhesion. *Cell*, *184*, 3702–3716.

Weinstein, M., Xu, X., Ohyama, K., & Deng, C. X. (1998). FGFR-3 and FGFR-4 function cooperatively to direct alveogenesis in the murine lung. *Development*, *125*, 3615–3623.

Williams, E. J., Furness, J., Walsh, F. S., & Doherty, P. (1994). Activation of the FGF receptor underlies neurite outgrowth stimulated by L1, N-CAM, and N-cadherin. *Neuron*, *13*, 583–594.

Wong, A., Bogni, S., Kotka, P., de Graaff, E., D'Agati, V., Costantini, F., et al. (2005). Phosphotyrosine 1062 is critical for the in vivo activity of the Ret9 receptor tyrosine kinase isoform. *Molecular and Cellular Biology*, *25*, 9661–9673.

Yayon, A., Klagsbrun, M., Esko, J. D., Leder, P., & Ornitz, D. M. (1991). Cell surface, heparin-like molecules are required for binding of basic fibroblast growth factor to its high affinity receptor. *Cell*, *64*, 841–848.

Zhang, L., Zhou, F., Han, W., Shen, B., Luo, J., Shibuya, M., et al. (2010). VEGFR-3 ligand-binding and kinase activity are required for lymphangiogenesis but not for angiogenesis. *Cell Research*, *20*, 1319–1331.

CHAPTER FIVE

Cellular and molecular mechanisms of EPH/EPHRIN signaling in evolution and development

Jeffrey O. Bush[a,b,c,d,*]
[a]Department of Cell and Tissue Biology, University of California San Francisco, San Francisco, CA, United States
[b]Program in Craniofacial Biology, University of California San Francisco, San Francisco, CA, United States
[c]Institute for Human Genetics, University of California San Francisco, San Francisco, CA, United States
[d]Eli and Edythe Broad Center of Regeneration Medicine and Stem Cell Research, University of California San Francisco, San Francisco, CA, United States
*Corresponding author: e-mail address: Jeffrey.bush@ucsf.edu

Contents

Current Topics in Developmental Biology, Volume 149
ISSN 0070-2153
https://doi.org/10.1016/bs.ctdb.2022.02.005

Abstract

The EPH receptor tyrosine kinases and their signaling partners, the EPHRINS, comprise a large class of cell signaling molecules that plays diverse roles in development. As cell membrane-anchored signaling molecules, they regulate cellular organization by modulating the strength of cellular contacts, usually by impacting the actin cytoskeleton or cell adhesion programs. Through these cellular functions, EPH/EPHRIN signaling often regulates tissue shape. Indeed, recent evidence indicates that this signaling family is ancient and associated with the origin of multicellularity. Though extensively studied, our understanding of the signaling mechanisms employed by this large family of signaling proteins remains patchwork, and a truly "canonical" EPH/EPHRIN signal transduction pathway is not known and may not exist. Instead, several foundational evolutionarily conserved mechanisms are overlaid by a myriad of tissue -specific functions, though common themes emerge from these as well. Here, I review recent advances and the related contexts that have provided new understanding of the conserved and varied molecular and cellular mechanisms employed by EPH/EPHRIN signaling during development.

1. Introduction

The EPH/EPHRIN signaling family is the largest receptor tyrosine kinase system in the vertebrate genome and has myriad roles in development, bone homeostasis, the immune system, and cancer (Arthur & Gronthos, 2021; Darling & Lamb, 2019; Kania & Klein, 2016; Pasquale, 2010). The most conserved and reiterated functions of EPH/EPHRIN signaling in development relate to guidance of axon or cell movements, and to the establishment of boundaries for tissue separation, but it is clear that EPH/EPHRIN signaling can also regulate cell proliferation, cell fate specification, apoptosis, cytokinesis, and mitotic spindle orientation in various contexts (Kania & Klein, 2016). It should not be surprising that such a large signaling family, highly elaborated through gene duplication in vertebrates, would have acquired widely-varied cellular functions during development. Further, the diversity of cellular function also comes into play in a number of different cancers where deregulation of EPH/EPHRIN signaling has been implicated in growth, migration and invasion of tumor cells, as well as in tumor angiogenesis and in immune response to cancer, and for these reasons have been considered as attractive therapeutic targets (Barquilla & Pasquale, 2015; Janes, Vail, Ernst, & Scott, 2021; Nievergall, Lackmann, & Janes, 2012; Noberini, Lamberto, & Pasquale, 2012; Noberini & Pasquale, 2009; Pasquale, 2010). It is clear that many of these functions reflect the developmental roles of EPH/EPHRIN signaling, but our understanding of the basic molecular and cellular mechanisms employed remains incomplete. In this chapter, I will review and highlight recent discoveries that reveal both conserved and divergent functions for EPH/EPHRIN signaling in development across evolutionary time and across distinct tissue contexts during development. I will focus on the cellular and molecular mechanisms by which EPH/EPHRIN signaling acts in development, and particularly look for opportunities to examine *in vivo* evidence in whole organisms supporting mechanisms proposed in cell culture.

1.1 Structure and specificity

The EPH receptors are the largest family of receptor tyrosine kinases and comprise EPHA and EPHB subfamilies, which are defined based on sequence similarity and on their binding preference to the two different types of EPHRIN signaling partner thus named EPHRIN-A and EPHRIN-B (Fig. 1A). Both classes of EPHRINs are cell membrane-bound and are defined by sequence homology as well as by their mode of membrane attachment: the B-type EPHRINs have a highly conserved transmembrane domain while the A-type EPHRINs are tethered to the membrane by a GPI-anchor. The intracellular domain of B-type EPHRINs includes several conserved tyrosines and a carboxy-terminal PSD95/Dlg/ZO1 (PDZ)-binding motif. The human genome encodes five EPHB receptors and nine EPHA receptors, including two receptors, EPHB6 and EPHA10, which lack intrinsic kinase catalytic activity. The domain structure of EPH receptors is conserved between subfamilies and includes an extracellular globular ligand binding domain, a cysteine rich region that includes sushi and epidermal growth factor (EGF)-like domains, two fibronectin type III (FN) domains, the transmembrane helix, a juxtamembrane region, a tyrosine kinase domain, a sterile α motif (SAM) domain, and PDZ-binding motif (Fig. 1A) (Himanen & Nikolov, 2003).

The human genome encodes five EPHB receptors, nine EPHA receptors, three EPHRINs-B and five EPHRINs-A. For the most part, EPH/EPHRIN interactions are promiscuous within sub-class, specificity likely enabled by the presence of a "class-specificity loop" of distinct sizes in the extracellular domain of the EPHA and EPHB receptors (Himanen, Henkemeyer, & Nikolov, 1998). Exceptions to sub-class specificity include EPHB2, which can bind to EPHRIN-A5, and EPHA4, which can bind to EPHRIN-B2 and EPHRIN-B3 (Gale et al., 1996; Himanen et al., 2004). Crystal structure studies have revealed that EPHA4 has conformational

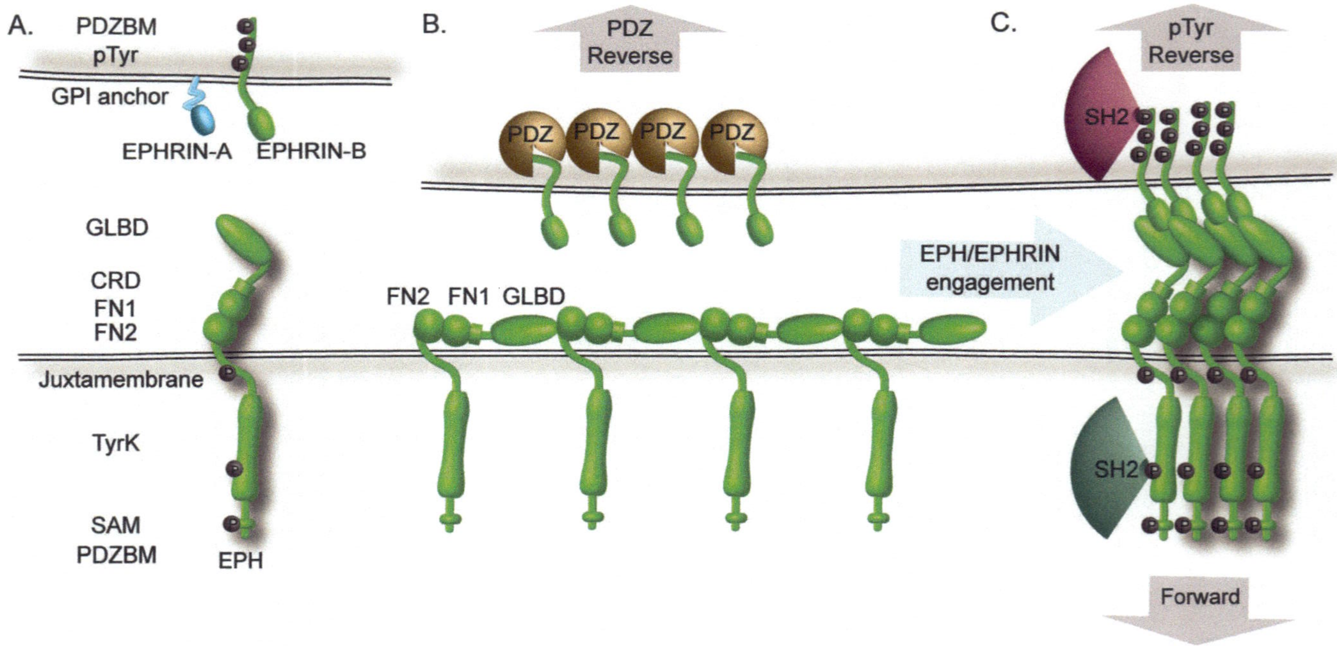

Fig. 1 Schematic representation of EPH/EPHRIN structure and activation. (A) Schematic depiction of the domain structure of EPHRIN-A, EPHRIN-B, and EPHB proteins (EPHA domain structure is similar). (B) Prior to EPH/EPHRIN signaling activation, EPH receptors make homotypic contacts between their GLBDs and FN2 domains. EPHRIN-B signaling partners can bind to PDZ-domain containing proteins independently of EPHB binding. (C) EPH/EPHRIN activation begins with heterodimerization followed by receptor autophosphorylation and recruitment of oligomeric signaling clusters which mediate maximal signaling. Binding of the EPH to EPHRIN-B signaling partners results in a conformational change that allows accessibility of the EPHRIN-B intracellular domain to a variety of tyrosine kinases that mediate its intracellular phosphorylation. SH2 domain-containing proteins bind to phosphorylated tyrosines on both the EPHRIN-B (mediating phosphorylation-dependent reverse signaling) and on the EPHB (mediating kinase-dependent forward signaling). Other modes of signaling are reviewed in (Kania & Klein, 2016). PDZBM, PSD95/Dlg/ZO1 (PDZ)-binding motif; pTyr, phospho-tyrosines; GPI, Glycosylphosphatidylinosil; GLBD, globular ligand binding domain; CRD, cysteine-rich domain; FN, Fibronectin type III domains; TyrK, tyrosine kinase; SAM, sterile α-motif.

plasticity in its ligand binding interface, such that when binding EPHRIN-A2, it resembles other A-class receptors, whereas when it binds EPHRIN-B2 or EPHRIN-B3 it takes on structural hallmarks of B-class receptors (Bowden et al., 2009; Guo & Lesk, 2014). EPHRIN-B1 binding to EPHA4 is not observed, likely because of a steric clash between a large tyrosine residue present at position 121 compared with a relatively small leucine found at the comparable position in EPHRIN-B2 and EPHRIN-B3 (Guo & Lesk, 2014). Interestingly, this amino acid difference enabling cross-class binding of these two EPHRINs with EPHA4 only exists in gnathostomes, and it is intriguing to consider the diversification of signaling functions that was enabled by this difference between otherwise well-conserved B-type EPHRINs (Arcas, Wilkinson, & Nieto, 2020).

In a signaling-off state, EPHs and EPHRINs are loosely homotypically clustered at cell membranes. Both un-liganded EPHA4 and EPHB2 can form head-to-tail interactions between the ligand binding and fibronectin domains of the extracellular domain that are autoinhibitory in nature (Fig. 1B) (Xu et al., 2021, 2013). Signaling interaction initiates through formation of a high-affinity EPH/EPHRIN heterodimer followed by the formation of higher-order oligomers and leading to full activation of kinase signaling (Fig. 1C) (Himanen et al., 2010; Ojosnegros et al., 2017; Seiradake, Harlos, Sutton, Aricescu, & Jones, 2010). A chemical genetic approach using a synthetic dimerizer with high affinity to the FK506 binding protein domain (FKBP) was used to drive the formation of EPHB2 receptor signaling complexes of varying sizes, and the strength of the resulting signal was measured by biochemical and cell biological assays, demonstrating that the degree of EPH receptor clustering determines the strength of kinase-dependent forward signaling and the strength of the cellular response (Schaupp et al., 2014). Unlike most receptor tyrosine kinases, which are activated upon dimerization, EPHB2 and EPHA4 were activated during the transition from dimers to trimers/tetramers. Smaller trimeric and tetrameric clusters were maximally effective in producing a cellular response, though increased activation was not observed upon driving formation of hexamers. Instead, graded responses in EPHB2 activation are achieved through the relative ratio between inactive clusters (monomers/dimers) and active multimers (trimers or greater). Surprisingly, this study also demonstrated that the PDZ binding motif and SAM domain negatively regulated EPHB2 oligomerization to throttle back EPHB2 signaling strength (Schaupp et al., 2014). It is worth noting, however, that crystal structures of EPHA2 in complex with EPHRIN-As indicated the ability to form large extended signaling arrays, whereas EPHA4 in complex with EPHRIN-B3 formed smaller clusters with a closed circular arrangement (Himanen et al., 2010; Seiradake et al., 2010, 2013). This clustered structure was determined by distinct ectodomain structure, suggesting that different EPH/EPHRIN signaling partners regulate cluster formation differently to achieve varying degrees of graded signaling responses. Further, EPH receptors can hetero-oligomerize both within and across sub-class, and, in cell culture, hetero-oligomerization can modulate signaling strength and cellular response (Freywald, Sharfe, & Roifman, 2002; Janes et al., 2011; Marquardt et al., 2005; Wimmer-Kleikamp, Janes, Squire, Bastiaens, & Lackmann, 2004). Hetero-oligomerization can occur through shared EPHRIN partnership, or independently of EPHRIN specificity through EPH-EPH interactions (Janes et al., 2011; Wimmer-Kleikamp et al., 2004). As multiple EPH/EPHRINs often exhibit overlapping expression within tissues during development, this property of EPH/EPHRIN signaling drastically increases possible signaling complexity beyond the already promiscuous EPH/EPHRIN signaling partnerships. Detailed mechanistic studies that manipulate the homo- and hetero-multimerization of EPH receptors *in vivo* will be required to more fully comprehend combinatorial EPH/EPHRIN signaling codes.

1.2 EPH/EPHRIN SH2-protein mediated signaling

Upon engagement of EPH receptor signaling, autophosphorylation of the juxtamembrane domain results in full kinase activation and phosphorylation of several intracellular tyrosines which act as binding sites for Src Homology 2 (SH2)-domain containing proteins, often resulting in their own tyrosine phosphorylation. Traditional biochemical approaches (Hock et al., 1998; Holland et al., 1997), together with mass spectrometry-based phosphoproteomic approaches, have revealed an extensive network of EPH receptor phosphorylation targets (Bush & Soriano, 2010, 2012; Jorgensen et al., 2009; Zhang, Fenyo, & Neubert, 2008; Zhang, Spellman, Skolnik, & Neubert, 2006). These include proteins with catalytic activity such as p120-RasGAP and α-chimaerin (Elowe, Holland, Kulkarni, & Pawson, 2001; Iwasato et al., 2007), as well as adaptors lacking catalytic activity such as NCK, SHC, and CRK (Hock et al., 1998). Compared with other receptor tyrosine kinases, understanding of the crucial proximal signal transduction mechanisms employed by EPH receptors has remained somewhat rudimentary. Recent work has confirmed and extended these findings by more detailed examination of the function of different SH2-domain containing proteins, mostly in cell culture, but in some cases, parallel phenotypes in genetic experiments shed light on relevant *in vivo* contexts (reviewed in Bush & Soriano, 2012).

The NCK SH2-adaptor proteins, NCK1 and NCK2, have long been known to interact with and be phosphorylated by EPH receptors, and genetic evidence supports their roles in mediating EPH receptor signaling-driven axon guidance functions in multiple contexts (Chang et al., 2018; Fawcett et al., 2007; Hock et al., 1998; Holland et al., 1997; Srivastava, Robichaux, Chenaux, Henkemeyer, & Cowan, 2013). Multiple biochemical functions for the NCK adaptor have been identified. NCK, together with p62DOK, forms a complex with p120-RasGAP, which negatively regulates RAS/mitogen activated protein kinase (MAPK) pathway activation (Elowe et al., 2001; Holland et al., 1997). NCK and p62DOK recruitment to EPHB1 and EPHB2 also mediates activation of the NCK interacting kinase (NIK) and modulates integrin-mediated attachment and activation of c-Jun Kinase (JNK), which may contribute to cytoskeletal regulation downstream of EPH/EPHRIN signaling (Becker et al., 2000; Stein, Huynh-Do, Lane, Cerretti, & Daniel, 1998). Indeed, NCK appears to facilitate EPH/EPHRIN actomyosin cytoskeletal regulation in multiple contexts. It is required for retraction of cellular protrusions and cell migration activated through EPHA3 (Hu et al., 2009). In *Xenopus,* EphA4 activation and phosphorylation provides sites for NCK2 binding and results in recruitment of p21 activated kinase (Pak1) which regulates levels of Rac, RhoA and Cdc42 GTPases and controls cell contact strength (Bisson, Poitras, Mikryukov, Tremblay, & Moss, 2007). A recent study has revealed that termination of EPH receptor tyrosine kinase downstream signaling depends on the phosphorylation of a conserved tyrosine in the SH3 domain of NCK, which prevents its association with multiple protein targets, including PAK1, and results in dismantling of the signaling network. This mechanism may therefore enable a rapid and reversible mechanism for EPH/EPHRIN transduction which is sufficiently dynamic to mediate rapid changes in the actomyosin cytoskeleton and its effects on cell shape (Dionne et al., 2018).

The SH2 domain-containing scaffold SHB was found to be highly phosphorylated by EPHB2 in HEK293 cells and in secondary palate mesenchyme cells, and appeared to have functional significance based on shRNA knockdown in an EPH/EPHRIN cell sorting assay (Bush & Soriano, 2010; Jorgensen et al., 2009). Recently, detailed examination by co-immunoprecipitation and mass spectrometry revealed that SHB scaffolds the formation of a multi-SH2-domain protein complex including NCK1/2, p120-RasGAP, and α- and β-chimaerin RacGAP proteins that depended on EPHB2 kinase activation (Wagner et al., 2020). Phosphorylation of distinct tyrosine residues on SHB enabled binding of each of these proteins, and mutation of these tyrosines to phenylalanine reduced EPH/EPHRIN cell sorting. Interestingly, interaction of SHB with EPHB2 was relatively weak, suggesting that it may be transient or indirect. Nevertheless, formation of this complex was also observed upon stimulation of signaling through EPHA4, EPHA8, or EPHB4, suggesting that multiple EPH receptors may utilize the SHB scaffold for simultaneous multi-protein interactions. Though some *Shb*$^{-/-}$ knockout mouse embryos exhibited tantalizingly parallel vascular or neural tube defects, most *Shb*$^{-/-}$ null mice were viable and fertile, suggesting partial compensation for SHB adapter function, for example, by related scaffolds SHD, SHE, or SHF (Kriz et al., 2007; Wagner et al., 2020).

The SH2-containing phosphoinositide phosphatase, SHIP2, is a conserved target of EPH receptor phosphorylation with functional significance in the regulation of cell segregation (Ashlin, Wu, Xu, & Wilkinson, 2019; Jorgensen et al., 2009). Upon EPHRIN-A1 stimulation, SHIP2 binds to the EPHA2 SAM domain and inhibits EPHA2 endocytosis, independent of EPHA2 kinase activity (Zhuang, Hunter, Hwang, & Chen, 2007). Interestingly, SHIP2 does not bind to EPHA3, EPHA4, or EPHB1, but does interact with activated EPHB2, albeit via its SH2 domain rather than its SAM domain (Ashlin et al., 2019). Further, SHIP2 does not appear to regulate EPHB2 endocytosis, but rather regulates directional migration of EPHB2 cells away from EPHRIN-B1 cells, possibly through effects on the actin cytoskeleton. Indeed, previous studies have shown that SHIP2 has an important role in cell polarization and migration through restriction of PI(3,4,5)P_3 localization at the leading edge of migrating cells and through interaction with RhoA, which also can mediate actomyosin contractility mediated effects on EPH/EPHRIN cell-cell contacts (Kato et al., 2012; Kindberg et al., 2021).

The related CRK, CRKII and CRK-L SH2 domain-containing adapter proteins bind to several activated EPH receptors, and CRKII has been shown to mediate EPHA3 signaling effects on cell adhesion and Rho family GTPase signaling in cell culture (Lawrenson et al., 2002). In human breast cancer cells, CRK is tyrosine phosphorylated upon EPHB4 stimulation, mediated at least in part by ABL and ARG, non-receptor tyrosine kinases that can also be phosphorylated by EPH receptor signaling (Huang, Wu, Jin, Stupack, & Wang, 2008; Noren, Foos, Hauser, & Pasquale, 2006).

Curiously, whereas ABL-CRK signaling enables inhibition of cell proliferation and migration by EPHB4 in breast cancer cells, ABL also acts downstream of EPHB2 to promote tumor growth during intestinal adenoma formation (Genander et al., 2009; Noren et al., 2006). These distinct activities of ABL hint at the broader paradoxical roles of the EPH/EPHRIN family in cancer, which probably reflect the broad array of cellular behaviors that are regulated by EPH/EPHRIN signaling, many of which become dysregulated in cancer (Pasquale, 2010).

1.3 Reverse signaling mechanisms

The signaling complexity of this pathway is further multiplied by an additional unique signaling property: famously, there is a capacity for bidirectional signaling, wherein an EPHRIN-expressing cell signals to an EPH-expressing cell (forward signaling) with the simultaneous signaling of the EPH-expressing cell to an EPHRIN-expressing cell (reverse signaling). Bidirectional signaling has been examined in numerous *in vitro* and *in vivo* studies and occurs in A- and B-type subfamilies, albeit by different mechanisms. Upon hetero-tetramerization with EPHB receptors, the EPHRIN-B transmembrane and cytoplasmic domains are repositioned to allow their phosphorylation by Src-family kinases, as well as PDGF and FGF receptors (Fig. 1C) (Brückner, Pasquale, & Klein, 1997; Chong, Park, Latimer, Friesel, & Daar, 2000; Himanen & Nikolov, 2003; Palmer et al., 2002). Tyrosine phosphorylation of EPHRIN-B allows the binding of SH2 domain-containing proteins and can impact cytoskeletal remodeling in cell culture (Brückner et al., 1997; Cowan & Henkemeyer, 2001; Holland et al., 1997). The EPHRIN-B C-terminal PDZ binding motif has also been shown to bind to a handful of PDZ-proteins independent of EPH-receptor binding to transduce a reverse signal (Lin, Gish, Songyang, & Pawson, 1999; Torres et al., 1998). Both phosphorylation-dependent and PDZ-dependent modes of reverse signaling have been well-studied, but other less-studied mechanisms, including serine phosphorylation of EPHRIN-B2, also exist (Bush & Soriano, 2012; Daar, 2012; Davy & Soriano, 2005; Niethamer & Bush, 2019). We recently published an extensive review of genetic interrogation of forward and reverse signaling function of B-type EPHRINs in mice (Niethamer & Bush, 2019), which led us to speculate that some of the *in vivo* functions currently attributed to B-type reverse signaling may instead be due to impacts of EPHRIN-B intracellular mutations on forward signaling;

here, I will not dwell on this challenging question beyond noting whether forward or reverse signaling has been implicated.

A-type EPHRINs also engage in reverse signaling. For example, activation of EPHRIN-A5 reverse signaling results in β1 integrin-dependent FAK tyrosine phosphorylation and MAPK activation as well as changes in cell morphology and adhesion (Davy & Robbins, 2000). Indeed, several studies have indicated that EPHRIN-A reverse signaling may generally regulate integrin-mediated cell adhesion or cell migration in distinct cellular contexts (Daar, 2012). In addition to integrins, EPHRIN-A reverse signaling often occurs through interaction with other membrane associated signaling molecules such as TrkB, p75NTR, and the ADAM10 membrane metalloprotease, all of which impacts axon pathfinding in multiple contexts (Xu & Henkemeyer, 2012). One challenge inherent to understanding the roles of EPHRIN-A reverse signaling *in vivo* is that EPHRIN-As can also interact with EPHAs within the same cell, resulting in cis-inhibition/attenuation of EPHA forward signaling (Carvalho et al., 2006; Marquardt et al., 2005). EPHRIN-As interact in cis with the membrane proximal fibronectin III domain of EPHA receptors, a different domain than the one utilized for trans-signaling, and reduce EPHA tyrosine phosphorylation and downstream signaling response (Carvalho et al., 2006). As such, modulating reverse signaling through increasing or decreasing EPHRIN-A expression also disrupts the balance between trans-activation and cis-inhibition (Kao & Kania, 2011). Indeed, the balance between trans-activation and cis-attenuation is likely to be a key property of EPHRIN-A mechanisms in axon guidance in the retina, tectum, and spinal motor neurons, and possibly in systems beyond axon guidance (Kao & Kania, 2011). Similar cis-attenuation mechanisms have not been identified for B-type EPH/EPHRIN signaling.

2. EPH/EPHRIN function through an evolutionary lens

It has been recognized for some time that EPH/EPHRIN signaling is evolutionarily ancient and is connected to the origin of animal multicellularity and to the elaboration of morphological complexity (Drescher, 2002; Mellott & Burke, 2008a). Recent work has refined the origin and evolution of this signaling family and has revealed new insights into its likely earliest functions (Fig. 2A–F) (Arcas et al., 2020; Krishnan et al., 2019). The exponentially increasing genome sequencing data in a wide variety

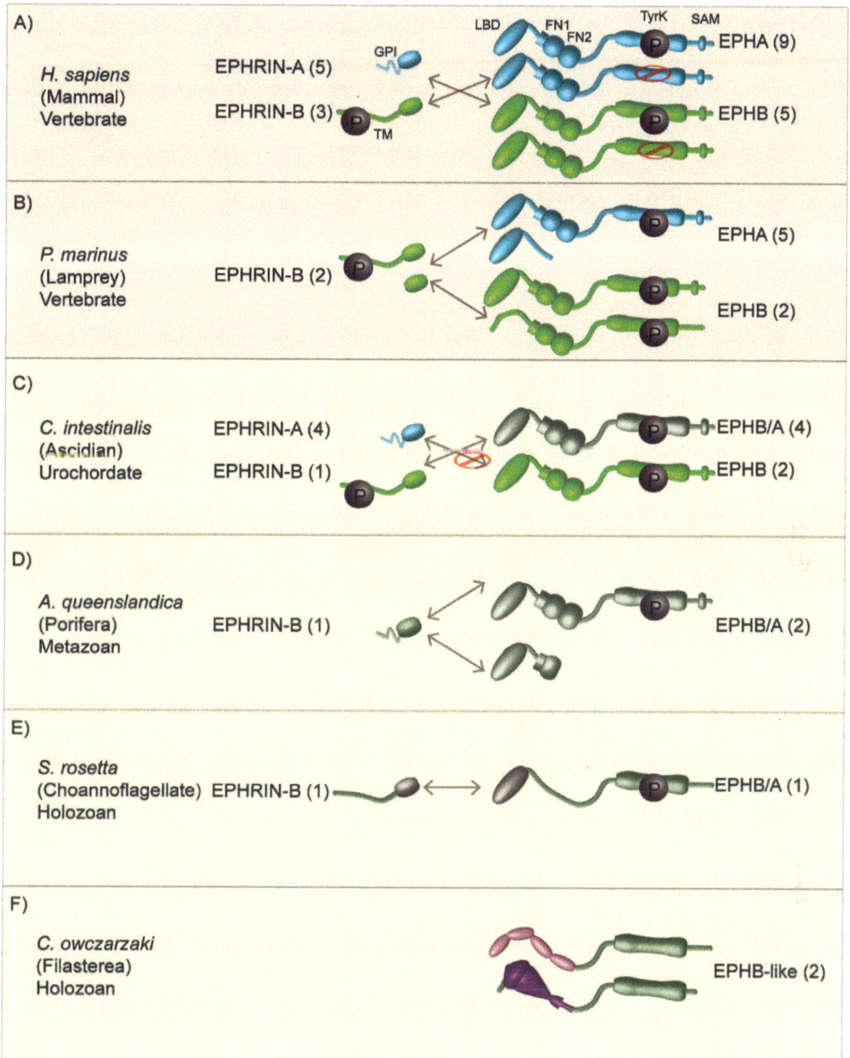

Fig. 2 Summary of evolutionary history of EPH/EPHRIN signaling. Schematic depiction of number of EPH/EPHRINs identified in different species (in parentheses) with an approximation of predicted EPH/EPHRIN binding relationships based on (Arcas et al., 2020; Krishnan, Degnan, & Degnan, 2019), which the reader is highly recommended to explore for more detailed information. (A) EPHA/EPHB subtypes in H. sapiens including kinase-active receptors which are found in EPHs from sponges to humans, and two kinase-dead EPH receptors. While EPHB6 is only kinase dead in mammals (prohibition symbol), EPHA10, which is also kinase dead, is the most ancient A-type receptor and arose in vertebrate lineages. EPHA receptors are only found in vertebrate lineages (peach boxes). (B) EPHB/EPHRIN-A Cross-subclass signaling likely arose in a common ancestor to the gnathastomes, but P. marinus lacks EPHRIN-As, EPHB/EPHRIN-A signaling does not occur in this species. (C) *C. intestinalis* have four EPHB-like receptors that also signal with EPHRIN-As while two receptors are closely related to vertebrate EPHB4,

(Continued)

of organisms has recently led to a more precise understanding of the diversification of the signaling networks that enabled the early evolution of cellular organization, proliferation, cell movement, and cell type specification. Unlike metazoan-specific modalities such as WNT signaling, tyrosine kinase signaling is ancient, arising in the earliest holozoans, and underwent extensive diversification in holozoans and metazoans (Suga, Torruella, Burger, Brown, & Ruiz-Trillo, 2014). Further, EPH receptors are amongst the most ancient receptor tyrosine kinases identified. Though it has been known for some time that bona fide EPHs and EPHRINs are found in sponges (*Porifera*) (Fig. 2D), one of the most evolutionarily basal animals, more recently, proteins with sequence homology to metazoan EPHs and EPHRINs were also identified in choanoflagellates, and structural predictions suggest they bind each other, enabling functional signaling (Arcas et al., 2020; Krishnan et al., 2019; Suga et al., 1999) (Fig. 2E). As the choanoflagellates are colonial unicellular eukaryotes and are the earliest ancestor that exhibited all metazoan EPH domain components together in one protein (Fig. 2E), EPH/EPHRIN signaling could already have regulated cellular organization in the multicellular colonies formed by these creatures.

Sequences corresponding to the domain architecture of part of metazoan EPH receptors, and with homology to the tyrosine kinase domain, have even been identified in unicellular Filasterea such as *C. owczarzaki* (Fig. 2F), a holozoan that shares a common ancestor with choanoflagellates and animals. This suggests the ancestral presence of a "proto-EPH-like" signaling molecule; as this protein lacks an EPHRIN binding domain, and an EPHRIN protein is not present in Filasterea, this "EPH" signals in an EPHRIN-independent manner (Arcas et al., 2020; Krishnan et al., 2019). An EPHRIN is first identifiable in choanoflagellates and has a B-type receptor binding domain (Fig. 2E). Sponges have EPHRIN orthologs that have either a transmembrane region, a GPI-anchor, or both in the same sequence, but are always predicted

Fig. 2—Cont'd suggesting specialized EPHRIN-B binding (prohibition symbol). (D) Bona-fide orthologs of EPHs and EPHRINs have been identified in Porifera, where all EPHRINs harbor a B-type ligand binding domain, though they may exhibit a transmembrane domain, a GPI-anchor, or both. (E) An EPH/EPHRIN-like pair was recently identified in choanoflagellates, which are closely related to those in Porifera and other ancient metazoan species. This appears to be the most ancient lineage to have the typical domain structure of metazoan EPH receptors. (F) Filasterea have sequences with distant homology to EPH receptors, which share part of the domain structure of metazoan EPH receptors. As Filasterea lack EPHRINs, these receptors may have bound other ligands, or acted as cell-cell adhesion molecules mediated by distinct extracellular domains (purple).

to be tethered to the membrane, indicating that EPH/EPHRIN signaling originated as a cell-contact communication system (Krishnan et al., 2019) (Fig. 2D). Whereas EPHRIN-A sequences originated in the common ancestor of tunicates and vertebrates (Fig. 2C), EPHA class receptors arose in the vertebrate lineage (Fig. 2A and B), indicating that EPH receptors in tunicates could bind to both EPHRIN-A and EPHRIN-B subtypes (Mellott & Burke, 2008a). As the sponge *A. queenslandica* and choanoflagellates also possess orthologs of downstream forward signaling effectors such as RhoA, Erk1/2, CDC42, it is possible that EPH/EPHRIN signals by similar core mechanisms in the earliest multicellular organisms. ROCK is present in *A. queenslandica* but not in choanoflagellates, suggesting that if EPH/EPHRIN signaling regulates actomyosin contractility in choanoflagellates, it does so through a different pathway. Interestingly, EPH and EPHRIN sequences are also found in some bacteria, suggesting that they acquired these sequences by horizontal gene transfer; whether they have a function in these prokaryotes is not yet known (Arcas et al., 2020).

2.1 Conserved functions for EPH/EPHRIN signaling in actomyosin contractility and cell positioning

Based on sequence similarities between orthologs, the ancestral EPH was a B-type receptor most similar to vertebrate EPHB4. Whereas the dramatic increase in the size of the EPH/EPHRIN signaling family in vertebrates does correlate loosely with the evolution of the nervous system, it is axiomatic, based on the existence of EPH/EPHRIN signaling in creatures lacking a nervous system, that ancestral function did not relate to neurodevelopment. Instead, it has been hypothesized that ancestral function may relate to regulation of cell movement through regulation of the actin cytoskeleton and cell adhesion, a hypothesis that accounts for the need for cell positioning information in the evolution of multicellular organisms (Drescher, 2002). A form of cell segregation in sponges achieves species-specific recognition during aggregation, and requires actomyosin contractility for cell motility (Arcas et al., 2020; Gaino, Bavestrello, & Magnino, 1999; Gaino & Burlando, 1990). Further support for an evolutionarily basal function of EPH/EPHRIN signaling in cell segregation and boundary formation is provided by the observation of complementary EPH/EPHRIN expression boundaries in the tentacles and developing buds of *Hydra vulgaris*, a simple cnidarian that evolved prior to the split of bilaterians from other metazoans. Functional experiments in an increasingly diverse array of organisms will be extremely informative to understanding the earliest functional roles for EPH/EPHRIN signaling in evolution.

A role for EPH/EPHRIN signaling in cell position through regulation of cell migration has also been discovered in the sea urchin, *S. purpuratus*, but here EPH/EPHRIN (Eph/Efn) signaling acts as an attractive cue rather than a repulsive one as it does in most scenarios in vertebrate development (Krupke, Zysk, Mellott, & Burke, 2016). Whereas Sp-Efn is expressed in the dorsal ectoderm, Sp-Eph is expressed within pigmented cells called immunocytes that normally migrate to and insert into the ectoderm. Blocking Eph/Efn signaling in sea urchins resulted in a failure of immunocyte migration and insertion into the dorsal ectoderm, and conversely, ectopic high levels of expression of Sp-Efn within the ventral ectoderm resulted in the aberrant attraction of immunocytes into these regions, where they also inserted. Interestingly, Sp-Efn localizes to long cellular protrusions called cytonemes, which deliver it to Sp-Eph-expressing immunocytes to facilitate long-distance attraction. These cytonemes are similar to filopodia-like protrusions that mediate long-distance EphrinB1/EphB3b signaling during hepatoblast positioning in zebrafish (Cayuso et al., 2016; Kornberg, 2019; Krupke et al., 2016). More recently, it was shown that EPHRIN-A1 is also localized to signaling filopodia which deliver it to EPHA2-expressing cancer cells to mediate a repulsive response, and filopodia also can facilitate EPHB2/EPHRIN-B1 repulsive signaling in HEK293 cells (Prospéri et al., 2015; Valenzuela & Perez, 2020). Together, these results indicate that cytoneme-mediated signaling may be an evolutionarily conserved mechanism of EPH/EPHRIN delivery at a distance. Further, these results indicate a basal function for EPH/EPHRIN signaling in cell positioning which has been dramatically expanded to accommodate cellular positioning in a wide array of processes in vertebrate embryos (Krupke et al., 2016).

2.2 An ancient and conserved EPH/EPHRIN signaling module

Even prior to the expansion of the EPH/EPHRIN signaling family in vertebrates, the pathway had already acquired multiple cellular functions. EPH/EPHRIN signaling function has been extensively examined in *Ciona intestinalis* ascidians, invertebrate urochordates that are a sister group to invertebrates. These studies have identified a repeatedly utilized general mechanism for cell fate specification that seems to be conserved throughout animals. In *Ciona* embryos, during early cleavage stages, an asymmetric cell division between two pairs of mother cells called A6.2 and A6.4 results in the generation of notochord and neural precursor cells. FGF signaling activation of Ras/MAPK drives cells to take on a notochord fate while its suppression results in neural fate. While FGF is expressed broadly and does not provide a positional cue for cell fate specification, Ci-ephrin-Ad is expressed in adjacent animal cells and signals to Eph2 and Eph3 in the mother cells prior to their asymmetric division. Following cell division, this cell-contact localized EPH/EPHRIN signal attenuates MAPK signaling and promotes neural fate within the daughter cells that arise nearer to the Ci-ephrin-Ad source. As one of only five signaling ligands expressed zygotically at the 16-cell stage, Ci-ephrin-Ad plays many parts. It restricts the neural cell fate in epidermal cells, again through repression of MAPK (Ohta & Satou, 2013). Asymmetric inhibition of MAPK by Ci-ephrin-Ad signaling from animal cells in direct contact with the A6.3 endomesodermal precursor mother cell, whose daughters give rise to mesoderm or endoderm derivatives, promotes mesoderm over endoderm fate (Shi & Levine, 2008). Later stages of cell fate specification in *Ciona* are also governed by EPH/EPHRIN repression of MAPK. For example, EPH/EPHRIN signaling is required to restrict MAPK activation during the neural plate stage to prevent the formation of too many pigment cells, as well as during the development of the *Ciona* motor ganglion (MG), which controls rhythmic swimming behavior in the tadpole stage (Haupaix et al., 2014; Stolfi, Wagner, Taliaferro, Chou, & Levine, 2011). In MG neuron specification, Ephrin-Ab is expressed in a cell lineage positioned directly posterior to the A9.30 cell lineage, which ultimately gives rise to the four anterior motor neuron cell types. Signaling by Ephrin-Ab to cells directly anterior to the A9.30 cell represses MAPK activation in this lineage, increasing expression of the Delta2 ligand, which in turn relays Notch activation to the next-most anterior cell, thereby elaborating downstream cell fate selection (Stolfi et al., 2011).

It is notable that in every scenario described above, EPH receptor signal transduction is mediated through the activation of p120RasGAP, a multi-domain GTPase-activating protein that inhibits Ras by increasing the rate of GTP hydrolysis (Haupaix et al., 2014). EPH/EPHRIN signaling also acts through p120RasGAP to inhibit Ras/MAPK in vertebrate cells, resulting in cellular responses such as neurite retraction, myogenic differentiation, and reduced endothelial cell migration and proliferation, suggesting possible evolutionary conservation of this signaling module (Elowe et al., 2001; Kim et al., 2002; Minami, Koyama, Wakayama, Fukuhara, & Mochizuki, 2011).

Indeed, EPH/EPHRIN signaling can repress Ras/MAPK signaling in a wide variety of mammalian cell culture cell types and antagonize Ras/MAPK activation by a variety of growth factor pathways (Miao et al., 2001). The evolutionary conservation of this signaling module and its deployment in so many different cell types make a strong case for its utilization in vertebrate development. In fact, treatment of human microvascular endothelial cells (HDMEC) with an EPHB4 inhibitor resulted in increased pERK1/2 and increased proliferation, whereas induction by addition of pre-clustered EPHRIN-B2-Fc resulted in a decrease in pERK1/2 and decreased proliferation, supporting, at least, that EPH/EPHRIN signaling can repress Ras/MAPK in endothelial cells (Groppa et al., 2018). Intriguingly, loss of function of p120RasGAP in mice results in vascular defects and embryonic lethality by E9.5 (Henkemeyer et al., 1995), and mutations in the *RASA1* gene that encodes p120RasGAP causes capillary malformation-arteriovenous malformation (CM-AVM) in humans (Eerola et al., 2003; Gerety, Wang, Chen, & Anderson, 1999). Similar vascular phenotypes to these are also observed upon loss of EPHRIN-B2 or EPHB4 in mice, and zebrafish with loss of function of EPHB4 and p120RasGAP also exhibit highly similar abnormalities in blood vessel formation (Gerety et al., 1999; Kawasaki et al., 2014). Moreover, recent studies have shown that germline mutations in the *EPHB4* gene also result in CM-AVM and other vascular anomalies in humans (Amyere et al., 2017; Zeng et al., 2019). Broader mechanisms for EPH/EPHRIN signaling in angiogenesis have been extensively studied and recently reviewed (Du, Li, He, Li, & He, 2020; Kania & Klein, 2016).

Parallel phenotypes have also been observed upon loss of function of EPHB4/EPHRIN-B2 and p120RasGAP in lymphatic valve development in mice, where kinase-dependent forward signaling through the EPHB4 receptor is crucial (Lapinski et al., 2017; Makinen et al., 2005; Zhang et al., 2015). Kinase inactivating mutations in EPHB4 cause lymphatic abnormalities in humans, which are also observed in CM-AVM (Li et al., 2018; Martin-Almedina et al., 2016, 2021). Further studies are likely to provide definitive *in vivo* genetic evidence of a shared EPHRIN-B2/EPHB4/p120RasGAP pathway and further illuminate the cellular mechanisms by which it regulates vascular and lymphangiogenic development. Together, these studies make a compelling case for conserved signaling function of the most ancestral EPH receptor EPHB4, which seems to utilize a similar signaling mechanism beginning prior to the emergence of vertebrates, conserved in humans, and relevant to human disease.

Tantalizingly, p120RasGAP is also found in the genome of sponges and choanoflagellates, suggesting the intriguing possibility that this signaling relationship may even be conserved from the earliest origins of EPH/ EPHRIN signaling (Jaber Chehayeb, Wang, Stiegler, & Boggon, 2020).

2.3 Conserved and divergent roles for EPH/EPHRIN signaling in vertebrate gastrulation

EPH/EPHRIN signaling has both evolutionarily conserved and divergent functions in the cellular movements and fate-induction signaling that occur during vertebrate gastrulation. Initial evidence for a role of EPH/EPHRIN signaling in gastrulation came from studies in zebrafish, in which exogenous expression of human EPHA3 or EPHRIN-A5 resulted in a failure of convergent extension of mesodermal cells and a severely deformed notochord (Oates et al., 1999). These results were interpreted as dominant negative effects because simultaneous overexpression of EPHA3 and EPHRIN-A5 reduced the frequency and severity of these phenotypes. Similarly, dominant negative disruption of B-type EPHRINs caused anteroposterior shortening and lateral expansion of the notochord and prechordal plate, likely through defects in convergent extension cell movements (Chan et al., 2001). During gastrulation in *Xenopus*, convergent extension of the involuting mesoderm elongates the antero-posterior axis and is regulated by signaling between Ephrin-A1 in the ectoderm and EphA4 in the involuting mesoderm (Park, Cho, Kim, Choi, & Han, 2011). Morpholino knockdown of either of these, or dominant negative disruption by expressing a kinase-dead EphA4, resulted in a failure to undergo convergent extension. Neither disruption of EphA4 signaling in *Xenopus*, nor Ephrin-B signaling in zebrafish disrupted fate specification of the mesoderm (Chan et al., 2001; Park et al., 2011). Instead, EphA4/Ephrin-A1 signaling was found to regulate tissue separation of the involuting mesoderm and non-involuting ectoderm; failure of tissue separation and disruption of convergent extension resulting from EphA4 disruption could be rescued by expressing a constitutively active form of the low molecular weight GTPase RhoA (Park et al., 2011). Interestingly, similar tissue separation phenotypes were also observed upon expression of a dominant negative ADAM10, consistent with cleavage enabling conversion of the EphA4/ ephrin-A1 adhesive interaction to a repulsive outcome (Janes et al., 2005; Park et al., 2011). Numerous EPH/EPHRIN signaling partners contribute to tissue separation during *Xenopus* gastrulation. Additive antiparallel signaling by Ephrin-B1 from the ectoderm to EphB receptors in the mesoderm, and Ephrin-B2 from the mesoderm to EphB receptors in the ectoderm are required for ectoderm/mesoderm separation (Rohani, Canty, Luu, Fagotto, & Winklbauer, 2011). As above, Ephrin-B/EphB signaling also utilizes RhoA signaling, though the low molecular weight GTPase Rac is required as well, possibly to regulate trans-endocytosis of EPH/EPHRIN signaling complexes to remove this physical tether from the interface (Gaitanos, Koerner, & Klein, 2016) (see also subsequent section on cellular mechanisms of cell sorting for more details on RhoA in tissue separation). A further role for EPH/EPHRIN signaling in *Xenopus* gastrulation is observed in the regulation of *Brachyury* expression by signaling between EphA4 and Ephrin-B2. This drives a cellular process called "peak involution" wherein dorsal mesoderm undergo bending and anteroposterior elongation to contribute to the earliest phase convergent extension (Evren et al., 2014).

A recent study showed that pan-inhibition of Eph receptor signaling in *Ciona* and in the distantly related ascidian *Phallusia mammillata* resulted in a disruption of endoderm invagination at the beginning of gastrulation without disruption of germ layer fate specification (Fiuza, Negishi, Rouan, Yasuo, & Lemaire, 2020). Following endoderm apical constriction and flattening of the embryo, Nodal signaling drives transcriptional upregulation of the Eph1 receptor in endodermal precursors, which is required for relocalization of phospho-myosin enrichment from apical to basolateral and for subsequent apicobasal shortening of the endoderm. Nodal signaling also activates expression of EphA4 in the mesoderm of *Xenopus*, and EphA4 is required for mesendoderm internalization, suggesting that this may be a shared ancestral mechanism utilized during gastrulation (Evren et al., 2014; Wills & Baker, 2015). However, whereas invagination drives internalization of endoderm in cephalochordates and tunicates such as *Ciona*, vertebrates undergo cellular ingression by migration (Wen & Winklbauer, 2017). Live imaging studies of slice explants in *Xenopus* revealed that ingression is driven by endodermal cell lengthening and rearrangement that enables their amoeboid-like migration. Ephrin-B1 exhibits highly localized enrichment at the trailing edge of migrating endoderm cells where it undergoes trans-endocytosis to permit the retraction of the trailing edge during migration, though it is intriguing to speculate that EPH/EPHRIN signaling at this interface may also enable increased actomyosin contractility to terminate this cell contact (see also below discussion on cell sorting) (Wen & Winklbauer, 2017).

Though EPH/EPHRIN signaling clearly has numerous roles in gastrulation in multiple chordate species, no known roles have been identified in higher vertebrate species. Gastrulation defects have not been observed upon targeted disruption of *Efnb1*, *Efnb2*, or *Epha4* in mice, and indeed, it is surprising to note that despite expression of multiple EPH/EPHRIN family members during gastrulation in mice, none have yet been reported to have functional roles in this process. This difference may point to functional redundancy in mammals, or subtle defects in gastrulation may have been missed because of the relative inaccessibility of gastrulation cell behaviors in a viviparous organism; alternatively, this difference may reflect evolutionarily divergent functions amongst vertebrates.

2.4 EPH/EPHRIN signaling family as a substrate for evolutionary variation

The dramatic expansion of the EPH/EPHRIN signaling family throughout vertebrate evolution, and the considerable inherent functional redundancy that resulted, may have enabled this pathway to serve as a substrate for evolutionary variation. One example comes from crested pigeons, which display a head crest: neck feathers that grow toward the top of the head instead of down the neck as in other pigeon breeds. The head crest exhibits simple Mendelian inheritance, and the same genetic locus appears to be responsible in multiple crested pigeon breeds. This phenotype is caused by a variant in the catalytic loop of EPHB2, which is predicted to affect kinase activity resulting in disruption in the patterning of feather placode polarity and inversion of feather bud outgrowth (Shapiro et al., 2013). A similar feather phenotype in ringneck doves is also caused by an amino acid substitution that reduces EPHB2 kinase activity, indicating that evolutionary convergence impacting EPH/EPHRIN signaling drives phenotypic variation in distinct species, suggesting a possible evolutionarily conserved function of EPHB2 in feather bud patterning in these species, and possibly in other birds (Vickrey, Domyan, Horvath, & Shapiro, 2015). As compound loss of EPHB2 and EPHB3 results in skeletal and axon guidance phenotypes, function of these receptors in key developmental processes is also likely to be overlapping in avians, allowing the adoption of additional functions and contributing to phenotypic diversity.

Our recent morphometric studies of an allelic series of *Ephb1*; *Ephb2*; *Ephb3* compound mutant mouse embryos, revealed both quantitatively additive shared effects on head shape as well as qualitatively distinct effects of loss of function of each receptor (Niethamer et al., 2020; Mincer, Niethamer, Teng, Bush, & Percival, 2022). *Ephb2*$^{-/-}$; *Ephb3*$^{-/-}$ compound homozygotes display a severe cleft palate phenotype with high penetrance in mice which is incompatible with life (Orioli, Henkemeyer, Lemke, Klein, & Pawson, 1996; Risley, Garrod, Henkemeyer, & McLean, 2009), these results suggest that partial functional redundancy between EPH receptors ensures viability allowing partial divergence of receptor function, which may contribute to face shape variation in mammals.

3. Cellular mechanisms underlying EPH/EPHRIN-mediated cell sorting and boundary formation

Active mechanisms are employed by the embryo to establish and maintain boundaries between cell populations to withstand constant cell movements that would otherwise cause cell intermingling. Indeed, EPH/EPHRIN signaling often acts to establish and/or enforce boundaries in numerous contexts in development, including gastrulation, rhombomeres and elsewhere in the nervous system, somites, craniofacial primordia, and in the development of endodermal organs, and has been reviewed extensively (Fagotto, Winklbauer, & Rohani, 2014; Kindberg & Bush, 2019; Niethamer & Bush, 2019). In addition to these roles in early morphogenesis, EPH/EPHRIN signaling contributes to boundary formation and cellular organization of germinal centers in the immune system (Laidlaw et al., 2017; Lu, Shih, & Qi, 2017).

Our understanding of the cellular mechanisms by which EPH/EPHRIN signaling regulates cell sorting and tissue separation has continually evolved together with our understanding of the general processes of cell sorting. Extensive reviews of general mechanisms governing cell sorting can be found elsewhere (Cayuso, Xu, & Wilkinson, 2015; Fagotto, 2014; Kindberg & Bush, 2019), but in brief, cell sorting has long been known as a crucial driver of cellular self-organization during development (Townes & Holtfreter, 1955). This process can be driven by qualitative or quantitative differences in cell-cell adhesion strength between two populations of cells (termed selective and differential adhesion, respectively) (Duguay, Foty, & Steinberg, 2003), or by negative effects on the strength of cell-cell contacts exerted by cortical actomyosin cytoskeletal contractility (cortical tension) (Krieg et al., 2008; Maitre et al., 2012). Together, these intercellular forces are thought to regulate the interfacial tension (the force of attraction at an interface) to achieve differences in cell-cell contact strength, a unifying summation that is termed the differential interfacial tenson hypothesis (DITH) (Brodland, 2002; Winklbauer, 2015). These biophysical concepts constitute the founding principles upon which our understanding of EPH/EPHRIN-driven cellular organization are based.

The discovery of the importance of EPH/EPHRIN signaling in cell sorting originated with studies of hindbrain rhombomere segmentation in zebrafish (Xu, Alldus, Holder, & Wilkinson, 1995; Xu, Mellitzer, Robinson, & Wilkinson, 1999). The hindbrain rhombomeres are distinct compartments that are initially patterned by morphogen signaling and then refined, and physically separated, by cell sorting behaviors at rhombomere boundaries. EPHs and EPHRINs are expressed in a reciprocal pattern of expression in alternate rhombomeres, creating EPH/EPHRIN heterotypic interfaces at each rhombomere boundary; disruption of EPH/EPHRIN signaling results in inappropriate intermixing of cells from neighboring rhombomeres (Kindberg & Bush, 2019; Krumlauf & Wilkinson, 2021; Xu et al., 1995). It was recognized from the very earliest studies that EPH/EPHRIN signaling could drive cell sorting by repulsion (Xu et al., 1999). When ephrin-B2 expression was driven in cells of rhombomeres r3/r5, which normally express EphA4 and not ephrin-B2, ephrin-B2 expressing cells were repelled to the boundaries of these rhombomeres; conversely, cell sorting also occurred when the extracellular domain of EphA4 was overexpressed to activate ephrin-B2 reverse signaling (Xu et al., 1999). These, and other similar results suggested that bidirectional signaling drives cell sorting at rhombomere boundaries through cell repulsion (Cayuso et al., 2015).

3.1 EPH/EPHRIN signaling and actomyosin tension in cell sorting

The broad concept of a repulsive EPH/EPHRIN positional signal comprises more specific underlying mechanisms, such as an effect of EPH/EPHRIN signaling on repolarization and migration of cells upon contact, an impact on strength of cell contact in the context of random cell movement, or some combination of the two (Cayuso et al., 2015; Kindberg & Bush, 2019). As these detailed cellular mechanisms are difficult to distinguish *in vivo*, we recently took a reductionist approach in which we placed two HEK293 cells in low adhesion agarose plates and assessed their extent of contact by measuring their cell-cell contact angles as a proxy for their interfacial tension in the absence of cell-substrate contact (Kindberg et al., 2021). We found that while EPHB2-EPHB2 or EPHRIN-B1-EPHRIN-B1 homotypic cell pairs exhibited extensive cell-cell contacts reflecting low

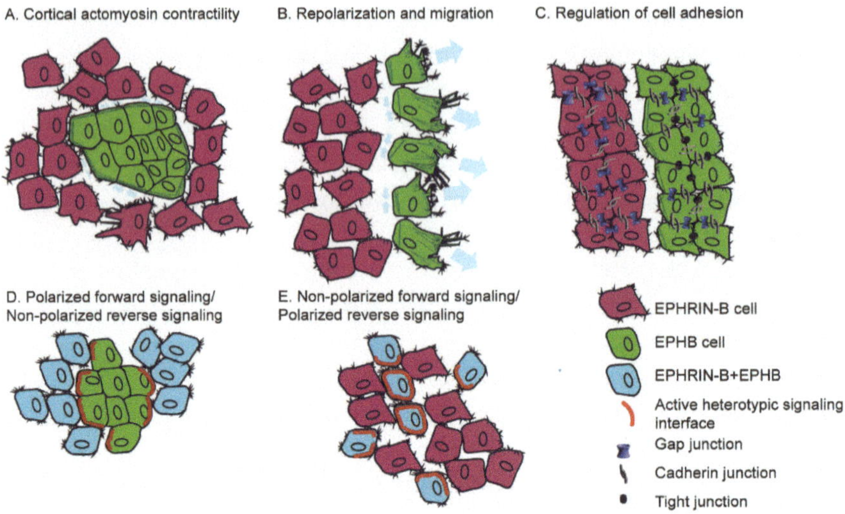

Fig. 3 Cellular behaviors driving EPH/EPHRIN-mediated cell sorting. Schematic simplification of non-exclusive cellular mechanisms that have been proposed to drive cell sorting. (A) Cortical actomyosin contractility within EPHB cells at heterotypic cell interfaces reduces the strength of heterotypic cell contacts and simultaneously drives increased homotypic cell contacts. (B) Localized collapse of EPHB cells at heterotypic cell contacts results in cell repolarization and directed migration of EPHB cells away from EPHRIN-B cells. (C) Cell-cell adhesion is regulated by EPHRIN-B through multiple mechanisms including forward signaling recruitment of ADAM10 metalloprotease and cleavage of E-cadherin at the heterotypic interface and EPHRIN-B binding to Cx43 and prevention of the formation of gap junctions at heterotypic interfaces. EPHRIN-B binding to Par-3 (not shown) prevents formation of tight junctions within EPHRIN-B cells. (D, E) Schematic representation of the concept of "Polarized" and "Non-polarized" signaling. (D) When all cells express EPHB, but are mosaic for expression of EPHRIN-B, polarized forward signaling, and non-polarized reverse signaling occur. (E) When all cells express EPHRIN-B, but are mosaic for expression of EPHB, non-polarized forward signaling, and polarized reverse signaling occur. In (E, F), productive forward signaling interfaces are highlighted in yellow to demonstrate that polarized forward signaling allows EPHB cells to differentiate between heterotypic and homotypic interfaces, whereas non-polarized forward signaling does not.

interfacial tension, EPHB2-EPHRIN-B1 heterotypic cell pairs exhibited extremely reduced cell contacts, reflecting increased interfacial tension. Inhibition of actomyosin contractility increased heterotypic cell contacts, and heterotypic cell cultures had increased stiffness when measured by atomic force microscopy, consistent with the notion that EPHB2/EPHRIN-B1 heterotypic interfacial tension is increased through effects on the cortical actomyosin cytoskeleton (Fig. 3A).

Larger scale cell mixing experiments, where EPHRIN-B1 cells are mixed with EPHB2 cells, resulting in cell sorting, also reflects the requirement of actomyosin contractility. As for individual cell-cell contacts, inhibition of key regulators of actomyosin contractility including RhoA, Rho kinase (ROCK), Myosin light chain kinase (MLCK), or non-muscle myosin II (NMII), drastically reduced cell sorting in culture, and dominant negative disruption of ROCK, or genetic loss of function of non-muscle myosin resulted in a loss of cell sorting *in vivo*, indicating that actomyosin contractility is indeed crucial for this process (Kindberg et al., 2021; O'Neill et al., 2016; Tanaka, Kamo, Ota, & Sugimura, 2003). These results also fit with observations made during tissue separation in *Xenopus* gastrulation. In this system the ectoderm and mesoderm both express both EPHs and EPHRINs, and antiparallel EPH/EPHRIN forward signaling between them results in repeated rounds of attachment and detachment to facilitate migration of the mesoderm on the ectoderm, while maintaining tissue separation (Rohani et al., 2011). In this study, detachment required EPH/EPHRIN regulation of actomyosin contractility by RhoA at the tissue interface and was therefore one of the first studies to indicate that EPH/EPHRIN boundary formation is regulated by EPH/EPHRIN effects on cortical contractility. Interestingly, in our two-cell system, we also observed heterotypic EPHB2-EPHRIN-B1 cell pairs moving through phases of sparse or close contact over time, indicating that EPH/EPHRIN impacts on interfacial tension are dynamic in nature (Kindberg et al., 2021). This may occur through impacts of EPH/EPHRIN signaling on cadherin clustering (see also below), or through feedback regulation wherein close cell contact promoted by cadherin-based adhesion allows increased EPH/EPHRIN activation, thereby promoting actomyosin contractility-driven minimization of contact, which may in turn be terminated at the heterotypic interface through endocytosis (Fagotto, 2014; Gaitanos et al., 2016; Kindberg et al., 2021; Rohani et al., 2011; Zimmer, Palmer, Köhler, & Klein, 2003).

We proposed that EPH/EPHRIN-mediated cortical actomyosin contractility regulates cellular organization through its impact on cell-contact strength, but it is alternatively or additionally possible that disruption of actomyosin contractility might prevent cell sorting by preventing cell migration, thus preventing repulsive reorganization (see below). However, adding actomyosin contractility inhibitors Y27632 and ML7 after completion of cell sorting resulted in dramatic re-intermixing of EPHB2 and EPHRIN-B1 cells, indicating that cells still move in these conditions and that actomyosin contractility is instead required to maintain EPH/EPHRIN interfacial boundaries (Kindberg et al., 2021). These findings fit well with the previous observations of actomyosin enrichment at

EPH/EPHRIN boundaries in zebrafish rhombomeres, the *Drosophila* wing disc, the mouse secondary palate mesenchyme, and at the *Xenopus* ectoderm/mesoderm boundary during gastrulation (Calzolari, Terriente, & Pujades, 2014; Monier, Pélissier-Monier, Brand, & Sanson, 2010; O'Neill et al., 2016; Rohani et al., 2011; Umetsu, Dunst, & Dahmann, 2014).

Of course, as a predominantly cell-cell contact mediated signaling system, productive EPH/EPHRIN signaling occurs at the heterotypic interface of cells expressing an EPHRIN and EPH; as such, it is the differential between homotypic and heterotypic interfacial tensions as established by cortical actomyosin contractility that governs cell sorting through this model. This property of EPH/EPHRIN-driven cell sorting was demonstrated by Canty et al. in 2017, who expressed that a differential in homotypic interfacial tension between two cell populations need not exist to achieve heterotypic interfacial tension between the populations (Canty, Zarour, Kashkooli, François, & Fagotto, 2017). It is notable, though, that in our studies in which we reduced cadherin-based cell-cell adhesion through use of calcium-free media, we revealed a distinct cortical tension mechanism that mediates EPHB2-EPHB2 cell contact, suggesting that differential homotypic cell affinities could also contribute to cellular self-organization, at least in this cell system (Kindberg et al., 2021). Consistent with this finding, increased homotypic cell density in EPH receptor expressing cells when mixed with EPHRIN-B1 expressing cells has been observed in multiple studies (Kindberg et al., 2021; Taylor et al., 2017). Further, when we performed mixing of 3 cell types: EPHB2-expressing, EPHRIN-B1 expressing, and wtHEK293 cells, not only were the EPHB2 cells sorted out from EPHRIN-B1 cells, but wtHEK293 cells were also eliminated from EPHB2-expressing cell groups, suggesting that forward signaling into EPHB2 cells also drove non-autonomous exclusive sorting of signaling-inert cells (Kindberg et al., 2021). Further, EPHB2 kinase activity is required for the compaction of EPHB2 cell groups when mixed with EPHRIN-B1 expressing cells (Taylor et al., 2017). Together these findings suggest the possibility that a forward signaling effect on cellular cohesion might contribute to elimination of EPHRIN-expressing cells from EPH-expressing compartments, for example, as observed for ephrin-B2 expressing cells in the zebrafish hindbrain (Kemp, Cooke, & Moens, 2009; Taylor et al., 2017; Xu et al., 1999). *In vivo* experiments interrogating EPH/EPHRIN cell segregation in endoderm/mesoderm separation during *Xenopus* gastrulation, in the neural folds of mouse embryos, and in cell culture support that EPH receptor forward signaling can be sufficient to drive cell sorting (Fuller, 2003; O'Neill et al., 2016; Rohani, Parmeggiani, Winklbauer, & Fagotto, 2014; Wu, Ashlin, Xu, & Wilkinson, 2019), though a more minor role from phosphorylation-dependent reverse signaling also contributes (Jorgensen et al., 2009; Rohani et al., 2011; Wu et al., 2019).

3.2 EPH/EPHRIN signaling and repulsive cell migration in cell sorting

In the above studies, EPH/EPHRIN may drive cell sorting through cortical actomyosin contractility or cadherin-based effects on cell-cell contact strength in the presence of random movement of cells. However, repolarization and directed migration has also been predicted by mathematical modeling to cause cell sorting (Fig. 3B) (Aharon et al., 2014). In fact, effects on cell contact strength may occur in combination with repulsive migration, defined as a cell repolarizing its protrusions and migrating away from a heterotypic cell type, akin to contact inhibition of locomotion (Kindberg & Bush, 2019; Taylor et al., 2017). This scenario fits well with the repeated demonstration of the importance of EPH/EPHRIN signaling in regulation of cell and axon migration during development (Krull, 2010). Heterotypic repulsive migration has been observed for cells cultured at low density, wherein EPHB2-expressing cells repolarize and increase their speed of migration away from an EPHRIN-B1 expressing cell (Poliakov, Cotrina, Pasini, & Wilkinson, 2008; Taylor et al., 2017). EPH/EPHRIN heterotypic boundaries can have different effects on cell polarity when cells are at high density. When human epithelial ARPE-19 cells expressing EPHRIN-B1 or EPHB3 were micropatterned into distinct regions and then allowed to intermix, they maintained their boundaries and polarized their actin cytoskeleton, microtubule organizing centers and cell shapes parallel to the EPHRIN-B1/EPHB3 interface, and cells moved along the interface, rather than away from it (Javaherian et al., 2017). These data suggest the intriguing possibility that in addition to regulating cell intermixing, EPH/EPHRIN expression boundaries may also contribute to coordinating planar cell polarity across a tissue.

Heterotypic repulsive migration, driven by phosphorylation-dependent reverse signaling, can also occur in EPHRIN-B1 expressing cells, and contributes to preventing intermixing of EPHB2 and EPHRIN-B1 expressing cells in less potent manner than forward signaling (Wu et al., 2019). In a recent study, the ability of "polarized" vs "non-polarized" forward and reverse signaling was examined. Polarized forward signaling was defined as the situation in which EPH receptor expressing cells encounter both EPHRIN-expressing and non-expressing cells, whereas non-polarized forward signaling was defined as the situation in which EPH receptor expressing cells only encounter EPHRIN-B1-expressing cells (Fig. 3D and E). In this scenario, polarization of the reverse signal is the inverse of forward; polarized forward signaling occurs together with non-polarized reverse signaling; non-polarized forward signaling occurs with polarized reverse signaling (Fig. 3D and E). Non-polarized forward signaling and polarized reverse signaling (i.e. achieved by combining EPHB2; EPHRIN-B1-expressing cells signaling with EPHRIN-B1-expressing cells), did not drive sorting as robustly as polarized forward signaling (i.e. EPHB2; EPHRIN-B1-expressing cells signaling with EPHB2-expressing cells) (Fig. 3E) (Wu et al., 2019). Together, these results suggest that forward signaling allows EPHB2 cells to detect and respond to a heterotypic interface. It is worth pointing out that maintaining EPHB2 and EPHRIN-B1 co-expression in culture also resulted in reduced EPHB2 expression, probably through endocytosis of constantly activated receptor, a situation that has also been observed *in vivo* (Bush & Soriano, 2010; Wu et al., 2019). As such, these co-expressing cells were likely desensitized to EPHRIN-B1 at the time of mixing with EPHRIN-B1 cells. Nevertheless, based on the accumulated evidence, "non-polarized forward signaling" is not capable of driving cell sorting because a differential is not generated, as all EPHB2 cell-cell contacts are made with other EPHRIN-B1 expressing cells (Fig. 3E). This forward signaling differential may reflect the impact of cortical contractility on cell contact strength, a repulsive migratory cue, or both. Further, the above results indicate that reverse signaling is less potent than forward signaling, which drives repulsion at heterotypic relative to homotypic interfaces as a central mechanism for EPH/EPHRIN-driven cell sorting.

3.3 EPH/EPHRIN signaling and differential adhesion in cell sorting

EPH/EPHRIN signaling has also been proposed to regulate cell sorting by modulating cadherin-based intercellular adhesion, which in turn drives cell segregation through a differential adhesion mechanism (Fig. 3C) (Fagotto, Rohani, Touret, & Li, 2013; Solanas, Cortina, Sevillano, & Batlle, 2011). In colorectal cancer epithelial cells, E-cadherin accumulation was decreased at an EPHB3/EPHRIN-B1 heterotypic interface, and this effect was reversed by pan-metalloproteinase inhibition (Solanas et al., 2011). Further, EPHB3/EPHRIN-B1 cell sorting was reduced by knockdown of ADAM10 or ADAM15 metalloproteinases, or by expressing a dominant-negative ADAM10 mutant lacking metalloproteinase activity. In the small intestine, EPHRIN-B1 expression in the villus opposes expression of EPHB2 in the crypt base and EPHB3 expression in the Paneth cells, and loss of EPHB2 and EPHB3 function result in Paneth cells becoming inappropriately positioned within the intestinal epithelium (Batlle et al., 2002). Dominant-negative loss of ADAM10 metalloprotease activity resulted in a Paneth cell intermixing phenotype that greatly mimicked the loss of EPHB3 in mice. These results formed the basis for a model in which EPHB3/EPHRIN-B1 signaling recruits ADAM10 to heterotypic interfaces to reduce cadherin-based cell adhesion (Solanas et al., 2011). A similar molecular mechanism has been recently proposed to regulate EPHA4/EPHRIN-B2-driven separation of inner pillar and outer pillar cells during the development of the cochlear sensory epithelium (Defourny, Peuckert, Kullander, & Malgrange, 2019). It is important to point out that ADAM metalloproteases can also cleave both A- and B-type EPHRINs and function to convert an adhesive molecular interaction to a repulsive cellular outcome in many contexts (Atapattu, Lackmann, & Janes, 2014; Hattori, Osterfield, & Flanagan, 2000; Janes et al., 2005; Ji et al., 2014; Wei et al., 2010). In *Xenopus*, ADAM13 cleavage of Ephrin-B1 and Ephrin-B2 is critical for Neural Crest Cell induction, and ADAM10 can cleave Ephrin-B2 upon loss of the interacting protein flotillin-1 during neural tube closure (Ji et al., 2014; Wei et al., 2010). Recently, it was shown that ADAM10 mediates cleavage of EPHRIN-B2 in human myofibroblasts to promote fibroblast chemotaxis skin and lung fibrosis (Lagares et al., 2017). Expression of dominant negative ADAM10 in a Co115 colorectal cancer cell line did not reveal dramatic differences in EPHB3 phosphorylation when mixed with EPHRIN-B1-expressing cells, however, so it was concluded that ADAM10 in that system functions specifically to mediate EPH/EPHRIN regulation of E-cadherin-based cell adhesion (Solanas et al., 2011).

EPH/EPHRIN signaling can also indirectly regulate cadherin-mediated cell adhesion through its impact on actomyosin contractility. In *Xenopus*, separation of the notochord and presomitic mesoderm is mediated by EphB4/ephrinB2 expression in the presomitic mesoderm and notochord, respectively, and morpholino knockdown of either resulted in inappropriate intermixing of the two cell types (Fagotto et al., 2013). EPH/EPHRIN signaling at this boundary mediates extensive actomyosin contractility that results in cell blebbing, but presomitic mesoderm and notochord cells remain in contact, causing the authors to conclude that tissue separation is not directly caused by actomyosin contractility effects on cell contacts. Instead, the authors hypothesize that actomyosin contractility prevented the formation of cadherin clusters, thereby diminishing heterotypic cadherin-based adhesion relative to homotypic contacts amongst either cell type (Fagotto et al., 2013). In EPHB2/EPHRIN-B1 cell mixing experiments in HEK293 cells, knockdown of N-cadherin or depletion of cadherin-based adhesion through culture in low calcium media reduced cell-cell contacts generally, but neither disrupted cell sorting (Kindberg et al., 2021; Taylor et al., 2017). Further, N-cadherin knockdown decreased both homotypic and heterotypic repulsion as measured by cell contact time and increased protrusion collapse frequency, indicating that, at least in these cells, differential adhesion due to cadherins is not likely a major effector of EPH/EPHRIN-driven cell sorting in these cells (Taylor et al., 2017). These results suggest that EPH/EPHRIN-driven actomyosin contractility may impact interfacial tension and boundary formation through multiple mechanisms and the role of cadherin-based adhesion may be context-dependent.

EPH/EPHRIN signaling can regulate cell contact strength through multiple other mechanisms. In *Xenopus* early development, overexpression of Ephrin-B1 or its intracellular domain, leads to dissociation of blastomeres (Jones et al., 1998). This is likely because Ephrin-B1 binds to the cell polarity protein Par-6, sequestering it from the Par polarity complex (Par-3, Par-6, Cdc42, aPKC) and resulting in a loss of localized ZO-1 and a loss of epithelial tight junctions (Lee, Nishanian, Mood, Bong, & Daar, 2008). This event is regulated by tyrosine phosphorylation of the intracellular domain of Ephrin-B1, which prevents binding of Ephrin-B1 to Par-6 and restores tight junctions. In mouse craniofacial development, EPHRIN-B1 also regulates the formation of gap junctions through its interaction with Connexin 43 (Davy, Bush, & Soriano, 2006). At EPHB2/EPHRIN-B1 heterotypic interfaces, gap junction formation is prevented, effectively de-coupling EPHRIN-B1 expressing and non-expressing tissues and impacting craniofacial osteogenic differentiation. Similar regulation of gap junction formation occurs in zebrafish animal caps, where EPHB/EPHRIN-B interfaces show reduced gap junction formation and restricted communication across these boundaries (Mellitzer, Xu, & Wilkinson, 1999). As additional modes of cell contact regulation, EPH/EPHRIN regulation of tight and gap junctions, together with impacts on adherens junctions and cortical contractility, determine the tension differential between heterotypic and homotypic interfaces, a critical part of EPH/EPHRIN boundary formation.

4. Nervous system development

EPH/EPHRIN signaling plays a vital role in the development of the vast complexity of the central nervous system architecture. In early development, EPH/EPHRIN signaling is critical for early neural patterning and neural tube morphogenesis. Its roles include EPH/EPHRIN-mediated cell sorting, as discussed above, which is crucial for processes including rhombomere boundary formation and proper distribution of Cajal-Retzius cells, which distribute throughout the cerebral cortex through EPH/EPHRIN contact-mediated repulsion (Krumlauf & Wilkinson, 2021; Villar-Cerviño et al., 2013). Later in development, through spatially nested and differing temporal levels of expression, and through use of multiple modes of signaling, EPH/EPHRIN signaling directs neuronal migration, differentiation, axon guidance and synaptic connectivity. As in other contexts, multiple signal transduction pathways, including Ras and Rho family GTPases are regulated by numerous signaling intermediaries, including guanine exchange factors GEFs such as EPHEXIN, INTERSECTIN, KALIRIN, VAV, and TIAM1 and GTPase activating proteins such as the chimaerins. These signaling pathways have been recently reviewed (Kania & Klein, 2016) and I will focus on recent studies that have further illuminated the diverse modes that EPH/EPHRIN signaling utilizes in nervous system development.

Two of the earliest known and most highly conserved vertebrate functions for EPH/EPHRIN signaling are the regulation of midline axonal crossing and axonal targeting. Several studies have revealed the importance of EPH/EPHRIN signaling in formation of the corpus callosum (CC), the largest interhemispheric axon tract in mammals (Bush & Soriano, 2009; Mendes, Henkemeyer, & Liebl, 2006; Orioli et al., 1996; Robichaux et al., 2016; Soskis et al., 2012). Whereas our previous study showed that EPHRIN-B1 expression in CC axons is required to receive a PDZ-dependent reverse signal from midline glial guidepost cells, a very recent study performing deletion of *Efnb1* within the dorsal telencephalon did not observe agenesis of the CC. Though the basis for this discrepancy is not clear, the more recent study revealed that EPHRIN-B1 is dynamically regulated to prevent axonal midline re-crossing (Mire et al., 2018). EPHRIN-B1 increases in expression in axons following midline crossing and interacts with NRP1, a receptor for the axonal attractant SEMA3C, to silence SEMA3C/NRP1 signaling (Mire et al., 2018). This silencing allows growth of CC axons away from the midline following their crossing by effectively blinding them to the declining attractive SEMA3C gradient in the contralateral hemisphere. Whether this silencing activity requires the PDZ binding motif is not known, but N-glycosylation of the extracellular domain of EPHRIN-B1 was required, indicating this mechanism may be distinct (Mire et al., 2018).

A recent study further illuminated the role of EPHRIN-B1/EPHB signaling in axonal targeting following midline crossing of the CC. As receptors for the primary excitatory neurotransmitter in the brain, NMDA receptors (NMDAR), were presumed to play a critical role in CC formation through excitatory synaptic activity. However, in mice with cortex-specific knockout of GluN1, the essential subunit of the NMDAR, loss of NMDAR function led to loss of targeted callosal innervation of the S1/S2 border and an increase in callosal innervation in S1 during CC development. This was not due to excitatory activity and ion channel function of the receptor, but instead through physical interaction of EPHB2 with the NMDAR which resulted in stabilization of both proteins in the target neurons. EPHRIN-B1 loss specifically in projecting neurons resulted in axonal mistargeting in the contralateral cortex. It remains to be determined if the mechanism by which the callosal axons mis-target is through loss of repulsive cues from EPHB/EPHRIN signals, and if there is disruption of forward or reverse signaling or both. Though multiple studies have shown that NMDA and EPHB co-localize and physically interact (Dalva et al., 2000; Nolt et al., 2011), this is the first report demonstrating their collaborative function in neural circuit formation during development.

Neuronal target selection and determination of whether to initiate synaptogenesis must involve interpretation of repulsive and attractive cues by dendritic filopodia. Through use of a dual-color ratiometric indicator of EPHB kinase activity and manipulation of activation through use of a photoactivatable EPHB2, a recent study proposed that a dendrite's choice of whether to retract or to generate stable axo-dendritic contacts is governed by differences in the kinetics of EPHB2 activation (Mao et al., 2018). Dendritic filopodial retraction is initiated by a rapid increase in EPHB activation, while slower activation results in stabilization of contacts and synapse formation. In addition to providing the exciting insight that

EPH/EPHRIN signaling may serve both repulsive and attractive purposes in synaptogenesis, it also provides powerful new reagents for the spatiotemporal observation and manipulation of EPHB2 kinase activity that can be applied in diverse contexts.

In addition to regulating cellular position and proper neuronal connections, EPH/EPHRIN signaling also regulates cell fate specification and progenitor identity in neural development (Laussu, Khuong, Gautrais, & Davy, 2014). For example, in the vertebrate spinal cord, EPHRIN-B2 and EPHRIN-B3 are expressed in distinct domains: EPHRIN-B2 is expressed in progenitors of motor neurons (pMN) whereas EPHRIN-B3 is expressed in p3 progenitors that give rise to interneurons. Tissue-specific conditional loss of function of *Efnb2* resulted in normal initial pMN specification, but reduced maintenance of pMN neurons at later stages of development (Laussu et al., 2017). Conversely, *Efnb3* loss of function led to fewer p3 progenitor cells and increased pMN/p3 intermixing. Instead of an effect on relative cell proliferation rate, fate specification by EPHRIN-B2:EPHRIN-B3 appears to work in collaboration with SHH signaling, a key regulator of dorsoventral cell fates in the spinal cord, because compound $Efnb2^{+/-}$; $Shh^{+/-}$ mutant embryos also exhibited reduced pMN relative to p3 cell numbers while neither $Efnb2^{+/-}$ or $Shh^{+/-}$ heterozygosity had an effect on its own. This role in morphogen-dependent tissue patterning is especially interesting when considered in light of roles for EPH/EPHRINs in refining cell position following morphogenetic patterning, and suggests an additional way that these two events may be coupled.

During neural stem cell differentiation, genetic and metabolic changes are coordinated to specify distinct cellular properties that actualize distinct cell states. Activation of EPHB signaling was previously shown to promote adult neural stem cell differentiation, but the downstream signaling pathways were unclear (Ashton et al., 2012; Ottone et al., 2014). It has been recently shown that EPHRIN-B1/EPHB signaling inhibits expression of dihydrofolate reductase (DHFR), which is a key enzyme in the 1C folate metabolic pathway (Fawal et al., 2018). Lack of folate has long been linked to developmental abnormalities including neural tube defects, as well as other neurodevelopmental disorders, and 1C metabolism impacts epigenetic programs through histone and DNA methylation. Inhibition of DHFR by EPHB forward signaling leads to increased neural stem cell differentiation during development through altered 1C metabolism which modifies H2K4 methylation of key progenitor genes. These epigenetic changes lock cells into a differentiated state that is maintained and inherited in future generations. As such, impacts of EPH/EPHRIN on 1C metabolism are propagated from cell signaling to transcriptional and epigenetic changes and may be relevant to neurological disorders involving folate deficiency and DHFR dysfunction.

5. Craniofacial and musculoskeletal development

Multiple EPH/EPHRINs regulate development of the skeleton, including roles in craniofacial morphogenesis. Numerous studies have demonstrated that EPH/EPHRIN signaling is crucial for guidance of the migratory cranial NCCs that compose most of the skeleton of the viscerocranium in vertebrates (Kindberg & Bush, 2019; Mellott & Burke, 2008b; Theveneau & Mayor, 2012). Notably, *EFNB1* is mutated in an X-linked condition called craniofrontonasal syndrome (CFNS) that affects craniofacial, axial skeletal, and neurological development (Compagni, Logan, Klein, & Adams, 2003; Twigg et al., 2004; Wieland et al., 2004). Patients with CFNS exhibit a shorter midface (frontonasal dysplasia), inappropriate fusion of calvaria of the skull (coronal synostosis), wide-set eyes (hypertelorism), and occasionally also exhibit cleft lip/palate, limb anomalies, and agenesis of the CC. CFNS is unusual in that despite being an X-linked condition, it affects females with heterozygous loss of function of *EFNB1* severely, whereas males with hemizygous loss remain mildly affected or unaffected. Targeted disruption of *Efnb1* in mice results in most of the same phenotypes as in humans, including increased severity in *Efnb1*$^{+/-}$ females, but with some differences; *Efnb1*$^{+/-}$ mutant mice do not exhibit coronal synostosis, but instead exhibit nasofrontal suture synostosis, and exhibit cleft palate only rather than cleft lip and palate (Bush & Soriano, 2010; Compagni et al., 2003; Davy, Aubin, & Soriano, 2004). It was immediately recognized that this increased heterozygous severity was likely related to cell sorting of mosaic EPHRIN-B1 expressing and non-expressing cells generated by random X chromosome inactivation, which we have since confirmed using a human induced pluripotent stem cell (hiPSC) model (Compagni et al., 2003; Davy et al., 2004; Niethamer et al., 2017).

How cell sorting occurred in this context, and how it caused dysmorphology, was more mysterious. It was initially proposed that cell sorting upon heterozygous loss of *EFNB1* caused craniofacial phenotypes by disrupting the NCC/mesoderm boundary or by disrupting NCC migration at early stages (Davy et al., 2004; Twigg et al., 2004). Recently, we found that inducing *Efnb1* heterozygosity in the post-migratory NCC-derived craniofacial mesenchyme can drive cell sorting and craniofacial dysmorphology similar to what is observed upon germline deletion of *Efnb1*, suggesting that CFNS pathology may be caused by later effects on craniofacial shape (Niethamer et al., 2020). Using the secondary palate as a model for EPHRIN-B1 driven tissue shape change, we found that boundaries between EPHRIN-B1 expressing and non-expressing domains correlate with dysmorphology in a manner that is reminiscent of cell aggregates generated in culture. Specifically, when EPHRIN-B1 expressing and EPHB2 expressing cells are allowed to aggregate, they minimize contact with each other in accordance with a minimization of interfacial tension, causing them to be irregular in shape, just as the secondary palatal shelves are in *Efnb1*$^{+/-}$ mutants (Bush & Soriano, 2010; Kindberg et al., 2021; Niethamer et al., 2020). We speculate that polydactyly, which is also consistently observed in *Efnb1*$^{+/-}$ mutants, may also relate to tissue shape changes initiated at EPHRIN-B1 expression boundaries.

These data suggest that tissue dysmorphology in CFNS may also be caused by EPH/EPHRIN-driven heterotypic interfacial tension. We found that blocking ROCK and MLCK activity could relax tortuous tissue shape changes in aggregates, consistent with these changes also being driven by effects on cortical actomyosin contractility (Kindberg et al., 2021). We and others also demonstrated that EPHB2; EPHB3 combined loss of function recapitulated most of the *Efnb1*$^{Y/-}$ hemizygous craniofacial phenotype. Interestingly, using quantitative three-dimensional morphometric analysis, we found that even compound loss of function of all of the EPH receptors that bind to EPHRIN-B1 in *Ephb1*$^{-/-}$; *Ephb2*$^{-/-}$; *Ephb3*$^{-/-}$ mutant mice does not recapitulate *Efnb1*$^{+/-}$ phenotypic severity, consistent with cell sorting causing these more severe phenotypes (Niethamer et al., 2020; Risley et al., 2009). Though *Efnb1*$^{+/-}$; *Ephb1*$^{-/-}$; *Ephb2*$^{-/-}$; *Ephb3*$^{-/-}$ mutant embryos still exhibited cell sorting, it was dramatically reduced, indicating either that EPHRIN-B1 has EPH-independent functions in segregation, or that EPH receptors that lack substantial biochemical affinity may also contribute (Niethamer et al., 2020). It remains mysterious why CFNS phenotypes are so highly stereotyped. Although this may relate to constraint of dysmorphology to the domains of expression of EPHRIN-B1, we demonstrated using quantitative morphometric approaches that many *Efnb1*$^{+/-}$ craniofacial shape changes were present in *Efnb1*$^{Y/-}$ embryos as well, albeit less severe, which would not necessarily be expected if CFNS were a result of random dysmorphology constrained to specific structures. It is therefore likely that additional morphogenetic mechanisms contribute to CFNS dysmorphology.

EPH/EPHRIN signaling also has later roles in skeletal development, and skeletal phenotypes in CFNS, the only human skeleton condition known to be caused by EPH/EPHRIN dysfunction, may relate to these functions (Arthur & Gronthos, 2021). Whereas cartilage segmentation phenotypes observed upon loss of function of *Efnb1* in the limbs and ribs in mice may relate to early patterning and progenitor cellular position, its disruption in osteoblast progenitor cells in *Osx-cre; Efnb1*$^{lox/lox}$ mice resulted in perturbed long-bone growth, reduction in osteoblast progenitor cells and osteoclasts, and disruption of growth plate structure (Compagni et al., 2003; Davy et al., 2004; Nguyen et al., 2016). Disruption of *Efnb1* function in osteoblasts using a Col1α2-Cre also resulted in skull and long bone defects and reduced bone formation and mineralization, whereas overexpression of EPHRIN-B1 within osteoblast progenitors resulted in increased bone thickness (Cheng et al., 2013; Xing, Kim, Wergedal, Chen, & Mohan, 2010). The function of EPHRIN-B1 in bone formation was found to relate to EPHRIN-B1 reverse signaling-driven dephosphorylation of the Taz transcription factor, resulting in an increase in expression of Osterix/Sp7, the master regulator of bone formation. Genetic disruption of the PDZ domain binding motif in *Efnb1*$^{\Delta V}$ mice did not result in apparent skeletal or bone defects, though the quality of bone was not examined in detail in that study (Bush & Soriano, 2009). Disruption of EPHRIN-B1 in osteoprogenitor cells also resulted in disruption of bone homeostasis with increased osteoclasts resulting in osteoporosis and delayed fracture healing (Arthur et al., 2018; Arthur, Paton, Zannettino, & Gronthos, 2020). The function of EPHRIN-B1 in regulating osteoclast numbers appears to relate to a forward signaling function, indicating that bidirectional EPHRIN-B1/EPHB2 signaling between multiple osteogenic cell populations may regulate bone homeostasis (Arthur & Gronthos, 2021).

In contrast, disruption of *Efnb2* in the same osteoprogenitor cell types in *Osx-Cre; Efnb2*$^{lox/lox}$ mutant mice resulted in impaired osteoclastogenesis at the growth plate, which in turn resulted in increased trabecular bone formation and greater cortical bone thickness (Tonna et al., 2016). This Cre drives recombination in both osteoblasts and chondrocytes, where loss of EPHRIN-B2 resulted in hypertrophic chondrocytes that failed to promote cartilage degradation. The impaired destruction of cartilage during endochondral ossification is a hallmark of the neonatal osteopetrosis phenotype that these mice exhibited, though this phenotype resolved by six weeks of age. EPHRIN-B2/EPHB4 signaling also regulates bone homeostasis in adult mice by controlling the balance between bone resorption by osteoclasts and bone formation by osteoblasts (Zhao et al., 2006). Osteoclast progenitors primarily express EPHRIN-B2, where it is regulated by NFATC1, a master regulator of osteoclast differentiation, and signal with osteoblast progenitors that express EPHB4. PDZ-dependent reverse signaling from osteoblasts to osteoclast progenitors leads to inhibition of osteoclast differentiation through negative feedback on *Nfatc1* transcription. Forward signaling from osteoclasts to osteoblast progenitors increases osteoblast differentiation and enhances bone formation. In this way, direct cell-cell interaction between osteoblasts and osteoclasts may facilitate the dynamic transition between bone resorption and formation states. Subsequent studies refined our understanding of EPHB4/EPHRIN-B2 in bone formation and revealed that insulin-like growth factor (IGF-I) signaling is required for expression of EPHRIN-B2 in osteoclasts and expression of EPHRIN-B2 and EPHB4 in osteoblasts and chondrocytes and that EPHRIN-B2/EPHB4 mediate the effects of IGF-I signaling on endochondral bone formation (Wang et al., 2014). Furthermore, EPHRIN-B2 loss from chondrocytes resulted in reduced chondrocyte-osteoblast trans-differentiation and decreased bone formation during fracture repair suggesting a new avenue for therapeutic benefit (Wang et al., 2020).

Recent studies have revealed that EPH/EPHRIN signaling also regulates myogenic differentiation (Arnold et al., 2020). Targeted disruption of *Epha7* in mice resulted in hindlimb muscles with fewer and smaller myofibers, that exhibited slower muscle regeneration upon injury. EPHA7 is expressed by differentiated myocytes during muscle regeneration, whereas proliferating myoblasts express the signaling partner EPHRIN-A5. Treatment of satellite muscle stem cells with exogenous EPHA7 extracellular domain resulted in induction of myogenic differentiation in culture, suggesting that EPHA7 signals from differentiated myocytes to myoblasts to regulate their differentiation. Though the reverse signaling mechanisms utilized in this context are not yet known, this work suggests that contact-mediated EPH/EPHRIN signaling contributes to a "community effect" that drives coordinated myogenic differentiation to ensure a sufficient number of myocytes for fusion and formation of a muscle fiber.

6. Conclusions and perspectives

Through the continual discovery of new and surprising developmental functions of EPH/EPHRIN signaling as well as the identification of both unique and overlapping signaling mechanisms employed, our awareness of the complexity of the EPH/EPHRIN signaling family continues to increase. Though mechanisms such as suppression of RAS/MAPK to modulate differentiation, and activation of RhoA to regulate actomyosin dynamics are shared across multiple contexts, signaling functions are just as often cell type-specific and context-dependent. As such, consideration of EPH/EPHRIN signaling from a universal perspective may have marginal utility, and continued, detailed *in vivo* study of signal transduction mechanisms is warranted. For example, while beautiful structural biology and cell culture work has provided great insight into the multimerization of EPH/EPHRIN signaling complexes and the potential for homo- and hetero-oligomers, future work that manipulates multimerization *in vivo* will be required for an understanding of how this pathway signals in the embryo, and whether context-specific regulation of these properties is at play. Approaches for the *in vivo* examination of homo- vs. hetero-oligomeric states, for example through applying bimolecular fluorescence complementation, may begin to provide insights into when and how distinct oligomeric states are utilized (Rogers & Fantauzzo, 2021). The *in vivo* context-specific utilization of proximal signaling partners also remains a major gap in knowledge. New spatial and single-cell proteomic methods applied to embryonic tissues may enlighten our understanding of overlapping functions as well as the shared and distinct context-specific downstream signal transduction pathways that allow specificity of signaling output to emerge from this complex signaling family.

Acknowledgments

I apologize to many colleagues whose work I could not discuss due to space limitations. I am grateful to Nancy Ann Oberheim and my laboratory colleagues, Yoon-Gu Jang, Abigail Kindberg, Ace Lewis, Terren Niethamer, Marisol O'Neill, and Camilla Teng for helpful comments on the manuscript. J.O.B was supported by R01DE023337, R01DE025877, and R01HL144785 to J.O.B.

References

Aharon, R., Janes, P. W., Burgess, A. W., Hamza, K., Klebaner, F., & Lackmann, M. (2014). A mathematical model for Eph/Ephrin-directed segregation of intermingled cells. *PLoS One, 9*, e111803. https://doi.org/10.1371/journal.pone.0111803.

Amyere, M., Revencu, N., Helaers, R., Pairet, E., Baselga, E., Cordisco, M., et al. (2017). Germline loss-of-function mutations in EPHB4 cause a second form of capillary malformation-arteriovenous malformation (CM-AVM2) deregulating RAS-MAPK signaling. *Circulation, 136*, 1037–1048. https://doi.org/10.1161/CIRCULATIONAHA.116.026886.

Arcas, A., Wilkinson, D. G., & Nieto, M.Á. (2020). The evolutionary history of Ephs and ephrins: Toward multicellular organisms. *Molecular Biology and Evolution*, *37*, 379–394. https://doi.org/10.1093/molbev/msz222.

Arnold, L. L., Cecchini, A., Stark, D. A., Ihnat, J., Craigg, R. N., Carter, A., et al. (2020). EphA7 promotes myogenic differentiation via cell-cell contact. *Elife*, *9*, e53689. https://doi.org/10.7554/eLife.53689.

Arthur, A., & Gronthos, S. (2021). Eph-ephrin signaling mediates cross-talk within the bone microenvironment. *Frontiers in Cell and Developmental Biology*, *9*, 598612. https://doi.org/10.3389/fcell.2021.598612.

Arthur, A., Nguyen, T. M., Paton, S., Klisuric, A., Zannettino, A. C. W., & Gronthos, S. (2018). The osteoprogenitor-specific loss of ephrinB1 results in an osteoporotic phenotype affecting the balance between bone formation and resorption. *Science Reports*, *8*, 12756. https://doi.org/10.1038/s41598-018-31190-2.

Arthur, A., Paton, S., Zannettino, A. C. W., & Gronthos, S. (2020). Conditional knockout of ephrinB1 in osteogenic progenitors delays the process of endochondral ossification during fracture repair. *Bone*, *132*, 115189. https://doi.org/10.1016/j.bone.2019.115189.

Ashlin, T. G., Wu, Z., Xu, Q., & Wilkinson, D. G. (2019). *Role of SHIP2 in cell repulsion regulated by Eph receptor and ephrin signaling*. https://doi.org/10.1101/872846.

Ashton, R. S., Conway, A., Pangarkar, C., Bergen, J., Lim, K.-I., Shah, P., et al. (2012). Astrocytes regulate adult hippocampal neurogenesis through ephrin-B signaling. *Nature Neuroscience*, *15*, 1399–1406. https://doi.org/10.1038/nn.3212.

Atapattu, L., Lackmann, M., & Janes, P. W. (2014). The role of proteases in regulating Eph/ephrin signaling. *Cell Adhesion & Migration*, *8*, 294–307. https://doi.org/10.4161/19336918.2014.970026.

Barquilla, A., & Pasquale, E. B. (2015). Eph receptors and ephrins: therapeutic opportunities. *Annual Review of Pharmacology and Toxicology*, *55*, 465–487. https://doi.org/10.1146/annurev-pharmtox-011112-140226.

Batlle, E., Henderson, J. T., Beghtel, H., van den Born, M. M., Sancho, E., Huls, G., et al. (2002). Beta-catenin and TCF mediate cell positioning in the intestinal epithelium by controlling the expression of EphB/ephrinB. *Cell*, *111*, 251–263.

Becker, E., Huynh-Do, U., Holland, S., Pawson, T., Daniel, T. O., & Skolnik, E. Y. (2000). Nck-interacting Ste20 kinase couples Eph receptors to c-Jun N-terminal kinase and integrin activation. *Molecular and Cellular Biology*, *20*, 1537–1545. https://doi.org/10.1128/MCB.20.5.1537-1545.2000.

Bisson, N., Poitras, L., Mikryukov, A., Tremblay, M., & Moss, T. (2007). EphA4 signaling regulates blastomere adhesion in the Xenopus embryo by recruiting Pak1 to suppress Cdc42 function. *Molecular Biology of the Cell*, *18*, 1030–1043. https://doi.org/10.1091/mbc.e06-04-0294.

Bowden, T. A., Aricescu, A. R., Nettleship, J. E., Siebold, C., Rahman-Huq, N., Owens, R. J., et al. (2009). Structural plasticity of Eph receptor A4 facilitates cross-class ephrin signaling. *Structure*, *17*, 1386–1397. https://doi.org/10.1016/j.str.2009.07.018.

Brodland, G. W. (2002). The differential interfacial tension hypothesis (DITH): A comprehensive theory for the self-rearrangement of embryonic cells and tissues. *Journal of Biomechanical Engineering*, *124*, 188–197. https://doi.org/10.1115/1.1449491.

Brückner, K., Pasquale, E. B., & Klein, R. (1997). Tyrosine phosphorylation of transmembrane ligands for Eph receptors. *Science*, *275*, 1640–1643. https://doi.org/10.1126/science.275.5306.1640.

Bush, J. O., & Soriano, P. (2009). Ephrin-B1 regulates axon guidance by reverse signaling through a PDZ-dependent mechanism. *Genes Development*, *23*, 1586–1599. https://doi.org/10.1101/gad.1807209.

Bush, J. O., & Soriano, P. (2010). Ephrin-B1 forward signaling regulates craniofacial morphogenesis by controlling cell proliferation across Eph-ephrin boundaries. *Genes Development*, *24*, 2068–2080. https://doi.org/10.1101/gad.1963210.

Bush, J. O., & Soriano, P. (2012). Eph/ephrin signaling: Genetic, phosphoproteomic, and transcriptomic approaches. *Seminars in Cell & Developmental Biology*, *23*, 26–34. https://doi.org/10.1016/j.semcdb.2011.10.018.

Calzolari, S., Terriente, J., & Pujades, C. (2014). Cell segregation in the vertebrate hindbrain relies on actomyosin cables located at the interhombomeric boundaries. *The EMBO Journal*, *33*, 686–701. https://doi.org/10.1002/embj.201386003.

Canty, L., Zarour, E., Kashkooli, L., François, P., & Fagotto, F. (2017). Sorting at embryonic boundaries requires high heterotypic interfacial tension. *Nature Communications*, *8*, 157. https://doi.org/10.1038/s41467-017-00146-x.

Carvalho, R. F., Beutler, M., Marler, K. J., Knoll, B., Becker-Barroso, E., Heintzmann, R., et al. (2006). Silencing of EphA3 through a cis interaction with ephrinA5. *Nature Neuroscience*, *9*, 322–330. https://doi.org/10.1038/nn1655.

Cayuso, J., Dzementsei, A., Fischer, J. C., Karemore, G., Caviglia, S., Bartholdson, J., et al. (2016). EphrinB1/EphB3b coordinate bidirectional epithelial-mesenchymal interactions controlling liver morphogenesis and laterality. *Developmental Cell*, *39*, 316–328. https://doi.org/10.1016/j.devcel.2016.10.009.

Cayuso, J., Xu, Q., & Wilkinson, D. G. (2015). Mechanisms of boundary formation by Eph receptor and ephrin signaling. *Developmental Biology, Cell Adhesion in Development*, *401*, 122–131. https://doi.org/10.1016/j.ydbio.2014.11.013.

Chan, J., Mably, J. D., Serluca, F. C., Chen, J. N., Goldstein, N. B., Thomas, M. C., et al. (2001). Morphogenesis of prechordal plate and notochord requires intact Eph/ephrin B signaling. *Developmental Biology*, *234*, 470–482. https://doi.org/10.1006/dbio.2001.0281.

Chang, C.-J., Chang, M.-Y., Lee, Y.-C., Chen, K.-Y., Hsu, T.-I., Wu, Y.-H., et al. (2018). Nck2 is essential for limb trajectory selection by spinal motor axons. *Developmental Dynamics*, *247*, 1043–1056. https://doi.org/10.1002/dvdy.24656.

Cheng, S., Kesavan, C., Mohan, S., Qin, X., Alarcon, C. M., Wergedal, J., et al. (2013). Transgenic overexpression of Ephrin B1 in bone cells promotes bone formation and an anabolic response to mechanical loading in mice. *PLoS One*, *8*, e69051. https://doi.org/10.1371/journal.pone.0069051.

Chong, L. D., Park, E. K., Latimer, E., Friesel, R., & Daar, I. O. (2000). Fibroblast growth factor receptor-mediated rescue of x-ephrin B1-induced cell dissociation in Xenopus embryos. *Molecular Cell Biology*, *20*, 724–734. https://doi.org/10.1128/MCB.20.2.724-734.2000.

Compagni, A., Logan, M., Klein, R., & Adams, R. H. (2003). Control of skeletal patterning by ephrinB1-EphB interactions. *Developmental Cell*, *5*, 217–230.

Cowan, C. A., & Henkemeyer, M. (2001). The SH2/SH3 adaptor Grb4 transduces B-ephrin reverse signals. *Nature*, *413*, 174–179. https://doi.org/10.1038/35093123.

Daar, I. O. (2012). Non-SH2/PDZ reverse signaling by ephrins. *Seminars in Cell & Developmental Biology*, *23*, 65–74. https://doi.org/10.1016/j.semcdb.2011.10.012.

Dalva, M. B., Takasu, M. A., Lin, M. Z., Shamah, S. M., Hu, L., Gale, N. W., et al. (2000). EphB receptors interact with NMDA receptors and regulate excitatory synapse formation. *Cell*, *103*, 945–956.

Darling, T. K., & Lamb, T. J. (2019). Emerging roles for Eph receptors and ephrin ligands in immunity. *Frontiers in Immunology*, *10*, 1473. https://doi.org/10.3389/fimmu.2019.01473.

Davy, A., Aubin, J., & Soriano, P. (2004). Ephrin-B1 forward and reverse signaling are required during mouse development. *Genes Development*, *18*, 572–583. https://doi.org/10.1101/gad.1171704.

Davy, A., Bush, J. O., & Soriano, P. (2006). Inhibition of gap junction communication at ectopic Eph/ephrin boundaries underlies craniofrontonasal syndrome. *PLoS Biology*, *4*, e315. https://doi.org/10.1371/journal.pbio.0040315.

Davy, A., & Robbins, S. M. (2000). Ephrin-A5 modulates cell adhesion and morphology in an integrin-dependent manner. *The EMBO Journal*, *19*, 5396–5405. https://doi.org/10.1093/emboj/19.20.5396.

Davy, A., & Soriano, P. (2005). Ephrin signaling in vivo: Look both ways. *Developmental Dynamics*, *232*, 1–10. https://doi.org/10.1002/dvdy.20200.

Defourny, J., Peuckert, C., Kullander, K., & Malgrange, B. (2019). EphA4-ADAM10 interplay patterns the cochlear sensory epithelium through local disruption of adherens junctions. *iScience*, *11*, 246–257. https://doi.org/10.1016/j.isci.2018.12.017.

Dionne, U., Chartier, F. J. M., de Los, L., Santos, Y., Lavoie, N., Bernard, D. N., et al. (2018). Direct phosphorylation of SRC homology 3 domains by tyrosine kinase receptors disassembles ligand-induced signaling networks. *Molecular Cell*, *70*, 995–1007.e11. https://doi.org/10.1016/j.molcel.2018.05.013.

Drescher, U. (2002). Eph family functions from an evolutionary perspective. *Current Opinion in Genetics & Development*, *12*, 397–402.

Du, E., Li, X., He, S., Li, X., & He, S. (2020). The critical role of the interplays of EphrinB2/EphB4 and VEGF in the induction of angiogenesis. *Molecular Biology Reports*, *47*, 4681–4690. https://doi.org/10.1007/s11033-020-05470-y.

Duguay, D., Foty, R. A., & Steinberg, M. S. (2003). Cadherin-mediated cell adhesion and tissue segregation: Qualitative and quantitative determinants. *Developmental Biology*, *253*, 309–323. https://doi.org/10.1016/S0012-1606(02)00016-7.

Eerola, I., Boon, L. M., Mulliken, J. B., Burrows, P. E., Dompmartin, A., Watanabe, S., et al. (2003). Capillary malformation-arteriovenous malformation, a new clinical and genetic disorder caused by RASA1 mutations. *American Journal of Human Genetics*, *73*, 1240–1249. https://doi.org/10.1086/379793.

Elowe, S., Holland, S. J., Kulkarni, S., & Pawson, T. (2001). Downregulation of the Ras-mitogen-activated protein kinase pathway by the EphB2 receptor tyrosine kinase is required for ephrin-induced neurite retraction. *Molecular Cell Biology*, *21*, 7429–7441. https://doi.org/10.1128/MCB.21.21.7429-7441.2001.

Evren, S., Wen, J. W. H., Luu, O., Damm, E. W., Nagel, M., & Winklbauer, R. (2014). EphA4-dependent Brachyury expression is required for dorsal mesoderm involution in the Xenopus gastrula. *Development*, *141*, 3649–3661. https://doi.org/10.1242/dev.111880.

Fagotto, F. (2014). The cellular basis of tissue separation. *Development*, *141*, 3303–3318. https://doi.org/10.1242/dev.090332.

Fagotto, F., Rohani, N., Touret, A.-S., & Li, R. (2013). A molecular base for cell sorting at embryonic boundaries: Contact inhibition of cadherin adhesion by ephrin/Eph-dependent contractility. *Developmental Cell*, *27*, 72–87. https://doi.org/10.1016/j.devcel.2013.09.004.

Fagotto, F., Winklbauer, R., & Rohani, N. (2014). Ephrin-Eph signaling in embryonic tissue separation. *Cell Adhesion & Migration*, *8*, 308–326. https://doi.org/10.4161/19336918.2014.970028.

Fawal, M.-A., Jungas, T., Kischel, A., Audouard, C., Iacovoni, J. S., & Davy, A. (2018). Cross talk between one-carbon metabolism, Eph signaling, and histone methylation promotes neural stem cell differentiation. *Cell Reports*, *23*, 2864–2873.e7. https://doi.org/10.1016/j.celrep.2018.05.005.

Fawcett, J. P., Georgiou, J., Ruston, J., Bladt, F., Sherman, A., Warner, N., et al. (2007). Nck adaptor proteins control the organization of neuronal circuits important for walking. *Proceedings of the National Academy of Sciences of the United States of America*, *104*, 20973–20978. https://doi.org/10.1073/pnas.0710316105.

Fiuza, U.-M., Negishi, T., Rouan, A., Yasuo, H., & Lemaire, P. (2020). A Nodal/Eph signalling relay drives the transition from apical constriction to apico-basal shortening in ascidian endoderm invagination. *Development, 147*, dev186965. https://doi.org/10.1242/dev.186965.

Freywald, A., Sharfe, N., & Roifman, C. M. (2002). The kinase-null EphB6 receptor undergoes transphosphorylation in a complex with EphB1. *Journal of Biological Chemistry, 277*, 3823–3828. https://doi.org/10.1074/jbc.M108011200.

Fuller, T. (2003). Forward EphB4 signaling in endothelial cells controls cellular repulsion and segregation from ephrinB2 positive cells. *Journal of Cell Science, 116*, 2461–2470. https://doi.org/10.1242/jcs.00426.

Gaino, E., Bavestrello, G., & Magnino, G. (1999). Self/non-self recognition in sponges. *Italian Journal of Zoology, 66*, 299–315. https://doi.org/10.1080/11250009909356270.

Gaino, E., & Burlando, B. (1990). Sponge cell motility: A model system for the study of morphogenetic processes. *Bollettino di zoologia, 57*, 109–118. https://doi.org/10.1080/11250009009355684.

Gaitanos, T. N., Koerner, J., & Klein, R. (2016). Tiam–Rac signaling mediates trans-endocytosis of ephrin receptor EphB2 and is important for cell repulsion. *Journal of Cellular Biology*. https://doi.org/10.1083/jcb.201512010.

Gale, N. W., Holland, S. J., Valenzuela, D. M., Flenniken, A., Pan, L., Ryan, T. E., et al. (1996). Eph receptors and ligands comprise two major specificity subclasses and are reciprocally compartmentalized during embryogenesis. *Neuron, 17*, 9–19.

Genander, M., Halford, M. M., Xu, N. J., Eriksson, M., Yu, Z., Qiu, Z., et al. (2009). Dissociation of EphB2 signaling pathways mediating progenitor cell proliferation and tumor suppression. *Cell, 139*, 679–692. https://doi.org/10.1016/j.cell.2009.08.048.

Gerety, S. S., Wang, H. U., Chen, Z. F., & Anderson, D. J. (1999). Symmetrical mutant phenotypes of the receptor EphB4 and its specific transmembrane ligand ephrin-B2 in cardiovascular development. *Molecular Cell, 4*, 403–414.

Groppa, E., Brkic, S., Uccelli, A., Wirth, G., Korpisalo-Pirinen, P., Filippova, M., et al. (2018). EphrinB2/EphB4 signaling regulates non-sprouting angiogenesis by VEGF. *The EMBO Reports, 19*, e45054. https://doi.org/10.15252/embr.201745054.

Guo, F.-Y., & Lesk, A. M. (2014). Sizes of interface residues account for cross-class binding affinity patterns in Eph receptor-ephrin families. *Proteins, 82*, 349–353. https://doi.org/10.1002/prot.24414.

Hattori, M., Osterfield, M., & Flanagan, J. G. (2000). Regulated cleavage of a contact-mediated axon repellent. *Science, 289*, 1360–1365. https://doi.org/10.1126/science.289.5483.1360.

Haupaix, N., Abitua, P. B., Sirour, C., Yasuo, H., Levine, M., & Hudson, C. (2014). Ephrin-mediated restriction of ERK1/2 activity delimits the number of pigment cells in the Ciona CNS. *Developmental Biology, 394*, 170–180. https://doi.org/10.1016/j.ydbio.2014.07.010.

Henkemeyer, M., Rossi, D. J., Holmyard, D. P., Puri, M. C., Mbamalu, G., Harpal, K., et al. (1995). Vascular system defects and neuronal apoptosis in mice lacking ras GTPase-activating protein. *Nature, 377*, 695–701. https://doi.org/10.1038/377695a0.

Himanen, J. P., Chumley, M. J., Lackmann, M., Li, C., Barton, W. A., Jeffrey, P. D., et al. (2004). Repelling class discrimination: Ephrin-A5 binds to and activates EphB2 receptor signaling. *Nature Neuroscience*, 7, 501–509. https://doi.org/10.1038/nn1237.

Himanen, J.-P., Henkemeyer, M., & Nikolov, D. B. (1998). Crystal structure of the ligand-binding domain of the receptor tyrosine kinase EphB2. *Nature, 396*, 486–491. https://doi.org/10.1038/24904.

Himanen, J. P., & Nikolov, D. B. (2003). Eph signaling: A structural view. *Trends in Neuroscience, 26*, 46–51.

Himanen, J. P., Yermekbayeva, L., Janes, P. W., Walker, J. R., Xu, K., Atapattu, L., et al. (2010). Architecture of Eph receptor clusters. *Proceedings of the National Academy of Sciences of the United States of America*, *107*, 10860–10865. https://doi.org/10.1073/pnas.1004148107.

Hock, B., Böhme, B., Karn, T., Feller, S., Rübsamen-Waigmann, H., & Strebhardt, K. (1998). Tyrosine-614, the major autophosphorylation site of the receptor tyrosine kinase HEK2, functions as multi-docking site for SH2-domain mediated interactions. *Oncogene*, *17*, 255–260. https://doi.org/10.1038/sj.onc.1201907.

Holland, S. J., Gale, N. W., Gish, G. D., Roth, R. A., Songyang, Z., Cantley, L. C., et al. (1997). Juxtamembrane tyrosine residues couple the Eph family receptor EphB2/Nuk to specific SH2 domain proteins in neuronal cells. *The EMBO Journal*, *16*, 3877–3888. https://doi.org/10.1093/emboj/16.13.3877.

Hu, T., Shi, G., Larose, L., Rivera, G. M., Mayer, B. J., & Zhou, R. (2009). Regulation of process retraction and cell migration by EphA3 is mediated by the adaptor protein Nck1. *Biochemistry*, *48*, 6369–6378. https://doi.org/10.1021/bi900831k.

Huang, X., Wu, D., Jin, H., Stupack, D., & Wang, J. Y. J. (2008). Induction of cell retraction by the combined actions of Abl-CrkII and Rho-ROCK1 signaling. *The Journal of Cell Biology*, *183*, 711–723. https://doi.org/10.1083/jcb.200801192.

Iwasato, T., Katoh, H., Nishimaru, H., Ishikawa, Y., Inoue, H., Saito, Y. M., et al. (2007). Rac-GAP alpha-chimerin regulates motor-circuit formation as a key mediator of EphrinB3/EphA4 forward signaling. *Cell*, *130*, 742–753. https://doi.org/10.1016/j.cell.2007.07.022.

Jaber Chehayeb, R., Wang, J., Stiegler, A. L., & Boggon, T. J. (2020). The GTPase-activating protein p120RasGAP has an evolutionarily conserved "FLVR-unique" SH2 domain. *The Journal of Biological Chemistry*, *295*, 10511–10521. https://doi.org/10.1074/jbc.RA120.013976.

Janes, P. W., Griesshaber, B., Atapattu, L., Nievergall, E., Hii, L. L., Mensinga, A., et al. (2011). Eph receptor function is modulated by heterooligomerization of A and B type Eph receptors. *The Journal of Cell Biology*, *195*, 1033–1045. https://doi.org/10.1083/jcb.201104037.

Janes, P. W., Saha, N., Barton, W. A., Kolev, M. V., Wimmer-Kleikamp, S. H., Nievergall, E., et al. (2005). Adam meets Eph: An ADAM substrate recognition module acts as a molecular switch for ephrin cleavage in trans. *Cell*, *123*, 291–304. https://doi.org/10.1016/j.cell.2005.08.014.

Janes, P. W., Vail, M. E., Ernst, M., & Scott, A. M. (2021). Eph receptors in the immunosuppressive tumor microenvironment. *Cancer Research*, *81*, 801–805. https://doi.org/10.1158/0008-5472.CAN-20-3047.

Javaherian, S., D'Arcangelo, E., Slater, B., Londono, C., Xu, B., & McGuigan, A. P. (2017). Modulation of cellular polarization and migration by ephrin/Eph signal-mediated boundary formation. *Integrative Biology*. https://doi.org/10.1039/C7IB00176B.

Ji, Y. J., Hwang, Y.-S., Mood, K., Cho, H.-J., Lee, H.-S., Winterbottom, E., et al. (2014). EphrinB2 affects apical constriction in Xenopus embryos and is regulated by ADAM10 and flotillin-1. *Nature Communications*, *5*, 3516. https://doi.org/10.1038/ncomms4516.

Jones, T. L., Chong, L. D., Kim, J., Xu, R. H., Kung, H. F., & Daar, I. O. (1998). Loss of cell adhesion in Xenopus laevis embryos mediated by the cytoplasmic domain of XLerk, an erythropoietin-producing hepatocellular ligand. *Proceedings of the National Academy of Sciences of the United States of America*, *95*, 576–581. https://doi.org/10.1073/pnas.95.2.576.

Jorgensen, C., Sherman, A., Chen, G. I., Pasculescu, A., Poliakov, A., Hsiung, M., et al. (2009). Cell-specific information processing in segregating populations of Eph receptor ephrin-expressing cells. *Science, 326*, 1502–1509. https://doi.org/10.1126/science.1176615.

Kania, A., & Klein, R. (2016). Mechanisms of ephrin-Eph signalling in development, physiology and disease. *Nature Reviews. Molecular Cell Biology, 17*, 240–256. https://doi.org/10.1038/nrm.2015.16.

Kao, T.-J., & Kania, A. (2011). Ephrin-mediated cis-attenuation of Eph receptor signaling is essential for spinal motor axon guidance. *Neuron, 71*, 76–91. https://doi.org/10.1016/j.neuron.2011.05.031.

Kato, K., Yazawa, T., Taki, K., Mori, K., Wang, S., Nishioka, T., et al. (2012). The inositol 5-phosphatase SHIP2 is an effector of RhoA and is involved in cell polarity and migration. *Molecular Biology of the Cell, 23*, 2593–2604. https://doi.org/10.1091/mbc.e11-11-0958.

Kawasaki, J., Aegerter, S., Fevurly, R. D., Mammoto, A., Mammoto, T., Sahin, M., et al. (2014). RASA1 functions in EPHB4 signaling pathway to suppress endothelial mTORC1 activity. *Journal of Clinical Investigation, 124*, 2774–2784. https://doi.org/10.1172/JCI67084.

Kemp, H. A., Cooke, J. E., & Moens, C. B. (2009). EphA4 and EfnB2a maintain rhombomere coherence by independently regulating intercalation of progenitor cells in the zebrafish neural keel. *Developmental Biology, 327*, 313–326. https://doi.org/10.1016/j.ydbio.2008.12.010.

Kim, I., Ryu, Y. S., Kwak, H. J., Ahn, S. Y., Oh, J.-L., Yancopoulos, G. D., et al. (2002). EphB ligand, ephrinB2, suppresses the VEGF- and angiopoietin 1-induced Ras/mitogen-activated protein kinase pathway in venous endothelial cells. *FASEB Journal, 16*, 1126–1128. https://doi.org/10.1096/fj.01-0805fje.

Kindberg, A. A., & Bush, J. O. (2019). Cellular organization and boundary formation in craniofacial development. *Genesis, 57*, e23271. https://doi.org/10.1002/dvg.23271.

Kindberg, A. A., Srivastava, V., Muncie, J. M., Weaver, V. M., Gartner, Z. J., & Bush, J. O. (2021). EPH/EPHRIN regulates cellular organization by actomyosin contractility effects on cell contacts. *Journal of Cell Biology, 220*, e202005216. https://doi.org/10.1083/jcb.202005216.

Kornberg, T. B. (2019). Scripting a place in time. *Developmental Biology, 447*, 24–27. https://doi.org/10.1016/j.ydbio.2017.09.014.

Krieg, M., Arboleda-Estudillo, Y., Puech, P.-H., Käfer, J., Graner, F., Müller, D. J., et al. (2008). Tensile forces govern germ-layer organization in zebrafish. *Nature Cell Biology, 10*, 429–436. https://doi.org/10.1038/ncb1705.

Krishnan, A., Degnan, B. M., & Degnan, S. M. (2019). The first identification of complete Eph-ephrin signalling in ctenophores and sponges reveals a role for neofunctionalization in the emergence of signalling domains. *BMC Evolutionary Biology, 19*, 96. https://doi.org/10.1186/s12862-019-1418-z.

Kriz, V., Mares, J., Wentzel, P., Funa, N. S., Calounova, G., Zhang, X.-Q., et al. (2007). Shb null allele is inherited with a transmission ratio distortion and causes reduced viability in utero. *Developmental Dynamics, 236*, 2485–2492. https://doi.org/10.1002/dvdy.21257.

Krull, C. E. (2010). Neural crest cells and motor axons in avians: Common and distinct migratory molecules. *Cell Adhesion & Migration, 4*, 631–634. https://doi.org/10.4161/cam.4.4.13594.

Krumlauf, R., & Wilkinson, D. G. (2021). Segmentation and patterning of the vertebrate hindbrain. *Development, 148*, dev186460. https://doi.org/10.1242/dev.186460.

Krupke, O. A., Zysk, I., Mellott, D. O., & Burke, R. D. (2016). Eph and Ephrin function in dispersal and epithelial insertion of pigmented immunocytes in sea urchin embryos. *eLife*, *5*, e16000. https://doi.org/10.7554/eLife.16000.

Lagares, D., Ghassemi-Kakroodi, P., Tremblay, C., Santos, A., Probst, C. K., Franklin, A., et al. (2017). ADAM10-mediated ephrin-B2 shedding promotes myofibroblast activation and organ fibrosis. *Nature Medicine*, *23*, 1405–1415. https://doi.org/10.1038/nm.4419.

Laidlaw, B. J., Schmidt, T. H., Green, J. A., Allen, C. D. C., Okada, T., & Cyster, J. G. (2017). The Eph-related tyrosine kinase ligand Ephrin-B1 marks germinal center and memory precursor B cells. *Journal of Experimental Medicine*, *214*, 639–649. https://doi.org/10.1084/jem.20161461.

Lapinski, P. E., Lubeck, B. A., Chen, D., Doosti, A., Zawieja, S. D., Davis, M. J., et al. (2017). RASA1 regulates the function of lymphatic vessel valves in mice. *Journal of Clinical Investigation*, *127*, 2569–2585. https://doi.org/10.1172/JCI89607.

Laussu, J., Audouard, C., Kischel, A., Assis-Nascimento, P., Escalas, N., Liebl, D. J., et al. (2017). Eph/Ephrin signaling controls progenitor identities in the ventral spinal cord. *Neural Development*, *12*, 10. https://doi.org/10.1186/s13064-017-0087-0.

Laussu, J., Khuong, A., Gautrais, J., & Davy, A. (2014). Beyond boundaries- -Eph:ephrin signaling in neurogenesis. *Cell Adhesion & Migration*, *8*, 349–359. https://doi.org/10.4161/19336918.2014.969990.

Lawrenson, I. D., Wimmer-Kleikamp, S. H., Lock, P., Schoenwaelder, S. M., Down, M., Boyd, A. W., et al. (2002). Ephrin-A5 induces rounding, blebbing and de-adhesion of EphA3-expressing 293T and melanoma cells by CrkII and Rho-mediated signalling. *Journal of Cell Science*, *115*, 1059–1072.

Lee, H.-S., Nishanian, T. G., Mood, K., Bong, Y.-S., & Daar, I. O. (2008). EphrinB1 controls cell-cell junctions through the Par polarity complex. *Nature Cell Biology*, *10*, 979–986. https://doi.org/10.1038/ncb1758.

Li, D., Wenger, T. L., Seiler, C., March, M. E., Gutierrez-Uzquiza, A., Kao, C., et al. (2018). Pathogenic variant in EPHB4 results in central conducting lymphatic anomaly. *Human Molecular Genetics*, *27*, 3233–3245. https://doi.org/10.1093/hmg/ddy218.

Lin, D., Gish, G. D., Songyang, Z., & Pawson, T. (1999). The carboxyl terminus of B class ephrins constitutes a PDZ domain binding motif. *Journal of Biological Chemistry*, *274*, 3726–3733.

Lu, P., Shih, C., & Qi, H. (2017). Ephrin B1-mediated repulsion and signaling control germinal center T cell territoriality and function. *Science*, *356*, eaai9264. https://doi.org/10.1126/science.aai9264.

Maitre, J.-L., Berthoumieux, H., Krens, S. F. G., Salbreux, G., Julicher, F., Paluch, E., et al. (2012). Adhesion functions in cell sorting by mechanically coupling the cortices of adhering cells. *Science*, *338*, 253–256. https://doi.org/10.1126/science.1225399.

Makinen, T., Adams, R. H., Bailey, J., Lu, Q., Ziemiecki, A., Alitalo, K., et al. (2005). PDZ interaction site in ephrinB2 is required for the remodeling of lymphatic vasculature. *Genes Development*, *19*, 397–410. https://doi.org/10.1101/gad.330105.

Mao, Y.-T., Zhu, J. X., Hanamura, K., Iurilli, G., Datta, S. R., & Dalva, M. B. (2018). Filopodia conduct target selection in cortical neurons using differences in signal kinetics of a single kinase. *Neuron*, *98*, 767–782.e8. https://doi.org/10.1016/j.neuron.2018.04.011.

Marquardt, T., Shirasaki, R., Ghosh, S., Andrews, S. E., Carter, N., Hunter, T., et al. (2005). Coexpressed EphA receptors and ephrin-a ligands mediate opposing actions on growth cone navigation from distinct membrane domains. *Cell*, *121*, 127–139. https://doi.org/10.1016/j.cell.2005.01.020.

Martin-Almedina, S., Martinez-Corral, I., Holdhus, R., Vicente, A., Fotiou, E., Lin, S., et al. (2016). EPHB4 kinase-inactivating mutations cause autosomal dominant lymphatic-related hydrops fetalis. *The Journal of Clinical Investigation*, *126*, 3080–3088. https://doi.org/10.1172/JCI85794.

Martin-Almedina, S., Ogmen, K., Sackey, E., Grigoriadis, D., Karapouliou, C., Nadarajah, N., et al. (2021). Janus-faced EPHB4-associated disorders: Novel pathogenic variants and unreported intrafamilial overlapping phenotypes. *Genetics in Medicine*, *23*, 1315–1324. https://doi.org/10.1038/s41436-021-01136-7.

Mellitzer, G., Xu, Q., & Wilkinson, D. G. (1999). Eph receptors and ephrins restrict cell intermingling and communication. *Nature*, *400*, 77–81. https://doi.org/10.1038/21907.

Mellott, D. O., & Burke, R. D. (2008a). The molecular phylogeny of eph receptors and ephrin ligands. *BMC Cell Biology*, *9*, 27. https://doi.org/10.1186/1471-2121-9-27.

Mellott, D. O., & Burke, R. D. (2008b). Divergent roles for Eph and ephrin in avian cranial neural crest. *BMC Developmental Biology*, *8*, 56. https://doi.org/10.1186/1471-213X-8-56.

Mendes, S. W., Henkemeyer, M., & Liebl, D. J. (2006). Multiple Eph receptors and B-class ephrins regulate midline crossing of corpus callosum fibers in the developing mouse forebrain. *Journal of Neuroscience*, *26*, 882–892. https://doi.org/10.1523/JNEUROSCI.3162-05.2006.

Miao, H., Wei, B. R., Peehl, D. M., Li, Q., Alexandrou, T., Schelling, J. R., et al. (2001). Activation of EphA receptor tyrosine kinase inhibits the Ras/MAPK pathway. *Nature Cell Biology*, *3*, 527–530. https://doi.org/10.1038/35074604.

Minami, M., Koyama, T., Wakayama, Y., Fukuhara, S., & Mochizuki, N. (2011). EphrinA/EphA signal facilitates insulin-like growth factor-I–induced myogenic differentiation through suppression of the Ras/extracellular signal–regulated kinase 1/2 cascade in myoblast cell lines. *Molecular Biology of the Cell*, *22*, 3508–3519. https://doi.org/10.1091/mbc.e11-03-0183.

Mincer, S. T., Niethamer, T. K., Teng, T., Bush, J. O., & Percival, C. J. (2022). Investigating the effects of compound paralogous EPHB receptor mutations on mouse facial development. *Developmental Dynamics*. https://doi.org/10.1002/dvdy.454.

Mire, E., Hocine, M., Bazellières, E., Jungas, T., Davy, A., Chauvet, S., et al. (2018). Developmental upregulation of Ephrin-B1 silences Sema3C/neuropilin-1 signaling during post-crossing navigation of corpus callosum axons. *Current Biology*, *28*, 1768–1782.e4. https://doi.org/10.1016/j.cub.2018.04.026.

Monier, B., Pélissier-Monier, A., Brand, A. H., & Sanson, B. (2010). An actomyosin-based barrier inhibits cell mixing at compartmental boundaries in Drosophila embryos. *Nature Cell Biology*, *12*, 60–65. https://doi.org/10.1038/ncb2005.

Nguyen, T. M., Arthur, A., Paton, S., Hemming, S., Panagopoulos, R., Codrington, J., et al. (2016). Loss of ephrinB1 in osteogenic progenitor cells impedes endochondral ossification and compromises bone strength integrity during skeletal development. *Bone*, *93*, 12–21. https://doi.org/10.1016/j.bone.2016.09.009.

Niethamer, T. K., & Bush, J. O. (2019). Getting direction(s): The Eph/ephrin signaling system in cell positioning. *Developmental Biology*, *447*, 42–57. https://doi.org/10.1016/j.ydbio.2018.01.012.

Niethamer, T. K., Larson, A. R., O'Neill, A. K., Bershteyn, M., Hsiao, E. C., Klein, O. D., et al. (2017). EPHRIN-B1 mosaicism drives cell segregation in craniofrontonasal syndrome hiPSC-derived neuroepithelial cells. *Stem Cell Reports*, *8*, 529–537. https://doi.org/10.1016/j.stemcr.2017.01.017.

Niethamer, T. K., Teng, T., Franco, M., Du, Y. X., Percival, C. J., & Bush, J. O. (2020). Aberrant cell segregation in the craniofacial primordium and the emergence of facial dysmorphology in craniofrontonasal syndrome. *PLoS Genetics*, *16*, e1008300. https://doi.org/10.1371/journal.pgen.1008300.

Nievergall, E., Lackmann, M., & Janes, P. W. (2012). Eph-dependent cell-cell adhesion and segregation in development and cancer. *Cellular & Molecular Life Sciences*, *69*, 1813–1842. https://doi.org/10.1007/s00018-011-0900-6.

Noberini, R., Lamberto, I., & Pasquale, E. B. (2012). Targeting Eph receptors with peptides and small molecules: Progress and challenges. *Seminars in Cell & Developmental Biology*, *23*, 51–57. https://doi.org/10.1016/j.semcdb.2011.10.023.

Noberini, R., & Pasquale, E. B. (2009). Proliferation and tumor suppression: Not mutually exclusive for eph receptors. *Cancer Cell*, *16*, 452–454. https://doi.org/10.1016/j.ccr.2009.11.008.

Nolt, M. J., Lin, Y., Hruska, M., Murphy, J., Sheffler-Colins, S. I., Kayser, M. S., et al. (2011). EphB controls NMDA receptor function and synaptic targeting in a subunit-specific manner. *Journal of Neuroscience*, *31*, 5353–5364. https://doi.org/10.1523/JNEUROSCI.0282-11.2011.

Noren, N. K., Foos, G., Hauser, C. A., & Pasquale, E. B. (2006). The EphB4 receptor suppresses breast cancer cell tumorigenicity through an Abl-Crk pathway. *Nature Cell Biology*, *8*, 815–825. https://doi.org/10.1038/ncb1438.

O'Neill, A. K., Kindberg, A. A., Niethamer, T. K., Larson, A. R., Ho, H.-Y. H., Greenberg, M. E., et al. (2016). Unidirectional Eph/ephrin signaling creates a cortical actomyosin differential to drive cell segregation. *The Journal of Cell Biology*, *215*, 217–229. https://doi.org/10.1083/jcb.201604097.

Oates, A. C., Lackmann, M., Power, M. A., Brennan, C., Down, L. M., Do, C., et al. (1999). An early developmental role for eph-ephrin interaction during vertebrate gastrulation. *Mechanisms of Development*, *83*, 77–94. https://doi.org/10.1016/s0925-4773(99)00036-2.

Ohta, N., & Satou, Y. (2013). Multiple signaling pathways coordinate to induce a threshold response in a chordate embryo. *PLoS Genetics*, *9*, e1003818. https://doi.org/10.1371/journal.pgen.1003818.

Ojosnegros, S., Cutrale, F., Rodríguez, D., Otterstrom, J. J., Chiu, C. L., Hortigüela, V., et al. (2017). Eph-ephrin signaling modulated by polymerization and condensation of receptors. *Proceedings of the National Academy of Sciences of the United States of America*, *114*, 13188–13193. https://doi.org/10.1073/pnas.1713564114.

Orioli, D., Henkemeyer, M., Lemke, G., Klein, R., & Pawson, T. (1996). Sek4 and Nuk receptors cooperate in guidance of commissural axons and in palate formation. *The EMBO Journal*, *15*, 6035–6049.

Ottone, C., Krusche, B., Whitby, A., Clements, M., Quadrato, G., Pitulescu, M. E., et al. (2014). Direct cell-cell contact with the vascular niche maintains quiescent neural stem cells. *Nature Cell Biology*, *16*, 1045–1056. https://doi.org/10.1038/ncb3045.

Palmer, A., Zimmer, M., Erdmann, K. S., Eulenburg, V., Porthin, A., Heumann, R., et al. (2002). EphrinB phosphorylation and reverse signaling: Regulation by Src kinases and PTP-BL phosphatase. *Molecular Cell*, *9*, 725–737.

Park, E. C., Cho, G.-S., Kim, G.-H., Choi, S.-C., & Han, J.-K. (2011). The involvement of Eph–Ephrin signaling in tissue separation and convergence during Xenopus gastrulation movements. *Developmental Biology*, *350*, 441–450. https://doi.org/10.1016/j.ydbio.2010.12.012.

Pasquale, E. B. (2010). Eph receptors and ephrins in cancer: Bidirectional signalling and beyond. *Nature Reviews. Cancer*, *10*, 165–180. https://doi.org/10.1038/nrc2806.

Poliakov, A., Cotrina, M. L., Pasini, A., & Wilkinson, D. G. (2008). Regulation of EphB2 activation and cell repulsion by feedback control of the MAPK pathway. *The Journal of Cell Biology*, *183*, 933–947. https://doi.org/10.1083/jcb.200807151.

Prospéri, M.-T., Lépine, P., Dingli, F., Paul-Gilloteaux, P., Martin, R., Loew, D., et al. (2015). Myosin 1b functions as an effector of EphB signaling to control cell repulsion. *The Journal of Cell Biology*, *210*, 347–361. https://doi.org/10.1083/jcb.201501018.

Risley, M., Garrod, D., Henkemeyer, M., & McLean, W. (2009). EphB2 and EphB3 forward signalling are required for palate development. *Mechanisms of Development, 126*, 230–239. https://doi.org/10.1016/j.mod.2008.10.009.

Robichaux, M. A., Chenaux, G., Ho, H.-Y. H., Soskis, M. J., Greenberg, M. E., Henkemeyer, M., et al. (2016). EphB1 and EphB2 intracellular domains regulate the formation of the corpus callosum and anterior commissure. *Developmental Neurobiology, 76*, 405–420. https://doi.org/10.1002/dneu.22323.

Rogers, M. A., & Fantauzzo, K. A. (2021). *PDGFR dimer-specific activation, trafficking and downstream signaling dynamics*. https://doi.org/10.1101/2021.10.26.465978.

Rohani, N., Canty, L., Luu, O., Fagotto, F., & Winklbauer, R. (2011). EphrinB/EphB signaling controls embryonic germ layer separation by contact-induced cell detachment. *PLoS Biology, 9*, e1000597. https://doi.org/10.1371/journal.pbio.1000597.

Rohani, N., Parmeggiani, A., Winklbauer, R., & Fagotto, F. (2014). Variable combinations of specific ephrin ligand/Eph receptor pairs control embryonic tissue separation. *PLoS Biology, 12*, e1001955. https://doi.org/10.1371/journal.pbio.1001955.

Schaupp, A., Sabet, O., Dudanova, I., Ponserre, M., Bastiaens, P., & Klein, R. (2014). The composition of EphB2 clusters determines the strength in the cellular repulsion response. *The Journal of Cell Biology, 204*, 409–422. https://doi.org/10.1083/jcb.201305037.

Seiradake, E., Harlos, K., Sutton, G., Aricescu, A. R., & Jones, E. Y. (2010). An extracellular steric seeding mechanism for Eph-ephrin signaling platform assembly. *Nature Structural & Molecular Biology, 17*, 398–402. https://doi.org/10.1038/nsmb.1782.

Seiradake, E., Schaupp, A., del Toro Ruiz, D., Kaufmann, R., Mitakidis, N., Harlos, K., et al. (2013). Structurally encoded intraclass differences in EphA clusters drive distinct cell responses. *Nature Structural & Molecular Biology, 20*, 958–964. https://doi.org/10.1038/nsmb.2617.

Shapiro, M. D., Kronenberg, Z., Li, C., Domyan, E. T., Pan, H., Campbell, M., et al. (2013). Genomic diversity and evolution of the head crest in the rock pigeon. *Science, 339*, 1063–1067. https://doi.org/10.1126/science.1230422.

Shi, W., & Levine, M. (2008). Ephrin signaling establishes asymmetric cell fates in an endomesoderm lineage of the Ciona embryo. *Development, 135*, 931–940. https://doi.org/10.1242/dev.011940.

Solanas, G., Cortina, C., Sevillano, M., & Batlle, E. (2011). Cleavage of E-cadherin by ADAM10 mediates epithelial cell sorting downstream of EphB signalling. *Nature Cell Biology, 13*, 1100–1107. https://doi.org/10.1038/ncb2298.

Soskis, M. J., Ho, H.-Y. H., Bloodgood, B. L., Robichaux, M. A., Malik, A. N., Ataman, B., et al. (2012). A chemical genetic approach reveals distinct mechanisms of EphB signaling during brain development. *Nature Neuroscience, 15*, 1645–1654. https://doi.org/10.1038/nn.3249.

Srivastava, N., Robichaux, M. A., Chenaux, G., Henkemeyer, M., & Cowan, C. W. (2013). EphB2 receptor forward signaling controls cortical growth cone collapse via Nck and Pak. *Molecular and Cellular Neurosciences, 52*, 106–116. https://doi.org/10.1016/j.mcn.2012.11.003.

Stein, E., Huynh-Do, U., Lane, A. A., Cerretti, D. P., & Daniel, T. O. (1998). Nck recruitment to Eph receptor, EphB1/ELK, couples ligand activation to c-Jun kinase. *Journal of Biological Chemistry, 273*, 1303–1308. https://doi.org/10.1074/jbc.273.3.1303.

Stolfi, A., Wagner, E., Taliaferro, J. M., Chou, S., & Levine, M. (2011). Neural tube patterning by Ephrin, FGF and Notch signaling relays. *Development, 138*, 5429–5439. https://doi.org/10.1242/dev.072108.

Suga, H., Koyanagi, M., Hoshiyama, D., Ono, K., Iwabe, N., Kuma, K., et al. (1999). Extensive gene duplication in the early evolution of animals before the parazoan-eumetazoan split demonstrated by G proteins and protein tyrosine kinases from sponge and hydra. *Journal of Molecular Evolution, 48*, 646–653.

Suga, H., Torruella, G., Burger, G., Brown, M. W., & Ruiz-Trillo, I. (2014). Earliest holozoan expansion of phosphotyrosine signaling. *Molecular Biology and Evolution*, *31*, 517–528. https://doi.org/10.1093/molbev/mst241.

Tanaka, M., Kamo, T., Ota, S., & Sugimura, H. (2003). Association of Dishevelled with Eph tyrosine kinase receptor and ephrin mediates cell repulsion. *EMBO Journal*, *22*, 847–858. https://doi.org/10.1093/emboj/cdg088.

Taylor, H. B., Khuong, A., Wu, Z., Xu, Q., Morley, R., Gregory, L., et al. (2017). Cell segregation and border sharpening by Eph receptor-ephrin-mediated heterotypic repulsion. *Journal of the Royal Society Interface*, *14*, 20170338. https://doi.org/10.1098/rsif.2017.0338.

Theveneau, E., & Mayor, R. (2012). Neural crest migration: interplay between chemorepellents, chemoattractants, contact inhibition, epithelial–mesenchymal transition, and collective cell migration. *Wiley Interdisciplinary Reviews. Developmental Biology*, *1*, 435–445. https://doi.org/10.1002/wdev.28.

Tonna, S., Poulton, I. J., Taykar, F., Ho, P. W. M., Tonkin, B., Crimeen-Irwin, B., et al. (2016). Chondrocytic ephrin B2 promotes cartilage destruction by osteoclasts in endochondral ossification. *Development*, *143*, 648–657. https://doi.org/10.1242/dev.125625.

Torres, R., Firestein, B. L., Dong, H., Staudinger, J., Olson, E. N., Huganir, R. L., et al. (1998). PDZ proteins bind, cluster, and synaptically colocalize with Eph receptors and their ephrin ligands. *Neuron*, *21*, 1453–1463.

Townes, P. L., & Holtfreter, J. (1955). Directed movements and selective adhesion of embryonic amphibian cells. *Journal of Experimental Zoology*, *128*, 53–120.

Twigg, S. R. F., Kan, R., Babbs, C., Bochukova, E. G., Robertson, S. P., Wall, S. A., et al. (2004). Mutations of ephrin-B1 (EFNB1), a marker of tissue boundary formation, cause craniofrontonasal syndrome. *Proceedings of the National Academy of Sciences of the United States of America*, *101*, 8652–8657. https://doi.org/10.1073/pnas.0402819101.

Umetsu, D., Dunst, S., & Dahmann, C. (2014). An RNA interference screen for genes required to shape the anteroposterior compartment boundary in Drosophila identifies the Eph receptor. *PLoS One*, *9*, e114340. https://doi.org/10.1371/journal.pone.0114340.

Valenzuela, J. I., & Perez, F. (2020). Localized intercellular transfer of ephrin-As by trans-endocytosis enables long-term signaling. *Developmental Cell*, *52*, 104–117.e5. https://doi.org/10.1016/j.devcel.2019.11.013.

Vickrey, A. I., Domyan, E. T., Horvath, M. P., & Shapiro, M. D. (2015). Convergent evolution of head crests in two domesticated columbids is associated with different missense mutations in EphB2. *Molecular Biology and Evolution*, *32*, 2657–2664. https://doi.org/10.1093/molbev/msv140.

Villar-Cerviño, V., Molano-Mazón, M., Catchpole, T., Valdeolmillos, M., Henkemeyer, M., Martínez, L. M., et al. (2013). Contact repulsion controls the dispersion and final distribution of Cajal-Retzius cells. *Neuron*, *77*, 457–471. https://doi.org/10.1016/j.neuron.2012.11.023.

Wagner, M. J., Hsiung, M. S., Gish, G. D., Bagshaw, R. D., Doodnauth, S. A., Soliman, M. A., et al. (2020). The Shb scaffold binds the Nck adaptor protein, p120 RasGAP, and Chimaerins and thereby facilitates heterotypic cell segregation by the receptor EphB2. *The Journal of Biological Chemistry*, *295*, 3932–3944. https://doi.org/10.1074/jbc.RA119.009276.

Wang, Y., Ling, L., Tian, F., Won Kim, S. H., Ho, S., & Bikle, D. D. (2020). Ablation of ephrin B2 in Col2 expressing cells delays fracture repair. *Endocrinology*, *161*, bqaa179. https://doi.org/10.1210/endocr/bqaa179.

Wang, Y., Menendez, A., Fong, C., ElAlieh, H. Z., Chang, W., & Bikle, D. D. (2014). Ephrin B2/EphB4 mediates the actions of IGF-I signaling in regulating endochondral bone formation. *Journal of Bone and Mineral Research*, *29*, 1900–1913. https://doi.org/10.1002/jbmr.2196.

Wei, S., Xu, G., Bridges, L. C., Williams, P., White, J. M., & DeSimone, D. W. (2010). ADAM13 induces cranial neural crest by cleaving class B Ephrins and regulating Wnt signaling. *Developmental Cell*, *19*, 345–352. https://doi.org/10.1016/j.devcel.2010.07.012.

Wen, J. W., & Winklbauer, R. (2017). Ingression-type cell migration drives vegetal endoderm internalisation in the Xenopus gastrula. *Elife*, *6*, e27190. https://doi.org/10.7554/eLife.27190.

Wieland, I., Jakubiczka, S., Muschke, P., Cohen, M., Thiele, H., Gerlach, K. L., et al. (2004). Mutations of the ephrin-B1 gene cause craniofrontonasal syndrome. *American Journal of Human Genetics*, *74*, 1209–1215. https://doi.org/10.1086/421532.

Wills, A. E., & Baker, J. C. (2015). E2a is necessary for Smad2/3-dependent transcription and the direct repression of lefty during gastrulation. *Developmental Cell*, *32*, 345–357. https://doi.org/10.1016/j.devcel.2014.11.034.

Wimmer-Kleikamp, S. H., Janes, P. W., Squire, A., Bastiaens, P. I. H., & Lackmann, M. (2004). Recruitment of Eph receptors into signaling clusters does not require ephrin contact. *The Journal of Cell Biology*, *164*, 661–666. https://doi.org/10.1083/jcb.200312001.

Winklbauer, R. (2015). Cell adhesion strength from cortical tension—An integration of concepts. *Journal of Cell Science*, *128*, 3687–3693. https://doi.org/10.1242/jcs.174623.

Wu, Z., Ashlin, T. G., Xu, Q., & Wilkinson, D. G. (2019). Role of forward and reverse signaling in Eph receptor and ephrin mediated cell segregation. *Experimental Cell Research*, *381*, 57–65. https://doi.org/10.1016/j.yexcr.2019.04.040.

Xing, W., Kim, J., Wergedal, J., Chen, S.-T., & Mohan, S. (2010). Ephrin B1 regulates bone marrow stromal cell differentiation and bone formation by influencing TAZ transactivation via complex formation with NHERF1. *Molecular and Cellular Biology*, *30*, 711–721. https://doi.org/10.1128/MCB.00610-09.

Xu, Q., Alldus, G., Holder, N., & Wilkinson, D. G. (1995). Expression of truncated Sek-1 receptor tyrosine kinase disrupts the segmental restriction of gene expression in the Xenopus and zebrafish hindbrain. *Development*, *121*, 4005–4016.

Xu, N.-J., & Henkemeyer, M. (2012). Ephrin reverse signaling in axon guidance and synaptogenesis. *Seminars in Cell & Developmental Biology*, *23*, 58–64. https://doi.org/10.1016/j.semcdb.2011.10.024.

Xu, Q., Mellitzer, G., Robinson, V., & Wilkinson, D. G. (1999). In vivo cell sorting in complementary segmental domains mediated by Eph receptors and ephrins. *Nature*, *399*, 267–271. https://doi.org/10.1038/20452.

Xu, Y., Robev, D., Saha, N., Wang, B., Dalva, M. B., Xu, K., et al. (2021). The Ephb2 receptor uses homotypic, head-to-tail interactions within its ectodomain as an autoinhibitory control mechanism. *International Journal of Molecular Sciences*, *22*, 10473. https://doi.org/10.3390/ijms221910473.

Xu, K., Tzvetkova-Robev, D., Xu, Y., Goldgur, Y., Chan, Y.-P., Himanen, J. P., et al. (2013). Insights into Eph receptor tyrosine kinase activation from crystal structures of the EphA4 ectodomain and its complex with ephrin-A5. *Proceedings of the National Academy of Sciences of the United States of America*, *110*, 14634–14639. https://doi.org/10.1073/pnas.1311000110.

Zeng, X., Hunt, A., Jin, S. C., Duran, D., Gaillard, J., & Kahle, K. T. (2019). EphrinB2-EphB4-RASA1 signaling in human cerebrovascular development and disease. *Trends in Molecular Medicine*, *25*, 265–286. https://doi.org/10.1016/j.molmed.2019.01.009.

Zhang, G., Brady, J., Liang, W.-C., Wu, Y., Henkemeyer, M., & Yan, M. (2015). EphB4 forward signalling regulates lymphatic valve development. *Nature Communications*, *6*, 6625. https://doi.org/10.1038/ncomms7625.

Zhang, G., Fenyo, D., & Neubert, T. A. (2008). Screening for EphB signaling effectors using SILAC with a linear ion trap-orbitrap mass spectrometer. *Journal of Proteome Research*, *7*, 4715–4726. https://doi.org/10.1021/pr800255a.

Zhang, G., Spellman, D. S., Skolnik, E. Y., & Neubert, T. A. (2006). Quantitative phosphotyrosine proteomics of EphB2 signaling by stable isotope labeling with amino acids in cell culture (SILAC). *Journal of Proteome Research*, *5*, 581–588. https://doi.org/10.1021/pr050362b.

Zhao, C., Irie, N., Takada, Y., Shimoda, K., Miyamoto, T., Nishiwaki, T., et al. (2006). Bidirectional ephrinB2-EphB4 signaling controls bone homeostasis. *Cell Metabolism*, *4*, 111–121. https://doi.org/10.1016/j.cmet.2006.05.012.

Zhuang, G., Hunter, S., Hwang, Y., & Chen, J. (2007). Regulation of EphA2 receptor endocytosis by SHIP2 lipid phosphatase via phosphatidylinositol 3-Kinase-dependent Rac1 activation. *Journal of Biological Chemistry*, *282*, 2683–2694. https://doi.org/10.1074/jbc.M608509200.

Zimmer, M., Palmer, A., Köhler, J., & Klein, R. (2003). EphB–ephrinB bi-directional endocytosis terminates adhesion allowing contact mediated repulsion. *Nature Cell Biology*, *5*, 869–878. https://doi.org/10.1038/ncb1045.

CHAPTER SIX

Integrins, anchors and signal transducers of hematopoietic stem cells during development and in adulthood

Peter W. Krenn[a,b,*], Eloi Montanez[c], Mercedes Costell[d,e], and Reinhard Fässler[a]

[a]Department of Molecular Medicine, Max Planck Institute of Biochemistry, Martinsried, Germany
[b]Department of Biosciences and Medical Biology, Cancer Cluster Salzburg, Paris-Lodron University of Salzburg, Salzburg, Austria
[c]Department of Physiological Sciences, Faculty of Medicine and Health Sciences, University of Barcelona and Bellvitge Biomedical Research Institute, L'Hospitalet del Llobregat, Barcelona, Spain
[d]Departamento de Bioquímica y Biología Molecular, Facultad de Ciencias Biológicas, Universitat de València, Burjassot, Spain
[e]Institut Universitari de Biotecnologia i Biomedicina, Universitat de València, Burjassot, Spain
*Corresponding author: e-mail address: peter.krenn@plus.ac.at

Contents

Abstract

Hematopoietic stem cells (HSCs), the apex of the hierarchically organized blood cell production system, are generated in the yolk sac, aorta-gonad-mesonephros region and placenta of the developing embryo. To maintain life-long hematopoiesis, HSCs emigrate from their site of origin and seed in distinct microenvironments, called niches, of fetal liver and bone marrow where they receive supportive signals for self-renewal, expansion and production of hematopoietic progenitor cells (HPCs), which in turn orchestrate the production of the hematopoietic effector cells. The interactions of hematopoietic stem and progenitor cells (HSPCs) with niche components are to a large

Current Topics in Developmental Biology, Volume 149
ISSN 0070-2153
https://doi.org/10.1016/bs.ctdb.2022.02.009

part mediated by the integrin superfamily of adhesion molecules. Here, we summarize the current knowledge regarding the functional properties of integrins and their activators, Talin-1 and Kindlin-3, for HSPC generation, function and fate decisions during development and in adulthood. In addition, we discuss integrin-mediated mechanosensing for HSC-niche interactions, *ex vivo* protocols aimed at expanding HSCs for therapeutic use, and recent approaches targeting the integrin-mediated adhesion in leukemia-inducing HSCs in their protecting, malignant niches.

1. Integrins

Cell-cell and cell-matrix adhesion is essential for the development and homeostasis of multicellular organisms. Integrins are αβ heterodimeric transmembrane receptors that attach cells to the extracellular matrix (ECM) or neighboring cells, and in addition probe the biochemical and biophysical properties of the ECM substrate and transmit this information *via* large signaling hubs that are linked to the contractile actomyosin cytoskeleton to various intracellular compartments. Since integrin-mediated signaling cooperates with and is modified by various signaling pathways, particularly those elicited by growth factors, the response to integrin signals can differ from cell type to cell type and includes induction of cell polarity, migration, proliferation, survival and differentiation. Before summarizing the current knowledge on how integrins impact on the development of hematopoietic stem and progenitor cells (HSPCs), we briefly introduce basics of integrin biology.

In mammals, the integrin superfamily comprises 24 heterodimeric integrins that are generated from 18 α- and 8 β-subunits, many of which come in different splice variants affecting the extracellular or cytoplasmic domain (Fig. 1A). A single integrin can bind several ligands and a single ligand can be bound by several integrins, which results in overlapping functions and makes integrin analyses difficult. Integrin ligands include ECM proteins such as collagens, laminins, fibronectin, or cellular counter receptors such as VCAM-1, ICAM-1 or MAdCAM-1 (Humphries et al., 2006; Hynes, 2002).

Integrin α- and β-subunits are type I transmembrane proteins consisting of a large ectodomain, a single pass transmembrane domain and, with the exception of the β4 integrin subunit, a short cytoplasmic domain that assembles hundreds of adaptor, cytoskeletal and signaling proteins into a gigantic signaling hub termed adhesome. The ectodomains of the α- and β-subunits are non-covalently associated and together form the ligand binding site.

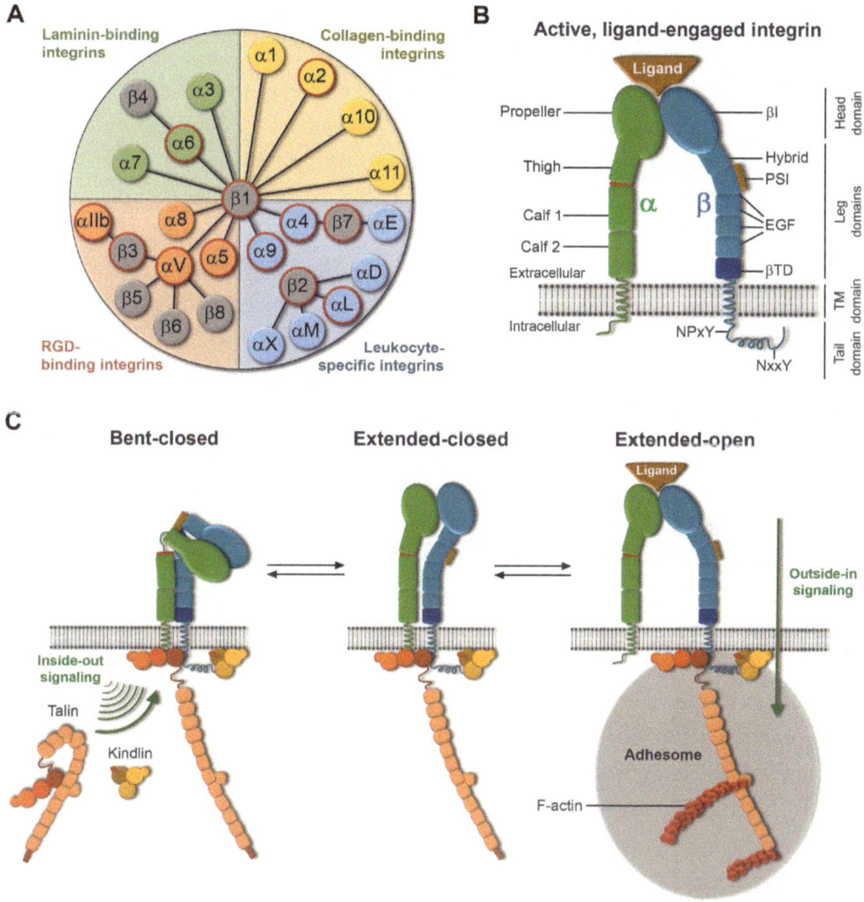

Fig. 1 Integrin superfamily. (A) Classification of the mammalian integrins according to their ligand-binding specificity. Integrins shown to be expressed by hematopoietic stem cells are circled red. (B) Domain composition of an active, ligand-engaged αβ heterodimeric integrin: head domain, knee, leg domains, transmembrane (TM) domain, cytoplasmic or tail domain. Tails of β-subunits harbor the membrane proximal NPxY and membrane distal NxxY motif, which bind Talin and Kindlin, respectively. The ectodomain of α-subunits consists of α-I (α-inserted) von Willebrand factor-like domain; Propeller, β propeller domain; Thigh, thigh domain; Calf, calf domain. The ectodomain of β-subunits consists of βI, β-inserted von Willebrand factor-like domain; Hybrid, hybrid domain; PSI, plexin-semaphorin-integrin domain; EGF, epidermal growth factor domain; βTD, β-tail domain. (C) In hematopoietic cells, agonist induced "inside-out" signaling triggers Talin and Kindlin binding to β-tails leading to conformational changes from inactive (closed-bent) to ligand-primed (extended-closed) and finally active, ligand-engaged (extended-open) integrins. Alternatively, integrins fluctuate between conformations and rarely adopt, due to high energy requirements, the extended-closed and extended-open conformations at thermodynamic equilibrium. If ligand binds, the conformation is stabilized by Talin and Kindlin binding to the β-tail followed by the linkage to the actomyosin cytoskeleton, force transduction *via* the transmembrane domain and stabilization of the integrin-ligand bond.

Integrin α-subunits consist of β-propeller, thigh and two calf domains. Nine integrin α-subunits contain an additional α-I (α-inserted) domain with a von Willebrand-like fold that is inserted in the β-propeller domain and functions as ligand binding site controlled by the β-integrin subunits. These α-I domains are mainly but not exclusively found in the "leukocyte integrins" αL-, αM-, αD- and αX-subunits that associate with the β2-subunit, the E-cadherin-binding αE-subunit that associates with the β7-subunit and collagen-binding α1-, α2-, α10- and α11-subunits that associate with the β1-subunit. The von Willebrand factor-like β-I domain in the β-integrin subunit is inserted into a hybrid domain, which in turn is followed by a plexin-semaphorin-integrin (PSI) domain, four epidermal growth factor (EGF)-like domains and a β-tail domain. The β cytoplasmic tails harbor an NPxY and NxxY motif, which recruit Talins and Kindlins and hence, play major regulatory roles for integrin activation (Campbell & Humphries, 2011; Gahmberg et al., 2019) (Fig. 1B).

A hallmark of integrins is their ability to reversibly switch between different affinity states: inactive (bent-closed), ligand-primed (extended-closed) and active, ligand-engaged (extended-open) states, in which closed and open refers to the functional status of the ligand binding pocket (Luo, Carman, & Springer, 2007; Nishida et al., 2006) (Fig. 1C). In hematopoietic cells the concept of "inside-out" activation of integrins is well established and suggests that agonists such as chemokines, adenosine diphosphate, thrombin or antigens induce intracellular signaling pathways leading to the activation and binding of Talin-1 and Kindlin-3 to integrin β-tails, which in turn results in the separation of integrin legs, transmembrane and tail domains, swing away of the hybrid domain from the α-subunit and finally opening of the integrin-ligand pocket (Abram & Lowell, 2009; Hu & Luo, 2013; Moser, Nieswandt, Ussar, Pozgajova, & Fässler, 2008; Springer & Dustin, 2012; Tadokoro et al., 2003; Ye, Snider, & Ginsberg, 2014). Springer and colleagues have also shown that integrins thermodynamically fluctuate between the three conformations. The authors also reported that in a process of "outside-in" activation of integrins the rare extended-open state binds ligand whose bonds are stabilized by actin-mediated force transduction *via* Talin-1 and Kindlin-3 (Li et al., 2017; Li & Springer, 2017). In hematopoietic cells both activation modes very likely act in parallel to ensure that a high number of integrins become ligand-engaged and resist the particularly high shear forces when cells are exposed to the blood flow (Sun, Costell, & Fässler, 2019). Whereas ligand binding in non-hematopoietic cells is followed by integrin clustering and the assembly of hundreds of adaptor, signaling and actin-binding proteins, which are collectively called the adhesome (Campbell & Humphries, 2011; Horton et al., 2016; Li & Springer, 2017; Winograd-Katz, Fässler, Geiger, & Legate, 2014), the adhesion sites of hematopoietic cells form small integrin clusters, are highly dynamic and short-lived and their adhesomes are still ill-defined (Horton et al., 2015; Jankowska et al., 2018).

In this review we summarize the current knowledge regarding the role of integrins and the integrin activators, Talin-1 and Kindlin-3, for homing, engraftment, maintenance (self-renewal and stemness), and differentiation of hematopoietic stem and progenitor cells (HSPCs) in their microenvironments during development and in adulthood and discuss current advances and implications on how integrin-mediated signals could be exploited for *ex vivo* hematopoietic stem cell (HSC) culture and expansion, and therapeutic targeting of leukemia-inducing HSCs.

2. A common origin of vascular and hematopoietic cells

A prerequisite for vertebrate embryogenesis is the establishment of a transporting system that supports cells and organs with oxygen and nutrition over large distances. This process starts shortly after gastrulation with the development of a specific subpopulation of mesodermal cells in the posterior primitive streak that express Flk1 (VEGFR-2) in response to fibroblast growth factor (FGF) and bone morphogenic protein 4 (BMP4) (Flamme, Breier, & Risau, 1995; Park et al., 2004). At mouse embryonic day (E) 7 the Flk1$^+$ cells migrate to extraembryonic sites including yolk sac and allantois and intraembryonic sites (Kinder et al., 1999; Saykali et al., 2019) (Fig. 2).

In the yolk sac, Flk1$^+$ cells aggregate into proliferative cell clusters, differentiate and establish blood islands consisting of an outer endothelial cell layer enclosing unipotent and short-lived HPCs. The concomitant development, spatial proximity and expression of certain marker genes in outer and inner cells of the blood islands suggests a common origin, or at least, a direct lineage relationship between endothelial and HPCs (Boucher & Pedersen, 1996; Dumont et al., 1995; Elefanty, Begley, Hartley, Papaevangeliou, & Robb, 1999; Fennie, Cheng, Dowbenko, Young, & Lasky, 1995; Fong, Klingensmith, Wood, Rossant, & Breitman, 1996; Goldie, Nix, & Hirschi, 2008; Kallianpur, Jordan, & Brandt, 1994; Palis, McGrath, & Kingsley, 1995). *In vitro* differentiation and colony-forming assays with Flk1$^+$ mesodermal cells isolated from the

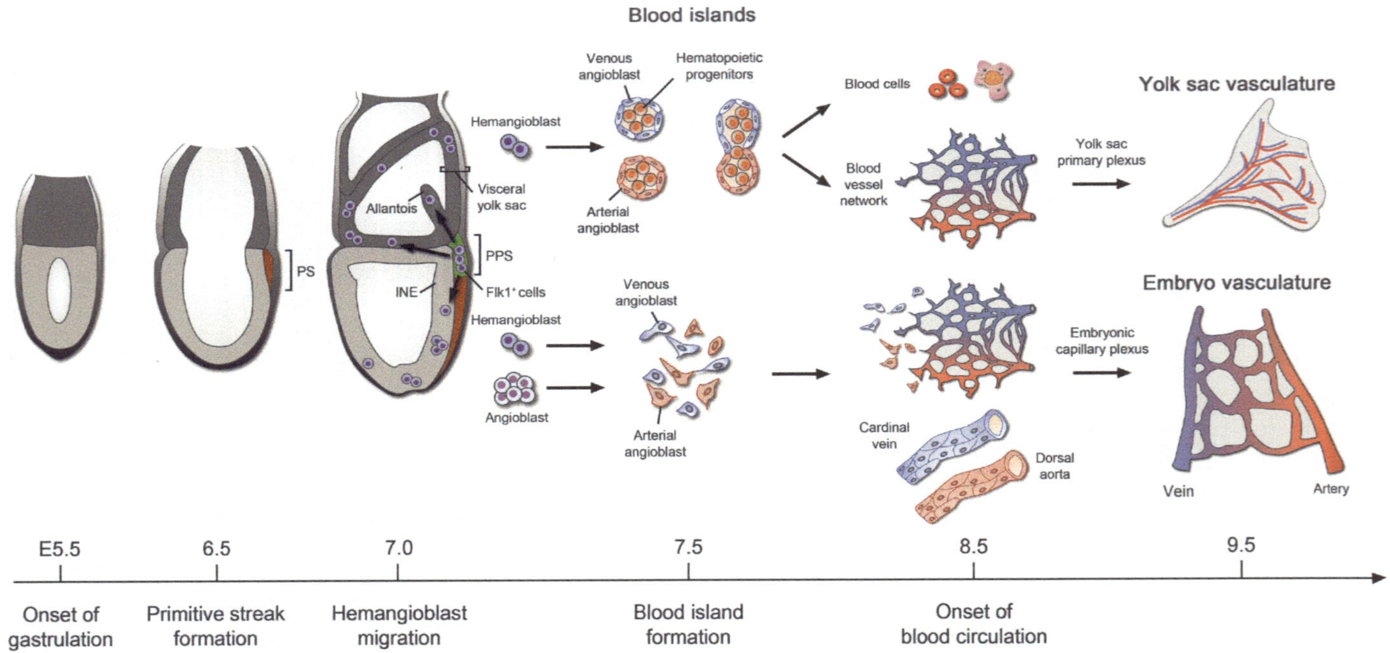

Fig. 2 The hemangioblast, common progenitor for endothelial and hematopoietic cells. Upon gastrulation and formation of the primitive streak (PS) in mouse development, $Flk1^+$ hemangioblasts and angioblasts form in the posterior primitive streak (PPS) region and migrate to extraembryonic and/or intraembryonic (INE) sites. In the yolk sac, hemangioblasts cluster and establish blood islands giving rise to the first endothelial and hematopoietic progenitor cells. Intraembryonic hemangioblasts and angioblasts assemble into the dorsal aorta and cardinal veins connected by a capillary network. At around mouse embryonic day (E) 8.75 the extraembryonic vascular plexus fuses with intraembryonic vessels marking the onset of blood circulation in the embryo.

primitive streak or early blood islands corroborated this notion and showed that single cells can give rise to both, adherent endothelial and non-adherent HPCs. These findings fostered the idea of a "monoclonal" origin of blood islands and coined the name hemangioblast (Choi, Kennedy, Kazarov, Papadimitriou, & Keller, 1998; Huber, Kouskoff, Fehling, Palis, & Keller, 2004; Stefanska et al., 2017). *In vivo* lineage tracing experiments confirmed that hemangioblasts develop and expand in the primitive streak, migrate to the yolk sac, cluster and form blood islands (Padrón-Barthe et al., 2014; Ueno & Weissman, 2006). Integrins seem dispensable for these processes as genetic loss-of-function studies resulting in the constitutive ablation of either the β1-, β2-, β3- or αV-class integrins revealed normal blood island development in the yolk sac (Bader, Rayburn, Crowley, & Hynes, 1998; Hirsch, Iglesias, Potocnik, Hartmann, & Fässler, 1996; Hodivala-Dilke et al., 1999; Scharffetter-Kochanek et al., 1998).

Flk1$^+$ mesodermal cells that migrate to the allantois or intraembryonic sites do not form blood islands but differentiate into endothelial cells. It is currently believed that these intraembryonic Flk1$^+$ cells produce angioblasts and to a lesser extent hemangioblasts (reviewed by Bautch (2011)). The newly generated endothelial cells assemble into primitive vessels of the umbilical vascular system and the primary embryonic circulatory system including that of the endocardium, the dorsal aorta and head vasculature (Collart et al., 2021; Downs, Gifford, Blahnik, & Gardner, 1998; Drake & Fleming, 2000). At around E8.75 the extraembryonic vascular plexus anastomoses with intraembryonic vessels leading to the formation of a functional vascular network that connects to the developing heart and establishes the blood circulation with the initiation of heart beats (Collart et al., 2021; Risau & Flamme, 1995) (Fig. 2). So far antibody-based inhibition studies and genetic deletion of α4, α5, αV, β1, β3 or β8 integrins, respectively, have been shown to interfere with vascular development and/or integrity at this stage of embryonic development (reviewed by Avraamides, Garmy-Susini, and Varner (2008)).

3. Involvement of integrins in the establishment of the hematopoietic system

Blood cells fulfill numerous vital functions in mammals (Magner, 1992; Menche & Raichle, 2020). It was, however, only until the 1950s that a series of key experiments identified the bone marrow as the site of postnatal blood production. Jacobson, Ford and colleagues showed for the first time that the intravenous injection of adult mouse bone marrow cells into lethally irradiated mice replaces the destroyed hematopoietic system by faithfully establishing multilineage hematopoiesis (Ford, Hamerton, Barnes, & Loutit, 1956; Jacobson, Simmons, Marks, & Eldredge, 1951). These pioneering experiments initiated a long scientific search until HSCs were identified and defined as the apex of the hierarchal blood cell producing system that gives rise to multipotent hematopoietic progenitor cells (HPCs), unipotent HPCs (as defined in Table 1) and effector cells. To maintain hematopoiesis over the entire lifespan of an organism and also to rapidly respond to stress situations such as blood loss or infections with HPCs and effector cell production, HSCs receive cell-extrinsic cues from the different components of the microenvironment and possess cell-intrinsic determinants such as epigenetic marks like histone modifications, which together define HSC lineage commitment and tightly balance dormancy, asymmetric divisions, which produce one HSC and one HPC, or symmetric divisions, which yield either two HSCs or two HPCs (reviewed by Olson, Kang, and Passegué (2020) and Haas, Trumpp, and Milsom (2018)). With the identification of the hierarchically organized hematopoietic system and its dependency on bone marrow-derived signals for maintenance many laboratories began to decipher the contributions of integrin-mediated interactions of HSCs with the hematopoiesis-supporting microenvironments.

3.1 Early hematopoietic progenitor cells

The development of the hematopoietic system is established in spatially and temporally distinguishable waves. The first wave of hematopoiesis is induced at E7.0–8.25 by the hemangioblast-derived blood islands in the yolk sac, which leads to the establishment of transient and unipotent HPCs that express PECAM-1 (CD31), Tie-2 and low levels of αIIbβ3 integrin and produce short-lived cell populations of primitive nucleated, embryonic α- and β-globin expressing erythroblasts (EryPs), macrophages, and megakaryocytes (Ema et al., 2006; Ferkowicz et al., 2003; Li, Ferkowicz, Johnson, Shelley, & Yoder, 2005; Palis, 2016). The EryP population expands in the yolk sac, seeds into the circulation upon initiation of cardiac contractions, transports oxygen and scavenges reactive oxygen species. The increasing number of circulating EryPs results in elevated blood viscosity and shear stress, which in turn supports the development of the vascular system (Blatnik, Schmid-Schönbein, & Sung, 2005; Kingsley et al., 2013; Li et al., 2005; Palis, Robertson, Kennedy, Wall, & Keller, 1999; Wells, 1979; Wong, Chung, Reicheld, & Chui, 1986).

Table 1 Assays to characterize HSPCs. HSCs are defined by the ability to re-establish multilineage (multipotency) and life-long (self-renewal) hematopoiesis upon transplantation into lethally irradiated recipients. Several additional assays have been established to study HSC properties. However, many of them address multipotent or unipotent HPCs rather than HSCs. The most frequently used assays are listed below.

Assay	Procedure	Readout	Additional information	Referred to in this review as	References
Transplantation assay	Transplantation of, e.g., adult bone marrow (BM) cells into lethally irradiated adult recipients	Restoration of multilineage hematopoiesis indicates the presence of definitive HSCs		HSCs	Arora et al. (2014); McCulloch and Till (1960); Yoder and Hiatt (1997)
	Transplantation of, e.g., embryonic tissue-derived cells into busulfan-pretreated or irradiated neonatal recipients	Restoration of multilineage hematopoiesis demonstrates the presence of immature or pre-HSCs		Pre-HSCs	
Serial transplantation assay	Transplantation in series of, e.g., adult BM cells into lethally irradiated adult recipients	Restoration of hematopoiesis in series demonstrates the presence of long-term repopulating and self-renewing HSCs		HSCs	Iscove and Nawa (1997)
Competitive transplantation/repopulation/reconstitution assay	Transplantation of two distinguishable donor cell populations into lethally irradiated recipients, or transplantation of one donor cell population into recipients with reduced HSC numbers and/or activity	Comparison of the repopulation efficiencies of two competing HSC populations within one recipient		HSCs	Harrison, Jordan, Zhong, and Astle (1993); Pawliuk, Eaves, and Humphries (1996)
Limiting dilution assay (LDA)	Serial dilutions of, e.g., BM donor cells are co-transplanted with effector cells into lethally irradiated recipients	Estimation of HSPCs or HSCs present in donor cell populations	4.7–6.0 short-term repopulating HSPCs per 10^5 C57BL6 BM cells; 0.9–1.3 long-term repopulating HSCs per 10^5 C57BL6 BM cells	Analysis $\leq$20 weeks after transplantation allows estimating HSPCs; analysis $\geq$20 weeks after transplantation allows estimating HSCs	McCulloch and Till (1960); Osawa, Hanada, Hamada, and Nakauchi (1996); Szilvassy, Humphries, Lansdorp, Eaves, and Eaves (1990); Zhong, Astle, and Harrison (1996)

Continued

Table 1 Assays to characterize HSPCs. HSCs are defined by the ability to re-establish multilineage (multipotency) and life-long (self-renewal) hematopoiesis upon transplantation into lethally irradiated recipients. Several additional assays have been established to study HSC properties. However, many of them address multipotent or unipotent HPCs rather than HSCs. The most frequently used assays are listed below.—cont'd

Assay	Procedure	Readout	Additional information	Referred to in this review as	References
Colony-forming unit spleen (CFU-S) assay	BM cells transplanted into lethally irradiated recipients form colonies in spleen	Numbers of colonies in spleen 9–14 days after transplantation correlate with numbers of transplanted self-renewing HSCs and multi- or unipotent HPCs; spleen colonies represent mixture of short-term and long-term repopulating cells		HSPCs	Magli, Iscove, and Odartchenko (1982); Till and McCulloch (1961)
In vitro colony forming unit (CFU-GM, CFU-M, CFU-G, BFU-E, CFU-E, CFU GEMM, Pre-B) or long-term culture-initiating cell (LTC-IC) assays	Hematopoietic tissue derived cells are cultured in semi-solid media supplemented with cytokines and growth factors, or on supportive cell layers	*In vitro* differentiation and proliferation of HPCs; number of lineage specific colonies correlate with abundance of distinct multi- or unipotent HPCs		HPCs	Bradley and Metcalf (1966); Iscove, Sieber, and Winterhalter (1974); Pamphilon et al. (2013); Ploemacher, van der Sluijs, Voerman, and Brons (1989); Pluznik and Sachs (1965)
HSC/HPC identification and enrichment	Antibody staining of murine adult BM, spleen, or blood cells	lineage– ($Mac\text{-}1^-$, $Gr\text{-}1^-$, $B220^-$, $Ter119^-$, $CD4^-$, $CD8^-$, $CD3^-$), $Sca\text{-}1^+$, $c\text{-}kit^+$ (LSK) cells	1 HSC in 10 mouse BM-derived LSK cells	HSPCs	Bhatia, Wang, Kapp, Bonnet, and Dick (1997); Camargo, Chambers, Drew, McNagny, and Goodell (2006); Goodell et al. (1997); Kiel, Yilmaz, Iwashita, Terhorst, and Morrison (2005); Kim, He, Yilmaz, Kiel, and Morrison (2006); Pietras et al. (2015)
		LSK, $CD48^+$, $CD150^-$		HPCs	
		LSK, $CD48^-$, $CD150^+$	1 HSC in 2 mouse BM-derived LSK, $CD48^-$, $CD150^+$ cells	HSCs	

		$CD48^-$, $CD150^+$	1 HSC in 5 mouse BM-derived $CD48^-$, $CD150^+$ cells	HSCs
		Thy^{low}, $Sca\text{-}1^+$, $lineage^-$, $Mac\text{-}1^+$	1 HSC in 8.9 fetal liver-derived Thy^{low}, $Sca\text{-}1^+$, $lineage^-$, $Mac\text{-}1^+$ cells	HSPCs
	Antibody staining of murine fetal liver (FL) cells	Thy^{low}, $Sca\text{-}1^+$, $lineage^-$, $Mac\text{-}1^+$, $CD150^+$, $CD48^-$	1 HSC in 2.7 fetal liver-derived Thy^{low}, $Sca\text{-}1^+$, $lineage^-$, $Mac\text{-}1^+$ cells	HSCs
	Antibody staining of human hematopoietic tissue or blood	$lineage^-$, $CD34^+$, $CD38^-$	1 in 617 cord blood derived lineage-, $CD34^+$, $CD38^-$ cells engraft the BM of immune deficient mice	HSPCs
	Staining of, e.g., BM cells with Hoechst 33342 dye	Cells with highest Hoechst 33342 fluorescent efflux and lowest dye staining are referred to as side population (SP) cells	1 in 3 dye^{low} murine BM-derived cells shows long-term repopulating capacity	HSCs

Depending on localization and stage of differentiation, EryPs differ in their integrin expression profiles: yolk sac-resident EryPs express high levels of αIIb, α4, α5, αV, α6, β1, β2, and β3 integrins, circulating EryPs downregulate expression of almost all integrins, and upon homing to the developing fetal liver EryPs undergo final maturation and re-express and/or upregulate α4, α5 and β1 integrins suggesting that these integrins may influence EryP circulation, and maturation (Isern et al., 2011; Isern, Fraser, He, & Baron, 2008). Direct evidence for this hypothesis, however, is limited to α4β1 integrin, which enables EryPs to attach to VCAM-1-expressing macrophages in the fetal liver resulting in the formation of so-called erythroblastic islands. In erythroblastic islands, EryPs enucleate and complete their differentiation into reticulocytes. During the final maturation steps α4, α5 and β1 integrins are distributed to the plasma membrane encapsulating the extruded nuclei, which allows reticulocytes to detach from macrophages and egress into the circulation (Arroyo, Yang, Rayburn, & Hynes, 1999; Fraser, Isern, & Baron, 2007; Isern et al., 2008; McGrath et al., 2008) (Fig. 3). Of note, a similar integrin-driven process also operates during definitive erythropoiesis in the adult spleen and bone marrow (Lee et al., 2006; Sadahira, Yoshino, & Monobe, 1995).

The second wave of hematopoiesis is induced at E8.25 by a subset of intraluminal endothelial cells lining the developing yolk sac capillary network, the so-called hemogenic endothelium, which upregulate Scl and Runx1 transcription factors leading to an endothelial-to-hematopoietic-transition by rounding up, detaching from the endothelial layer and differentiating into bipotent erythro-myeloid progenitors (EMPs) with the potential to differentiate into definitive erythroid (EryD) or myeloid cells (Li et al., 2005; Palis et al., 1999, 2001; Wong et al., 1986; Yokomizo et al., 2001). The hemogenic endothelial cells as well as the EMPs are identified by αIIb and α4 integrins, whose expression is induced by Scl and Runx1 (Ferkowicz et al., 2003; Li et al., 2005; Mikkola, Fujiwara, Schlaeger, Traver, & Orkin, 2003; Ogawa et al., 1999). Functionally, however, neither αIIb nor α4 integrins seem required for the endothelial-to-hematopoietic transition or the formation of EMPs (Domínguez-Soto, Relloso, Vega, Corbí, & Puig-Kröger, 2005; Lie-A-Ling et al., 2014; Mikkola et al., 2003; Tanaka et al., 2012). Recent data obtained from single cell transcriptomics of E8.25 embryos suggest that also α5, αV, α6 and β1, β2, and β3 integrins are expressed in hemogenic endothelium and EMPs (Ibarra-Soria et al., 2018). Whether they play a functional role, or compensate for αIIb and/or α4 integrins in the above-mentioned genetic-loss-of-function experiments remains to be shown.

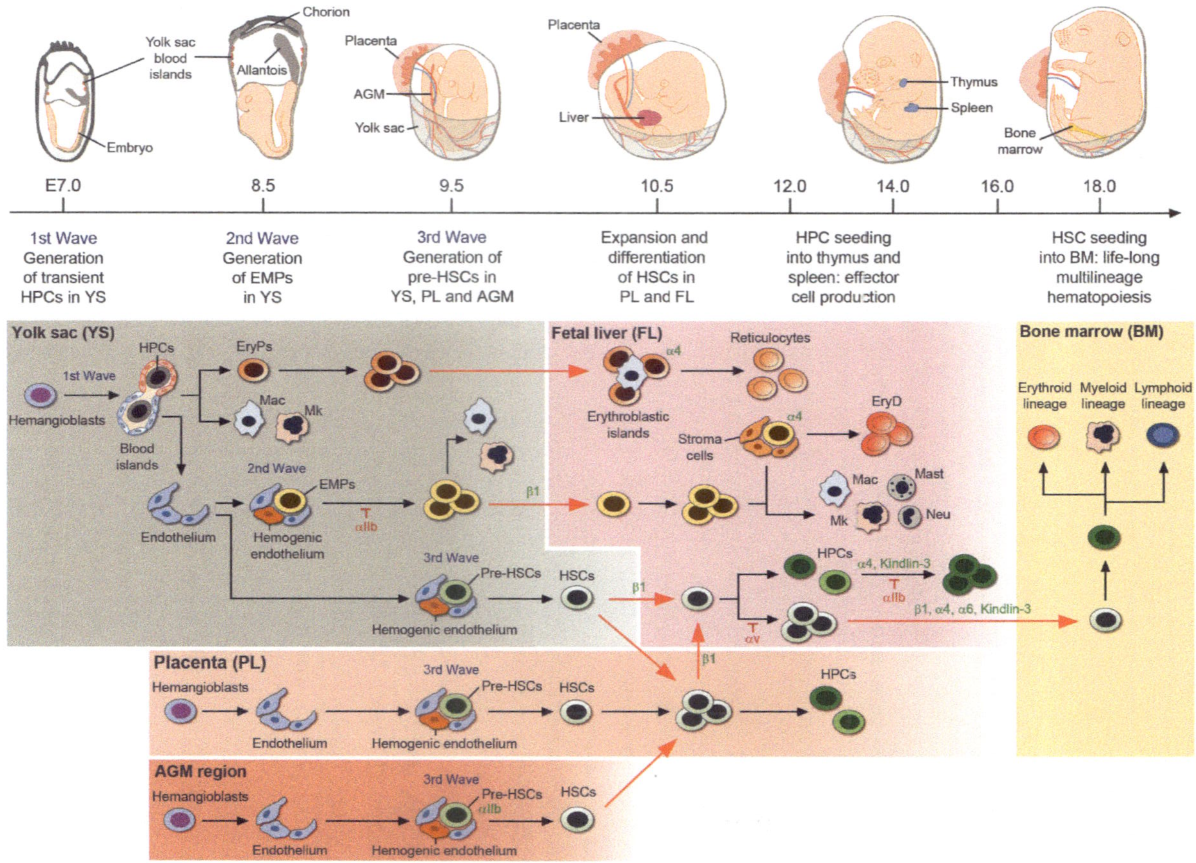

Fig. 3 Integrins and the development of hematopoiesis. Hematopoiesis evolves in three successive waves resulting in the production of hematopoietic progenitor cells (HPCs), erythro-myeloid progenitors (EMPs), hemogenic endothelium, and finally hematopoietic stem cells (HSCs). A simplified and incomplete hematopoietic tree is depicted for HSC-dependent hematopoiesis in the bone marrow. Integrins promoting HSPC functions are marked in green and integrins inhibiting HSCP functions are marked in red. Migration and homing steps *via* blood circulation are indicated by red arrows. EryPs: primitive erythroblasts; Mac: macrophages; Mk: megakaryocytes; EryD: definitive erythroid cells; Mast: mast cells; Neu: neutrophils; E: mouse embryonic day.

Once cardiac function and blood circulation are established between E8.25 and E10, yolk sac EMPs migrate to extraembryonic sites including allantois and placenta, and to intraembryonic sites such as the developing fetal liver where they rapidly expand, differentiate and establish the major pool of fetal erythrocytes and tissue-resident macrophages (McGrath et al., 2015; McGrath, Koniski, Malik, & Palis, 2003; Soares-da-Silva, Burlen-Defranoux, et al., 2020; Soares-da-Silva, Peixoto, et al., 2020; Wu & Hirschi, 2020).

Whereas genetic deletion of the αIIb integrin-encoding *ItgaIIb* gene allowed differentiation of yolk sac-derived EMPs *in vitro*, the numbers of EMPs in yolk sac and fetal liver were 2-fold increased pointing to an αIIb-mediated inhibition of EMP cell proliferation, particularly *in vivo*. Interestingly, the increase in EMP numbers was without consequences for embryonic or adult erythro- or myelopoiesis (Emambokus & Frampton, 2003).

Chimeric mice generated by complementing wildtype blastocysts with *Itgb1*- or *Itga4*-deficient embryonic stem (ES) cells have normal EMP numbers in yolk sacs but decreased numbers in fetal livers. In the case of *Itgb1* loss, the colonization of the fetal liver by EMPs was abrogated, whereas in the absence of *Itga4* EMPs can seed the fetal liver but cannot associate with, transmigrate beneath and obtain support from stromal cells required for EMP differentiation and subsequent erythrocyte production (Arroyo et al., 1999; Arroyo, Yang, Rayburn, & Hynes, 1996; Hirsch et al., 1996; Potocnik, Brakebusch, & Fässler, 2000). It is currently not clear whether *Itga4* on EMPs receive supportive signals from VCAM-1 on stromal cells or fibronectin (Eshghi, Vogelezang, Hynes, Griffith, & Lodish, 2007; Yanai, Sekine, Yagita, & Obinata, 1994). Since *Itgb7*-null mice display normal hematopoiesis, the EMP defects seem to be mediated by α4β1 and not α4β7 integrin (Wagner et al., 1996).

The different requirements of β1 and α4 integrin for EMP homing suggests that in addition to α4β1 also other members of the β1 integrin-class contribute to EMP-fetal liver homing. Since deletion of neither *Itga5* nor *Itgav* result in reduced EMP numbers, expansion or differentiation in the fetal liver (Arroyo et al., 1996, 1999, 1996, 1999), consequences of *Itga6* deletions or combinations of multiple deletions of β1 integrin-associating α-subunits in EMPs would have to be analyzed to unravel possible compensatory mechanisms (Fig. 3).

3.2 Hematopoietic stem cells, the insurance for life-long hematopoiesis

The third wave of hematopoiesis produces self-renewing and multipotent HSCs, which are the source for life-long hematopoiesis. Despite their importance for the entire life, HSCs are generated from hemogenic endothelium during a very narrow time window in development and therefore, the organism has to operate very economically with the consumption of the HSC pool (Bautch, 2011; Boisset et al., 2010; Emambokus & Frampton, 2003; Robin, Ottersbach, Boisset, Oziemlak, & Dzierzak, 2011; Swiers, Rode, Azzoni, & de Bruijn, 2013; Yokomizo & Dzierzak, 2010; Zovein et al., 2008). Using *Ncx1*-deficient embryos, which have non-beating hearts and hence, lack blood circulation, it was demonstrated that the E8.5–9.5 yolk sac, placenta and paraaortic splanchnopleure/aorta-gonad-mesonephros (AGM) region produce independently of each other cells with multi-lineage hematopoietic potential when cultured under supportive and differentiation-inducing conditions *ex vivo* (Corbel, Salaün, Belo-Diabangouaya, & Dieterlen-Lièvre, 2007; Cumano, Ferraz, Klaine, Di Santo, & Godin, 2001; Rhodes et al., 2008). Although these short-term *in vitro* differentiation assays strongly indicate that yolk sac, placenta and AGM harbor a cell population with HSC potential, they failed to engraft upon transplantation into primary adult recipient mice. Surprisingly, however, when first transplanted into newborn and from them into adult recipient mice, HSCs from yolk sac, placenta and AGM clearly demonstrated their self-renewal and multilineage repopulation capacity (Ganuza et al., 2018; Gekas, Dieterlen-Lièvre, Orkin, & Mikkola, 2005; Medvinsky & Dzierzak, 1996; Müller, Medvinsky, Strouboulis, Grosveld, & Dzierzak, 1994; Yamane, Washino, & Yamazaki, 2013; Yoder, Hiatt, Dutt, et al., 1997; Yoder, Hiatt, & Mukherjee, 1997). The observations showing that neonatal recipients preferentially support engraftment of early embryo-derived immature precursor HSCs (pre-HSCs) compared to fetal liver- or adult bone marrow-derived definitive HSCs (Arora et al., 2014) suggest that neonates harbor hematopoietic niches that provide an environment primarily for pre-HSCs. Neither the biochemical composition nor biophysical properties of pre-HSC supporting niches nor integrin profiles of pre-HSCs have been defined so far. Hence, it is not known whether and how these embryonic niches differ from fetal liver and adult bone marrow niches and to which extent integrin-mediated adhesion and signaling are involved in the maturation from pre-HSCs to HSCs.

The production of HSCs does not only occur in a narrow time window but is also a rare event and results in as few as 0.1, 0.5, and 0.3 long-term repopulating HSCs from E11 yolk sac, placenta and AGM region, respectively. Therefore, it is probably necessary that the more numerous pre-HSCs gradually mature, expand and contribute to the definitive HSC pool that sustains hematopoiesis after birth and consists of 4, 50, 3, and 93 long-term repopulating HSCs from E12.5 yolk sac, placenta, AGM region and fetal liver, respectively. It has been shown that pre-HSCs enter the circulation at E11.5 and extravasate to the placenta prior to re-entering the circulation and homing to the fetal liver (Gekas et al., 2005; Ottersbach & Dzierzak, 2005; Rybtsov, Ivanovs, Zhao, & Medvinsky, 2016). Whereas the placenta microenvironment supports maturation of pre-HSCs into HSCs and the expansion of the HSCs by symmetric divisions, the fetal liver microenvironment enables expansion and differentiation of HSCs by both, symmetric and asymmetric division (Chen et al., 2020; Gekas et al., 2005, 2010; Kumaravelu et al., 2002; Ottersbach & Dzierzak, 2005). Following the period of expansion and differentiation in the fetal liver, HSCs and/or HPCs colonize the thymus at around E12, spleen at around E15 and bone marrow at around E17.5 to secure long-term HSPC maintenance, differentiation and effector cell production (Christensen, Wright, Wagers, & Weissman, 2004; Itoi, Kawamoto, Katsura, & Amagai, 2001). Shortly after birth, the liver undergoes a remodeling step characterized by the formation of portal veins and loss of HSC-supporting Nestin^{+}NG2^{+} pericytes, which leads to the emigration of the HSC population into the bone marrow and the establishment of the bone marrow as primary site for adult hematopoiesis (Khan et al., 2016).

A combination of experiments including *in vivo* imaging, flow cytometry analysis and transplantation assays showed that αIIb integrins are expressed on all AGM- and a small fraction of yolk sac- but not on placenta-derived (pre-)HSCs (Boisset, Clapes, Van Der Linden, Dzierzak, & Robin, 2013; Robin et al., 2011). This observation together with *in vivo* lineage tracing studies, which revealed that less than 5% of adult bone marrow cells are derived from αIIb integrin-positive (pre-)HSCs (Emambokus & Frampton, 2003) suggest that the pool of αIIb integrin-negative yolk sac and/or placenta (pre-)HSCs play a more important role for adult hematopoiesis than those expressing αIIb and derived from the AGM. In support of these findings, genetic deletion of the αIIb integrin-encoding *ItgaIIb* gene exclusively reduced the number of long-term repopulating HSCs from the AGM region but not those from yolk sac or placenta (Boisset et al., 2013). Although these findings point to a role for αIIb integrins in maintaining the AGM-derived HSC pool, the consequences for postnatal hematopoiesis are not entirely clear. *ItgaIIb*- and *Itgb3*-deficient mice and patients suffering from Glanzmann's thrombasthenia, who—like the mutant mice—lack expression of αIIbβ3 integrins, display severe bleedings but no obvious HSC defects postnatally (Hodivala-Dilke et al., 1999; Nurden, Fiore, Nurden, & Pillois, 2011; Tronik-Le Roux et al., 2000). Very likely, the embryonic defect is compensated by other integrins, later in development.

Whereas several laboratories have shown with snapshot analyses that fetal liver- and adult bone marrow-derived HSPCs express α2, α4, α5, α6, α9, αV, αL, β1, β2 and β3 integrins on their surface (Biswas et al., 2020; Qian et al., 2007; Ruppert et al., 2015), a complete and unbiased measurement of integrin expression profiles throughout HSC development was obtained by single cell RNA sequencing (Zhou et al., 2016). The integrin mRNA expression levels of AGM hemogenic endothelial cells, AGM pre-HSCs, fetal liver HSCs and adult bone marrow HSCs revealed that they all share the expression of several integrins including α4, α5, α6, α9, αV, and β1. Interestingly, αIIb and β3 integrins are downregulated and αL, αM, and β2 integrin levels increase upon transition from pre-HSC to fetal liver HSC (Zhou et al., 2016). The mRNA sequencing study by Biswas et al. (2020) suggested that fetal liver but not adult bone marrow HSCs express α1, α8, αX, α11, β4 and β6 integrins. Although the mRNA data have not been confirmed at the protein level, they indicate that integrin expression is dynamic and changes during or directly after HSC homing to the fetal liver and/or bone marrow. How integrin expression is regulated and whether changes of integrin surface profiles are required for successful HSC homing and expansion remains to be analyzed.

Genetic studies identified the β1-class integrins as essential adhesion molecules for fetal liver homing of HSCs. Blastocyst complementation assays, in which wildtype blastocysts were complemented with *Itgb1*-null ES cells revealed that HSCs accumulate in the circulation and are absent in fetal liver, spleen and bone marrow. The origin of the accumulating *Itgb1*-null HSCs has not been determined. However, the normal HSC production and differentiation in the absence of the *Itgb1* gene suggests that particularly HSC homing into the fetal liver is β1-class integrin-dependent (Hirsch et al., 1996; Potocnik et al., 2000). Since none of the genetic loss-of-function experiments abrogating the expression of α-subunits able to associate with the β1-subunit recapitulates this severe phenotype, it was concluded that more than one member of the β1-class integrins mediates fetal liver homing. Based on integrin profile analyses of fetal liver HSCs, the integrins α2β1, α4β1, α5β1, αVβ1, α6β1 and α9β1 are candidates (Biswas et al., 2020; Qian, Georges-Labouesse, et al., 2007; Ruppert et al., 2015). These integrins indeed share ligands to a certain extent and have partially overlapping functions (see Table 2).

The α4-subunit, which can associate with β1- and β7-subunits, has been extensively analyzed using blastocyst complementation assays. HSPCs lacking the *Itga4* gene home to and expand in fetal liver and bone marrow, but differentiate less efficiently into erythroid, myeloid and lymphoid cells and fail to provide hematopoiesis after birth. The normal *in vitro* differentiation capacity of fetal *Itga4*-null HSPCs excludes cell intrinsic defects but rather points to defective interactions with and support by the maturing bone marrow microenvironment after birth (Arroyo et al., 1996, 1999). This hypothesis would also explain the severe bone marrow homing and engraftment defect of fetal liver HSPCs that were treated with anti-α4 integrin blocking antibodies (Qian, Georges-Labouesse, et al., 2007). Potential α4β1 integrin ligands mediating these processes include VCAM-1, fibronectin, osteopontin and MAdCAM-1. Whereas MAdCAM-1 is not required for hematopoiesis, osteopontin, which can be bound by α4β1 and in addition by α9β1 integrins, regulates HPC differentiation and maintains HSC quiescence in fetal and neonatal bone marrow (Cao et al., 2019; Schippers et al., 2009). Although fibronectin as well as VCAM-1 are expressed in fetal liver (Ayres-Silva et al., 2011; Schweitzer et al., 1996; Soares-da-Silva, Burlen-Defranoux, et al., 2020; Soares-da-Silva, Peixoto, et al., 2020; Sugiyama et al., 2013; Sugiyama, Koike, & Shiojiri, 2010), their early embryonic lethality in genetic loss-of-function experiments did not allow investigating their role for fetal liver HSCs *in vivo* (George, Georges-Labouesse, Patel-King, Rayburn, & Hynes, 1993; Kwee et al., 1995).

Deletion of the *Itgav* gene with an inducible *Vav-Cre* transgene revealed an expansion of short-term repopulating HSPCs and an elevation of total fetal liver cell numbers without altering the numbers of long-term repopulating HSCs. This phenotype was attributed to the inability of HSPCs to bind periostin (POSTN), an ECM protein produced by the fetal liver vascular endothelium (Biswas et al., 2020) and shown to bind αVβ3 and αVβ5 integrins (Gillan et al., 2002).

Fetal liver HSPCs lacking the expression of the laminin-binding α6β1 integrin display impaired adhesion to laminin-411, -511 and -521 *in vitro*, and impaired homing of *Itga6*-null fetal liver HPCs but not HSCs to the adult

Table 2 Integrin ligands and integrin regulators operating in HSPC niches.

Integrin ligand	Expressed in	Expressed by	Function	Integrin involved	References
Fibronectin	YS, AGM	Endothelial cells	HSPC homing, migration, proliferation, differentiation	α5β1, α4β1, α4β7, αVβ3, αIIbβ3	Risau and Lemmon (1988)
	PL	Stromal cells, endothelial cells			Arai and Nishiyama (2007); Chen and Aplin (2003); Saylam, Ozdemir, Itil, Sendag, and Terek (2002)
	FL	Megakaryocytes, sinusoidal lining cells, stromal cells, hepatoblasts, hepatic stellate cells			Ayres-Silva, Manso, Madeira, Pelajo-Machado, and Lenzi (2011); Baloch, Klapper, Buchanan, Schwartz, & Amenta, 1992); Bockman and Gulati (1989); Hackney et al. (2002); Soares-da-Silva, Burlen-Defranoux, et al., 2020; Soares-da-Silva, Peixoto, et al., 2020; Sugiyama, Kulkeaw, and Mizuochi (2013); Tan, Kulkeaw, Nakanishi, and Sugiyama (2017)
	BM	Stromal cells, osteoblasts, megakaryocytes, endothelial cells			Carstanjen, Gross, Kosova, Fichtner, and Salama (2005); Craddock, Nakamoto, Andrews, Priestley, and Papayannopoulou (1997); Craddock, Nakamoto, Elices, and Papayannopoulou (1997); Gu, Nilsson, Eng, and Ekblom (2000); Malara et al. (2014); Nilsson et al. (1998); Tada and Fukuta (2010); Verfaillie, McCarthy, and McGlave (1991); Wilkinson et al. (2019); Williams, Rios, Stephens, and Patel (1991)
VCAM-1	YS, PL, FL, BM, Spleen	Endothelial cells, macrophages	HSPC homing and retention; erythropoiesis	α4β1	Gurtner et al. (1995); Kwee et al. (1995); Lichterfeld, Martin, Burkly, Haas, and Kronenwett (2000); Papayannopoulou, Priestley, and Nakamoto (1998); Soares-da-Silva, Burlen-Defranoux, et al., 2020; Soares-da-Silva, Peixoto, et al., 2020; Ulyanova et al. (2005); Yanai et al. (1994)

Continued

Table 2 Integrin ligands and integrin regulators operating in HSPC niches.—cont'd

Integrin ligand	Expressed in	Expressed by	Function	Integrin involved	References
ICAM-1	BM	Endothelial cells, mesenchymal stem cells, osteoblasts, CXCL12-abundant reticular cells	HSPC quiescence and retention	αLβ2	Liu et al. (2018)
MAdCAM-1	BM	Endothelial cells	HPC homing	α4β7	Katayama, Hidalgo, Peired, and Frenette (2004)
Laminins	PL, FL	not analyzed	not analyzed	α6β1	Saylam et al. (2002); Sugiyama et al. (2013); Tan et al. (2017); Yap, Tay, Nguyen, Tjin, and Tryggvason (2019)
	BM	Endothelial cells, pericytes, stromal cells	HSPC homing, maintenance, proliferation		Gu et al. (2003); Nilsson et al. (1998); Qian, Georges-Labouesse, et al. (2007); Siler et al. (2000); Susek et al. (2018)
Collagens	PL	not analyzed	not analyzed	α2β1	Arai and Nishiyama (2007); Oefner et al. (2015); Saylam et al. (2002)
	FL	Sinusoidal cells, hepatoblasts			Baloch et al. (1992)
	BM	Osteoblasts, megakaryocytes			Malara et al. (2014); Nilsson et al. (1998)
Periostin	FL	Endothelial cells	HSC quiescence and maintenance	αVβ3	Biswas et al. (2020)
	BM	Endothelial cells, HSPCs			Khurana et al. (2016)
	Spleen	Myofibroblasts, vascular smooth muscle cells			Mehatre et al. (2021)
Tenascin-C	FL	Stromal cells, hepatic stellate cells	not analyzed	α9β1	Papadopoulos et al. (2004); Tan et al. (2017)
	BM	Stromal cells, endothelial cells, osteoblasts	HSPC proliferation and differentiation		Mackie and Tucker (1992); Nakamura-Ishizu et al. (2012); Ohta, Sakai, Saga, Aizawa, and Saito (1998); Seiffert et al. (1998)

Osteopontin	PL	Epithelial cells	not analyzed	α9β1, α4β1	Cao et al. (2019); Grassinger et al. (2009); Nilsson et al. (2005); Stier et al. (2005)
	BM	Osteoblasts, stromal cells	HSPC quiescence, migration, retention		
Nephroblastoma overexpressed (Nov/CCN3)	BM	HSCs	HSC maintenance	αVβ3	Ishihara et al. (2014)
Del-1	BM	Arteriolar endothelial cells, CXCL12-abundant reticular cells and osteoblasts	HSPC retention and proliferation; myelopoiesis	αVβ3	Mitroulis et al. (2017)
Vitronectin	FL	Hepatoblasts, hepatic stellate cells	not analyzed	αVβ3	Sugiyama et al. (2013); Tan et al. (2017)
	BM	not analyzed	HSC maintenance		Seiffert (1996); Umemoto et al. (2012)
CXCL12 (SDF-1)	FL	Hepatoblasts, pericytes	HSPC homing and expansion; B-cell lymphopoiesis	α5β1, α4β1, αLβ2	Chou and Lodish (2010); Khan et al. (2016); Nagasawa et al. (1996); Zou, Kottmann, Kuroda, Taniuchi, and Littman (1998)
	BM	Perivascular stromal cells (mesenchymal stem and CXCL12-abundant reticular cells), endothelial cells, osteoblasts	HSPC maintenance, retention and homing; myelopoiesis		Ara et al. (2003); Ding and Morrison (2013); Greenbaum et al. (2013); Lai et al. (2014); Nagasawa et al. (1996); Sugiyama, Kohara, Noda, and Nagasawa (2006); Zhang et al. (2016)
Angiopoietin-1	YS, p-SP/AGM, FL	HSPCs, pericytes	Vessel sprouting and remodeling	β1	Takakura et al. (2000)
	BM	Osteoblasts	HSC quiescence		Arai et al. (2004)
G-CSF	YS, AGM, FL	not analyzed	not analyzed	α4β1	Okada et al. (1998)
	BM	Stromal cells, endothelial cells, (peripheral) hematopoietic cells	HSPC mobilization and proliferation		Demetri and Griffin (1991); Tamura et al. (1987)

Continued

Table 2 Integrin ligands and integrin regulators operating in HSPC niches.—cont'd

Integrin ligand	Expressed in	Expressed by	Function	Integrin involved	References
SCF/c-kitL	AGM, PL	Endothelial cells	HSC formation and expansion	β1	Sasaki et al. (2010)
	FL	Hepatoblasts	HSPC expansion		Chou and Lodish (2010)
	BM	Leptin receptor$^+$ stromal cells, endothelial cells, adipocytes	HSPC maintenance		Comazzetto et al. (2019); Ding, Saunders, Enikolopov, and Morrison (2012); Zhou et al. (2017)
FLT3L	FL, BM	Stromal cells, endothelial cells	HSPC mobilization and expansion	α4β1, α5β1	Brasel et al. (1996); Buza-Vidas et al. (2009); Lisovsky et al. (1996); Lyman et al. (1993); McKenna et al. (2000); Solanilla et al. (2000)
Calcium	BM	Osteoblasts	HSC quiescence	α4β1, α9β1	Cao et al. (2019); Luchsinger et al. (2019)
Thrombopoietin	FL	Hepatoblasts	HSC maintenance and expansion	αVβ3	Chou and Lodish (2010); Decker, Leslie, Liu, and Ding (2018)
	BM	Osteoblasts	HSC maintenance		Buza-Vidas et al. (2006); Decker et al. (2018); Ishihara et al. (2014); Qian et al. (2007); Umemoto et al. (2012); Yoshihara et al. (2007)

bone marrow in transplantation experiments. Despite these pronounced *in vitro* and *in vivo* defects, constitutive deletion of *Itga6* in mice revealed normal fetal liver and bone marrow hematopoiesis (Qian, Georges-Labouesse, et al., 2007). These conflicting results can only be explained by a differential requirement of HSCs and HPCs for α6 integrin-mediated adhesion to, probably, distinct niches and compensatory mechanisms established by HPCs during embryo development.

Mice lacking the collagen-binding α2β1 integrin establish hematopoiesis both, during development and postnatally (Holtkötter et al., 2002; Nieuwenhuis, Akkerman, Houdijk, & Sixma, 1985). Whether other collagen-binding integrins (α1β1, α10β1, α11β1) are expressed on HSCs and compensate for the loss of *Itga2* is not known.

This overview of integrin functions for the development of embryonic, fetal and postnatal hematopoiesis highlights their importance but also the difficulty to identify integrin-specific functions. This difficulty results from the ability of integrins to share ligands and hence, perform overlapping functions. This situation can only be overcome by simultaneously disrupting several integrin genes, or generating a mouse model, in which all 18 α and 8 β integrin-subunits are conditionally deleted in HSCs and individual integrins re-expressed one by one. Although tedious, this approach is probably achievable using the CRISPR-Cas9 technology. An alternative procedure is to impair the activation and hence function of many, if not all integrins, by constitutively or conditionally deleting the gene encoding Kindlin-3 or Talin-1 in mice or generating bone marrow chimeras with Kindlin-3- or Talin-1-deficient HSCs (Moretti et al., 2013; Moser et al., 2008, 2009; Petrich et al., 2007; Schmidt et al., 2011). The latter approach does not distinguish between individual integrins, but impairs the activation of compensatory, integrin-based mechanisms. This approach was pursued and came with surprises. In contrast to *Itgb1*-null HSCs, Kindlin-3-deficient HSCs were shown to colonize the fetal liver and the bone marrow, although inefficiently (Moser et al., 2008; Ruppert et al., 2015). This unexpected outcome could result either from the few poorly functioning integrins that adopt an extended-open, active conformation due to thermodynamics (Li et al., 2017; Li & Springer, 2017; Sun, Guo, & Fässler, 2016) or, less likely, the presence of low-levels of Kindlin-2 in (pre-)HSCs shortly after their endothelial-to-hematopoietic transition (Zhou et al., 2016). The analysis of the fetal liver HSPC compartment also revealed that the actual number of colony-forming HPCs but not HSCs was decreased in Kindlin-3-null fetal livers. Transplantation experiments in lethally irradiated recipient mice revealed that loss of Kindlin-3 significantly impairs homing and engraftment of fetal liver HSPCs to the bone marrow and eventually results in HSC exhaustion due to dysfunctions of almost all effector cells (Krüger et al., 2008; Moser et al., 2008; Ruppert et al., 2015) (Fig. 3).

4. Contribution of integrin-mediated adhesion to life-long hematopoiesis

In contrast to the high proliferation rates of fetal and neonatal HSCs, the majority of HSCs switch into a state of dormancy at around 4 weeks after birth, which is induced and maintained by the bone marrow microenvironment called hematopoietic stem cell niche (Arai et al., 2004; Bowie et al., 2006; Cheng et al., 2000; Cheshier, Morrison, Liao, & Weissman, 1999; Kiel et al., 2007; Morrison & Scadden, 2014; Pietras, Warr, & Passegué, 2011; Qian, Buza-Vidas, et al., 2007; Yamazaki et al., 2011; Yoshihara et al., 2007). Upon induction of hematopoietic stresses, such as infection or injury-induced blood loss, quiescent HSCs are activated, enter the cell cycle, divide asymmetrically and produce HPCs required for effector cell generation prior to re-entering the quiescent state again (Cheshier, Prohaska, & Weissman, 2007; Mirantes, Passegué, & Pietras, 2014; Schoedel et al., 2016; Schuettpelz & Link, 2013; Wilson et al., 2008). The balance between HSC dormancy, proliferation and differentiation is meticulously regulated to limit proliferation of HSCs, prevent HSC senescence, malfunctions and malignancies, and ensure life-long hematopoiesis (Bernitz, Kim, MacArthur, Sieburg, & Moore, 2016; Mirantes et al., 2014; Schoedel et al., 2016).

While site(s) and composition of hematopoietic stem cell niches are still debated and heavily investigated (see review Yu and Scadden (2016) and Table 2), (at least) two distinct HSC-supporting bone marrow microenvironments have been identified and characterized: (1) the endosteal niche, which promotes HSC quiescence, and (2) the vascular niche, which promotes HSC maintenance, regeneration and proliferation (Arai et al., 2004; Calvi et al., 2003; Ding et al., 2012; Kiel et al., 2005; Sugiyama et al., 2006; Winkler et al., 2012; Zhang et al., 2003). However, there is emerging evidence indicating that bone and vascular niches are adjoining or even intertwined and share cellular, ECM and signaling components resulting in considerable functional overlap (Kiel & Morrison, 2008; Lo Celso et al., 2009; Xie et al., 2009). This feature together with the high motility of HSCs on one hand, and the availability of integrin adhesion molecules on the other hand enables HSCs to select bone marrow microenvironments specifically to satisfy their constantly changing tasks ranging from proliferation, differentiation or entry into dormancy (De Grandis, Lhoumeau, Mancini, & Aurrand-Lions, 2016; Kokkaliaris et al., 2020; Lo Celso et al., 2009; Nilsson, Johnston, & Coverdale, 2001; Tokoyoda, Egawa, Sugiyama, Choi, & Nagasawa, 2004; Wright, Wagers, Gulati, Johnson, & Weissman, 2001; Xie et al., 2009).

The chemokine CXCL12, also known as SDF-1, is an important hematopoietic bone marrow niche factor with several key roles in hematopoiesis, some of them operate *via* the potential to activate integrins (Nagasawa et al., 1996). Mice carrying a *CXCL12-GFP* knock-in gene revealed strong CXCL12-GFP expression by perivascular stromal and endothelial cells and to a much lesser extent by endosteal osteoblasts. The differential CXCL12 expression is sensed by the chemokine receptor CXCR4 on HSPCs and directs HSPCs to bone marrow niches and promotes HSC quiescence (Nie, Han, & Zou, 2008; Sugiyama et al., 2006; Tzeng et al., 2011). Activation of the CXCL12-CXCR4 signaling axis polarizes HSPCs, activates integrins on HSPCs, and induces the formation of a forward-facing synapse-like adhesion structure termed magnopodium, which anchors HSPCs specifically to bone marrow osteoblasts and endothelial cells and allows the remaining HSPC cell body to dynamically scan the niche for factors that regulate HSPC integrin activity, self-renewal, quiescence, expansion or mobilization (see Table 2 and review Lee, Kim, & Cho, 2020 (Bessy et al., 2021; Ding et al., 2012; Peled, Petit, et al., 1999; Peled, Grabovsky, et al., 1999; Peled et al., 2000). Although integrins are major adhesion molecules expressed by HSPCs, HSPCs also produce a range of additional cell adhesion receptors including CD44, Notch, Eph/ephrins, selectins, selectin ligands, and N-cadherin, which contribute to HSC maintenance, HSPC homing to and retention in bone marrow niches, often by cooperating with integrins. The properties of these adhesion molecules for HSPCs have been reviewed elsewhere (Kulkarni & Kale, 2020) and will not be further discussed here.

Adult bone marrow HSPCs including long-term repopulating HSCs express α2, α4, α5, α6, α9, αL, αV, β1, β2, β3, and β7 integrin subunits and probably additional ones that have not been measured so far (Katayama et al., 2004; Ruppert et al., 2015; Umemoto et al., 2012; Wagers, Allsopp, & Weissman, 2002).

Abolishing the functions of β1-class integrins by genetic manipulations or function-blocking antibodies reduced adhesion of adult HSPC to endothelial or stromal cells *in vitro* and homing to the bone marrow in mice (Arai et al., 2004; Hirsch et al., 1996; Potocnik et al., 2000; Williams et al., 1991; Zhou, Ding, & Morrison, 2015). It was also reported that β1 integrins are upregulated on quiescent HSCs by Tie2/Angiopoietin-1 signaling, which promotes adhesion to osteoblasts in the endosteal niche (Arai et al., 2004). The relevance of these findings for HSC dormancy, however, could not be confirmed with conditional *Itgb1* deletion experiments in HSPCs using *Itgb1*$^{fl/fl; Mx\text{-}Cre}$ bone marrow chimeras, in which HSC dormancy, retention, and differentiation remained unaffected (Brakebusch et al., 2002). Simultaneous deletion of the *Itgb1* and *Itgb7* genes in HSPCs revealed a slight and transient increase of HPCs in bone marrow and peripheral blood pointing to impaired HPC retention, which is normalized over time by unknown mechanism(s) (Bungartz et al., 2006).

Deletion of *Itgb1* in HSPCs was accompanied by the reduction of α4 and loss of α5 and α6 integrin expression (Potocnik et al., 2000), whereas deletion of *Itga4* reduced but did not abolish β1 integrin expression (Scott, Priestley, & Papayannopoulou, 2003). These changes in integrin profile suggest that the severe endothelial transmigration and bone marrow homing/retention defects of β1 integrin-deficient HSPCs are probably caused by a combinatorial loss of α4β1, α5β1 and α6β1 integrins. Studies with function-blocking antibodies against α4, α5 and α6 integrins, respectively, reduced murine and human adult HSPC adhesion to their respective ligands (i.e., VCAM-1, fibronectin and laminins) *in vitro* and endothelial cells and/or stromal cells *ex vivo* (Hurley, McCarthy, & Verfaillie, 1995; Kerst et al., 1993; Qian, Tryggvason, Jacobsen, & Ekblom, 2006; Teixidó, Hemler, Greenberger, & Anklesaria, 1992; van der Loo et al., 1998; Williams et al., 1991).

Several experiments and clinical observations suggest that α4 integrins play an important role for HSPC functions. Inducible deletion of *Itga4* in HSPCs of adult mice caused a pronounced bone marrow retention defect with an accumulation of *Itga4*-null HSPCs in the circulation and spleen. The retention defect did not diminish the numbers of *Itga4*-null HSPCs in the bone marrow over time suggesting that the loss of *Itga4*-null HSPCs is compensated by increased proliferation, reduced apoptosis and/or increased re-homing of *Itga4*-null HSPCs to the bone marrow (Scott et al., 2003). A similar HSPC bone marrow retention and mobilization defect was also observed in multiple sclerosis patients treated with the anti-α4 integrin antibody natalizumab (Zohren et al., 2008). Furthermore, transplantation of murine as well as human adult HSPCs, in which α4 integrins were either genetically ablated or functionally blocked with anti-α4 integrin antibodies into lethally irradiated recipients or immunodeficient NOD/SCID mice, respectively, confirmed a role of α4 integrins for bone marrow homing, engraftment and bone marrow retention (Jiang, Bonig, Ulyanova, Chang, & Papayannopoulou, 2009; Katayama et al., 2003; Peled et al., 2000; Priestley, Scott, Ulyanova, & Papayannopoulou, 2006; Priestley, Ulyanova, & Papayannopoulou, 2007; Scott et al., 2003; Wagers et al., 2002; Williams et al., 1991). Due to strong homing defects, the involvement of α4 integrins for HSC stemness and self-renewal could not be tested in transplantation assays. In contrast to the experiments with bone marrow chimeras, the conditional deletion of *Itga4* with the *Mx-Cre* transgene in adult mice neither impaired HPC differentiation nor effector cell numbers (Arroyo et al., 1996; Scott et al., 2003). However, when challenged with 5-fluorouracil, which depletes HPCs and effector cells and leads to a strong activation and proliferation of HSCs (Lerner & Harrison, 1990; Wilson et al., 2008), the recovery of the peripheral blood cell pool is retarded due to loss of HPCs into the circulation (Scott et al., 2003). Treatment of wildtype mice with functionally blocking fibronectin peptides or anti-VCAM-1 and/or anti-α4 integrin antibodies revealed that the HPC retention defect is caused by an adhesion defect of α4 integrins to VCAM-1 rather than to fibronectin (Craddock, Nakamoto, Elices, et al., 1997; Papayannopoulou et al., 1998; Verfaillie et al., 1991). The HPC retention defect was also confirmed in mice with a conditional deletion of *VCAM-1* in hematopoietic and endothelial cells (Ulyanova et al., 2005). Despite the important role of the α4 integrin for the retention of HPCs in bone marrow, their mobilization is further increased by additionally abrogating CXCR4-CXCL12 signaling or co-administrating either G-CSF, SCF or FLT3 ligand (Christopher, Liu, Hilton, Long, & Link, 2009; Craddock, Nakamoto, Andrews, et al., 1997; Papayannopoulou et al., 1998; Qian et al., 2006), which points to a complex involvement of additional integrins and/or integrin-unrelated adhesion molecules for HPC retention under steady state and stress (Greenbaum & Link, 2011).

The studies investigating the role of α5 integrins for adult HSPC bone marrow homing, engraftment and hematopoiesis produced conflicting results. In one report it was shown that inhibition of α5 integrin with function-blocking antibodies does not impair HSPC homing to the bone marrow (Craddock, Nakamoto, Elices, et al., 1997), while several other studies reported a significant reduction of HSPC bone marrow homing and engraftment (Asaumi et al., 2001; Wierenga, Weersing, Dontje, de

Haan, & van Os, 2006). Although this conflict has not been solved yet and requires additional experiments, further insights on the role of α5 integrins for adult hematopoiesis were obtained from adult chimeric mice generated by complementing wildtype blastocysts with *Itga5*-null ES cells. These animals present normal numbers of lymphocytes, monocytes and granulocytes, which suggests that α5 integrins are not required for adult steady state hematopoiesis (Arroyo et al., 2000). Unfortunately, these mice were not exposed to hematopoietic stress.

Inhibition of α6 integrins on HSPCs with function-blocking antibodies reduced homing of long-term repopulating multilineage HSCs but not short-term repopulating HSPCs to bone marrow of mice. Combining anti-α6 integrin with anti-α4 integrin blocking antibodies reinforced HSPC homing defects to the bone marrow suggesting that the two integrins cooperate (Qian et al., 2006). The α6β1 integrin binds to several laminins. So far, deletion of the laminin α4 chain encoding *Lama4* gene in mice was tested in reciprocal bone marrow cell transplantation experiments, which revealed impaired homing of wildtype HPCs to the *Lama4*-deficient bone marrow. *Lama4*-deficient bone marrow-derived HSCs and HPCs displayed homing defects and reduced long-term engraftment, respectively, when transplanted into wildtype mice indicating HSPC intrinsic defects. However, *Lama4*-null mice present, if at all, only slightly reduced HSC, HPC and effector cell numbers in the bone marrow suggesting that the intrinsic defects only come into play upon stress-induced hematopoiesis (Susek et al., 2018). Although the functional studies of α6 integrins are far from clear, high α6 integrin surface expression on HSCs has been established as widely used marker that can distinguish long-term multilineage repopulating HSCs from HPCs in humans (Huntsman et al., 2015; Notta et al., 2011).

Whereas the majority of bone marrow-derived HSPCs express low levels of α2 integrin, a subset lacks α2 integrin expression (Wagers et al., 2002; Wagers & Weissman, 2006). Upon transplantation into lethally irradiated recipients only α2-negative HSPCs, but not $\alpha 2^{low}$ HSPCs, provide long-lasting multilineage hematopoiesis (Wagers & Weissman, 2006). This finding suggests that all self-renewing HSCs lack α2 integrin expression and that further differentiated HSPCs specifically induce α2 integrin expression by currently unknown mechanism(s). However, considering that *Itga2*-null mice present normal hematopoiesis, α2 integrin could be non-functional on these HSPCs and represent a mere surrogate marker for differentiation (Holtkötter et al., 2002).

Itga9-null mice die perinatally and suffer from respiratory defects and reduced fetal and neonatal neutrophil production (Chen et al., 2006; Huang et al., 2000). The role of the α9 integrin for HSPC functions has not been studied in conditional knockout mice or bone marrow chimeras. However, *in vitro* experiments using anti-α9 function-blocking antibodies revealed that α9β1 integrin promotes adhesion of HSPCs to osteoblasts, which express multiple α9β1 integrin ligands including VCAM-1, tenascin-C, and thrombin-cleaved osteopontin (Mackie & Tucker, 1992; Schreiber et al., 2009; Tanaka et al., 1995). Further *in vitro* studies revealed that HSPC adhesion to recombinant VCAM-1 and tenascin-C is not affected by anti-α9 integrin blocking antibodies (Schreiber et al., 2009) suggesting that osteopontin is probably the major α9β1 integrin ligand on osteoblasts or that additional adhesion receptors beyond α9β1 bind VCAM-1 and/or tenascin-C. *In vivo* studies defining a potential role for α9 integrin-mediated adhesion or signaling for HSPC function are limited to genetic loss-of-function studies targeting the non-exclusive α9β1 integrin ligands VCAM-1, tenascin-C and osteopontin and are therefore far from conclusive. Loss of VCAM-1 expression in hematopoietic and endothelial cells results in HPC retention defects (Ulyanova et al., 2005), loss of tenascin-C expression does not affect steady state hematopoiesis but impairs proliferation of HSPCs and regeneration of the hematopoietic system in stress situations such as 5-fluorouracil treatment (Nakamura-Ishizu et al., 2012) and loss of osteopontin expression diminishes HSPC proliferation and survival in the endosteal niche under steady state conditions (Grassinger et al., 2009; Nilsson et al., 2005; Stier et al., 2005).

Reports that investigated the roles of the four members of the "leukocyte-specific" β2-class integrins for HSPC function are scarce, despite their prominent roles for the extravasation of leukocytes (Bednarczyk, Stege, Grabbe, & Bros, 2020; Fagerholm, Guenther, Llort Asens, Savinko, & Uotila, 2019). The expression of αL integrin increases during HSC differentiation and shows highest expression level on HPCs with limited self-renewal and long-term reconstitution potential (Fathman et al., 2014). Mice lacking the *ItgαL* gene have normal HSPC numbers with the exception of the subpopulation of unipotent lymphoid-primed HPCs, also referred to as common lymphoid progenitors, which are reduced in the bone marrow resulting in partially impaired early T cell development in the thymus and slightly decreased B and T cell numbers in the circulation. Transplantation assays confirmed that αL-deficient HSPCs insufficiently generate lymphoid-primed HPCs in the presence of competing wild type cells (Bose, Colpitts, Pham, Puddington, & Lefrançois, 2014). In contrast to the αL-subunit deficiency, loss or reduced expression of the β2-subunit encoding *Itgb2* integrin gene increased HSPC numbers and improved HSC self-renewal capacity in bone marrow (Leon-Rico et al., 2014; Papayannopoulou et al., 2001). Although these observations seem contradicting, the latter studies did not investigate lymphoid-primed HPCs and can therefore not confirm or exclude a role for αLβ2 integrin for the production and/or maintenance of this specific HPC subpopulation in the bone marrow. Nevertheless, it could also be that the different members of the β2-class integrins exert either positive or negative effects on HSPC function and the latter prevails when all members of β2-class are removed in HSCs. Loss of ICAM-1, which binds all four members of the β2-class integrins, although with different affinities, also results in an expansion of HSPCs in the bone marrow and in addition, in a differentiation bias towards the myeloid lineage suggesting reduced HPC multipotency. Since reciprocal transplantation experiments excluded HSPC-intrinsic defects, the HSPC expansion seems to be caused by ICAM-1 deficiency in the bone marrow niche and probably by abrogating the interaction with β2-class integrins on HSPCs. In line with a niche role of ICAM-1, G-CSF treatment increases the mobilization of HSPCs from the bone marrow significantly more in ICAM-1-deficient mice than in wildtype control mice (Liu et al., 2018). Reports on the functional properties of the αM-, αD- or αX-subunits, which associate with β2 to generate the remaining three members of the β2-class integrins, for HSPCs are still missing.

HSPCs can be distinguished in high and low β3 integrin expressing populations (Umemoto et al., 2006). $\beta3^{high}$ HSPCs display a superior engraftment in competitive transplantation experiments over $\beta3^{low}$ HSPCs indicating that HSCs are enriched in the $\beta3^{high}$ HSPC population (Umemoto et al., 2006, 2008). The engraftment-promoting role of β3 integrin for HSCs was confirmed in transplantation assays with HSCs that constitutively lack β3 integrin expression or express an inactive, Talin-binding deficient β3 integrin, in which leucine-746 in the β3 integrin cytoplasmic domain was substituted for an alanine (Umemoto et al., 2012). The α-subunit that associates and forms a functional integrin with β3 on adult HSPCs is likely αV, as expression of the αIIb-subunit decreases at the end of development and becomes only re-expressed on HSCs at old age (Gekas & Graf, 2013; Umemoto et al., 2006). Competitive serial transplantation assays with αV-deficient HSPCs confirmed the important role of αVβ3 integrin for HSC engraftment and maintenance (Khurana et al., 2016). The ligands of αVβ3 integrins that have been associated with HSC maintenance include thrombopoietin, nephroblastoma overexpressed (Nov/CCN3), periostin (POSTN) and Del-1 (Ishihara et al., 2014; Khurana et al., 2016; Mitroulis et al., 2017; Qian, Buza-Vidas, et al., 2007; Umemoto et al., 2012; Yoshihara et al., 2007). Additional ligands such as fibronectin, vitronectin, osteopontin, tenascin-C and CD31 could also play a role, although they also bind to several other integrins on HSCs, which complicates the interpretation of their effects on HSCs (Malara et al., 2014; Nakamura-Ishizu et al., 2012; Nilsson et al., 2005; Ross et al., 2008; Umemoto et al., 2012).

Integrins are the principal components of the adhesome, which consists of hundreds of proteins, associates with the actomyosin cytoskeleton and transmits integrin signals to different subcellular compartments. Since it is impossible to review the roles of all adhesome proteins for HSPC functions, we decided to focus our attention to the two integrin-binding and -activating proteins, Talin-1 and Kindlin-3 in adult HSPCs. Genetic studies of Kindlin-3 identified several important functions for HSPCs including roles for HSPC homing to and retention of activated HSCPs in the bone marrow. Transplantation experiments demonstrated that Kindlin-3-deficient HSCs home poorly to the bone marrow, hyperproliferate, produce effector cells with dysfunctional integrin adhesion properties, are poorly retained in the bone marrow and eventually exhaust. However, when Kindlin-3-deficient HSCs are co-transplanted with wildtype HSPCs, they enter dormancy as long as they are not exposed to hematopoietic stress. In line with these observations, patients suffering from leukocyte adhesion deficiency-III caused by Kindlin-3 loss have also massively increased numbers of HSPCs in the circulation (Ruppert et al., 2015) (Fig. 4).

Depletion of Talin-1, which is co-expressed with Kindlin-3 in hematopoietic cells, with specific shRNAs impairs murine HSPCs adhesion to fibronectin *in vitro*, but still allows normal homing to the bone marrow and long-term reconstitution in transplantation assays (Ohmori et al., 2010). These relatively mild findings are in stark contrast to the strong phenotypes observed in Kindlin-3-deficient mice and are best explained by the incomplete depletion of Talin-1 in HSPCs using shRNAs. In line with this hypothesis a hypomorphic *Fermt3* gene mutation in mice that reduces Kindlin-3 levels by around 90% also demonstrates only very mild phenotypes under steady state conditions (Klapproth et al., 2015). Clearly, genetic studies with loss of Talin-1 expression in HSPCs will better disclose the important role of Talin-1 for HSPCs.

5. Integrin-mediated mechanosensing in the HSPC niche

Apart from the sole adhesion function of integrins required for bone marrow homing, retention and differentiation of HSCs to different supportive niches, HSCs likely use integrins to probe the biochemical and biophysical properties of the microenvironment and transduce this information into HSCs to support their lifelong maintenance (reviewed by Sun et al. (2016)). Although very important, little is known regarding the consequences, which HSCs experience upon settling in niches with different biophysical features. The tissue rigidity, for example, changes from 0.2 kPa in fetal liver to 3 kPa in the perivascular niche and >40 kPa in the endosteal niche of the bone marrow. Also flow-induced shear stress greatly differs between tissues and is around 0.5–5 dyne/cm^2 in the yolk sac, 5 dyne/cm^2 in the dorsal aorta, 0.1–10 dyne/cm^2 in the placenta and 0.1–50 dyne/cm^2 in the bone marrow (Brugger, Guettler, & Gauster, 2020; Choi & Harley, 2017; Chouinard-Pelletier, Jahnsen, & Jones, 2013; Engler, Sen, Sweeney, & Discher, 2006; Garcia-Abrego et al., 2020; Jansen, Birch, Schiffman, Crosby, & Peyton, 2015; Jones, Baron, Fraser, & Dickinson, 2004; Lee, Li, Evans, Diaz, & Wenzel, 2013; Ren et al., 2015). The few studies, which addressed substrate stiffness experimentally *in vitro* with ECM protein-coated polyacrylamide gels, observed an increased HSPC α5β1 integrin-mediated adhesion, migration and maintenance of multipotency on a ≥44 kPa fibronectin substrate and strong HSPC proliferation and differentiation into erythroid progenitors on a ≤3.7 kPa laminin substrate (Choi & Harley, 2017; Holst et al., 2010; Lee-Thedieck, Rauch, Fiammengo, Klein, & Spatz, 2012). Similarly, HSPCs cultured on mesenchymal stem cells or scaffolds coated with ECM and exposed to high shear stress proliferate and remain multipotent (Kang, Jeong, Lee, Lee, & Shin, 2018; Kim, Lee, Wu, Kang, & Shin, 2019). Further improvements and innovative design of artificial bone marrow niches on lab-on-a-chip devices with different supportive cell types embedded in 3-dimensional scaffolds or gels coated with ECM proteins such as fibronectin, laminins, tenascin-C, osteopontin and vitronectin and cell-presented integrin-ligands such as recombinant ICAM-1, periostin, osteopontin, nephroblastoma overexpressed and Del-1 and integrin activity regulating factors such as CXCL12, SCF, FLT3L, G-CSF and thrombopoietin (see Table 2) will help decipher how these niche components individually and in combination influence biochemical and biophysical signaling of integrins in HSPCs.

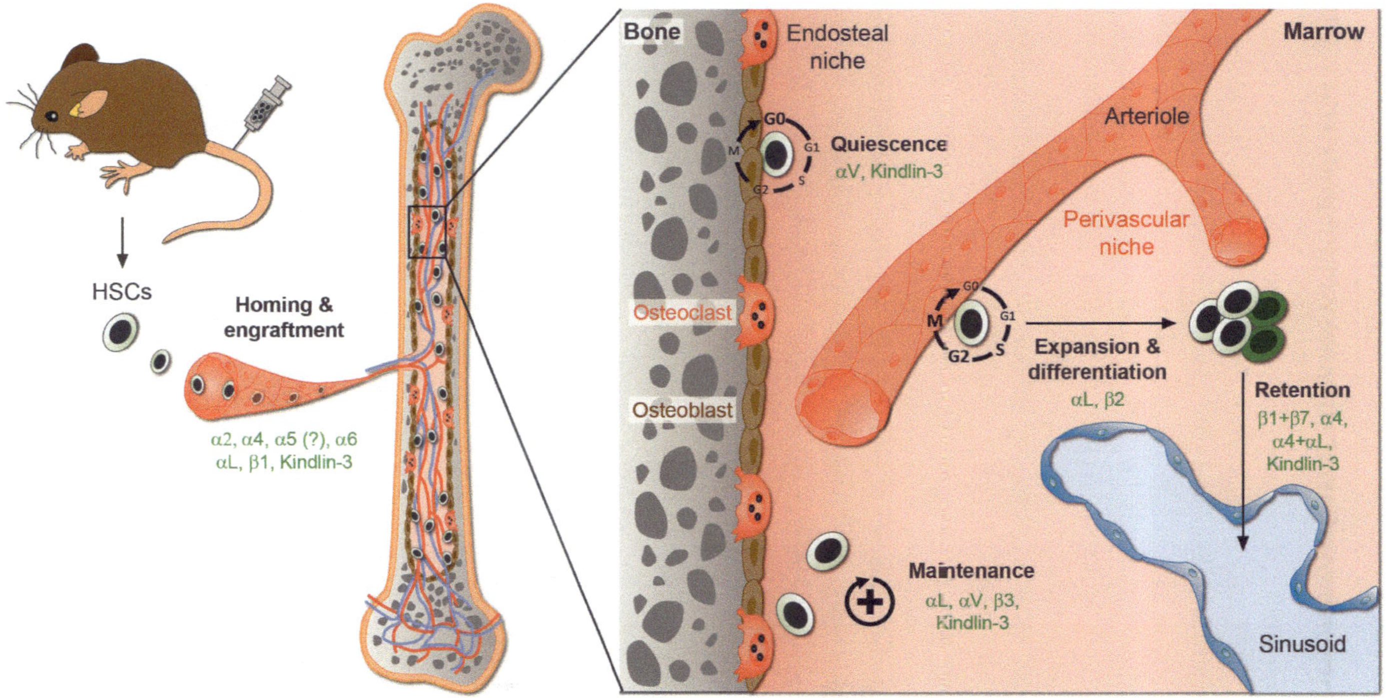

Fig. 4 Integrin-mediated adhesion and signaling for adult hematopoiesis. Integrins support HSPC homing to the bone marrow, positioning and retention in endosteal and perivascular niches, maintenance and cell cycle exit.

6. Integrin-mediated support of leukemic stem cells

The methods that are used to study hematopoiesis and HSC niches also found their way into studying hematological malignancies arising from transformed HSCs that are termed leukemic stem cells (LSCs). Similar to normal HSCs, LSCs require the support from bone marrow or spleen microenvironments for proliferation, differentiation, survival and self-renewal for leukemia initiation, progression and establishment of chemoresistance (Kim, Ruan, Ogana, & Kim, 2020; Krause & Scadden, 2015; Krenn, Koschmieder, & Fässler, 2020; Meads, Hazlehurst, & Dalton, 2008; Schemionek et al., 2012; Schepers, Campbell, & Passegué, 2015; Somervaille & Cleary, 2006). LSCs have the ability to substantially remodel the cellular and ECM composition of HSC niches to their advantage and at the expense of normal HSCs (Bhatia, McGlave, Dewald, Blazar, & Verfaillie, 1995; Krause & Scadden, 2015; Lane, Scadden, & Gilliland, 2009; Reynaud et al., 2011; Schemionek et al., 2012; Schepers et al., 2015; Somervaille & Cleary, 2006). These biochemical niche changes are sensed and transmitted by integrins and adhesome proteins of LSCs. Due to the absence of reliable markers that can clearly distinguish LSCs from HSCs, LSC-specific integrin expression profiles have still not been determined. *In vivo* studies using animal models of chronic myeloid leukemia (CML) and acute myeloid leukemia (AML), however, confirmed the importance of individual integrins, adhesome proteins and integrin activity regulators for LSCs with genetic deletion or application of function-blocking antibodies targeting α4, αV, or β3 integrins, Kindlin-3 and CXCR4, respectively. These studies revealed that abrogating integrin adhesion delays leukemia development, mobilizes LSCs in their supportive niches and increases their susceptibility to chemotherapy (Boyd et al., 2014; Krenn et al., 2020; Matsunaga et al., 2003; Miller et al., 2013; Nervi et al., 2009). In AML, increased β3 integrin mRNA expression has been associated with reduced patient overall survival (Yi et al., 2016). Although it is not entirely clear how the elevated β3 integrins impair patient survival, it was shown that increased αVβ3 integrin-induced signaling (*via* PI3K, AKT, GSK3, SYK) can enhance Wnt/β-catenin activity in AML cells, which in turn increases AML cell proliferation and chemoresistance (Johansen, Brenner, Bartaula-Brevik, Reikvam, & Bruserud, 2018; Miller et al., 2013; Yi et al., 2016). Increased β3 integrin expression and adhesion to fibronectin was also observed in chemoresistant CML-LSCs harboring the Bcr-Abl^{T315I}

point mutation. Interestingly, further overexpression of β3 integrins curbed CML development *in vivo*, indicating that the window of β3 expression capable of promoting leukemia is very narrow (Kumar et al., 2020).

In vitro studies with CML and AML cell lines have also underscored the role of integrin-mediated adhesion for the course of leukemia. The CML-inducing Bcr-Abl oncoprotein was shown to bind, modify or regulate the expression of actin, Talin, Kindlin-3, Focal Adhesion Kinase, Vinculin or Paxillin, and to increase CML cell adhesion to fibronectin, VCAM-1 and stromal cells (Fierro et al., 2008; Gotoh et al., 1995; Krenn et al., 2020; Kumar et al., 2020; Salgia et al., 1995; Salgia, Sattler, Pisick, Li, & Griffin, 1996; Wertheim et al., 2003). Also, the AML inducing FLT3-ITD oncoprotein can influence integrin functions by directly binding to β1 integrin tails and by phosphorylating and activating the focal adhesion kinase family member PYK2, which in turn reinforces inside-out integrin signaling, increases α4β1 integrin activation and AML cell adhesion to VCAM-1 expressing stromal cells (Katsumi et al., 2011). On the contrary to these observations, it was also reported that CML as well as AML patient-derived LSCs poorly adhere to fibronectin or stromal cells (Burger, Spoo, Dwenger, Burger, & Behringer, 2003; Gordon, Dowding, Riley, Goldman, & Greaves, 1987; Peled et al., 2002; Verfaillie, McCarthy, & McGlave, 1992). This different behavior of LSCs, particularly in AML and to a lesser extent in CML, may well depend on the mutations that result in the transformation of HSCs and produce disease with different outcome. In leukemia with poor prognosis, LSCs may no longer depend on niche-derived survival signals and loose the requirement for integrin-mediated niche retention and instead gain the ability to settle and grow in new territories. If this is indeed the case, integrin-mediated signaling should be blocked only at the onset of disease. Such a dual role of integrins in leukemia and also the mode of LSC adhesion in distant organs need to be clarified to make sure that attempts to therapeutically target LSCs by abrogating integrin-mediated adhesion do not end in disaster.

7. Conclusions

The majority of studies addressing the role of the integrin superfamily for the development of the mammalian hematopoietic system have mainly been studied in mice, in which individual integrin subunits or entire integrin classes were blocked with antibodies or constitutively deleted using gene targeting either in the entire organism or in the entire hematopoietic system.

These studies provided valuable information that highlight the important roles of integrins for HSPC homing to and retention in the different supportive niches to promote HSPC differentiation, expansion and self-renewal in development and adult homeostasis. A major obstacle of this approach, however, is that several integrins are expressed on a given cell type that share ligands and signaling pathways. A possible solution for this dilemma is the generation of a mouse model allowing the simultaneous disruption of all integrin classes in HSPCs. Although tedious, this approach should be achievable using CRISPR/Cas9 technology, and if combined with the generation of transgenes driving the expression of specific integrin heterodimers in specific HSCs or HPCs, would allow investigating individual integrins in a pan-integrin-null background. Similar studies have been carried out with immortalized fibroblasts *in vitro* (Schiller et al., 2013).

HSPCs home and are retained in specific niches, which provide niche-specific cocktails of growth factors, are equipped with varying integrin ligands and adopt distinct biophysical properties that orchestrate the well-being and destiny of HSPCs. In the past years, live cell imaging approaches and whole tissue bulk and single cell mRNA analysis spurred the identification and characterization of HSPCs niches. However, much more work is needed to identify and characterize the cellular composition and ECM assemblies that form the niches and the growth factors that are released by niches. Although tremendously helpful, these approaches are unable to spatially discriminate HSC niche cells from the bulk of the tissue. It will be necessary to establish new niche labeling strategies to isolate and then characterize the cells that are in direct proximity of individual HSC or HPC populations to better define the HSC and HPC niches and how niche interactions are mediated. Since integrin-niche interactions are important to guide, retain and protect HSPCs such studies would allow to design target-oriented genetic experiments for exploring niche properties and establish 2D or 3D artificial HSC niches for the *ex vivo* expansion of self-renewing and multipotent HSCs. Such studies would not only propel the knowledge of HSPC and niche biology and how HSPCs interact with niche components but also advance regenerative medicine. Furthermore, the establishment of a detailed map of healthy and malignant HSC niches will likely establish new clinical therapies with the aim to exclusively dislodge LSCs from their protective microenvironment and make them highly vulnerable to chemotherapies. Future work along these lines will further expand the understanding of integrin-mediated adhesion and signaling in the hematopoietic system.

Acknowledgments

This work was supported by the Max Planck Society, the Biomed Center Salzburg (Project 20102-F1901165-KZP), the University of Barcelona, the Spanish Ministry of Science, Innovation and Universities (PID2019-108902GB-I00, PID2020-120073GB-100, MCIN/AEI/10.13039/501100011033), and FEDER funds.

References

Abram, C. L., & Lowell, C. A. (2009). The ins and outs of leukocyte integrin signaling. *Annual Review of Immunology*, *27*, 339–362. https://doi.org/10.1146/annurev.immunol.021908.132554.

Ara, T., Tokoyoda, K., Sugiyama, T., Egawa, T., Kawabata, K., & Nagasawa, T. (2003). Long-term hematopoietic stem cells require stromal cell-derived factor-1 for colonizing bone marrow during ontogeny. *Immunity*, *19*(2), 257–267. https://doi.org/10.1016/s1074-7613(03)00201-2.

Arai, F., Hirao, A., Ohmura, M., Sato, H., Matsuoka, S., Takubo, K., et al. (2004). Tie2/angiopoietin-1 signaling regulates hematopoietic stem cell quiescence in the bone marrow niche. *Cell*, *118*(2), 149–161. https://doi.org/10.1016/j.cell.2004.07.004.

Arai, K. Y., & Nishiyama, T. (2007). Developmental changes in extracellular matrix messenger RNAs in the mouse placenta during the second half of pregnancy: Possible factors involved in the regulation of placental extracellular matrix expression. *Biology of Reproduction*, 77(6), 923–933. https://doi.org/10.1095/biolreprod.107.061382.

Arora, N., Wenzel, P. L., McKinney-Freeman, S. L., Ross, S. J., Kim, P. G., Chou, S. S., et al. (2014). Effect of developmental stage of HSC and recipient on transplant outcomes. *Developmental Cell*, *29*(5), 621–628. https://doi.org/10.1016/j.devcel.2014.04.013.

Arroyo, A. G., Taverna, D., Whittaker, C. A., Strauch, U. G., Bader, B. L., Rayburn, H., et al. (2000). In vivo roles of integrins during leukocyte development and traffic: Insights from the analysis of mice chimeric for alpha 5, alpha v, and alpha 4 integrins. *Journal of Immunology*, *165*(8), 4667–4675. https://doi.org/10.4049/jimmunol.165.8.4667.

Arroyo, A. G., Yang, J. T., Rayburn, H., & Hynes, R. O. (1996). Differential requirements for alpha4 integrins during fetal and adult hematopoiesis. *Cell*, *85*(7), 997–1008. https://doi.org/10.1016/s0092-8674(00)81301-x.

Arroyo, A. G., Yang, J. T., Rayburn, H., & Hynes, R. O. (1999). Alpha4 integrins regulate the proliferation/differentiation balance of multilineage hematopoietic progenitors in vivo. *Immunity*, *11*(5), 555–566. https://doi.org/10.1016/s1074-7613(00)80131-4.

Asaumi, N., Omoto, E., Mahmut, N., Katayama, Y., Takeda, K., Shinagawa, K., et al. (2001). Very late antigen-5 and leukocyte function-associated antigen-1 are critical for early stage hematopoietic progenitor cell homing. *Annals of Hematology*, *80*(7), 387–392. https://doi.org/10.1007/s002770100309.

Avraamides, C. J., Garmy-Susini, B., & Varner, J. A. (2008). Integrins in angiogenesis and lymphangiogenesis. *Nature Reviews. Cancer*, *8*(8), 604–617. https://doi.org/10.1038/nrc2353.

Ayres-Silva, J. E. P., Manso, P. P., Madeira, M. R., Pelajo-Machado, M., & Lenzi, H. L. (2011). Sequential morphological characteristics of murine fetal liver hematopoietic microenvironment in Swiss Webster mice. *Cell and Tissue Research*, *344*(3), 455–469. https://doi.org/10.1007/s00441-011-1170-1.

Bader, B. L., Rayburn, H., Crowley, D., & Hynes, R. O. (1998). Extensive vasculogenesis, angiogenesis, and organogenesis precede lethality in mice lacking all alpha v integrins. *Cell*, *95*(4), 507–519. https://doi.org/10.1016/s0092-8674(00)81618-9.

Baloch, Z., Klapper, J., Buchanan, L., Schwartz, M., & Amenta, P. S. (1992). Ontogenesis of the murine hepatic extracellular matrix: An immunohistochemical study. *Differentiation, 51*(3), 209–218. https://doi.org/10.1111/j.1432-0436.1992.tb00698.x.

Bautch, V. L. (2011). Stem cells and the vasculature. *Nature Medicine, 17*(11), 1437–1443. https://doi.org/10.1038/nm.2539.

Bednarczyk, M., Stege, H., Grabbe, S., & Bros, M. (2020). β2 Integrins-multi-functional leukocyte receptors in health and disease. *International Journal of Molecular Sciences, 21*(4). https://doi.org/10.3390/ijms21041402.

Bernitz, J. M., Kim, H. S., MacArthur, B., Sieburg, H., & Moore, K. (2016). Hematopoietic stem cells count and remember self-renewal divisions. *Cell, 167*(5), 1296–1309, e1210. https://doi.org/10.1016/j.cell.2016.10.022.

Bessy, T., Candelas, A., Souquet, B., Saadallah, K., Schaeffer, A., Vianay, B., et al. (2021). Hematopoietic progenitors polarize in contact with bone marrow stromal cells in response to SDF1. *The Journal of Cell Biology, 220*(11). https://doi.org/10.1083/jcb.202005085.

Bhatia, R., McGlave, P. B., Dewald, G. W., Blazar, B. R., & Verfaillie, C. M. (1995). Abnormal function of the bone marrow microenvironment in chronic myelogenous leukemia: Role of malignant stromal macrophages. *Blood, 85*(12), 3636–3645. https://www.ncbi.nlm.nih.gov/pubmed/7780147.

Bhatia, M., Wang, J. C., Kapp, U., Bonnet, D., & Dick, J. E. (1997). Purification of primitive human hematopoietic cells capable of repopulating immune-deficient mice. *Proceedings of the National Academy of Sciences of the United States of America, 94*(10), 5320–5325. https://doi.org/10.1073/pnas.94.10.5320.

Biswas, A., Roy, I. M., Babu, P. C., Manesia, J., Schouteden, S., Vijayakurup, V., et al. (2020). The Periostin/integrin-αv Axis regulates the size of hematopoietic stem cell Pool in the fetal liver. *Stem Cell Reports, 15*(2), 340–357. https://doi.org/10.1016/j.stemcr.2020.06.022.

Blatnik, J. S., Schmid-Schönbein, G. W., & Sung, L. A. (2005). The influence of fluid shear stress on the remodeling of the embryonic primary capillary plexus. *Biomechanics and Modeling in Mechanobiology, 4*(4), 211–220. https://doi.org/10.1007/s10237-005-0001-2.

Bockman, D. E., & Gulati, A. K. (1989). Localization of fibronectin in megakaryocytes of fetal liver. *The Anatomical Record, 223*(1), 90–94. https://doi.org/10.1002/ar.1092230113.

Boisset, J. C., Clapes, T., Van Der Linden, R., Dzierzak, E., & Robin, C. (2013). Integrin αIIb (CD41) plays a role in the maintenance of hematopoietic stem cell activity in the mouse embryonic aorta. *Biol Open, 2*(5), 525–532. https://doi.org/10.1242/bio.20133715.

Boisset, J. C., van Cappellen, W., Andrieu-Soler, C., Galjart, N., Dzierzak, E., & Robin, C. (2010). In vivo imaging of haematopoietic cells emerging from the mouse aortic endothelium. *Nature, 464*(7285), 116–120. https://doi.org/10.1038/nature08764.

Bose, T. O., Colpitts, S. L., Pham, Q. M., Puddington, L., & Lefrançois, L. (2014). CD11a is essential for normal development of hematopoietic intermediates. *Journal of Immunology, 193*(6), 2863–2872. https://doi.org/10.4049/jimmunol.1301820.

Boucher, D. M., & Pedersen, R. A. (1996). Induction and differentiation of extra-embryonic mesoderm in the mouse. *Reproduction, Fertility, and Development, 8*(4), 765–777. https://doi.org/10.1071/rd9960765.

Bowie, M. B., McKnight, K. D., Kent, D. G., McCaffrey, L., Hoodless, P. A., & Eaves, C. J. (2006). Hematopoietic stem cells proliferate until after birth and show a reversible phase-specific engraftment defect. *The Journal of Clinical Investigation, 116*(10), 2808–2816. https://doi.org/10.1172/JCI28310.

Boyd, A. L., Campbell, C. J., Hopkins, C. I., Fiebig-Comyn, A., Russell, J., Ulemek, J., et al. (2014). Niche displacement of human leukemic stem cells uniquely allows their competitive replacement with healthy HSPCs. *The Journal of Experimental Medicine, 211*(10), 1925–1935. https://doi.org/10.1084/jem.20140131.

Bradley, T. R., & Metcalf, D. (1966). The growth of mouse bone marrow cells in vitro. *The Australian Journal of Experimental Biology and Medical Science, 44*(3), 287–299. https://doi.org/10.1038/icb.1966.28.

Brakebusch, C., Fillatreau, S., Potocnik, A. J., Bungartz, G., Wilhelm, P., Svensson, M., et al. (2002). Beta1 integrin is not essential for hematopoiesis but is necessary for the T cell-dependent IgM antibody response. *Immunity, 16*(3), 465–477. https://doi.org/10.1016/s1074-7613(02)00281-9.

Brasel, K., McKenna, H. J., Morrissey, P. J., Charrier, K., Morris, A. E., Lee, C. C., et al. (1996). Hematologic effects of flt3 ligand in vivo in mice. *Blood, 88*(6), 2004–2012. https://www.ncbi.nlm.nih.gov/pubmed/8822919.

Brugger, B. A., Guettler, J., & Gauster, M. (2020). Go with the flow-trophoblasts in flow culture. *International Journal of Molecular Sciences, 21*(13). https://doi.org/10.3390/ijms21134666.

Bungartz, G., Stiller, S., Bauer, M., Müller, W., Schippers, A., Wagner, N., et al. (2006). Adult murine hematopoiesis can proceed without beta1 and beta7 integrins. *Blood, 108*(6), 1857–1864. https://doi.org/10.1182/blood-2005-10-007658.

Burger, J. A., Spoo, A., Dwenger, A., Burger, M., & Behringer, D. (2003). CXCR4 chemokine receptors (CD184) and alpha4beta1 integrins mediate spontaneous migration of human CD34+ progenitors and acute myeloid leukaemia cells beneath marrow stromal cells (pseudoemperipolesis). *British Journal of Haematology, 122*(4), 579–589. https://doi.org/10.1046/j.1365-2141.2003.04466.x.

Buza-Vidas, N., Antonchuk, J., Qian, H., Månsson, R., Luc, S., Zandi, S., et al. (2006). Cytokines regulate postnatal hematopoietic stem cell expansion: Opposing roles of thrombopoietin and LNK. *Genes & Development, 20*(15), 2018–2023. https://doi.org/10.1101/gad.385606.

Buza-Vidas, N., Cheng, M., Duarte, S., Charoudeh, H. N., Jacobsen, S. E., & Sitnicka, E. (2009). FLT3 receptor and ligand are dispensable for maintenance and posttransplantation expansion of mouse hematopoietic stem cells. *Blood, 113*(15), 3453–3460. https://doi.org/10.1182/blood-2008-08-174060.

Calvi, L. M., Adams, G. B., Weibrecht, K. W., Weber, J. M., Olson, D. P., Knight, M. C., et al. (2003). Osteoblastic cells regulate the haematopoietic stem cell niche. *Nature, 425*(6960), 841–846. https://doi.org/10.1038/nature02040.

Camargo, F. D., Chambers, S. M., Drew, E., McNagny, K. M., & Goodell, M. A. (2006). Hematopoietic stem cells do not engraft with absolute efficiencies. *Blood, 107*(2), 501–507. https://doi.org/10.1182/blood-2005-02-0655.

Campbell, I. D., & Humphries, M. J. (2011). Integrin structure, activation, and interactions. *Cold Spring Harbor Perspectives in Biology, 3*(3). https://doi.org/10.1101/cshperspect.a004994.

Cao, H., Cao, B., Heazlewood, C. K., Domingues, M., Sun, X., Debele, E., et al. (2019). Osteopontin is an important regulative component of the fetal bone marrow hematopoietic stem cell niche. *Cell, 8*(9). https://doi.org/10.3390/cells8090985.

Carstanjen, D., Gross, A., Kosova, N., Fichtner, I., & Salama, A. (2005). The alpha4beta1 and alpha5beta1 integrins mediate engraftment of granulocyte-colony-stimulating factor-mobilized human hematopoietic progenitor cells. *Transfusion, 45*(7), 1192–1200. https://doi.org/10.1111/j.1537-2995.2005.00172.x.

Chen, C. P., & Aplin, J. D. (2003). Placental extracellular matrix: Gene expression, deposition by placental fibroblasts and the effect of oxygen. *Placenta, 24*(4), 316–325. https://doi.org/10.1053/plac.2002.0904.

Chen, C., Huang, X., Atakilit, A., Zhu, Q. S., Corey, S. J., & Sheppard, D. (2006). The integrin alpha9beta1 contributes to granulopoiesis by enhancing granulocyte colony-stimulating factor receptor signaling. *Immunity*, *25*(6), 895–906. https://doi.org/10.1016/j.immuni.2006.10.013.

Chen, M. J., Lummertz da Rocha, E., Cahan, P., Kubaczka, C., Hunter, P., Sousa, P., et al. (2020). Transcriptome dynamics of hematopoietic stem cell formation revealed using a combinatorial Runx1 and Ly6a reporter system. *Stem Cell Reports*, *14*(5), 956–971. https://doi.org/10.1016/j.stemcr.2020.03.020.

Cheng, T., Rodrigues, N., Shen, H., Yang, Y., Dombkowski, D., Sykes, M., et al. (2000). Hematopoietic stem cell quiescence maintained by p21cip1/waf1. *Science*, *287*(5459), 1804–1808. https://doi.org/10.1126/science.287.5459.1804.

Cheshier, S. H., Morrison, S. J., Liao, X., & Weissman, I. L. (1999). In vivo proliferation and cell cycle kinetics of long-term self-renewing hematopoietic stem cells. *Proceedings of the National Academy of Sciences of the United States of America*, *96*(6), 3120–3125. https://doi.org/10.1073/pnas.96.6.3120.

Cheshier, S. H., Prohaska, S. S., & Weissman, I. L. (2007). The effect of bleeding on hematopoietic stem cell cycling and self-renewal. *Stem Cells and Development*, *16*(5), 707–717. https://doi.org/10.1089/scd.2007.0017.

Choi, J. S., & Harley, B. A. (2017). Marrow-inspired matrix cues rapidly affect early fate decisions of hematopoietic stem and progenitor cells. *Science Advances*, *3*(1), e1600455. https://doi.org/10.1126/sciadv.1600455.

Choi, K., Kennedy, M., Kazarov, A., Papadimitriou, J. C., & Keller, G. (1998). A common precursor for hematopoietic and endothelial cells. *Development*, *125*(4), 725–732. https://www.ncbi.nlm.nih.gov/pubmed/9435292.

Chou, S., & Lodish, H. F. (2010). Fetal liver hepatic progenitors are supportive stromal cells for hematopoietic stem cells. *Proceedings of the National Academy of Sciences of the United States of America*, *107*(17), 7799–7804. https://doi.org/10.1073/pnas.1003586107.

Chouinard-Pelletier, G., Jahnsen, E. D., & Jones, E. A. (2013). Increased shear stress inhibits angiogenesis in veins and not arteries during vascular development. *Angiogenesis*, *16*(1), 71–83. https://doi.org/10.1007/s10456-012-9300-2.

Christensen, J. L., Wright, D. E., Wagers, A. J., & Weissman, I. L. (2004). Circulation and chemotaxis of fetal hematopoietic stem cells. *PLoS Biology*, *2*(3), E75. https://doi.org/10.1371/journal.pbio.0020075.

Christopher, M. J., Liu, F., Hilton, M. J., Long, F., & Link, D. C. (2009). Suppression of CXCL12 production by bone marrow osteoblasts is a common and critical pathway for cytokine-induced mobilization. *Blood*, *114*(7), 1331–1339. https://doi.org/10.1182/blood-2008-10-184754.

Collart, C., Ciccarelli, A., Ivanovitch, K., Rosewell, I., Kumar, S., Kelly, G., et al. (2021). The migratory pathways of the cells that form the endocardium, dorsal aortae, and head vasculature in the mouse embryo. *BMC Developmental Biology*, *21*(1), 8. https://doi.org/10.1186/s12861-021-00239-3.

Comazzetto, S., Murphy, M. M., Berto, S., Jeffery, E., Zhao, Z., & Morrison, S. J. (2019). Restricted hematopoietic progenitors and erythropoiesis require SCF from leptin receptor+ niche cells in the bone marrow. *Cell Stem Cell*, *24*(3), 477–486.e476. https://doi.org/10.1016/j.stem.2018.11.022.

Corbel, C., Salaün, J., Belo-Diabangouaya, P., & Dieterlen-Lièvre, F. (2007). Hematopoietic potential of the pre-fusion allantois. *Developmental Biology*, *301*(2), 478–488. https://doi.org/10.1016/j.ydbio.2006.08.069.

Craddock, C. F., Nakamoto, B., Andrews, R. G., Priestley, G. V., & Papayannopoulou, T. (1997). Antibodies to VLA4 integrin mobilize long-term repopulating cells and augment cytokine-induced mobilization in primates and mice. *Blood*, *90*(12), 4779–4788. https://www.ncbi.nlm.nih.gov/pubmed/9389694.

Craddock, C. F., Nakamoto, B., Elices, M., & Papayannopoulou, T. (1997). The role of CS1 moiety of fibronectin in VLA mediated haemopoietic progenitor trafficking. *British Journal of Haematology*, *97*(1), 15–21. https://doi.org/10.1046/j.1365-2141.1997.d01-2120.x.

Cumano, A., Ferraz, J. C., Klaine, M., Di Santo, J. P., & Godin, I. (2001). Intraembryonic, but not yolk sac hematopoietic precursors, isolated before circulation, provide long-term multilineage reconstitution. *Immunity*, *15*(3), 477–485. https://doi.org/10.1016/s1074-7613(01)00190-x.

De Grandis, M., Lhoumeau, A. C., Mancini, S. J., & Aurrand-Lions, M. (2016). Adhesion receptors involved in HSC and early-B cell interactions with bone marrow microenvironment. *Cellular and Molecular Life Sciences*, *73*(4), 687–703. https://doi.org/10.1007/s00018-015-2064-2.

Decker, M., Leslie, J., Liu, Q., & Ding, L. (2018). Hepatic thrombopoietin is required for bone marrow hematopoietic stem cell maintenance. *Science*, *360*(6384), 106–110. https://doi.org/10.1126/science.aap8861.

Demetri, G. D., & Griffin, J. D. (1991). Granulocyte colony-stimulating factor and its receptor. *Blood*, *78*(11), 2791–2808. https://www.ncbi.nlm.nih.gov/pubmed/1720034.

Ding, L., & Morrison, S. J. (2013). Haematopoietic stem cells and early lymphoid progenitors occupy distinct bone marrow niches. *Nature*, *495*(7440), 231–235. https://doi.org/10.1038/nature11885.

Ding, L., Saunders, T. L., Enikolopov, G., & Morrison, S. J. (2012). Endothelial and perivascular cells maintain haematopoietic stem cells. *Nature*, *481*(7382), 457–462. https://doi.org/10.1038/nature10783.

Domínguez-Soto, A., Relloso, M., Vega, M. A., Corbí, A. L., & Puig-Kröger, A. (2005). RUNX3 regulates the activity of the CD11a and CD49d integrin gene promoters. *Immunobiology*, *210*(2–4), 133–139. https://doi.org/10.1016/j.imbio.2005.05.008.

Downs, K. M., Gifford, S., Blahnik, M., & Gardner, R. L. (1998). Vascularization in the murine allantois occurs by vasculogenesis without accompanying erythropoiesis. *Development*, *125*(22), 4507–4520. https://www.ncbi.nlm.nih.gov/pubmed/9778509.

Drake, C. J., & Fleming, P. A. (2000). Vasculogenesis in the day 6.5 to 9.5 mouse embryo. *Blood*, *95*(5), 1671–1679. https://www.ncbi.nlm.nih.gov/pubmed/10688823.

Dumont, D. J., Fong, G. H., Puri, M. C., Gradwohl, G., Alitalo, K., & Breitman, M. L. (1995). Vascularization of the mouse embryo: A study of flk-1, tek, tie, and vascular endothelial growth factor expression during development. *Developmental Dynamics*, *203*(1), 80–92. https://doi.org/10.1002/aja.1002030109.

Elefanty, A. G., Begley, C. G., Hartley, L., Papaevangeliou, B., & Robb, L. (1999). SCL expression in the mouse embryo detected with a targeted lacZ reporter gene demonstrates its localization to hematopoietic, vascular, and neural tissues. *Blood*, *94*(11), 3754–3763. https://www.ncbi.nlm.nih.gov/pubmed/10572089.

Ema, M., Yokomizo, T., Wakamatsu, A., Terunuma, T., Yamamoto, M., & Takahashi, S. (2006). Primitive erythropoiesis from mesodermal precursors expressing VE-cadherin, PECAM-1, Tie2, endoglin, and CD34 in the mouse embryo. *Blood*, *108*(13), 4018–4024. https://doi.org/10.1182/blood-2006-03-012872.

Emambokus, N. R., & Frampton, J. (2003). The glycoprotein IIb molecule is expressed on early murine hematopoietic progenitors and regulates their numbers in sites of hematopoiesis. *Immunity*, *19*(1), 33–45. https://doi.org/10.1016/s1074-7613(03)00173-0.

Engler, A. J., Sen, S., Sweeney, H. L., & Discher, D. E. (2006). Matrix elasticity directs stem cell lineage specification. *Cell*, *126*(4), 677–689. https://doi.org/10.1016/j.cell.2006.06.044.

Eshghi, S., Vogelezang, M. G., Hynes, R. O., Griffith, L. G., & Lodish, H. F. (2007). Alpha4beta1 integrin and erythropoietin mediate temporally distinct steps in erythropoiesis: Integrins in red cell development. *The Journal of Cell Biology*, *177*(5), 871–880. https://doi.org/10.1083/jcb.200702080.

Fagerholm, S. C., Guenther, C., Llort Asens, M., Savinko, T., & Uotila, L. M. (2019). Beta2-Integrins and interacting proteins in leukocyte trafficking, immune suppression, and immunodeficiency disease. *Frontiers in Immunology*, *10*, 254. https://doi.org/10.3389/fimmu.2019.00254.

Fathman, J. W., Fernhoff, N. B., Seita, J., Chao, C., Scarfone, V. M., Weissman, I. L., et al. (2014). Upregulation of CD11A on hematopoietic stem cells denotes the loss of long-term reconstitution potential. *Stem Cell Reports*, *3*(5), 707–715. https://doi.org/10.1016/j.stemcr.2014.09.007.

Fennie, C., Cheng, J., Dowbenko, D., Young, P., & Lasky, L. A. (1995). CD34+ endothelial cell lines derived from murine yolk sac induce the proliferation and differentiation of yolk sac CD34+ hematopoietic progenitors. *Blood*, *86*(12), 4454–4467. https://www.ncbi.nlm.nih.gov/pubmed/8541534.

Ferkowicz, M. J., Starr, M., Xie, X., Li, W., Johnson, S. A., Shelley, W. C., et al. (2003). CD41 expression defines the onset of primitive and definitive hematopoiesis in the murine embryo. *Development*, *130*(18), 4393–4403. https://doi.org/10.1242/dev.00632.

Fierro, F. A., Taubenberger, A., Puech, P. H., Ehninger, G., Bornhauser, M., Muller, D. J., et al. (2008). BCR/ABL expression of myeloid progenitors increases beta1-integrin mediated adhesion to stromal cells. *Journal of Molecular Biology*, *377*(4), 1082–1093. https://doi.org/10.1016/j.jmb.2008.01.085.

Flamme, I., Breier, G., & Risau, W. (1995). Vascular endothelial growth factor (VEGF) and VEGF receptor 2 (flk-1) are expressed during vasculogenesis and vascular differentiation in the quail embryo. *Developmental Biology*, *169*(2), 699–712. https://doi.org/10.1006/dbio.1995.1180.

Fong, G. H., Klingensmith, J., Wood, C. R., Rossant, J., & Breitman, M. L. (1996). Regulation of flt-1 expression during mouse embryogenesis suggests a role in the establishment of vascular endothelium. *Developmental Dynamics*, *207*(1), 1–10. https://doi.org/10.1002/(SICI)1097-0177(199609)207:1<1::AID-AJA1>3.0.CO;2-M.

Ford, C. E., Hamerton, J. L., Barnes, D. W., & Loutit, J. F. (1956). Cytological identification of radiation-chimaeras. *Nature*, *177*(4506), 452–454. https://doi.org/10.1038/177452a0.

Fraser, S. T., Isern, J., & Baron, M. H. (2007). Maturation and enucleation of primitive erythroblasts during mouse embryogenesis is accompanied by changes in cell-surface antigen expression. *Blood*, *109*(1), 343–352. https://doi.org/10.1182/blood-2006-03-006569.

Gahmberg, C. G., Grönholm, M., Madhavan, S., Jahan, F., Mikkola, E., Viazmina, L., et al. (2019). Regulation of cell adhesion: A collaborative effort of integrins, their ligands, cytoplasmic actors, and phosphorylation. *Quarterly Reviews of Biophysics*, *52*, e10. https://doi.org/10.1017/S0033583519000088.

Ganuza, M., Chabot, A., Tang, X., Bi, W., Natarajan, S., Carter, R., et al. (2018). Murine hematopoietic stem cell activity is derived from pre-circulation embryos but not yolk sacs. *Nature Communications*, *9*(1), 5405. https://doi.org/10.1038/s41467-018-07769-8.

Garcia-Abrego, C., Zaunz, S., Toprakhisar, B., Subramani, R., Deschaume, O., Jooken, S., et al. (2020). Towards mimicking the fetal liver niche: The influence of elasticity and oxygen tension on hematopoietic stem/progenitor cells cultured in 3D fibrin hydrogels. *International Journal of Molecular Sciences*, *21*(17). https://doi.org/10.3390/ijms21176367.

Gekas, C., Dieterlen-Lièvre, F., Orkin, S. H., & Mikkola, H. K. (2005). The placenta is a niche for hematopoietic stem cells. *Developmental Cell*, *8*(3), 365–375. https://doi.org/10.1016/j.devcel.2004.12.016.

Gekas, C., & Graf, T. (2013). CD41 expression marks myeloid-biased adult hematopoietic stem cells and increases with age. *Blood*, *121*(22), 4463–4472. https://doi.org/10.1182/blood-2012-09-457929.

Gekas, C., Rhodes, K. E., Van Handel, B., Chhabra, A., Ueno, M., & Mikkola, H. K. (2010). Hematopoietic stem cell development in the placenta. *The International Journal of Developmental Biology*, *54*(6–7), 1089–1098. https://doi.org/10.1387/ijdb.103070cg.

George, E. L., Georges-Labouesse, E. N., Patel-King, R. S., Rayburn, H., & Hynes, R. O. (1993). Defects in mesoderm, neural tube and vascular development in mouse embryos lacking fibronectin. *Development*, *119*(4), 1079–1091. https://www.ncbi.nlm.nih.gov/pubmed/8306876.

Gillan, L., Matei, D., Fishman, D. A., Gerbin, C. S., Karlan, B. Y., & Chang, D. D. (2002). Periostin secreted by epithelial ovarian carcinoma is a ligand for alpha(V)beta(3) and alpha(V)beta(5) integrins and promotes cell motility. *Cancer Research*, *62*(18), 5358–5364. https://www.ncbi.nlm.nih.gov/pubmed/12235007.

Goldie, L. C., Nix, M. K., & Hirschi, K. K. (2008). Embryonic vasculogenesis and hematopoietic specification. *Organogenesis*, *4*(4), 257–263. https://doi.org/10.4161/org.4.4.7416.

Goodell, M. A., Rosenzweig, M., Kim, H., Marks, D. F., DeMaria, M., Paradis, G., et al. (1997). Dye efflux studies suggest that hematopoietic stem cells expressing low or undetectable levels of CD34 antigen exist in multiple species. *Nature Medicine*, *3*(12), 1337–1345. https://doi.org/10.1038/nm1297-1337.

Gordon, M. Y., Dowding, C. R., Riley, G. P., Goldman, J. M., & Greaves, M. F. (1987). Altered adhesive interactions with marrow stroma of haematopoietic progenitor cells in chronic myeloid leukaemia. *Nature*, *328*(6128), 342–344. https://doi.org/10.1038/328342a0.

Gotoh, A., Miyazawa, K., Ohyashiki, K., Tauchi, T., Boswell, H. S., Broxmeyer, H. E., et al. (1995). Tyrosine phosphorylation and activation of focal adhesion kinase (p125FAK) by BCR-ABL oncoprotein. *Experimental Hematology*, *23*(11), 1153–1159. https://www.ncbi.nlm.nih.gov/pubmed/7556524.

Grassinger, J., Haylock, D. N., Storan, M. J., Haines, G. O., Williams, B., Whitty, G. A., et al. (2009). Thrombin-cleaved osteopontin regulates hemopoietic stem and progenitor cell functions through interactions with alpha9beta1 and alpha4beta1 integrins. *Blood*, *114*(1), 49–59. https://doi.org/10.1182/blood-2009-01-197988.

Greenbaum, A., Hsu, Y. M., Day, R. B., Schuettpelz, L. G., Christopher, M. J., Borgerding, J. N., et al. (2013). CXCL12 in early mesenchymal progenitors is required for haematopoietic stem-cell maintenance. *Nature*, *495*(7440), 227–230. https://doi.org/10.1038/nature11926.

Greenbaum, A. M., & Link, D. C. (2011). Mechanisms of G-CSF-mediated hematopoietic stem and progenitor mobilization. *Leukemia*, *25*(2), 211–217. https://doi.org/10.1038/leu.2010.248.

Gu, Y. C., Kortesmaa, J., Tryggvason, K., Persson, J., Ekblom, P., Jacobsen, S. E., et al. (2003). Laminin isoform-specific promotion of adhesion and migration of human bone marrow progenitor cells. *Blood*, *101*(3), 877–885. https://doi.org/10.1182/blood-2002-03-0796.

Gu, Y. C., Nilsson, K., Eng, H., & Ekblom, M. (2000). Association of extracellular matrix proteins fibulin-1 and fibulin-2 with fibronectin in bone marrow stroma. *British Journal of Haematology*, *109*(2), 305–313. https://doi.org/10.1046/j.1365-2141.2000.02011.x.

Gurtner, G. C., Davis, V., Li, H., McCoy, M. J., Sharpe, A., & Cybulsky, M. I. (1995). Targeted disruption of the murine VCAM1 gene: Essential role of VCAM-1 in chorioallantoic fusion and placentation. *Genes & Development*, *9*(1), 1–14. https://doi.org/10.1101/gad.9.1.1.

Haas, S., Trumpp, A., & Milsom, M. D. (2018). Causes and consequences of hematopoietic stem cell heterogeneity. *Cell Stem Cell*, *22*(5), 627–638. https://doi.org/10.1016/j.stem.2018.04.003.

Hackney, J. A., Charbord, P., Brunk, B. P., Stoeckert, C. J., Lemischka, I. R., & Moore, K. A. (2002). A molecular profile of a hematopoietic stem cell niche. *Proceedings of the National Academy of Sciences of the United States of America*, *99*(20), 13061–13066. https://doi.org/10.1073/pnas.192124499.

Harrison, D. E., Jordan, C. T., Zhong, R. K., & Astle, C. M. (1993). Primitive hemopoietic stem cells: Direct assay of most productive populations by competitive repopulation with simple binomial, correlation and covariance calculations. *Experimental Hematology*, *21*(2), 206–219. https://www.ncbi.nlm.nih.gov/pubmed/8425559.

Hirsch, E., Iglesias, A., Potocnik, A. J., Hartmann, U., & Fässler, R. (1996). Impaired migration but not differentiation of haematopoietic stem cells in the absence of beta1 integrins. *Nature*, *380*(6570), 171–175. https://doi.org/10.1038/380171a0.

Hodivala-Dilke, K. M., McHugh, K. P., Tsakiris, D. A., Rayburn, H., Crowley, D., Ullman-Cullere, M., et al. (1999). Beta3-integrin-deficient mice are a model for Glanzmann thrombasthenia showing placental defects and reduced survival. *The Journal of Clinical Investigation*, *103*(2), 229–238. https://doi.org/10.1172/JCI5487.

Holst, J., Watson, S., Lord, M. S., Eamegdool, S. S., Bax, D. V., Nivison-Smith, L. B., et al. (2010). Substrate elasticity provides mechanical signals for the expansion of hemopoietic stem and progenitor cells. *Nature Biotechnology*, *28*(10), 1123–1128. https://doi.org/10.1038/nbt.1687.

Holtkötter, O., Nieswandt, B., Smyth, N., Müller, W., Hafner, M., Schulte, V., et al. (2002). Integrin alpha 2-deficient mice develop normally, are fertile, but display partially defective platelet interaction with collagen. *The Journal of Biological Chemistry*, *277*(13), 10789–10794. https://doi.org/10.1074/jbc.M112307200.

Horton, E. R., Byron, A., Askari, J. A., Ng, D. H. J., Millon-Frémillon, A., Robertson, J., et al. (2015). Definition of a consensus integrin adhesome and its dynamics during adhesion complex assembly and disassembly. *Nature Cell Biology*, *17*(12), 1577–1587. https://doi.org/10.1038/ncb3257.

Horton, E. R., Humphries, J. D., James, J., Jones, M. C., Askari, J. A., & Humphries, M. J. (2016). The integrin adhesome network at a glance. *Journal of Cell Science*, *129*(22), 4159–4163. https://doi.org/10.1242/jcs.192054.

Hu, P., & Luo, B. H. (2013). Integrin bi-directional signaling across the plasma membrane. *Journal of Cellular Physiology*, *228*(2), 306–312. https://doi.org/10.1002/jcp.24154.

Huang, X. Z., Wu, J. F., Ferrando, R., Lee, J. H., Wang, Y. L., Farese, R. V., et al. (2000). Fatal bilateral chylothorax in mice lacking the integrin alpha9beta1. *Molecular and Cellular Biology*, *20*(14), 5208–5215. https://doi.org/10.1128/mcb.20.14.5208-5215.2000.

Huber, T. L., Kouskoff, V., Fehling, H. J., Palis, J., & Keller, G. (2004). Haemangioblast commitment is initiated in the primitive streak of the mouse embryo. *Nature*, *432*(7017), 625–630. https://doi.org/10.1038/nature03122.

Humphries, J. D., Byron, A., & Humphries, M. J. (2006). Integrin ligands at a glance. *Journal of Cell Science*, *119*(Pt 19), 3901–3903. https://doi.org/10.1242/jcs.03098.

Huntsman, H. D., Bat, T., Cheng, H., Cash, A., Cheruku, P. S., Fu, J. F., et al. (2015). Human hematopoietic stem cells from mobilized peripheral blood can be purified based on CD49f integrin expression. *Blood*, *126*(13), 1631–1633. https://doi.org/10.1182/blood-2015-07-660670.

Hurley, R. W., McCarthy, J. B., & Verfaillie, C. M. (1995). Direct adhesion to bone marrow stroma via fibronectin receptors inhibits hematopoietic progenitor proliferation. *The Journal of Clinical Investigation*, *96*(1), 511–519. https://doi.org/10.1172/JCI118063.

Hynes, R. O. (2002). Integrins: Bidirectional, allosteric signaling machines. *Cell*, *110*(6), 673–687. https://doi.org/10.1016/s0092-8674(02)00971-6.

Ibarra-Soria, X., Jawaid, W., Pijuan-Sala, B., Ladopoulos, V., Scialdone, A., Jörg, D. J., et al. (2018). Defining murine organogenesis at single-cell resolution reveals a role for the leukotriene pathway in regulating blood progenitor formation. *Nature Cell Biology*, *20*(2), 127–134. https://doi.org/10.1038/s41556-017-0013-z.

Iscove, N. N., & Nawa, K. (1997). Hematopoietic stem cells expand during serial transplantation in vivo without apparent exhaustion. *Current Biology*, 7(10), 805–808. https://doi.org/10.1016/s0960-9822(06)00341-1.

Iscove, N. N., Sieber, F., & Winterhalter, K. H. (1974). Erythroid colony formation in cultures of mouse and human bone marrow: Analysis of the requirement for erythropoietin by gel filtration and affinity chromatography on agarose-concanavalin A. *Journal of Cellular Physiology*, *83*(2), 309–320. https://doi.org/10.1002/jcp.1040830218.

Isern, J., Fraser, S. T., He, Z., & Baron, M. H. (2008). The fetal liver is a niche for maturation of primitive erythroid cells. *Proceedings of the National Academy of Sciences of the United States of America*, *105*(18), 6662–6667. https://doi.org/10.1073/pnas.0802032105.

Isern, J., He, Z., Fraser, S. T., Nowotschin, S., Ferrer-Vaquer, A., Moore, R., et al. (2011). Single-lineage transcriptome analysis reveals key regulatory pathways in primitive erythroid progenitors in the mouse embryo. *Blood*, *117*(18), 4924–4934. https://doi.org/10.1182/blood-2010-10-313676.

Ishihara, J., Umemoto, T., Yamato, M., Shiratsuchi, Y., Takaki, S., Petrich, B. G., et al. (2014). Nov/CCN3 regulates long-term repopulating activity of murine hematopoietic stem cells via integrin αvβ3. *International Journal of Hematology*, *99*(4), 393–406. https://doi.org/10.1007/s12185-014-1534-x.

Itoi, M., Kawamoto, H., Katsura, Y., & Amagai, T. (2001). Two distinct steps of immigration of hematopoietic progenitors into the early thymus anlage. *International Immunology*, *13*(9), 1203–1211. https://doi.org/10.1093/intimm/13.9.1203.

Jacobson, L. O., Simmons, E. L., Marks, E. K., & Eldredge, J. H. (1951). Recovery from radiation injury. *Science*, *113*(2940), 510–511. https://doi.org/10.1126/science.113.2940.510.

Jankowska, K. I., Williamson, E. K., Roy, N. H., Blumenthal, D., Chandra, V., Baumgart, T., et al. (2018). Integrins modulate T cell receptor signaling by constraining actin flow at the immunological synapse. *Frontiers in Immunology*, *9*, 25. https://doi.org/10.3389/fimmu.2018.00025.

Jansen, L. E., Birch, N. P., Schiffman, J. D., Crosby, A. J., & Peyton, S. R. (2015). Mechanics of intact bone marrow. *Journal of the Mechanical Behavior of Biomedical Materials*, *50*, 299–307. https://doi.org/10.1016/j.jmbbm.2015.06.023.

Jiang, Y., Bonig, H., Ulyanova, T., Chang, K., & Papayannopoulou, T. (2009). On the adaptation of endosteal stem cell niche function in response to stress. *Blood*, *114*(18), 3773–3782. https://doi.org/10.1182/blood-2009-05-219840.

Johansen, S., Brenner, A. K., Bartaula-Brevik, S., Reikvam, H., & Bruserud, Ø. (2018). The possible importance of β3 Integrins for Leukemogenesis and Chemoresistance in acute myeloid leukemia. *International Journal of Molecular Sciences*, *19*(1). https://doi.org/10.3390/ijms19010251.

Jones, E. A., Baron, M. H., Fraser, S. E., & Dickinson, M. E. (2004). Measuring hemodynamic changes during mammalian development. *American Journal of Physiology. Heart and Circulatory Physiology*, *287*(4), H1561–H1569. https://doi.org/10.1152/ajpheart.00081.2004.

Kallianpur, A. R., Jordan, J. E., & Brandt, S. J. (1994). The SCL/TAL-1 gene is expressed in progenitors of both the hematopoietic and vascular systems during embryogenesis. *Blood*, *83*(5), 1200–1208. https://www.ncbi.nlm.nih.gov/pubmed/8118024.

Kang, Y. G., Jeong, J. Y., Lee, T. H., Lee, H. S., & Shin, J. W. (2018). Synergistic integration of mesenchymal stem cells and hydrostatic pressure in the expansion and maintenance of human hematopoietic/progenitor cells. *Stem Cells International*, *2018*, 4527929. https://doi.org/10.1155/2018/4527929.

Katayama, Y., Hidalgo, A., Furie, B. C., Vestweber, D., Furie, B., & Frenette, P. S. (2003). PSGL-1 participates in E-selectin-mediated progenitor homing to bone marrow: Evidence for cooperation between E-selectin ligands and alpha4 integrin. *Blood*, *102*(6), 2060–2067. https://doi.org/10.1182/blood-2003-04-1212.

Katayama, Y., Hidalgo, A., Peired, A., & Frenette, P. S. (2004). Integrin alpha4beta7 and its counterreceptor MAdCAM-1 contribute to hematopoietic progenitor recruitment into bone marrow following transplantation. *Blood, 104*(7), 2020–2026. https://doi.org/10.1182/blood-2003-12-4157.

Katsumi, A., Kiyoi, H., Abe, A., Tanizaki, R., Iwasaki, T., Kobayashi, M., et al. (2011). FLT3/ITD regulates leukaemia cell adhesion through α4β1 integrin and Pyk2 signalling. *European Journal of Haematology, 86*(3), 191–198. https://doi.org/10.1111/j.1600-0609.2010.01556.x.

Kerst, J. M., Sanders, J. B., Slaper-Cortenbach, I. C., Doorakkers, M. C., Hooibrink, B., van Oers, R. H., et al. (1993). Alpha 4 beta 1 and alpha 5 beta 1 are differentially expressed during myelopoiesis and mediate the adherence of human CD34+ cells to fibronectin in an activation-dependent way. *Blood, 81*(2), 344–351. https://www.ncbi.nlm.nih.gov/pubmed/7678511.

Khan, J. A., Mendelson, A., Kunisaki, Y., Birbrair, A., Kou, Y., Arnal-Estapé, A., et al. (2016). Fetal liver hematopoietic stem cell niches associate with portal vessels. *Science, 351*(6269), 176–180. https://doi.org/10.1126/science.aad0084.

Khurana, S., Schouteden, S., Manesia, J. K., Santamaria-Martínez, A., Huelsken, J., Lacy-Hulbert, A., et al. (2016). Outside-in integrin signalling regulates haematopoietic stem cell function via Periostin-Itgav axis. *Nature Communications*, 7, 13500. https://doi.org/10.1038/ncomms13500.

Kiel, M. J., He, S., Ashkenazi, R., Gentry, S. N., Teta, M., Kushner, J. A., et al. (2007). Haematopoietic stem cells do not asymmetrically segregate chromosomes or retain BrdU. *Nature, 449*(7159), 238–242. https://doi.org/10.1038/nature06115.

Kiel, M. J., & Morrison, S. J. (2008). Uncertainty in the niches that maintain haematopoietic stem cells. *Nature Reviews. Immunology, 8*(4), 290–301. https://doi.org/10.1038/nri2279.

Kiel, M. J., Yilmaz, O. H., Iwashita, T., Terhorst, C., & Morrison, S. J. (2005). SLAM family receptors distinguish hematopoietic stem and progenitor cells and reveal endothelial niches for stem cells. *Cell, 121*(7), 1109–1121. https://doi.org/10.1016/j.cell.2005.05.026.

Kim, I., He, S., Yilmaz, O. H., Kiel, M. J., & Morrison, S. J. (2006). Enhanced purification of fetal liver hematopoietic stem cells using SLAM family receptors. *Blood, 108*(2), 737–744. https://doi.org/10.1182/blood-2005-10-4135.

Kim, J. E., Lee, E. J., Wu, Y., Kang, Y. G., & Shin, J. W. (2019). The combined effects of hierarchical scaffolds and mechanical stimuli on ex vivo expansion of haematopoietic stem/progenitor cells. *Artificial Cells, Nanomedicine, and Biotechnology, 47*(1), 586–593. https://doi.org/10.1080/21691401.2019.1573180.

Kim, H. N., Ruan, Y., Ogana, H., & Kim, Y. M. (2020). Cadherins, selectins, and Integrins in CAM-DR in leukemia. *Frontiers in Oncology, 10*, 592733. https://doi.org/10.3389/fonc.2020.592733.

Kinder, S. J., Tsang, T. E., Quinlan, G. A., Hadjantonakis, A. K., Nagy, A., & Tam, P. P. (1999). The orderly allocation of mesodermal cells to the extraembryonic structures and the anteroposterior axis during gastrulation of the mouse embryo. *Development, 126*(21), 4691–4701. https://www.ncbi.nlm.nih.gov/pubmed/10518487.

Kingsley, P. D., Greenfest-Allen, E., Frame, J. M., Bushnell, T. P., Malik, J., McGrath, K. E., et al. (2013). Ontogeny of erythroid gene expression. *Blood, 121*(6), e5–e13. https://doi.org/10.1182/blood-2012-04-422394.

Klapproth, S., Moretti, F. A., Zeiler, M., Ruppert, R., Breithaupt, U., Mueller, S., et al. (2015). Minimal amounts of kindlin-3 suffice for basal platelet and leukocyte functions in mice. *Blood, 126*(24), 2592–2600. https://doi.org/10.1182/blood-2015-04-639310.

Kokkaliaris, K. D., Kunz, L., Cabezas-Wallscheid, N., Christodoulou, C., Renders, S., Camargo, F., et al. (2020). Adult blood stem cell localization reflects the abundance of reported bone marrow niche cell types and their combinations. *Blood, 136*(20), 2296–2307. https://doi.org/10.1182/blood.2020006574.

Krause, D. S., & Scadden, D. T. (2015). A hostel for the hostile: The bone marrow niche in hematologic neoplasms. *Haematologica, 100*(11), 1376–1387. https://doi.org/10.3324/haematol.2014.113852.

Krenn, P. W., Koschmieder, S., & Fässler, R. (2020). Kindlin-3 loss curbs chronic myeloid leukemia in mice by mobilizing leukemic stem cells from protective bone marrow niches. *Proceedings of the National Academy of Sciences of the United States of America, 117*(39), 24326–24335. https://doi.org/10.1073/pnas.2009078117.

Krüger, M., Moser, M., Ussar, S., Thievessen, I., Luber, C. A., Forner, F., et al. (2008). SILAC mouse for quantitative proteomics uncovers kindlin-3 as an essential factor for red blood cell function. *Cell, 134*(2), 353–364. https://doi.org/10.1016/j.cell.2008.05.033.

Kulkarni, R., & Kale, V. (2020). Physiological cues involved in the regulation of adhesion mechanisms in hematopoietic stem cell fate decision. *Frontiers in Cell and Development Biology, 8*, 611. https://doi.org/10.3389/fcell.2020.00611.

Kumar, R., Pereira, R. S., Zanetti, C., Minciacchi, V. R., Merten, M., Meister, M., et al. (2020). Specific, targetable interactions with the microenvironment influence imatinib-resistant chronic myeloid leukemia. *Leukemia, 34*(8), 2087–2101. https://doi.org/10.1038/s41375-020-0866-1.

Kumaravelu, P., Hook, L., Morrison, A. M., Ure, J., Zhao, S., Zuyev, S., et al. (2002). Quantitative developmental anatomy of definitive haematopoietic stem cells/long-term repopulating units (HSC/RUs): Role of the aorta-gonad-mesonephros (AGM) region and the yolk sac in colonisation of the mouse embryonic liver. *Development, 129*(21), 4891–4899. https://www.ncbi.nlm.nih.gov/pubmed/12397098.

Kwee, L., Baldwin, H. S., Shen, H. M., Stewart, C. L., Buck, C., Buck, C. A., et al. (1995). Defective development of the embryonic and extraembryonic circulatory systems in vascular cell adhesion molecule (VCAM-1) deficient mice. *Development, 121*(2), 489–503. https://www.ncbi.nlm.nih.gov/pubmed/7539357.

Lai, C. Y., Yamazaki, S., Okabe, M., Suzuki, S., Maeyama, Y., Iimura, Y., et al. (2014). Stage-specific roles for CXCR4 signaling in murine hematopoietic stem/progenitor cells in the process of bone marrow repopulation. *Stem Cells, 32*(7), 1929–1942. https://doi.org/10.1002/stem.1670.

Lane, S. W., Scadden, D. T., & Gilliland, D. G. (2009). The leukemic stem cell niche: Current concepts and therapeutic opportunities. *Blood, 114*(6), 1150–1157. https://doi.org/10.1182/blood-2009-01-202606.

Lee, D., Kim, D. W., & Cho, J. Y. (2020). Role of growth factors in hematopoietic stem cell niche. *Cell Biology and Toxicology, 36*(2), 131–144. https://doi.org/10.1007/s10565-019-09510-7.

Lee, H. J., Li, N., Evans, S. M., Diaz, M. F., & Wenzel, P. L. (2013). Biomechanical force in blood development: Extrinsic physical cues drive pro-hematopoietic signaling. *Differentiation, 86*(3), 92–103. https://doi.org/10.1016/j.diff.2013.06.004.

Lee, G., Lo, A., Short, S. A., Mankelow, T. J., Spring, F., Parsons, S. F., et al. (2006). Targeted gene deletion demonstrates that the cell adhesion molecule ICAM-4 is critical for erythroblastic island formation. *Blood, 108*(6), 2064–2071. https://doi.org/10.1182/blood-2006-03-006759.

Lee-Thedieck, C., Rauch, N., Fiammengo, R., Klein, G., & Spatz, J. P. (2012). Impact of substrate elasticity on human hematopoietic stem and progenitor cell adhesion and motility. *Journal of Cell Science, 125*(Pt 16), 3765–3775. https://doi.org/10.1242/jcs.095596.

Leon-Rico, D., Aldea, M., Sanchez, R., Segovia, J. C., Weiss, L. A., Hidalgo, A., et al. (2014). Brief report: Reduced expression of CD18 leads to the in vivo expansion of hematopoietic stem cells in mouse bone marrow. *Stem Cells, 32*(10), 2794–2798. https://doi.org/10.1002/stem.1762.

Lerner, C., & Harrison, D. E. (1990). 5-fluorouracil spares hemopoietic stem cells responsible for long-term repopulation. *Experimental Hematology, 18*(2), 114–118. https://www.ncbi.nlm.nih.gov/pubmed/2303103.

Li, W., Ferkowicz, M. J., Johnson, S. A., Shelley, W. C., & Yoder, M. C. (2005). Endothelial cells in the early murine yolk sac give rise to CD41-expressing hematopoietic cells. *Stem Cells and Development*, *14*(1), 44–54. https://doi.org/10.1089/scd.2005.14.44.

Li, J., & Springer, T. A. (2017). Integrin extension enables ultrasensitive regulation by cytoskeletal force. *Proceedings of the National Academy of Sciences of the United States of America*, *114*(18), 4685–4690. https://doi.org/10.1073/pnas.1704171114.

Li, J., Su, Y., Xia, W., Qin, Y., Humphries, M. J., Vestweber, D., et al. (2017). Conformational equilibria and intrinsic affinities define integrin activation. *The EMBO Journal*, *36*(5), 629–645. https://doi.org/10.15252/embj.201695803.

Lichterfeld, M., Martin, S., Burkly, L., Haas, R., & Kronenwett, R. (2000). Mobilization of CD34+ haematopoietic stem cells is associated with a functional inactivation of the integrin very late antigen 4. *British Journal of Haematology*, *110*(1), 71–81. https://doi.org/10.1046/j.1365-2141.2000.02130.x.

Lie-A-Ling, M., Marinopoulou, E., Li, Y., Patel, R., Stefanska, M., Bonifer, C., et al. (2014). RUNX1 positively regulates a cell adhesion and migration program in murine hemogenic endothelium prior to blood emergence. *Blood*, *124*(11), e11–e20. https://doi.org/10.1182/blood-2014-04-572958.

Lisovsky, M., Braun, S. E., Ge, Y., Takahira, H., Lu, L., Savchenko, V. G., et al. (1996). Flt3-ligand production by human bone marrow stromal cells. *Leukemia*, *10*(6), 1012–1018. https://www.ncbi.nlm.nih.gov/pubmed/8667636.

Liu, Y. F., Zhang, S. Y., Chen, Y. Y., Shi, K., Zou, B., Liu, J., et al. (2018). ICAM-1 deficiency in the bone marrow niche impairs quiescence and repopulation of hematopoietic stem cells. *Stem Cell Reports*.. https://doi.org/10.1016/j.stemcr.2018.05.016.

Lo Celso, C., Fleming, H. E., Wu, J. W., Zhao, C. X., Miake-Lye, S., Fujisaki, J., et al. (2009). Live-animal tracking of individual haematopoietic stem/progenitor cells in their niche. *Nature*, *457*(7225), 92–96. https://doi.org/10.1038/nature07434.

Luchsinger, L. L., Strikoudis, A., Danzl, N. M., Bush, E. C., Finlayson, M. O., Satwani, P., et al. (2019). Harnessing hematopoietic stem cell low intracellular calcium improves their maintenance in vitro. *Cell Stem Cell*, *25*(2), 225–240.e227. https://doi.org/10.1016/j.stem.2019.05.002.

Luo, B. H., Carman, C. V., & Springer, T. A. (2007). Structural basis of integrin regulation and signaling. *Annual Review of Immunology*, *25*, 619–647. https://doi.org/10.1146/annurev.immunol.25.022106.141618.

Lyman, S. D., James, L., Vanden Bos, T., de Vries, P., Brasel, K., Gliniak, B., et al. (1993). Molecular cloning of a ligand for the flt3/flk-2 tyrosine kinase receptor: A proliferative factor for primitive hematopoietic cells. *Cell*, *75*(6), 1157–1167. https://doi.org/10.1016/0092-8674(93)90325-k.

Mackie, E. J., & Tucker, R. P. (1992). Tenascin in bone morphogenesis: Expression by osteoblasts and cell type-specific expression of splice variants. *Journal of Cell Science*, *103*(Pt 3), 765–771. https://www.ncbi.nlm.nih.gov/pubmed/1282516.

Magli, M. C., Iscove, N. N., & Odartchenko, N. (1982). Transient nature of early haematopoietic spleen colonies. *Nature*, *295*(5849), 527–529. https://doi.org/10.1038/295527a0.

Magner, L. N. (1992). *A history of medicine*. Dekker. https://archive.org/details/historyofmedicin0000magn.

Malara, A., Currao, M., Gruppi, C., Celesti, G., Viarengo, G., Buracchi, C., et al. (2014). Megakaryocytes contribute to the bone marrow-matrix environment by expressing fibronectin, type IV collagen, and laminin. *Stem Cells*, *32*(4), 926–937. https://doi.org/10.1002/stem.1626.

Matsunaga, T., Takemoto, N., Sato, T., Takimoto, R., Tanaka, I., Fujimi, A., et al. (2003). Interaction between leukemic-cell VLA-4 and stromal fibronectin is a decisive factor for minimal residual disease of acute myelogenous leukemia. *Nature Medicine*, *9*(9), 1158–1165. https://doi.org/10.1038/nm909.

McCulloch, E. A., & Till, J. E. (1960). The radiation sensitivity of normal mouse bone marrow cells, determined by quantitative marrow transplantation into irradiated mice. *Radiation Research, 13*, 115–125. https://www.ncbi.nlm.nih.gov/pubmed/13858509.

McGrath, K. E., Frame, J. M., Fegan, K. H., Bowen, J. R., Conway, S. J., Catherman, S. C., et al. (2015). Distinct sources of hematopoietic progenitors emerge before HSCs and provide functional blood cells in the mammalian embryo. *Cell Reports, 11*(12), 1892–1904. https://doi.org/10.1016/j.celrep.2015.05.036.

McGrath, K. E., Kingsley, P. D., Koniski, A. D., Porter, R. L., Bushnell, T. P., & Palis, J. (2008). Enucleation of primitive erythroid cells generates a transient population of "pyrenocytes" in the mammalian fetus. *Blood, 111*(4), 2409–2417. https://doi.org/10.1182/blood-2007-08-107581.

McGrath, K. E., Koniski, A. D., Malik, J., & Palis, J. (2003). Circulation is established in a stepwise pattern in the mammalian embryo. *Blood, 101*(5), 1669–1676. https://doi.org/10.1182/blood-2002-08-2531.

McKenna, H. J., Stocking, K. L., Miller, R. E., Brasel, K., De Smedt, T., Maraskovsky, E., et al. (2000). Mice lacking flt3 ligand have deficient hematopoiesis affecting hematopoietic progenitor cells, dendritic cells, and natural killer cells. *Blood, 95*(11), 3489–3497. https://www.ncbi.nlm.nih.gov/pubmed/10828034.

Meads, M. B., Hazlehurst, L. A., & Dalton, W. S. (2008). The bone marrow microenvironment as a tumor sanctuary and contributor to drug resistance. *Clinical Cancer Research, 14*(9), 2519–2526. https://doi.org/10.1158/1078-0432.CCR-07-2223.

Medvinsky, A., & Dzierzak, E. (1996). Definitive hematopoiesis is autonomously initiated by the AGM region. *Cell, 86*(6), 897–906. https://doi.org/10.1016/s0092-8674(00)80165-8.

Mehatre, S. H., Roy, I. M., Biswas, A., Prit, D., Schouteden, S., Huelsken, J., et al. (2021). Niche-mediated integrin signaling supports steady-state hematopoiesis in the spleen. *Journal of Immunology, 206*(7), 1549–1560. https://doi.org/10.4049/jimmunol.2001066.

Menche, N., & Raichle, G. (2020). *Biologie Anatomie Physiologie*. Elsevier Health Sciences.

Mikkola, H. K., Fujiwara, Y., Schlaeger, T. M., Traver, D., & Orkin, S. H. (2003). Expression of CD41 marks the initiation of definitive hematopoiesis in the mouse embryo. *Blood, 101*(2), 508–516. https://doi.org/10.1182/blood-2002-06-1699.

Miller, P. G., Al-Shahrour, F., Hartwell, K. A., Chu, L. P., Järås, M., Puram, R. V., et al. (2013). In vivo RNAi screening identifies a leukemia-specific dependence on integrin beta 3 signaling. *Cancer Cell, 24*(1), 45–58. https://doi.org/10.1016/j.ccr.2013.05.004.

Mirantes, C., Passegué, E., & Pietras, E. M. (2014). Pro-inflammatory cytokines: Emerging players regulating HSC function in normal and diseased hematopoiesis. *Experimental Cell Research, 329*(2), 248–254. https://doi.org/10.1016/j.yexcr.2014.08.017.

Mitroulis, I., Chen, L. S., Singh, R. P., Kourtzelis, I., Economopoulou, M., Kajikawa, T., et al. (2017). Secreted protein Del-1 regulates myelopoiesis in the hematopoietic stem cell niche. *The Journal of Clinical Investigation, 127*(10), 3624–3639. https://doi.org/10.1172/JCI92571.

Moretti, F. A., Moser, M., Lyck, R., Abadier, M., Ruppert, R., Engelhardt, B., et al. (2013). Kindlin-3 regulates integrin activation and adhesion reinforcement of effector T cells. *Proceedings of the National Academy of Sciences of the United States of America, 110*(42), 17005–17010. https://doi.org/10.1073/pnas.1316032110.

Morrison, S. J., & Scadden, D. T. (2014). The bone marrow niche for haematopoietic stem cells. *Nature, 505*(7483), 327–334. https://doi.org/10.1038/nature12984.

Moser, M., Bauer, M., Schmid, S., Ruppert, R., Schmidt, S., Sixt, M., et al. (2009). Kindlin-3 is required for beta2 integrin-mediated leukocyte adhesion to endothelial cells. *Nature Medicine, 15*(3), 300–305. https://doi.org/10.1038/nm.1921.

Moser, M., Nieswandt, B., Ussar, S., Pozgajova, M., & Fässler, R. (2008). Kindlin-3 is essential for integrin activation and platelet aggregation. *Nature Medicine, 14*(3), 325–330. https://doi.org/10.1038/nm1722.

Müller, A. M., Medvinsky, A., Strouboulis, J., Grosveld, F., & Dzierzak, E. (1994). Development of hematopoietic stem cell activity in the mouse embryo. *Immunity*, *1*(4), 291–301. https://doi.org/10.1016/1074-7613(94)90081-7.

Nagasawa, T., Hirota, S., Tachibana, K., Takakura, N., Nishikawa, S., Kitamura, Y., et al. (1996). Defects of B-cell lymphopoiesis and bone-marrow myelopoiesis in mice lacking the CXC chemokine PBSF/SDF-1. *Nature*, *382*(6592), 635–638. https://doi.org/10.1038/382635a0.

Nakamura-Ishizu, A., Okuno, Y., Omatsu, Y., Okabe, K., Morimoto, J., Uede, T., et al. (2012). Extracellular matrix protein tenascin-C is required in the bone marrow microenvironment primed for hematopoietic regeneration. *Blood*, *119*(23), 5429–5437. https://doi.org/10.1182/blood-2011-11-393645.

Nervi, B., Ramirez, P., Rettig, M. P., Uy, G. L., Holt, M. S., Ritchey, J. K., et al. (2009). Chemosensitization of acute myeloid leukemia (AML) following mobilization by the CXCR4 antagonist AMD3100. *Blood*, *113*(24), 6206–6214. https://doi.org/10.1182/blood-2008-06-162123.

Nie, Y., Han, Y. C., & Zou, Y. R. (2008). CXCR4 is required for the quiescence of primitive hematopoietic cells. *The Journal of Experimental Medicine*, *205*(4), 777–783. https://doi.org/10.1084/jem.20072513.

Nieuwenhuis, H. K., Akkerman, J. W., Houdijk, W. P., & Sixma, J. J. (1985). Human blood platelets showing no response to collagen fail to express surface glycoprotein Ia. *Nature*, *318*(6045), 470–472. https://doi.org/10.1038/318470a0.

Nilsson, S. K., Debatis, M. E., Dooner, M. S., Madri, J. A., Quesenberry, P. J., & Becker, P. S. (1998). Immunofluorescence characterization of key extracellular matrix proteins in murine bone marrow in situ. *The Journal of Histochemistry and Cytochemistry*, *46*(3), 371–377. https://doi.org/10.1177/002215549804600311.

Nilsson, S. K., Johnston, H. M., & Coverdale, J. A. (2001). Spatial localization of transplanted hemopoietic stem cells: Inferences for the localization of stem cell niches. *Blood*, *97*(8), 2293–2299. https://doi.org/10.1182/blood.v97.8.2293.

Nilsson, S. K., Johnston, H. M., Whitty, G. A., Williams, B., Webb, R. J., Denhardt, D. T., et al. (2005). Osteopontin, a key component of the hematopoietic stem cell niche and regulator of primitive hematopoietic progenitor cells. *Blood*, *106*(4), 1232–1239. https://doi.org/10.1182/blood-2004-11-4422.

Nishida, N., Xie, C., Shimaoka, M., Cheng, Y., Walz, T., & Springer, T. A. (2006). Activation of leukocyte beta2 integrins by conversion from bent to extended conformations. *Immunity*, *25*(4), 583–594. https://doi.org/10.1016/j.immuni.2006.07.016.

Notta, F., Doulatov, S., Laurenti, E., Poeppl, A., Jurisica, I., & Dick, J. E. (2011). Isolation of single human hematopoietic stem cells capable of long-term multilineage engraftment. *Science*, *333*(6039), 218–221. https://doi.org/10.1126/science.1201219.

Nurden, A. T., Fiore, M., Nurden, P., & Pillois, X. (2011). Glanzmann thrombasthenia: A review of ITGA2B and ITGB3 defects with emphasis on variants, phenotypic variability, and mouse models. *Blood*, *118*(23), 5996–6005. https://doi.org/10.1182/blood-2011-07-365635.

Oefner, C. M., Sharkey, A., Gardner, L., Critchley, H., Oyen, M., & Moffett, A. (2015). Collagen type IV at the fetal-maternal interface. *Placenta*, *36*(1), 59–68. https://doi.org/10.1016/j.placenta.2014.10.012.

Ogawa, M., Kizumoto, M., Nishikawa, S., Fujimoto, T., Kodama, H., & Nishikawa, S. I. (1999). Expression of alpha4-integrin defines the earliest precursor of hematopoietic cell lineage diverged from endothelial cells. *Blood*, *93*(4), 1168–1177. https://www.ncbi.nlm.nih.gov/pubmed/9949159.

Ohmori, T., Kashiwakura, Y., Ishiwata, A., Madoiwa, S., Mimuro, J., Furukawa, Y., et al. (2010). Vinculin is indispensable for repopulation by hematopoietic stem cells, independent of integrin function. *The Journal of Biological Chemistry*, *285*(41), 31763–31773. https://doi.org/10.1074/jbc.M109.099085.

Ohta, M., Sakai, T., Saga, Y., Aizawa, S., & Saito, M. (1998). Suppression of hematopoietic activity in tenascin-C-deficient mice. *Blood*, *91*(11), 4074–4083. https://www.ncbi.nlm.nih.gov/pubmed/9596652.

Okada, H., Watanabe, T., Niki, M., Takano, H., Chiba, N., Yanai, N., et al. (1998). AML1(−/−) embryos do not express certain hematopoiesis-related gene transcripts including those of the PU.1 gene. *Oncogene*, *17*(18), 2287–2293. https://doi.org/10.1038/sj.onc.1202151.

Olson, O. C., Kang, Y. A., & Passegué, E. (2020). Normal hematopoiesis is a balancing act of self-renewal and regeneration. *Cold Spring Harbor Perspectives in Medicine*, *10*(12). https://doi.org/10.1101/cshperspect.a035519.

Osawa, M., Hanada, K., Hamada, H., & Nakauchi, H. (1996). Long-term lymphohematopoietic reconstitution by a single CD34-low/negative hematopoietic stem cell. *Science*, *273*(5272), 242–245. https://doi.org/10.1126/science.273.5272.242.

Ottersbach, K., & Dzierzak, E. (2005). The murine placenta contains hematopoietic stem cells within the vascular labyrinth region. *Developmental Cell*, *8*(3), 377–387. https://doi.org/10.1016/j.devcel.2005.02.001.

Padrón-Barthe, L., Temiño, S., Villa del Campo, C., Carramolino, L., Isern, J., & Torres, M. (2014). Clonal analysis identifies hemogenic endothelium as the source of the blood-endothelial common lineage in the mouse embryo. *Blood*, *124*(16), 2523–2532. https://doi.org/10.1182/blood-2013-12-545939.

Palis, J. (2016). Hematopoietic stem cell-independent hematopoiesis: Emergence of erythroid, megakaryocyte, and myeloid potential in the mammalian embryo. *FEBS Letters*, *590*(22), 3965–3974. https://doi.org/10.1002/1873-3468.12459.

Palis, J., Chan, R. J., Koniski, A., Patel, R., Starr, M., & Yoder, M. C. (2001). Spatial and temporal emergence of high proliferative potential hematopoietic precursors during murine embryogenesis. *Proceedings of the National Academy of Sciences of the United States of America*, *98*(8), 4528–4533. https://doi.org/10.1073/pnas.071002398.

Palis, J., McGrath, K. E., & Kingsley, P. D. (1995). Initiation of hematopoiesis and vasculogenesis in murine yolk sac explants. *Blood*, *86*(1), 156–163. https://www.ncbi.nlm.nih.gov/pubmed/7795222.

Palis, J., Robertson, S., Kennedy, M., Wall, C., & Keller, G. (1999). Development of erythroid and myeloid progenitors in the yolk sac and embryo proper of the mouse. *Development*, *126*(22), 5073–5084. https://www.ncbi.nlm.nih.gov/pubmed/10529424.

Pamphilon, D., Selogie, E., McKenna, D., Cancelas-Peres, J. A., Szczepiorkowski, Z. M., Sacher, R., et al. (2013). Current practices and prospects for standardization of the hematopoietic colony-forming unit assay: A report by the cellular therapy team of the biomedical excellence for safer transfusion (BEST) collaborative. *Cytotherapy*, *15*(3), 255–262. https://doi.org/10.1016/j.jcyt.2012.11.013.

Papadopoulos, N., Simopoulos, C., Sigalas, J., Kotini, A., Cheva, A., & Tamiolakis, D. (2004). Induction of hepatic hematopoiesis with tenascin-C expression during the second trimester of development. *European Journal of Obstetrics, Gynecology, and Reproductive Biology*, *113*(1), 56–60. https://doi.org/10.1016/j.ejogrb.2003.05.006.

Papayannopoulou, T., Priestley, G. V., & Nakamoto, B. (1998). Anti-VLA4/VCAM-1-induced mobilization requires cooperative signaling through the kit/mkit ligand pathway. *Blood*, *91*(7), 2231–2239. https://www.ncbi.nlm.nih.gov/pubmed/9516120.

Papayannopoulou, T., Priestley, G. V., Nakamoto, B., Zafiropoulos, V., Scott, L. M., & Harlan, J. M. (2001). Synergistic mobilization of hemopoietic progenitor cells using concurrent beta1 and beta2 integrin blockade or beta2-deficient mice. *Blood*, *97*(5), 1282–1288. https://doi.org/10.1182/blood.v97.5.1282.

Park, C., Afrikanova, I., Chung, Y. S., Zhang, W. J., Arentson, E., Fong, G., et al. (2004). A hierarchical order of factors in the generation of FLK1- and SCL-expressing hematopoietic and endothelial progenitors from embryonic stem cells. *Development*, *131*(11), 2749–2762. https://doi.org/10.1242/dev.01130.

Pawliuk, R., Eaves, C., & Humphries, R. K. (1996). Evidence of both ontogeny and transplant dose-regulated expansion of hematopoietic stem cells in vivo. *Blood*, *88*(8), 2852–2858. https://www.ncbi.nlm.nih.gov/pubmed/8874181.

Peled, A., Grabovsky, V., Habler, L., Sandbank, J., Arenzana-Seisdedos, F., Petit, I., et al. (1999). The chemokine SDF-1 stimulates integrin-mediated arrest of CD34(+) cells on vascular endothelium under shear flow. *The Journal of Clinical Investigation*, *104*(9), 1199–1211. https://doi.org/10.1172/JCI7615.

Peled, A., Hardan, I., Trakhtenbrot, L., Gur, E., Magid, M., Darash-Yahana, M., et al. (2002). Immature leukemic CD34+CXCR4+ cells from CML patients have lower integrin-dependent migration and adhesion in response to the chemokine SDF-1. *Stem Cells*, *20*(3), 259–266. https://doi.org/10.1634/stemcells.20-3-259.

Peled, A., Kollet, O., Ponomaryov, T., Petit, I., Franitza, S., Grabovsky, V., et al. (2000). The chemokine SDF-1 activates the integrins LFA-1, VLA-4, and VLA-5 on immature human CD34(+) cells: Role in transendothelial/stromal migration and engraftment of NOD/SCID mice. *Blood*, *95*(11), 3289–3296. https://www.ncbi.nlm.nih.gov/pubmed/10828007.

Peled, A., Petit, I., Kollet, O., Magid, M., Ponomaryov, T., Byk, T., et al. (1999). Dependence of human stem cell engraftment and repopulation of NOD/SCID mice on CXCR4. *Science*, *283*(5403), 845–848. https://doi.org/10.1126/science.283.5403.845.

Petrich, B. G., Marchese, P., Ruggeri, Z. M., Spiess, S., Weichert, R. A., Ye, F., et al. (2007). Talin is required for integrin-mediated platelet function in hemostasis and thrombosis. *The Journal of Experimental Medicine*, *204*(13), 3103–3111. https://doi.org/10.1084/jem.20071800.

Pietras, E. M., Reynaud, D., Kang, Y. A., Carlin, D., Calero-Nieto, F. J., Leavitt, A. D., et al. (2015). Functionally distinct subsets of lineage-biased multipotent progenitors control blood production in Normal and regenerative conditions. *Cell Stem Cell*, *17*(1), 35–46. https://doi.org/10.1016/j.stem.2015.05.003.

Pietras, E. M., Warr, M. R., & Passegué, E. (2011). Cell cycle regulation in hematopoietic stem cells. *The Journal of Cell Biology*, *195*(5), 709–720. https://doi.org/10.1083/jcb.201102131.

Ploemacher, R. E., van der Sluijs, J. P., Voerman, J. S., & Brons, N. H. (1989). An in vitro limiting-dilution assay of long-term repopulating hematopoietic stem cells in the mouse. *Blood*, *74*(8), 2755–2763. https://www.ncbi.nlm.nih.gov/pubmed/2819245.

Pluznik, D. H., & Sachs, L. (1965). The cloning of normal "mast" cells in tissue culture. *Journal of Cellular Physiology*, *66*(3), 319–324. https://doi.org/10.1002/jcp.1030660309.

Potocnik, A. J., Brakebusch, C., & Fässler, R. (2000). Fetal and adult hematopoietic stem cells require beta1 integrin function for colonizing fetal liver, spleen, and bone marrow. *Immunity*, *12*(6), 653–663. https://www.ncbi.nlm.nih.gov/pubmed/10894165.

Priestley, G. V., Scott, L. M., Ulyanova, T., & Papayannopoulou, T. (2006). Lack of alpha4 integrin expression in stem cells restricts competitive function and self-renewal activity. *Blood*, *107*(7), 2959–2967. https://doi.org/10.1182/blood-2005-07-2670.

Priestley, G. V., Ulyanova, T., & Papayannopoulou, T. (2007). Sustained alterations in biodistribution of stem/progenitor cells in Tie2Cre+ alpha4(f/f) mice are hematopoietic cell autonomous. *Blood*, *109*(1), 109–111. https://doi.org/10.1182/blood-2006-06-026427.

Qian, H., Buza-Vidas, N., Hyland, C. D., Jensen, C. T., Antonchuk, J., Månsson, R., et al. (2007). Critical role of thrombopoietin in maintaining adult quiescent hematopoietic stem cells. *Cell Stem Cell*, *1*(6), 671–684. https://doi.org/10.1016/j.stem.2007.10.008.

Qian, H., Georges-Labouesse, E., Nyström, A., Domogatskaya, A., Tryggvason, K., Jacobsen, S. E., et al. (2007). Distinct roles of integrins alpha6 and alpha4 in homing of fetal liver hematopoietic stem and progenitor cells. *Blood*, *110*(7), 2399–2407. https://doi.org/10.1182/blood-2006-10-051276.

Qian, H., Tryggvason, K., Jacobsen, S. E., & Ekblom, M. (2006). Contribution of alpha6 integrins to hematopoietic stem and progenitor cell homing to bone marrow and collaboration with alpha4 integrins. *Blood*, *107*(9), 3503–3510. https://doi.org/10.1182/blood-2005-10-3932.

Ren, L., Yang, P., Wang, Z., Zhang, J., Ding, C., & Shang, P. (2015). Biomechanical and biophysical environment of bone from the macroscopic to the pericellular and molecular level. *Journal of the Mechanical Behavior of Biomedical Materials*, *50*, 104–122. https://doi.org/10.1016/j.jmbbm.2015.04.021.

Reynaud, D., Pietras, E., Barry-Holson, K., Mir, A., Binnewies, M., Jeanne, M., et al. (2011). IL-6 controls leukemic multipotent progenitor cell fate and contributes to chronic myelogenous leukemia development. *Cancer Cell*, *20*(5), 661–673. https://doi.org/10.1016/j.ccr.2011.10.012.

Rhodes, K. E., Gekas, C., Wang, Y., Lux, C. T., Francis, C. S., Chan, D. N., et al. (2008). The emergence of hematopoietic stem cells is initiated in the placental vasculature in the absence of circulation. *Cell Stem Cell*, *2*(3), 252–263. https://doi.org/10.1016/j.stem.2008.01.001.

Risau, W., & Flamme, I. (1995). Vasculogenesis. *Annual Review of Cell and Developmental Biology*, *11*, 73–91. https://doi.org/10.1146/annurev.cb.11.110195.000445.

Risau, W., & Lemmon, V. (1988). Changes in the vascular extracellular matrix during embryonic vasculogenesis and angiogenesis. *Developmental Biology*, *125*(2), 441–450. https://doi.org/10.1016/0012-1606(88)90225-4.

Robin, C., Ottersbach, K., Boisset, J. C., Oziemlak, A., & Dzierzak, E. (2011). CD41 is developmentally regulated and differentially expressed on mouse hematopoietic stem cells. *Blood*, *117*(19), 5088–5091. https://doi.org/10.1182/blood-2011-01-329516.

Ross, E. A., Freeman, S., Zhao, Y., Dhanjal, T. S., Ross, E. J., Lax, S., et al. (2008). A novel role for PECAM-1 (CD31) in regulating haematopoietic progenitor cell compartmentalization between the peripheral blood and bone marrow. *PLoS One*, *3*(6), e2338. https://doi.org/10.1371/journal.pone.0002338.

Ruppert, R., Moser, M., Sperandio, M., Rognoni, E., Orban, M., Liu, W. H., et al. (2015). Kindlin-3-mediated integrin adhesion is dispensable for quiescent but essential for activated hematopoietic stem cells. *The Journal of Experimental Medicine*, *212*(9), 1415–1432. https://doi.org/10.1084/jem.20150269.

Rybtsov, S., Ivanovs, A., Zhao, S., & Medvinsky, A. (2016). Concealed expansion of immature precursors underpins acute burst of adult HSC activity in foetal liver. *Development*, *143*(8), 1284–1289. https://doi.org/10.1242/dev.131193.

Sadahira, Y., Yoshino, T., & Monobe, Y. (1995). Very late activation antigen 4-vascular cell adhesion molecule 1 interaction is involved in the formation of erythroblastic islands. *The Journal of Experimental Medicine*, *181*(1), 411–415. https://doi.org/10.1084/jem.181.1.411.

Salgia, R., Brunkhorst, B., Pisick, E., Li, J. L., Lo, S. H., Chen, L. B., et al. (1995). Increased tyrosine phosphorylation of focal adhesion proteins in myeloid cell lines expressing p210BCR/ABL. *Oncogene*, *11*(6), 1149–1155. https://www.ncbi.nlm.nih.gov/pubmed/7566975.

Salgia, R., Sattler, M., Pisick, E., Li, J. L., & Griffin, J. D. (1996). p210BCR/ABL induces formation of complexes containing focal adhesion proteins and the protooncogene product p120c-Cbl. *Experimental Hematology*, *24*(2), 310–313. https://www.ncbi.nlm.nih.gov/pubmed/8641358.

Sasaki, T., Mizuochi, C., Horio, Y., Nakao, K., Akashi, K., & Sugiyama, D. (2010). Regulation of hematopoietic cell clusters in the placental niche through SCF/kit signaling in embryonic mouse. *Development*, *137*(23), 3941–3952. https://doi.org/10.1242/dev.051359.

Saykali, B., Mathiah, N., Nahaboo, W., Racu, M. L., Hammou, L., Defrance, M., et al. (2019). Distinct mesoderm migration phenotypes in extra-embryonic and embryonic regions of the early mouse embryo. *eLife*, *8*. https://doi.org/10.7554/eLife.42434.

Saylam, C., Ozdemir, N., Itil, I. M., Sendag, F., & Terek, M. C. (2002). Distribution of fibronectin, laminin and collagen type IV in the materno-fetal boundary zone of the developing mouse placenta. Experimental study. *Arch Gynecol Obstet*, *266*(2), 83–85. https://doi.org/10.1007/s004040100204.

Scharffetter-Kochanek, K., Lu, H., Norman, K., van Nood, N., Munoz, F., Grabbe, S., et al. (1998). Spontaneous skin ulceration and defective T cell function in CD18 null mice. *The Journal of Experimental Medicine*, *188*(1), 119–131. https://doi.org/10.1084/jem.188.1.119.

Schemionek, M., Spieker, T., Kerstiens, L., Elling, C., Essers, M., Trumpp, A., et al. (2012). Leukemic spleen cells are more potent than bone marrow-derived cells in a transgenic mouse model of CML. *Leukemia*, *26*(5), 1030–1037. https://doi.org/10.1038/leu.2011.366.

Schepers, K., Campbell, T. B., & Passegué, E. (2015). Normal and leukemic stem cell niches: Insights and therapeutic opportunities. *Cell Stem Cell*, *16*(3), 254–267. https://doi.org/10.1016/j.stem.2015.02.014.

Schiller, H. B., Hermann, M. R., Polleux, J., Vignaud, T., Zanivan, S., Friedel, C. C., et al. (2013). β1- and αv-class integrins cooperate to regulate myosin II during rigidity sensing of fibronectin-based microenvironments. *Nature Cell Biology*, *15*(6), 625–636. https://doi.org/10.1038/ncb2747.

Schippers, A., Leuker, C., Pabst, O., Kochut, A., Prochnow, B., Gruber, A. D., et al. (2009). Mucosal addressin cell-adhesion molecule-1 controls plasma-cell migration and function in the small intestine of mice. *Gastroenterology*, *137*(3), 924–933. https://doi.org/10.1053/j.gastro.2009.05.039.

Schmidt, S., Nakchbandi, I., Ruppert, R., Kawelke, N., Hess, M. W., Pfaller, K., et al. (2011). Kindlin-3-mediated signaling from multiple integrin classes is required for osteoclast-mediated bone resorption. *The Journal of Cell Biology*, *192*(5), 883–897. https://doi.org/10.1083/jcb.201007141.

Schoedel, K. B., Morcos, M. N. F., Zerjatke, T., Roeder, I., Grinenko, T., Voehringer, D., et al. (2016). The bulk of the hematopoietic stem cell population is dispensable for murine steady-state and stress hematopoiesis. *Blood*, *128*(19), 2285–2296. https://doi.org/10.1182/blood-2016-03-706010.

Schreiber, T. D., Steinl, C., Essl, M., Abele, H., Geiger, K., Müller, C. A., et al. (2009). The integrin alpha9beta1 on hematopoietic stem and progenitor cells: Involvement in cell adhesion, proliferation and differentiation. *Haematologica*, *94*(11), 1493–1501. https://doi.org/10.3324/haematol.2009.006072.

Schuettpelz, L. G., & Link, D. C. (2013). Regulation of hematopoietic stem cell activity by inflammation. *Frontiers in Immunology*, *4*, 204. https://doi.org/10.3389/fimmu.2013.00204.

Schweitzer, K. M., Dräger, A. M., van der Valk, P., Thijsen, S. F., Zevenbergen, A., Theijsmeijer, A. P., et al. (1996). Constitutive expression of E-selectin and vascular cell adhesion molecule-1 on endothelial cells of hematopoietic tissues. *The American Journal of Pathology*, *148*(1), 165–175. https://www.ncbi.nlm.nih.gov/pubmed/8546203.

Scott, L. M., Priestley, G. V., & Papayannopoulou, T. (2003). Deletion of alpha4 integrins from adult hematopoietic cells reveals roles in homeostasis, regeneration, and homing. *Molecular and Cellular Biology*, *23*(24), 9349–9360. https://doi.org/10.1128/mcb.23.24.9349-9360.2003.

Seiffert, D. (1996). Detection of vitronectin in mineralized bone matrix. *The Journal of Histochemistry and Cytochemistry*, *44*(3), 275–280. https://doi.org/10.1177/44.3.8648088.

Seiffert, M., Beck, S. C., Schermutzki, F., Müller, C. A., Erickson, H. P., & Klein, G. (1998). Mitogenic and adhesive effects of tenascin-C on human hematopoietic cells are mediated by various functional domains. *Matrix Biology*, *17*(1), 47–63. https://doi.org/10.1016/s0945-053x(98)90124-x.

Siler, U., Seiffert, M., Puch, S., Richards, A., Torok-Storb, B., Müller, C. A., et al. (2000). Characterization and functional analysis of laminin isoforms in human bone marrow. *Blood*, *96*(13), 4194–4203. https://www.ncbi.nlm.nih.gov/pubmed/11110691.

Soares-da-Silva, F., Burlen-Defranoux, O., Elsaid, R., Iturri, L., Freyer, L., Sismeiro, O., et al. (2020). Yolk sac erythromyeloid progenitors sustain erythropoiesis throughout embryonic life. *bioRxiv*. https://doi.org/10.1101/2020.02.27.968230.

Soares-da-Silva, F., Peixoto, M., Cumano, A., & Pinto-do-Ó, P. (2020). Crosstalk between the hepatic and hematopoietic systems during embryonic development. *Frontiers in Cell and Development Biology*, *8*, 612. https://doi.org/10.3389/fcell.2020.00612.

Solanilla, A., Grosset, C., Lemercier, C., Dupouy, M., Mahon, F. X., Schweitzer, K., et al. (2000). Expression of Flt3-ligand by the endothelial cell. *Leukemia*, *14*(1), 153–162. https://doi.org/10.1038/sj.leu.2401635.

Somervaille, T. C., & Cleary, M. L. (2006). Identification and characterization of leukemia stem cells in murine MLL-AF9 acute myeloid leukemia. *Cancer Cell*, *10*(4), 257–268. https://doi.org/10.1016/j.ccr.2006.08.020.

Springer, T. A., & Dustin, M. L. (2012). Integrin inside-out signaling and the immunological synapse. *Current Opinion in Cell Biology*, *24*(1), 107–115. https://doi.org/10.1016/j.ceb.2011.10.004.

Stefanska, M., Batta, K., Patel, R., Florkowska, M., Kouskoff, V., & Lacaud, G. (2017). Primitive erythrocytes are generated from hemogenic endothelial cells. *Scientific Reports*, 7(1), 6401. https://doi.org/10.1038/s41598-017-06627-9.

Stier, S., Ko, Y., Forkert, R., Lutz, C., Neuhaus, T., Grünewald, E., et al. (2005). Osteopontin is a hematopoietic stem cell niche component that negatively regulates stem cell pool size. *The Journal of Experimental Medicine*, *201*(11), 1781–1791. https://doi.org/10.1084/jem.20041992.

Sugiyama, T., Kohara, H., Noda, M., & Nagasawa, T. (2006). Maintenance of the hematopoietic stem cell pool by CXCL12-CXCR4 chemokine signaling in bone marrow stromal cell niches. *Immunity*, *25*(6), 977–988. https://doi.org/10.1016/j.immuni.2006.10.016.

Sugiyama, Y., Koike, T., & Shiojiri, N. (2010). Developmental changes of cell adhesion molecule expression in the fetal mouse liver. *Anat Rec (Hoboken)*, *293*(10), 1698–1710. https://doi.org/10.1002/ar.21204.

Sugiyama, D., Kulkeaw, K., & Mizuochi, C. (2013). TGF-beta-1 up-regulates extra-cellular matrix production in mouse hepatoblasts. *Mechanisms of Development*, *130*(2–3), 195–206. https://doi.org/10.1016/j.mod.2012.09.003.

Sun, Z., Costell, M., & Fässler, R. (2019). Integrin activation by Talin, kindlin and mechanical forces. *Nature Cell Biology*, *21*(1), 25–31. https://doi.org/10.1038/s41556-018-0234-9.

Sun, Z., Guo, S. S., & Fässler, R. (2016). Integrin-mediated mechanotransduction. *The Journal of Cell Biology*, *215*(4), 445–456. https://doi.org/10.1083/jcb.201609037.

Susek, K. H., Korpos, E., Huppert, J., Wu, C., Savelyeva, I., Rosenbauer, F., et al. (2018). Bone marrow laminins influence hematopoietic stem and progenitor cell cycling and homing to the bone marrow. *Matrix Biology*, *67*, 47–62. https://doi.org/10.1016/j.matbio.2018.01.007.

Swiers, G., Rode, C., Azzoni, E., & de Bruijn, M. F. (2013). A short history of hemogenic endothelium. *Blood Cells, Molecules & Diseases*, *51*(4), 206–212. https://doi.org/10.1016/j.bcmd.2013.09.005.

Szilvassy, S. J., Humphries, R. K., Lansdorp, P. M., Eaves, A. C., & Eaves, C. J. (1990). Quantitative assay for totipotent reconstituting hematopoietic stem cells by a competitive repopulation strategy. *Proceedings of the National Academy of Sciences of the United States of America*, *87*(22), 8736–8740. https://doi.org/10.1073/pnas.87.22.8736.

Tada, T., & Fukuta, K. (2010). Expression of cell adhesion molecules at the collapse and recovery of haematopoiesis in bone marrow of mouse. *Anatomia, Histologia, Embryologia*, *39*(5), 403–410. https://doi.org/10.1111/j.1439-0264.2010.01009.x.

Tadokoro, S., Shattil, S. J., Eto, K., Tai, V., Liddington, R. C., de Pereda, J. M., et al. (2003). Talin binding to integrin beta tails: A final common step in integrin activation. *Science*, *302*(5642), 103–106. https://doi.org/10.1126/science.1086652.

Takakura, N., Watanabe, T., Suenobu, S., Yamada, Y., Noda, T., Ito, Y., et al. (2000). A role for hematopoietic stem cells in promoting angiogenesis. *Cell*, *102*(2), 199–209. https://doi.org/10.1016/s0092-8674(00)00025-8.

Tamura, M., Hattori, K., Nomura, H., Oheda, M., Kubota, N., Imazeki, I., et al. (1987). Induction of neutrophilic granulocytosis in mice by administration of purified human native granulocyte colony-stimulating factor (G-CSF). *Biochemical and Biophysical Research Communications*, *142*(2), 454–460. https://doi.org/10.1016/0006-291x(87)90296-8.

Tan, K. S., Kulkeaw, K., Nakanishi, Y., & Sugiyama, D. (2017). Expression of cytokine and extracellular matrix mRNAs in fetal hepatic stellate cells. *Genes to Cells*, *22*(9), 836–844. https://doi.org/10.1111/gtc.12517.

Tanaka, Y., Joshi, A., Wilson, N. K., Kinston, S., Nishikawa, S., & Göttgens, B. (2012). The transcriptional programme controlled by Runx1 during early embryonic blood development. *Developmental Biology*, *366*(2), 404–419. https://doi.org/10.1016/j.ydbio.2012.03.024.

Tanaka, Y., Morimoto, I., Nakano, Y., Okada, Y., Hirota, S., Nomura, S., et al. (1995). Osteoblasts are regulated by the cellular adhesion through ICAM-1 and VCAM-1. *Journal of Bone and Mineral Research*, *10*(10), 1462–1469. https://doi.org/10.1002/jbmr.5650101006.

Teixidó, J., Hemler, M. E., Greenberger, J. S., & Anklesaria, P. (1992). Role of beta 1 and beta 2 integrins in the adhesion of human CD34hi stem cells to bone marrow stroma. *The Journal of Clinical Investigation*, *90*(2), 358–367. https://doi.org/10.1172/JCI115870.

Till, J. E., & McCulloch, E. A. (1961). A direct measurement of the radiation sensitivity of normal mouse bone marrow cells. *Radiation Research*, *14*, 213–222. https://www.ncbi.nlm.nih.gov/pubmed/13776896.

Tokoyoda, K., Egawa, T., Sugiyama, T., Choi, B. I., & Nagasawa, T. (2004). Cellular niches controlling B lymphocyte behavior within bone marrow during development. *Immunity*, *20*(6), 707–718. https://doi.org/10.1016/j.immuni.2004.05.001.

Tronik-Le Roux, D., Roullot, V., Poujol, C., Kortulewski, T., Nurden, P., & Marguerie, G. (2000). Thrombasthenic mice generated by replacement of the integrin alpha(IIb) gene: Demonstration that transcriptional activation of this megakaryocytic locus precedes lineage commitment. *Blood*, *96*(4), 1399–1408. https://www.ncbi.nlm.nih.gov/pubmed/10942384.

Tzeng, Y. S., Li, H., Kang, Y. L., Chen, W. C., Cheng, W. C., & Lai, D. M. (2011). Loss of Cxcl12/Sdf-1 in adult mice decreases the quiescent state of hematopoietic stem/progenitor cells and alters the pattern of hematopoietic regeneration after myelosuppression. *Blood*, *117*(2), 429–439. https://doi.org/10.1182/blood-2010-01-266833.

Ueno, H., & Weissman, I. L. (2006). Clonal analysis of mouse development reveals a polyclonal origin for yolk sac blood islands. *Developmental Cell*, *11*(4), 519–533. https://doi.org/10.1016/j.devcel.2006.08.001.

Ulyanova, T., Scott, L. M., Priestley, G. V., Jiang, Y., Nakamoto, B., Koni, P. A., et al. (2005). VCAM-1 expression in adult hematopoietic and nonhematopoietic cells is controlled by tissue-inductive signals and reflects their developmental origin. *Blood*, *106*(1), 86–94. https://doi.org/10.1182/blood-2004-09-3417.

Umemoto, T., Yamato, M., Ishihara, J., Shiratsuchi, Y., Utsumi, M., Morita, Y., et al. (2012). Integrin-αvβ3 regulates thrombopoietin-mediated maintenance of hematopoietic stem cells. *Blood*, *119*(1), 83–94. https://doi.org/10.1182/blood-2011-02-335430.

Umemoto, T., Yamato, M., Shiratsuchi, Y., Terasawa, M., Yang, J., Nishida, K., et al. (2006). Expression of integrin beta3 is correlated to the properties of quiescent hemopoietic stem cells possessing the side population phenotype. *Journal of Immunology*, *177*(11), 7733–7739. https://doi.org/10.4049/jimmunol.177.11.7733.

Umemoto, T., Yamato, M., Shiratsuchi, Y., Terasawa, M., Yang, J., Nishida, K., et al. (2008). CD61 enriches long-term repopulating hematopoietic stem cells. *Biochemical and Biophysical Research Communications*, *365*(1), 176–182. https://doi.org/10.1016/j.bbrc.2007.10.168.

van der Loo, J. C., Xiao, X., McMillin, D., Hashino, K., Kato, I., & Williams, D. A. (1998). VLA-5 is expressed by mouse and human long-term repopulating hematopoietic cells and mediates adhesion to extracellular matrix protein fibronectin. *The Journal of Clinical Investigation*, *102*(5), 1051–1061. https://doi.org/10.1172/JCI3687.

Verfaillie, C. M., McCarthy, J. B., & McGlave, P. B. (1991). Differentiation of primitive human multipotent hematopoietic progenitors into single lineage clonogenic progenitors is accompanied by alterations in their interaction with fibronectin. *The Journal of Experimental Medicine*, *174*(3), 693–703. https://doi.org/10.1084/jem.174.3.693.

Verfaillie, C. M., McCarthy, J. B., & McGlave, P. B. (1992). Mechanisms underlying abnormal trafficking of malignant progenitors in chronic myelogenous leukemia. Decreased adhesion to stroma and fibronectin but increased adhesion to the basement membrane components laminin and collagen type IV. *The Journal of Clinical Investigation*, *90*(4), 1232–1241. https://doi.org/10.1172/JCI115985.

Wagers, A. J., Allsopp, R. C., & Weissman, I. L. (2002). Changes in integrin expression are associated with altered homing properties of Lin(−/lo)Thy1.1(lo)Sca-1(+)c-kit(+) hematopoietic stem cells following mobilization by cyclophosphamide/granulocyte colony-stimulating factor. *Experimental Hematology*, *30*(2), 176–185. https://doi.org/10.1016/s0301-472x(01)00777-9.

Wagers, A. J., & Weissman, I. L. (2006). Differential expression of alpha2 integrin separates long-term and short-term reconstituting Lin-/loThy1.1(lo)c-kit+ Sca-1+ hematopoietic stem cells. *Stem Cells*, *24*(4), 1087–1094. https://doi.org/10.1634/stemcells.2005-0396.

Wagner, N., Löhler, J., Kunkel, E. J., Ley, K., Leung, E., Krissansen, G., et al. (1996). Critical role for beta7 integrins in formation of the gut-associated lymphoid tissue. *Nature*, *382*(6589), 366–370. https://doi.org/10.1038/382366a0.

Wells, R. M. G. (1979). Haemoglobin-oxygen affinity in developing embryonic erythroid cells of the mouse. *Journal of Comparative Physiology*, *129*(4), 333–338. https://doi.org/10.1007/BF00686990.

Wertheim, J. A., Perera, S. A., Hammer, D. A., Ren, R., Boettiger, D., & Pear, W. S. (2003). Localization of BCR-ABL to F-actin regulates cell adhesion but does not attenuate CML development. *Blood*, *102*(6), 2220–2228. https://doi.org/10.1182/blood-2003-01-0062.

Wierenga, P. K., Weersing, E., Dontje, B., de Haan, G., & van Os, R. (2006). Differential role for very late antigen-5 in mobilization and homing of hematopoietic stem cells. *Bone Marrow Transplantation*, *38*(12), 789–797. https://doi.org/10.1038/sj.bmt.1705534.

Wilkinson, A. C., Ishida, R., Kikuchi, M., Sudo, K., Morita, M., Crisostomo, R. V., et al. (2019). Long-term ex vivo haematopoietic-stem-cell expansion allows nonconditioned transplantation. *Nature*, *571*(7763), 117–121. https://doi.org/10.1038/s41586-019-1244-x.

Williams, D. A., Rios, M., Stephens, C., & Patel, V. P. (1991). Fibronectin and VLA-4 in haematopoietic stem cell-microenvironment interactions. *Nature*, *352*(6334), 438–441. https://doi.org/10.1038/352438a0.

Wilson, A., Laurenti, E., Oser, G., van der Wath, R. C., Blanco-Bose, W., Jaworski, M., et al. (2008). Hematopoietic stem cells reversibly switch from dormancy to self-renewal during homeostasis and repair. *Cell, 135*(6), 1118–1129. https://doi.org/10.1016/j.cell.2008.10.048.

Winkler, I. G., Barbier, V., Nowlan, B., Jacobsen, R. N., Forristal, C. E., Patton, J. T., et al. (2012). Vascular niche E-selectin regulates hematopoietic stem cell dormancy, self renewal and chemoresistance. *Nature Medicine, 18*(11), 1651–1657. https://doi.org/10.1038/nm.2969.

Winograd-Katz, S. E., Fässler, R., Geiger, B., & Legate, K. R. (2014). The integrin adhesome: From genes and proteins to human disease. *Nature Reviews. Molecular Cell Biology, 15*(4), 273–288. https://doi.org/10.1038/nrm3769.

Wong, P. M., Chung, S. W., Reicheld, S. M., & Chui, D. H. (1986). Hemoglobin switching during murine embryonic development: Evidence for two populations of embryonic erythropoietic progenitor cells. *Blood, 67*(3), 716–721. https://www.ncbi.nlm.nih.gov/pubmed/3947744.

Wright, D. E., Wagers, A. J., Gulati, A. P., Johnson, F. L., & Weissman, I. L. (2001). Physiological migration of hematopoietic stem and progenitor cells. *Science, 294*(5548), 1933–1936. https://doi.org/10.1126/science.1064081.

Wu, Y., & Hirschi, K. K. (2020). Tissue-resident macrophage development and function. *Frontiers in Cell and Development Biology, 8*, 617879. https://doi.org/10.3389/fcell.2020.617879.

Xie, Y., Yin, T., Wiegraebe, W., He, X. C., Miller, D., Stark, D., et al. (2009). Detection of functional haematopoietic stem cell niche using real-time imaging. *Nature, 457*(7225), 97–101. https://doi.org/10.1038/nature07639.

Yamane, T., Washino, A., & Yamazaki, H. (2013). Common developmental pathway for primitive erythrocytes and multipotent hematopoietic progenitors in early mouse development. *Stem Cell Reports, 1*(6), 590–603. https://doi.org/10.1016/j.stemcr.2013.10.008.

Yamazaki, S., Ema, H., Karlsson, G., Yamaguchi, T., Miyoshi, H., Shioda, S., et al. (2011). Nonmyelinating Schwann cells maintain hematopoietic stem cell hibernation in the bone marrow niche. *Cell, 147*(5), 1146–1158. https://doi.org/10.1016/j.cell.2011.09.053.

Yanai, N., Sekine, C., Yagita, H., & Obinata, M. (1994). Roles for integrin very late activation antigen-4 in stroma-dependent erythropoiesis. *Blood, 83*(10), 2844–2850. https://www.ncbi.nlm.nih.gov/pubmed/7514048.

Yap, L., Tay, H. G., Nguyen, M. T. X., Tjin, M. S., & Tryggvason, K. (2019). Laminins in cellular differentiation. *Trends in Cell Biology, 29*(12), 987–1000. https://doi.org/10.1016/j.tcb.2019.10.001.

Ye, F., Snider, A. K., & Ginsberg, M. H. (2014). Talin and kindlin: The one-two punch in integrin activation. *Frontiers in Medicine, 8*(1), 6–16. https://doi.org/10.1007/s11684-014-0317-3.

Yi, H., Zeng, D., Shen, Z., Liao, J., Wang, X., Liu, Y., et al. (2016). Integrin alphavbeta3 enhances β-catenin signaling in acute myeloid leukemia harboring Fms-like tyrosine kinase-3 internal tandem duplication mutations: Implications for microenvironment influence on sorafenib sensitivity. *Oncotarget*, 7(26), 40387–40397. https://doi.org/10.18632/oncotarget.9617.

Yoder, M. C., & Hiatt, K. (1997). Engraftment of embryonic hematopoietic cells in conditioned newborn recipients. *Blood, 89*(6), 2176–2183. https://www.ncbi.nlm.nih.gov/pubmed/9058742.

Yoder, M. C., Hiatt, K., Dutt, P., Mukherjee, P., Bodine, D. M., & Orlic, D. (1997). Characterization of definitive lymphohematopoietic stem cells in the day 9 murine yolk sac. *Immunity*, 7(3), 335–344. https://doi.org/10.1016/s1074-7613(00)80355-6.

Yoder, M. C., Hiatt, K., & Mukherjee, P. (1997). In vivo repopulating hematopoietic stem cells are present in the murine yolk sac at day 9.0 postcoitus. *Proceedings of the National Academy of Sciences of the United States of America, 94*(13), 6776–6780. https://doi.org/10.1073/pnas.94.13.6776.

Yokomizo, T., & Dzierzak, E. (2010). Three-dimensional cartography of hematopoietic clusters in the vasculature of whole mouse embryos. *Development, 137*(21), 3651–3661. https://doi.org/10.1242/dev.051094.

Yokomizo, T., Ogawa, M., Osato, M., Kanno, T., Yoshida, H., Fujimoto, T., et al. (2001). Requirement of Runx1/AML1/PEBP2alphaB for the generation of haematopoietic cells from endothelial cells. *Genes to Cells, 6*(1), 13–23. https://doi.org/10.1046/j.1365-2443.2001.00393.x.

Yoshihara, H., Arai, F., Hosokawa, K., Hagiwara, T., Takubo, K., Nakamura, Y., et al. (2007). Thrombopoietin/MPL signaling regulates hematopoietic stem cell quiescence and interaction with the osteoblastic niche. *Cell Stem Cell, 1*(6), 685–697. https://doi.org/10.1016/j.stem.2007.10.020.

Yu, V. W., & Scadden, D. T. (2016). Hematopoietic stem cell and its bone marrow niche. *Current Topics in Developmental Biology, 118*, 21–44. https://doi.org/10.1016/bs.ctdb.2016.01.009.

Zhang, Y., Dépond, M., He, L., Foudi, A., Kwarteng, E. O., Lauret, E., et al. (2016). CXCR4/CXCL12 axis counteracts hematopoietic stem cell exhaustion through selective protection against oxidative stress. *Scientific Reports, 6*, 37827. https://doi.org/10.1038/srep37827.

Zhang, J., Niu, C., Ye, L., Huang, H., He, X., Tong, W. G., et al. (2003). Identification of the haematopoietic stem cell niche and control of the niche size. *Nature, 425*(6960), 836–841. https://doi.org/10.1038/nature02041.

Zhong, R. K., Astle, C. M., & Harrison, D. E. (1996). Distinct developmental patterns of short-term and long-term functioning lymphoid and myeloid precursors defined by competitive limiting dilution analysis in vivo. *Journal of Immunology, 157*(1), 138–145. https://www.ncbi.nlm.nih.gov/pubmed/8683107.

Zhou, B. O., Ding, L., & Morrison, S. J. (2015). Hematopoietic stem and progenitor cells regulate the regeneration of their niche by secreting Angiopoietin-1. *eLife, 4*, e05521. https://doi.org/10.7554/eLife.05521.

Zhou, F., Li, X., Wang, W., Zhu, P., Zhou, J., He, W., et al. (2016). Tracing haematopoietic stem cell formation at single-cell resolution. *Nature, 533*(7604), 487–492. https://doi.org/10.1038/nature17997.

Zhou, B. O., Yu, H., Yue, R., Zhao, Z., Rios, J. J., Naveiras, O., et al. (2017). Bone marrow adipocytes promote the regeneration of stem cells and haematopoiesis by secreting SCF. *Nature Cell Biology, 19*(8), 891–903. https://doi.org/10.1038/ncb3570.

Zohren, F., Toutzaris, D., Klärner, V., Hartung, H. P., Kieseier, B., & Haas, R. (2008). The monoclonal anti-VLA-4 antibody natalizumab mobilizes CD34+ hematopoietic progenitor cells in humans. *Blood, 111*(7), 3893–3895. https://doi.org/10.1182/blood-2007-10-120329.

Zou, Y. R., Kottmann, A. H., Kuroda, M., Taniuchi, I., & Littman, D. R. (1998). Function of the chemokine receptor CXCR4 in haematopoiesis and in cerebellar development. *Nature, 393*(6685), 595–599. https://doi.org/10.1038/31269.

Zovein, A. C., Hofmann, J. J., Lynch, M., French, W. J., Turlo, K. A., Yang, Y., et al. (2008). Fate tracing reveals the endothelial origin of hematopoietic stem cells. *Cell Stem Cell, 3*(6), 625–636. https://doi.org/10.1016/j.stem.2008.09.018.

CHAPTER SEVEN

Putting in the Erk: Growth factor signaling and mesoderm morphogenesis

Sarah E. McFann[a,b], Stanislav Y. Shvartsman[b,c,d], and Jared E. Toettcher[c,*]

[a]Department of Chemical and Biological Engineering, Princeton University, Princeton, NJ, United States
[b]Lewis Sigler Institute for Integrative Genomics, Princeton University, Princeton, NJ, United States
[c]Department of Molecular Biology, Princeton University, Princeton, NJ, United States
[d]Center for Computational Biology, Flatiron Institute, Simons Foundation, New York, NY, United States
*Corresponding author: e-mail address: toettcher@princeton.edu

Contents

Abstract

It has long been known that FGF signaling contributes to mesoderm formation, a germ layer found in triploblasts that is composed of highly migratory cells that give rise to muscles and to the skeletal structures of vertebrates. FGF signaling activates several

Current Topics in Developmental Biology, Volume 149
ISSN 0070-2153
https://doi.org/10.1016/bs.ctdb.2022.02.007

pathways in the developing mesoderm, including transient activation of the Erk pathway, which triggers mesodermal fate specification through the induction of the gene *brachyury* and activates morphogenetic programs that allow mesodermal cells to position themselves in the embryo. In this review, we discuss what is known about the generation and interpretation of transient Erk signaling in mesodermal tissues across species. We focus specifically on mechanisms that translate the level and duration of Erk signaling into cell fate and cell movement instructions and discuss strategies for further interrogating the role that Erk signaling dynamics play in mesodermal gastrulation and morphogenesis.

1. Introduction

"Time is a flat circle" in developmental signaling (Fukunaga, 2014): a small number of signaling pathways are used recurrently to dictate cell fates and movements throughout embryogenesis. Despite a relatively complete molecular parts list, there is still much to learn about how cell signaling pathways are activated at just the right positions and developmental times, and how pathway activity is interpreted into the gene expression programs required for each developmental fate choice.

One of the earliest and most fundamental developmental fate choices is germ layer specification during gastrulation. A collection of initially pluripotent cells differentiates into endoderm, ectoderm or mesoderm in a manner that depends on their position in the embryo and exposure to signaling cues. While the cues that establish each germ layer are well-established, it is still unclear how these cues are organized dynamically in space and time to control different aspects of cell fate specification and cell movements. Here, we explore these dynamics in one specific context: the role of transient Erk signaling in early mesoderm development. We note that we cover the first steps of mesoderm induction, prior to somitogenesis, which is already covered by excellent reviews (Aulehla & Pourquié, 2010).

Activation of the kinases Erk1 and Erk2 (henceforth grouped together as Erk) in the mesoderm is primarily triggered by activation of fibroblast growth factor receptor (FGFR), a receptor tyrosine kinase (RTK) that has long been implicated in mesoderm induction in vertebrates. Early experiments showed that ectopically injected fibroblast growth factor (FGF) is sufficient to induce mesoderm in *Xenopus* (Smith, 1989), and over the next decade, specific FGFs and FGFRs that participate in mesoderm induction were identified (Christen & Slack, 1997; Golub et al., 2000; Isaacs, Pownall, & Slack, 1995; Lombardo, Isaacs, & Slack, 1998). One of the primary roles FGF plays in mesoderm specification is induction of *brachyury* (Isaacs, Pownall, & Slack, 1994), a gene that marks the mesodermal precursor population in vertebrates. Experiments with small molecule inhibitors and morpholinos against Erk revealed that FGF triggers *brachyury* expression through Erk signaling in a manner that is conserved across vertebrates (Christen & Slack, 1999; Hardy, Yatskievych, Konieczka, Bobbs, & Antin, 2011; Krens, Corredor-Adámez, He, Snaar-Jagalska, & Spaink, 2008; LaBonne & Whitman, 1994; van Boxtel, Economou, Heliot, & Hill, 2018; Yao et al., 2003). Such experiments also revealed that FGF-induced Erk signaling plays a role in bringing the newly specified mesoderm to its proper location in the embryo (Hardy et al., 2011; Krens, He, et al., 2008; Sivak, Petersen, & Amaya, 2005).

Interestingly, Erk signaling dynamics in the mesoderm are also well-conserved. Early antibody stainings suggested that Erk signaling is transient in the vertebrate mesoderm (Christen & Slack, 1999; Krens, He, et al., 2008; Lunn, Fishwick, Halley, & Storey, 2007), as well as in certain invertebrate mesodermal subtypes (Mandal, Dumstrei, & Hartenstein, 2004; San Martin & Bate, 2001). Recently developed biosensors for live imaging of embryonic signaling have captured the dynamics of such signals at high temporal resolution (Wong, Akiyama, Bessho, & Matsui, 2018) and have detected transient Erk signals in tissues where older staining techniques failed (Corson, Yamanaka, Lai, & Rossant, 2003; Morgani et al., 2018). The fact that transient Erk signaling is so well-conserved begs the question: are the dynamics of Erk signaling playing a functional role in mesoderm specification?

Historically, this hypothesis has been difficult to test because Erk plays other important roles prior to gastrulation. In mouse, for example, Erk signaling is required for epiblast specification (Nakamura, Goto, Kondo, & Aoki, 2021), and in zebrafish, Erk signaling is required for epiboly, a morphogenetic movement that precedes mesodermal ingression (Krens, He, et al., 2008). With the advent of advanced signaling perturbation techniques like optogenetics, however, it is finally possible to test specific hypotheses regarding dynamic Erk signaling requirements for the induction of specific cell types and morphogenetic behaviors. Optogenetic tools developed to control Erk signaling in invertebrates have revealed that the *Drosophila* embryo depends on Erk signaling dynamics for several germ layer choices, including the choice between endoderm and neurogenic ectoderm and the choice between caudal visceral mesoderm and hindgut ectoderm (Johnson & Toettcher, 2019; McFann, Dutta, Toettcher, & Shvartsman, 2021). It is likely that vertebrates, too, possess molecular mechanisms for decoding Erk signaling dynamics during germ layer specification. Indeed, recent optogenetic experiments in zebrafish have shown that varying the duration of Erk signaling during germ layer specification affects cell movements during gastrulation (Patel et al., 2019).

In this review, we discuss what is known about the roles that Erk and its dynamics play in mesodermal specification. We draw examples from lower invertebrates (*Xenopus*, zebrafish) and higher vertebrates (chick, mouse) to highlight conserved features of Erk signaling during the earliest moments of mesodermal specification, like mechanisms for ensuring Erk transience at the margins and primitive streaks of embryos. Additionally, we include the *Drosophila* caudal visceral mesoderm as an example to highlight how some features of mesodermal Erk signaling are more widely conserved, like Erk's role in inducing and maintaining *brachyury* expression and its importance in highly migratory tissues.

We are also interested in contexts where mesodermal development diverges, like in gastrulation where different organisms employ different types of cell movement to position the mesoderm. Erk's continued presence in these divergent contexts suggests a flexibility in this signaling module that allows it to be coopted by evolution for a wide variety of morphogenetic purposes. On the other hand, it is important to note that Erk is only one process triggered by upstream RTK activation, and it remains possible that other RTK-dependent but Erk-independent processes are at least as important for coordinating cell movements. We spend the final section of this review discussing emerging techniques that may prove helpful in answering such questions going forward.

2. The fates of vertebrate mesodermal populations marked by *brachyury*

How can early mesodermal cells be identified across organisms and contexts? We focus particularly on the earliest cell populations that express *brachyury* and its homologs, as *brachyury* is expressed in every mesodermal subtype in vertebrates, and its expression marks the mesodermal precursor population (Amaya, Stein, Musci, & Kirschner, 1993; Burdsal, Flannery, & Pedersen, 1998; Griffin, Patient, & Holder, 1995; Isaacs et al., 1994; Kispert, Ortner, Cooke, & Herrmann, 1995; Krens, Corredor-Adámez, et al., 2008; LaBonne & Whitman, 1994; Schulte-Merker & Smith, 1995; Smith, Price, Green, Weigel, & Herrmann, 1991; Yao et al., 2003). This population is highly migratory. Upon receiving pro-mesodermal cues, mesodermal precursors loosen connections to their cellular neighbors and traverse great distances to reach their final embryonic positions. Even in invertebrates like *Drosophila*, in which mesoderm specification is initiated through different means than in vertebrates, the caudal visceral mesoderm, the only mesodermal sub-population to express *brachyury*, migrates the greatest distance of any *Drosophila* embryonic cell type (Bae, Trisnadi, Kadam, & Stathopoulos, 2012).

In lower vertebrates like *Xenopus* and zebrafish, mesodermal precursors arise from the margins of spherical embryos. Meanwhile, in higher vertebrates, the geometries of embryos vary from the cup-like mouse gastrula to the disc-like chick gastrula, with mesodermal precursors arising from a structure called the primitive streak. In this section, we discuss mesodermal gastrulation in *Xenopus*, zebrafish, chick, and mouse to set the groundwork for discussing general principles regarding Erk-dependent programming of mesodermal induction and morphogenesis across species.

In *Xenopus* and zebrafish, a *brachyury* homolog is expressed at the embryonic margin, a region that borders the blastopore (Fig. 1A, green region). Cells at the margin descend over the embryo through a process called epiboly. In *Xenopus*, epiboly involves the vegetal movement of cells over the endodermal precursor population, while in zebrafish, epiboly involves vegetal movement over the yolk (Solnica-Krezel & Sepich, 2012). Concurrent with the expression of *brachyury*, mesodermal precursors undergo gastrulation. In *Xenopus*, cells expressing *Xbra*, the *Xenopus brachyury* homolog, form the mesodermal layer by folding beneath the ectodermal layer, a process termed involution (Smith et al., 1991; Solnica-Krezel & Sepich, 2012). In zebrafish, cells expressing *no tail* (*ntl*), the zebrafish *brachyury* homolog, push past cells in the epiblast to form a mesoendodermal layer via ingression (Rodaway et al., 1999; Solnica-Krezel & Sepich, 2012).

In both *Xenopus* and zebrafish, the type of mesoderm each cell will become is determined by its dorsal-ventral location along the margin. The dorsal margin gives rise to axial mesoderm, while cells medially localized along the dorsal-ventral axis give rise to paraxial and lateral mesoderm (Dale & Wardle, 2015; Warga & Nüsslein-Volhard, 1999). In all vertebrates, the axial mesoderm goes on to form the prechordal plate and notochord. Meanwhile, paraxial mesoderm gives rise to the somites and tailbud, whereas lateral mesoderm contributes to the appendicular skeleton, heart, and other organs. In both *Xenopus* and zebrafish, upon entering the mesodermal layer, the axial mesoderm extends anteriorly. Meanwhile, the paraxial and lateral mesoderm extend anteriorly while also converging toward the dorsal midline (Solnica-Krezel & Sepich, 2012).

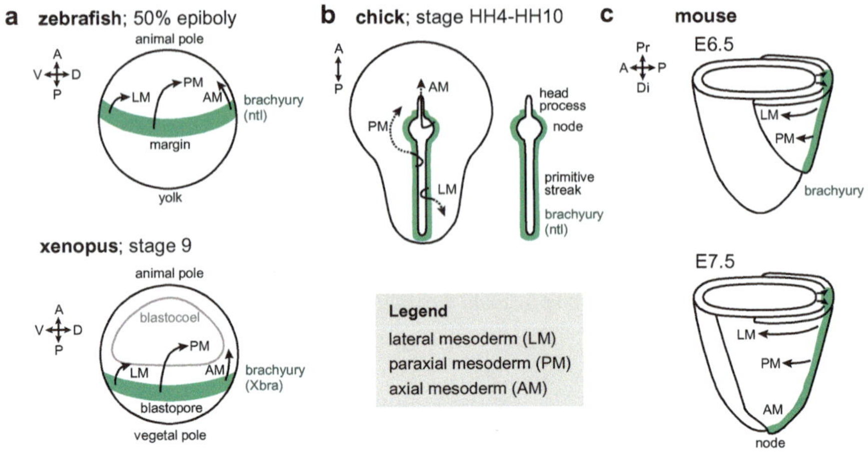

Fig. 1 Mesodermal cell populations across vertebrate species. (A) Mesodermal gastrulation in zebrafish and *Xenopus*. The dorsal-ventral and anterior-posterior axes are marked along with the animal cap, the yolk (zebrafish only) and the vegetal pole (*Xenopus* only). A *brachyury* homolog (*ntl* in zebrafish, *Xbra* in *Xenopus*) is expressed in a ring, marked in green, at the margin. Mesodermal precursors ingress at the margin. Axial mesoderm arises from the dorsal margin while paraxial and lateral mesoderm arise from medial regions along the dorsal-ventral axis. The axial mesoderm extends anteriorly, while the paraxial and lateral mesoderm converge dorsally while extending anteriorly. These cell movements are indicated by arrows. (B) Mesodermal gastrulation in chick. The anterior-posterior axis is marked. *brachyury*, marked in green, is expressed around the primitive streak and Hensen's node. Mesodermal precursors ingress in an anterior-to-posterior fashion, beginning at Hensen's node. Ingression is indicated by the solid portions of arrows, while migration occurring in the newly formed mesodermal layer is indicated by the dotted portions of arrows. Cells that ingress through Hensen's node migrate anteriorly to form the prechordal plate and head structures. As the primitive streak begins to retreat posteriorly, mesodermal precursors continue to ingress through the node and migrate anteriorly to form the notochord. Paraxial mesodermal precursors ingress through the anterior streak between HH4 and HH7 and travel a path first away from the streak and then toward the prechordal plate and head structures. Lateral mesoderm ingresses through the posterior streak between HH8 and HH10 and migrates away from the streak. (C) Mesodermal gastrulation in mouse. The anterior-posterior and proximal-distal axes are marked. *brachyury* is expressed along the primitive streak. Ingression into the streak is marked by the most proximal pair of arrows. Mesodermal precursors that ingress through the proximal streak migrate anteriorly away from the streak (marked by an arrow) and form lateral mesoderm. Precursors that ingress through the distal streak migrate anteriorly, but not as far as the lateral mesoderm, and form paraxial mesoderm (migration marked by an arrow). Axial mesoderm arises from cells that ingress through the node and migrate anteriorly.

In contrast to the circular margins of *Xenopus* and zebrafish, the primitive streak of birds and mammals resembles a slit and is built in a progressive manner. Primitive streak ingression in chick begins as a thickening of cells in the posterior blastodisc. The streak then extends anteriorly through convergence and extension. Mesodermal precursors enter the streak in an anterior-to-posterior manner and, upon internalization, travel to their final location via migration (Gilbert, 2000) (Fig. 1B). All mesodermal precursors that enter the streak express *brachyury* (Kispert et al., 1995). The most anteriorly positioned mesodermal precursors, the axial mesoderm population, enter Hensen's node and migrate anteriorly to become the head mesoderm and notochord. The next group of mesodermal precursors then enters the anterior streak to become paraxial mesoderm. Upon internalization, these cells are first directed away from the embryonic midline during migration and then anteriorly toward the head process. The final group of mesodermal precursors then enters the posterior streak. Upon internalization, these cells migrate posteriorly and away from the midline to form the lateral mesoderm (Gilbert, 2000).

In mouse, primitive streak ingression is initiated in the proximal posterior (Fig. 1C, top-right arrows), and the streak is extended throughs progressive induction of epithelial to mesenchymal transitions (EMT) along the proximal-distal axis (Williams, Burdsal, Periasamy, Lewandoski, & Sutherland, 2012). Upon ingression, mesodermal precursors migrate anteriorly from the midline to form paraxial and lateral mesoderm. Lateral mesoderm arises from the proximal streak and extends further anteriorly than the paraxial mesoderm, which arises from the distal streak and remains closer to the midline. Mesodermal precursors that remain at the midline form the axial mesoderm (Takada et al., 1994). All the mesodermal precursors express *brachyury* (Kispert & Herrmann, 1994).

Thus, while the specifics surrounding mesoderm induction and internalization differ among vertebrates, features like *brachyury* expression, internalization of the mesoderm through an opening, and patterning of the various mesodermal subtypes along that opening are ubiquitous.

3. FGF expression in the mesoderm: An upstream cue for Erk-dependent gene induction

The mechanisms by which *brachyury* expression is instantiated in the mesoderm are also well-conserved among vertebrates. At the time the embryo is being patterned for gastrulation, mesodermal precursors express receptor tyrosine kinases (RTKs), including members of the FGFR family. RTKs function by binding to extracellular ligands, which induce a conformational change that allows for phosphorylation of the RTK kinase domains and C-terminal tails, leading to recruitment of adaptor proteins and activation of multiple intracellular signaling pathways, including Erk. Depending on the type of RTK, the ligand to which the RTK binds, and which intracellular components are present to mediate adaptor recruitment, a wide range of different intracellular signaling pathways can be activated (Brewer, Mazot, & Soriano, 2016).

One RTK expressed by the mesodermal precursors of every model vertebrate we discuss here is FGFR1 (Golub et al., 2000; Lunn et al., 2007; Ota et al., 2010; Yamaguchi, Conlon, & Rossant, 1992). In addition to being present at the right place and right time to be involved in mesoderm induction and morphogenesis, inhibition or removal of FGFR downregulates *brachyury* expression in *Xenopus*, zebrafish, and chick (Chung et al., 2004; Hardy et al., 2011; Rodaway et al., 1999) and early mesodermal cell movements are disrupted in all four model vertebrates (Amaya et al., 1993; Griffin et al., 1995; Yamaguchi, Harpal, Henkemeyer, & Rossant, 1994; Yang, Dormann, Münsterberg, & Weijer, 2002).

Nevertheless, it is important to note that FGFR-to-Erk signaling is not one-to-one, but many-to-many. FGFRs activate a variety of canonical targets, including the Erk, PI3K, PLCγ, STAT, and Jnk pathways, and also participate in non-canonical interactions with cell-adhesion molecules, like cadherin and integrins (Brewer et al., 2016; Clark & Soriano, 2022). Conversely, there are many roads to Erk, including other RTKs.

In the lower vertebrates *Xenopus* and zebrafish, FGFR activation triggers Erk signaling at the margin (Christen & Slack, 1999; van Boxtel et al., 2018) and inhibiting Erk signaling via small molecule inhibitors or morpholinos disrupts both *brachyury* expression and mesodermal cell movements (Krens, Corredor-Adámez, et al., 2008; LaBonne & Whitman, 1994). Erk appears to control *brachyury* via phosphorylation of its canonical targets, Ets family transcription factors (Kawachi, Masuyama, & Nishida, 2003; Kjolby, Truchado-Garcia, Iruvanti, & Harland, 2019; McFaul, Hart, & Draper, 2020; Znosko et al., 2010). A similar situation is also found in the chick embryo, where FGFR activity induces *brachyury* through Erk in mesodermal precursors as they enter the primitive streak, as well as in mesodermal cells as they exit the primitive streak to migrate anteriorly (Lunn et al., 2007). Inhibiting Erk signaling during primitive streak formation disrupts *brachyury* expression and the expression of other genes required for proper cell movement (Hardy et al., 2011).

In mouse, the relationship between FGFRs, Erk pathway activation, and mesodermal development is less straightforward. Once again, the Erk pathway is transiently activated in cells entering the primitive streak and is activated a second time in mesodermal precursors as they migrate away from the streak to position themselves in the embryo (Morgani et al., 2018). Knockdown of Erk2, the predominant Erk isoform in the early embryo, also prevents both *brachyury* expression and mesoderm formation in the mouse, although Erk1 is able to compensate during *in vitro* mesoderm differentiation (Yao et al., 2003). However, some *brachyury* induction and initial mesoderm formation persists in the absence of a functional FGFR1, and a combination of FGFR1/2 variants that lack signal transduction to Erk are still able to produce *brachyury*-expressing mesoderm (Ray et al., 2020; Yamaguchi et al., 1994), suggesting that other upstream processes are involved in Erk pathway activation during mesoderm induction and primitive streak formation. In higher vertebrates, it has yet to be shown whether Erk signaling induces *brachyury* expression through phosphorylation of Ets family transcription factors as in lower vertebrates, although Ets family transcription factors are present in the proper location (the pre-mesodermal epiblast) for this to be the case (Chotteau-Lelievre et al., 2001; Lunn et al., 2007).

For an RTK to initiate signaling it must typically be stimulated by a ligand, and FGFRs are no exception to this rule. Vertebrate genomes tend to contain many different FGFs: the mammalian genome contains 22 and the zebrafish genome contains 31 (Ornitz & Itoh, 2001). Similarly, in *Xenopus*, at least 13 FGF ligands are expressed by stage 40 (Lea, Papalopulu, Amaya, & Dorey, 2009). However, only a few FGFs are expressed during mesodermal specification, and even fewer FGFs appear to be required. In mouse, for example, FGF3, FGF4, FGF5, and FGF8, and FGF17 are all expressed in the primitive streak at the time of gastrulation, but only FGF8 and FGF4 are required for embryonic development to proceed normally (Ciruna & Rossant, 2001; Maruoka et al., 1998; Sun, Meyers, Lewandoski, & Martin, 1999).

As touched on briefly above, mesodermal precursors in lower vertebrates experience a single pulse of Erk signaling as they pass through the margin, which induces *brachyury* in these cells and informs their convergence and extension movements. Conversely, many mesodermal precursors in higher vertebrates experience two pulses of Erk signaling: one as they enter the primitive streak and a second as they migrate away from the streak. These signaling events are triggered by FGFR activation in response to the spatiotemporally controlled release of FGF ligands. Interestingly, patterns of FGF expression and mechanisms for controlling the duration of FGF release differ among organisms, especially between lower and higher vertebrates.

Lower vertebrates with blastopores, like *Xenopus* and zebrafish, express FGF around the blastopore during mesodermal gastrulation, closely matching the domain of *brachyury* expression and mesoderm induction. In *Xenopus*, FGF3, FGF4, and FGF8 are expressed in a broad strip around the margin (Christen & Slack, 1997; Isaacs et al., 1995; Lombardo et al., 1998) (Fig. 2A). It is unclear exactly how these three FGFs are initially induced, as a complex cocktail of maternally deposited factors and signaling from the vegetal sphere are required for full FGF expression (Fletcher & Harland, 2008; Vonica & Gumbiner, 2002). The way in which FGF

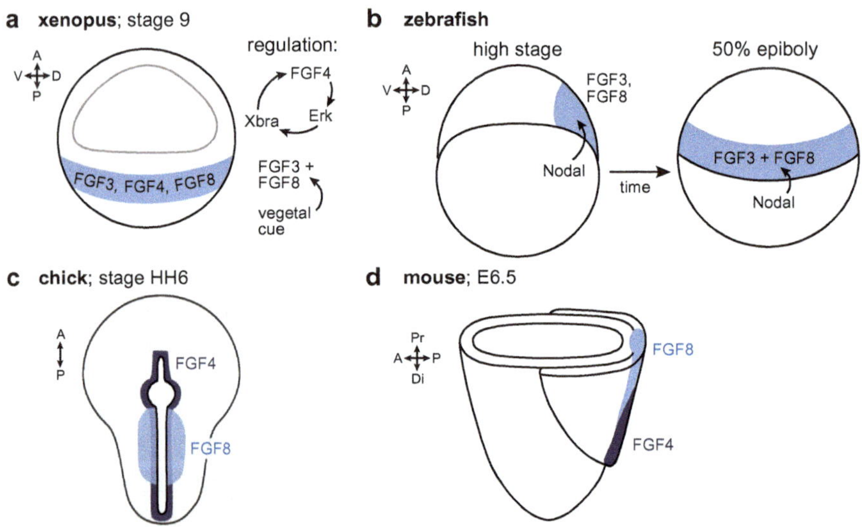

Fig. 2 FGF expression for mesoderm specification in vertebrates. (A) FGF expression in *Xenopus*. FGF3, FGF4, and FGF8 are expressed in similar patterns at the *Xenopus* margin, marked in light blue. FGF4 induces *Xbra* via Erk activation, and FGF4 expression is sustained by a positive feedback loop with *Xbra*. Meanwhile, FGF8 is sustained by signaling from the vegetal embryo. (B) FGF expression in zebrafish. FGF3 and FGF8 are expressed in similar patterns, first in the dorsal embryo and then at the margin in zebrafish downstream of Nodal signaling. (C) FGF expression in chick. FGF8 (light blue) and FGF4 (dark blue) are expressed in distinct patterns, with FGF8 expressed in the medial primitive streak and FGF4 expressed extending more anteriorly and more posteriorly along the streak. (D) FGF expression in mouse. FGF8 and FGF4 are expressed in distinct patterns, with FGF8 expressed in a proximal-to-distal gradient along the primitive streak and FGF4 in a distal-to-proximal gradient.

expression is sustained is better understood. FGF4 is maintained through a positive feedback loop with *Xbra*: maternally deposited FGF4 helps induce *Xbra*, which induces the zygotic transcription of FGF4. The zygotically induced FGF4 then induces more *Xbra* (Fletcher & Harland, 2008; Isaacs et al., 1994). Despite displaying a similar expression pattern to that of FGF4, FGF8 is sustained independently of prior FGFR activation (Fletcher & Harland, 2008), which suggests that FGF8 is not involved in a positive feedback loop with *Xbra*. Instead, FGF8 appears to be maintained by a separate signal, likely emitted by the vegetal hemisphere (Fig. 2A).

While similar in broad strokes, FGF expression in zebrafish differs in some important ways from that of *Xenopus*. First, FGF expression in the zebrafish embryo is initially asymmetric along the dorsoventral axis (Reifers et al., 1998): FGF is initially expressed only in dorsal regions of the embryo, and then its expression spreads to encompass the rest of the margin (Fürthauer, Reifers, Brand, Thisse, & Thisse, 2001) (Fig. 2B). This early dorsal-ventral asymmetry is important for specifying the axial mesoderm apart from the paraxial mesoderm, with high levels of FGF signaling at dorsal positions promoting axial mesodermal fates (Fürthauer, Van Celst, Thisse, & Thisse, 2004; Maegawa, Varga, & Weinberg, 2006). This is in marked contrast to *Xenopus*, in which FGF signaling promotes paraxial mesoderm specification at the expense of axial mesoderm (Fletcher & Harland, 2008).

A second difference involves the way in which FGF signaling is induced and maintained. In zebrafish, no positive feedback loop exists between FGF and expression of the *brachyury* homolog *ntl*, as exists in *Xenopus* (Lolas, Valenzuela, Tjian, & Liu, 2014). Instead, an orthogonal Nodal signal induces and sustains FGF3 and FGF8 at the margin (Mathieu et al., 2004; Rodaway et al., 1999; van Boxtel et al., 2018) (Fig. 2B). Thus, while FGF expression in *Xenopus* takes the cell's dynamic history of FGFR activation into account through positive feedback between FGF4 and *Xbra*, FGF expression in zebrafish does not, relying instead upon the continued supply of Nodal signaling. It is unknown whether these distinct modes of FGF maintenance are functionally equivalent or confer different downstream signaling properties or functions.

In both *Xenopus* and zebrafish, the different FGF subtypes are expressed in overlapping domains. This is not the case in vertebrates with primitive streaks, like chick and mouse. Both FGF8 and FGF4 are required for proper mesodermal development in these species, and in their absence, mesodermal precursors accumulate in the streak (Hardy et al., 2011; Sun et al., 1999).

In both chick and mouse, FGF4 is dependent upon FGF8 expression, so removing FGF8 in either organism is sufficient to prevent FGF4 expression as well (Hardy et al., 2011; Sun et al., 1999). Yet despite being dependent upon FGF8, FGF4 and FGF8 are expressed in different patterns. In chick, FGF8 is only expressed in the middle streak, while FGF4 is expressed in the middle streak, posterior streak, and anterior streak, including Hensen's node and the head process (Karabagli, Karabagli, Ladher, & Schoenwolf, 2002) (Fig. 2C). In mouse, FGF8 is expressed in a proximal-to-distal gradient along the streak, while FGF4 is expressed in a mirroring distal-to-proximal gradient (Sun et al., 1999) (Fig. 2D). It is still unclear how these non-overlapping domains of FGF4/8 expression are achieved. Two immediate models may be envisioned: a cell-migration model, where FGF4-expressing cells originate from the FGF8-expressing region but have migrated away from that region by the time FGF4 expression is in full swing; or a cell-communication model, where FGF8 expression drives FGF4 expression in a neighboring cell population through diffusion of a signal, perhaps FGF8 itself.

Although higher vertebrates share many common features with lower vertebrates, like FGF induction of *brachyury* in mesodermal precursors, one notable difference is that higher vertebrates express different FGF ligands in distinct, non-overlapping domains. An intriguing hypothesis is that FGF plays different morphogenetic roles in the two vertebrate classes and that these different roles require different ligand expression patterns. While lateral and paraxial mesodermal precursors employ cell intercalation to position themselves in *Xenopus* and zebrafish embryos, directed cell migration is used in chick and mouse (Solnica-Krezel & Sepich, 2012; Sun et al., 1999; Yang et al., 2002). It may be that distinct spatial patterns of FGF8 and FGF4 are required to direct migration, while overlapping FGF patterns are sufficient to program cell intercalation. In addition to evolutionary questions regarding how FGF expression patterns have evolved across species, questions remain regarding whether various FGF subtypes function redundantly. Can mesodermal precursors in chick or mouse still migrate properly if the expression patterns of FGF8 and FGF4 are switched but all else remains the same? We will further explore these questions in later sections of the review.

4. Generation and interpretation of transient Erk signaling in vertebrate mesoderm

We have already seen that mesodermal precursors are transiently localized to the margin or primitive streak before they pass through to a new layer to embark on their journey across the embryo. Prior to passing through the margin/streak, they are transiently exposed to FGF. Once they have passed through and left the margin/streak, this exposure ceases, implying that Erk may only be activated transiently in mesodermal precursors because of their highly mobile nature. This kind of transient Erk signaling during transit between tissues can be inferred from fixed staining of the chick primitive streak, where Erk phosphorylation is present in the FGF-expressing pre-ingression epiblast but absent in the primitive streak itself (Lunn et al., 2007). Likewise, in the mouse primitive streak, Erk signaling is only present in cells actively undergoing EMT, just as they enter the streak (Morgani et al., 2018). Further supporting this view, expression of Erk negative regulators is high in mesodermal precursors across vertebrate species, suggesting that Erk activity decreases rapidly as cells leave regions of continued FGF stimulation, which raises the question of how expression of specific negative regulators alters the dynamics of Erk activity (Branney, Faas, Steane, Pownall, & Isaacs, 2009; Fürthauer et al., 2001; Gómez et al., 2005; Hanafusa, Matsumoto, & Nishida, 2009; Hardy et al., 2011; Lolas et al., 2014; Lunn et al., 2007; Molina, Watkins, & Tsang, 2007; Morley et al., 2009; Sivak et al., 2005; Tsang et al., 2004). In this section, we will explore the origin and role of such negative regulation.

Several classes of negative regulators have been implicated in FGF signal interpretation in the mesoderm. Dual specificity phosphatases (DUSPs), which directly dephosphorylate Erk (Fig. 3A), are one such class of negative regulators (Caunt & Keyse, 2013; Theodosiou & Ashworth, 2002). In *Xenopus*, zebrafish, and mouse, a cytoplasmic DUSP, DUSP6, is induced downstream of *brachyury* in embryonic regions where FGF signaling is active (Gómez et al., 2005; Lolas et al., 2014; Molina et al., 2007; Morley et al., 2009; Tsang et al., 2004). This DUSP appears to primarily affect the amplitude of active Erk signaling in mesodermal tissues (Li, Scott, Hatch, Tian, & Mansour, 2007; Maillet et al., 2008; Tsang et al., 2004; Umbhauer, Marshall, Mason, Old, & Smith, 1995), as opposed to eliminating Erk signaling altogether or modulating Erk signal timing.

In contrast to DUSP6, other DUSPs have been implicated in the complete attenuation of Erk signaling. While Ntl is inducing DUSP6 at the zebrafish margin, Nodal signaling from the yolk induces DUSP4, a nuclear DUSP, in a subset of marginal cells. The presence of DUSP4 prevents Erk signaling from reaching the threshold for mesodermal gene induction, conferring an endodermal fate to cells expressing *DUSP4* (van Boxtel et al., 2018) (Fig. 3B). It is unclear whether a difference in localization, timing, or expression level is what confers a different fate to cells expressing *DUSP6* as

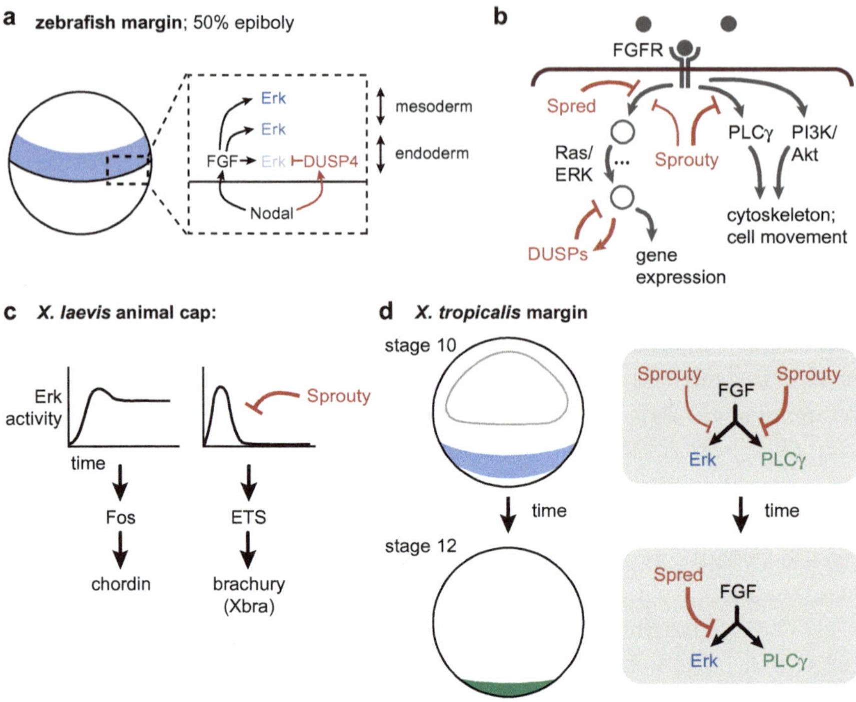

Fig. 3 Feedback loops on FGFR signaling regulate mesoderm specification. (A) DUSP4 downregulation of Erk for endoderm specification. In zebrafish, Nodal signaling from the yolk induces FGF and DUSP4 expression at the margin. While FGF as a ligand can spread significantly beyond where it is produced at the margin, DUSP4 only negatively regulates Erk signaling where DUSP4 is expressed. Different shades of blue represent the resulting dorsal-to-ventral Erk gradient at the margin. Endoderm is specified cell autonomously where DUSP4 is expressed (low Erk region), while mesoderm (high Erk region) is specified in a non cell-autonomous manner by FGF released from endoderm precursors. (B) Induction of Erk, PI3K, and PLCγ signaling by FGFR activation. FGFR is activated by FGF. Erk phosphorylation is initiated downstream of FGFR activation through the Ras/Erk signaling cascade. Phosphorylated Erk induces gene expression of pro-mesodermal genes and DUSPs, negative regulators that de-phosphorylate Erk. Ras/Erk signaling can be inhibited further upstream by negative regulators like Sprouty and Spred. Sprouty also inhibits PLCγ signaling. PI3K and PLCγ signaling promote cytoskeleton rearrangement and cell movement. The different widths of the Sprouty and Spred negative regulation arrows indicate that they inhibit different pathways to different extents. (C) Injecting *sprouty* into the *Xenopus* animal cap makes Erk signaling transient and prevents expression of genes like chordin. The *Xbra* enhancer contains Ets binding sites, and Erk induces *Xbra* expression via Ets phosphorylation. Meanwhile, the chordin enhancer contains Fos binding sites. FOS acts similarly in *Xenopus* as in mammalian cells, acting as an Erk duration sensor. (D) A switch from Erk to PLCγ signaling downstream of Sprouty and Spred in *X. tropicalis*. Sprouty expression blocks PLCγ signaling during gastrulation, causing FGF signaling to act through the Erk (blue) pathway. Around stage 12, Sprouty expression ceases and Spred is induced, blocking Erk signaling downstream of FGFR activation and allowing PLCγ (green) signaling to proceed.

opposed to those expressing *DUSP4*. Likewise, it has proven difficult to interpret what role individual DUSPs play in mesodermal development, as it is common for multiple DUSPs to be expressed in a cellular population at the same time. In *Xenopus*, for example, expression of *DUSP6* at the margin is accompanied by expression of the nuclear DUSPs *DUSP1* and *DUSP5* (Branney et al., 2009).

A second class of negative regulators, Sprouty-family proteins, function at the FGF receptor level to alter the balance of downstream signaling pathways, including the Ras/Erk pathway (Fig. 3A). While DUSPs have been implicated in controlling the level of Erk activity in cells, members of the Sprouty-family have been implicated in controlling Erk signal timing. The protein *Xsprouty2* is present in *Xenopus* and is itself induced by FGF and Erk signaling, forming a negative feedback loop (Nutt, Dingwell, Holt, & Amaya, 2001). When *Xenopus* animal caps are continuously exposed to FGF ligand, a transient Erk signal is produced. Injecting *Xsprouty2* makes the Erk signal more transient, while injecting a dominant negative version of *Xsprouty2* converts the Erk signal into a sustained one. Taken together, these results suggest that *Xsprouty2* helps dictate the duration of the native Erk signal (Hanafusa et al., 2009) (Fig. 3C).

Differences in Erk signaling dynamics may also contribute to differences in gene expression observed between different regions in the embryo. Active Erk is present in a dorsal-to-ventral gradient at the *Xenopus* margin, with sustained Erk signaling at the dorsal margin and more transient signaling at the ventral margin (Branney et al., 2009; Christen & Slack, 1999; Hanafusa et al., 2009). While *Xbra* is expressed throughout the margin, the gene *chordin* is only expressed dorsally. However, injecting a morpholino against *Xsprouty2* into the ventral margin induces *chordin* ventrally, suggesting that long duration Erk signaling is required for *chordin* induction, in contrast to *Xbra*, which can be induced by long or short duration Erk signals (Hanafusa et al., 2009). This difference in Erk-duration requirement can be explained by differences between the transcription factors that activate these genes. *Xbra* is induced by Ets-family transcription factors, while *chordin* is induced by XFos, a transcription factor whose mammalian homolog c-Fos is known to be stabilized through post-translational modification downstream of sustained Erk signaling (Hanafusa et al., 2009; Kawachi et al., 2003; Murphy & Blenis, 2006; Murphy, Smith, Chen, Fingar, & Blenis, 2002). This suggests a model where only long-duration Erk signaling is sufficient to both induce and stabilize XFos, leading to *chordin* induction.

Another example of Sprouty-family regulation of Erk signal timing is found in *X. tropicalis*. Sproutys are expressed at the *X. tropicalis* margin beginning in stage 10 during epiboly. At stage 12, however, *sprouty* expression is attenuated just as the expression of *spred*, another Sprouty-family member, begins (Sivak et al., 2005). As discussed above, Sprouty is a negative regulator of Erk signaling (Hanafusa et al., 2009). Spred, however, is a stronger repressor of Erk signaling than Sprouty (Sivak et al., 2005). Thus, the switch from *sprouty* so *spred* expression downregulates Erk signaling. Meanwhile, Sproutys are more effective than Spreds at negatively regulating PLCγ signaling. Thus, the switch from *sprouty* to *spred* expression marks a change in FGF signal interpretation; prior to the switch, FGFR activation preferentially triggers Erk signaling at the margin, but after the switch, FGFR activation preferentially triggers PLCγ signaling (Sivak et al., 2005) (Fig. 3D).

Sproutys and Spreds are expressed during mesodermal development in other species as well. In the chick preingression epiblast, Sproutys and Spreds are expressed downstream of FGF/Erk signaling in the primitive streak (Hardy et al., 2011), and in zebrafish, Sproutys are expressed downstream of FGF signaling at the margin (Fürthauer et al., 2001). It is currently unclear, however, whether Sproutys and Spreds in these scenarios are performing the same function as in *X. tropicalis*—regulating Erk signal timing and controlling the ratio of Erk to PLCγ signaling—or a different function entirely.

As we have just seen, Erk signaling can be made transient by negative regulation that targets the Ras/Erk pathway despite the presence of sustained FGFR activation upstream. Regulation of FGFR expression, however, is also commonly employed by mesodermal precursors in modulating the duration of Erk activation. In chick, DUSP6 is conspicuously absent from the primitive streak at HH3, but FGFR1 expression is actively regulated at this time; FGFR1 is not expressed in the primitive streak but is present in the pre-mesodermal epiblast. Erk signaling follows this pattern of FGFR1 expression, with active Erk present in the epiblast layer of the primitive streak but not in the emerging mesoderm (Lunn et al., 2007). Thus, it appears that controlled FGFR1 expression can dictate Erk signaling dynamics without the need for downstream negative regulation. In fact, tight spatiotemporal control of FGFR expression is common among gastrulation-stage embryos, with FGFR regulation present in *Xenopus*, chick, and mouse gastrulas as well as in the *Drosophila* caudal visceral mesoderm (Golub et al., 2000; Lunn et al., 2007; Mandal et al., 2004; Ota et al., 2010; San Martin & Bate, 2001; Yamaguchi et al., 1992).

To recap, Erk signaling dynamics during early mesoderm development are dictated by upstream patterns of FGF and FGFR expression, as well as through the induction of negative regulators. Negative regulation via DUSPs can modulate the level of Erk signaling or suppress signaling altogether. Other forms of negative regulation, like Sproutys and Spreds, can switch the mode through which FGF signaling operates, allowing for the separation of Erk signaling from other FGF-dependent signals in time. Each vertebrate species discussed here exhibits a unique combination of regulatory features during mesoderm induction, suggesting that regulatory features can be mixed and matched through evolution to elicit different cellular behaviors. In the sections that follow, we will explore these distinct cellular behaviors in greater depth.

5. The role of Erk signaling in initiating and sustaining *brachyury* expression

One recurring theme in mesoderm specification is the expression of a critical transcription factor, *brachyury*. In the preceding sections we presented evidence that Erk triggers *brachyury* expression, but that Erk activity is only transiently encountered prior to mesoderm precursors leaving the margin or primitive streak. These observations present interesting questions for *brachyury* dynamics. Is *brachyury* expression transient or sustained in mesodermal precursor cells, and how might it be maintained after Erk signaling is no longer present? In this section, we discuss the various Erk-dependent and Erk-independent feedback loops that have evolved for controlling the duration of *brachyury* expression and speculate on distinct functions that different durations of *brachyury* expression may play in mesodermal development.

In zebrafish, mesodermal *brachury* expression is sustained through a series of positive feedback loops. The *brachyury* homolog *no tail* (*ntl*) is expressed first in the dorsal epiblast and then at the margin (Schulte-Merker & Smith, 1995), closely tracking the timing and location of FGF signaling in Fig. 2B. Although Erk signaling induces *ntl* in both regions (Krens, Corredor-Adámez, et al., 2008), different mechanisms are responsible for sustaining it. At the margin, Ntl participates in a feedback loop with various Wnts. Ntl induces expression of Wnt3a and Wnt8 (Fig. 4A), with Wnt8 induced during early gastrulation at the time when the margin is giving rise to the anterior paraxial mesoderm, and Wnt3a induced during late gastrulation at the time when the tailbud, which will give rise to the posterior

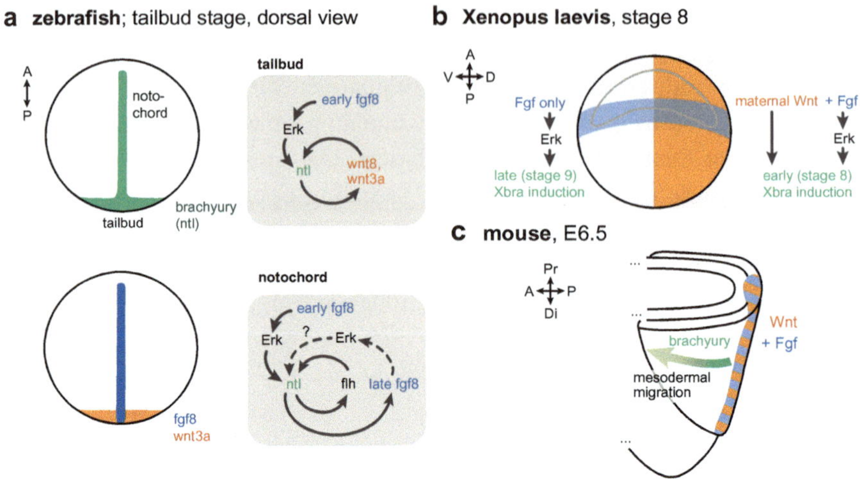

Fig. 4 Crosstalk between Erk and Wnt signaling regulates brachyury expression. (A) Mechanisms for *brachyury* induction and maintenance. Positive feedback loops maintain *ntl* (green) expression in the zebrafish notochord and future tailbud. In both the notochord and future tailbud region, FGF8 initially induces *ntl* via Erk activation. In the future tailbud region, *ntl* then induces *wnt8* during early gastrulation and *wnt3a* during late gastrulation (orange). Both Wnts maintain *ntl* expression through positive feedback loops. In the notochord, *ntl* induces the notochord-specific gene *flh*, maintaining *ntl* expression in a positive feedback loop. *ntl* also induces FGF8 (blue) in the notochord, which may help sustain *ntl* expression. (B) *Xbra* induced asynchronously in the dorsal vs. ventral *Xenopus* embryo due to Wnt signaling differences. While FGF (blue) is sufficient to induce *Xbra* (green) via Erk signaling in *Xenopus*, *Xbra* can also be induced via canonical Wnt signaling. The presence of maternal Wnt ligands (orange) in the dorsal embryo results in faster *Xbra* induction there (early, stage 8) than in the ventral embryo (late, stage 9). (C) Induction and maintenance of a posterior-to-anterior *brachyury* gradient in mouse. FGF (blue) and Wnt (orange) work synchronously to induce *brachyury* (green) expression. *brachyury* levels drop in mesodermal precursors as they migrate anteriorly away from the streak.

paraxial mesoderm, is beginning to form. Both ligands activate the canonical Wnt/β-catenin signaling pathway (Martin & Kimelman, 2008). In addition to the Ets sites discussed previously, the *ntl* enhancer possesses sites for Tcf, a transcription factor that is activated by Wnt/β-catenin signaling (van Noort & Clevers, 2002). Thus, through Tcf activation, Ntl-induced Wnt signaling goes on to induce more Ntl at the margin as paraxial mesoderm precursors ingress through it (Martin & Kimelman, 2008).

By contrast, *wnt3a* and *wnt8* expression is excluded from the zebrafish dorsal epiblast, which goes on to form axial mesoderm and then the notochord (Fig. 4A). Instead, *ntl* in the future axial mesoderm is induced by Erk signaling downstream of FGF and then maintained through a positive feedback loop with a notochord-specific transcription factor *floating head* (*flh*). It is noteworthy that FGF8 expression is itself a target of *ntl* in the notochord (Martin & Kimelman, 2008). It is thus tempting to speculate that this FGF8 might itself trigger additional *ntl* expression, similar to earlier embryonic stages and reminiscent of the Xbra/FGF4 feedback loop found in *Xenopus*. A Ntl/FGF8 feedback loop may thus exist that, alongside Flh, helps maintain notochordal *ntl* expression (Martin & Kimelman, 2008).

While Wnt signaling does not appear to play a role in sustaining *Xbra* expression in *Xenopus*, it does help control the timing of *Xbra* induction along the dorsal-ventral axis (Fig. 4B). As discussed previously, *Xbra* is sustained in *Xenopus* through a positive feedback loop involving *FGF4* (Isaacs et al., 1994). However, *Xbra* also possesses Tcf binding sites that make it sensitive to Wnt signaling. While Wnt signaling is not required for induction of *Xbra*, the presence of maternal Wnt signaling on the dorsal side of the embryo speeds up induction of *Xbra* there relative to the ventral embryo, resulting in early gastrulation of the dorsal margin relative to the ventral margin (Vonica & Gumbiner, 2002).

In contrast to the sustained *brachyury* expression observed in the vertebrate notochord and tailbud, *brachyury* expression in the lateral and trunk paraxial is transient, decaying with distance from the primitive streak. Like the *Xenopus* and zebrafish *brachyury* homologs, mouse *brachyury* possesses Tcf binding sites. Wnt signaling synergizes with FGF signaling in mouse to initiate *brachyury* expression (Ciruna & Rossant, 2001; Yamaguchi, Takada, Yoshikawa, Wu, & McMahon, 1999) (Fig. 4C), with *brachyury* expression being highest in the lateral mesoderm precursors, which express both *FGF8* and *wnt3a* (Yamaguchi et al., 1999). *wnt3a* is induced downstream of BMP signaling from the extraembryonic ectoderm and is locally enhanced by FGFR1 activation (Ben-Haim et al., 2006; Ciruna & Rossant, 2001). As lateral mesoderm cells migrate away from the streak, their *brachyury* levels decrease (Morgani et al., 2018), presumably because they are moving away from the source of FGF and Wnt ligands (Fig. 4C). Conversely, axial mesoderm develops under the continued influence of Wnt signaling, which maintains *brachyury* expression in the notochord (Yamaguchi et al., 1999). *wnt3a* maintains its own expression in the primitive streak through a self-sustaining positive feedback loop (Tortelote et al., 2013). While the role of Wnt signaling in *brachyury* maintenance is less clear in chick, superficially, the situation appears to be similar to that in mouse. *brachyury* in the migrating paraxial and lateral mesoderm is dependent on FGF signaling and attenuates once mesodermal precursors ingress, concomitant with the attenuation of FGFR expression in these cells (Hardy et al., 2011; Kispert et al., 1995; Lunn et al., 2007). In the notochord and Hensen's node, however, *brachyury* expression is sustained independently of FGF signaling (Hardy et al., 2011; Kispert et al., 1995).

To summarize, among vertebrates, *brachyury* is transiently induced in cells of the lateral and trunk paraxial mesoderm but sustained in cells of the axial mesoderm and future tailbud. While Erk signaling is ubiquitously involved in initiating *brachyury* expression among mesodermal subtypes, it is common for *brachyury* regulation to be handed off to other signaling pathways, particularly the Wnt pathway, in regions where *brachyury* expression will become sustained. Whether *brachyury* serves a different function when it is induced transiently as opposed to when it is expressed in a sustained manner remains an open question.

6. A role for Erk and a *brachyury* homology in *Drosophila* mesoderm specification

Widening the lens beyond vertebrates reveals dramatic differences in how signaling pathways control mesoderm specification. The *Drosophila* embryo presents a fascinating variation on our theme of spatial FGF patterns, Erk signaling, *brachyury* induction, and mesoderm specification that helps shed light on some shared principles. We will focus here on the signaling and morphogenetic mechanisms employed specifically by a mesodermal precursor population termed the caudal visceral mesoderm (CVM), a tissue that employs FGF signaling to embark on the longest migration in all of *Drosophila* embryogenesis, and the only mesodermal tissue subtype that the Drosophila *brachyury* homolog helps specify.

In *Drosophila*, mesoderm specification and invagination are triggered by the NF-κB homolog Dorsal (Dl) along the ventral surface of the embryo. This region will give rise to several mesodermal populations, including the trunk visceral mesoderm (TVM). Most *Drosophila* mesodermal cells do not express *brachyenteron* (*byn*), the *Drosophila* homolog of *brachyury* (Fig. 5A). Rather, mirroring the vertebrate scenario, *byn* expression is under control of the Erk pathway. At the stage at which the mesoderm is specified, Erk is active at the posterior pole, and *byn* expression takes the form of a ring in a subregion of the pole where Erk is active at a moderate level (Fig. 5A). Thus, while most *byn*-expressing cells do not form mesoderm, the ventral-most population of *byn*-expressing cells receive both a

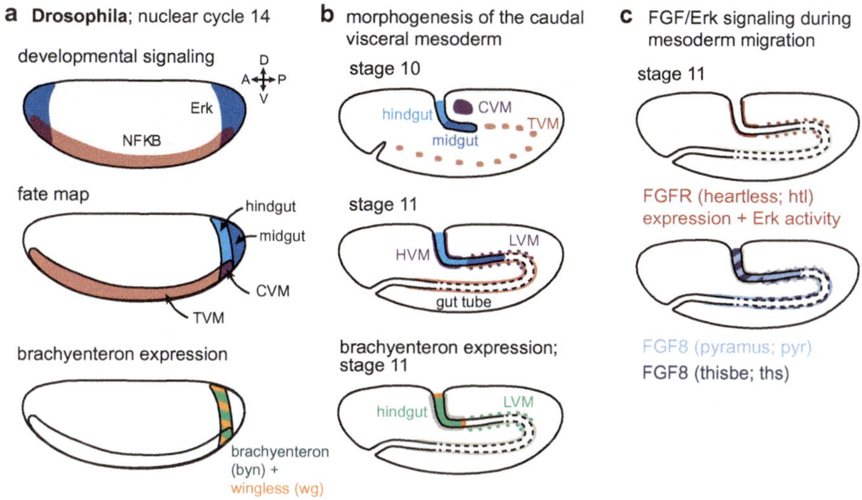

Fig. 5 Making invertebrate mesoderm: recurrent roles for *byn* and FGFR signaling. (A) NF-κB (pink) and Erk (blue) signaling specify the CVM in *Drosophila*. TVM (pink) is specified in the ventral embryo, where only NF-κB signaling is present. Hindgut (light blue) and midgut (dark blue) are specified in the posterior in regions where only Erk signaling is present. CVM (purple) is induced in the ventral posterior where Erk and NF-kB signaling intersect. *byn* (green) and *wg* (orange) are expressed in both the CVM and hindgut where Erk signaling is low, but not in the midgut where Erk signaling is high. (B) The CVM splits into HVM and LVM. The TVM (pink blobs at stage 10, pink midgut casing at stage 11) and CVM (purple blob) are internalized via ventral furrow invagination. The hindgut (light blue) and midgut (dark blue) are internalized via posterior midgut invagination (stage 10). Germband extension positions the CVM adjacent to the hindgut and midgut, and positions the TVM such that it marks where the gut tube will form (stage 10). At stage 11, the midgut uses the TVM as a guide for gut tube formation. Dotted lines represent regions where the gut tube has yet to form. The TVM forms circular fibers around the gut tube. The LVM (purple dots) migrates over the TVM, forming longitudinal fibers. The HVM (purple hindgut casing) forms circular fibers around the hindgut as the hindgut extends. *byn* (green) expression persists in the hindgut and LVM, but not in the HVM. *wg* (orange) persists at the anterior and posterior ends of the hindgut. (C) FGFR and FGF expression during HVM/LVM development. *htl* (red), an FGFR, is expressed strongly in the HVM and weakly in the LVM. Erk is active in the HVM and LVM where *htl* is expressed. FGFs *pyr* (light blue) and *ths* (dark blue) are expressed throughout the midgut, hindgut, and TVM. Each ligand is expressed in a distinct pattern. Stripes of *ths* and *pyr* represent regions where both ligands are expressed.

mesoderm-inducing NF-κB cue and posteriorizing Erk cue. These cells go on to form the CVM, a tissue that is internalized through ventral furrow invagination along with the rest of the mesoderm (Kusch & Reuter, 1999).

What is the fate of CVM cells? Around Stage 10, CVM cells begin migrating and differentiating into a tissue that gives rise to the circular muscles that encapsulate the *Drosophila* hindgut, termed the hindgut visceral mesoderm (HVM), as well as the longitudinal muscles that encapsulate the midgut, termed the longitudinal visceral mesoderm (LVM) (Frasch & Sink, 2006; San Martin & Bate, 2001) (Fig. 5B). Although it is unclear what cues the CVM to split into these distinct populations, *byn* in particular is expressed in the LVM at this stage but not in the HVM (Kusch & Reuter, 1999), suggesting that *byn* may serve as a marker for LVM specification.

In contrast to *byn*, FGF signaling is not involved in *Drosophila* mesoderm specification at all, but it does direct the migration of several mesodermal subpopulations, including the HVM and the LVM (Mandal et al., 2004; San Martin & Bate, 2001; Sun, Macabenta, Akos, & Stathopoulos, 2020). This makes *Drosophila* an excellent model system for exploring how FGF signaling dictates mesoderm morphogenesis independently of mesoderm induction. Expression of *heartless* (*htl*), the FGFR expressed in the CVM, coincides with the initiation of Erk activity and the beginning of migration in both the LVM and HVM (Mandal et al., 2004; San Martin & Bate, 2001). However, the mode of migration and level of FGFR expression varies between the two populations. In the HVM, migration is dependent on Connectin (Gould & White, 1992; Nose, Mahajan, & Goodman, 1992), a homophilic cell adhesion molecule that is induced downstream of *htl* and allows the HVM to spread over the hindgut. Conversely, cells in the LVM do not express Connectin and instead employ a different, yet-to-be-identified set of cell-cell adhesion mechanisms to travel in two, neat lines over the TVM via directed migration.

Cells in the HVM express a significantly higher level of *htl* than cells in the LVM (San Martin & Bate, 2001) (Fig. 5C). While the reason for this is still unclear, one possibility is that crosstalk between *byn* and *wingless* (*wg*), a *Drosophila* Wnt ligand, is responsible. *wg* is expressed downstream of Torso signaling in the posterior embryo in a similar ring pattern to that of *byn* (Hoch & Pankratz, 1996; Kispert, Herrmann, Leptin, & Reuter, 1994; Lengyel & Iwaki, 2002; Wu & Lengyel, 1998). *wg* continues to be expressed in the hindgut upon its invagination. Once the hindgut begins to extend, however, *wg* is restricted to the hindgut's anterior and posterior ends (Hoch & Pankratz, 1996; Lengyel & Iwaki, 2002; Wu & Lengyel, 1998). Wg released from the hindgut signals to the HVM, resulting in enhanced expression of *htl* through the induction of *twist* (*twi*) (San Martin & Bate, 2001), a gene associated with mesodermal development in both vertebrates and invertebrates (Castanon & Baylies, 2002).

In contrast to the HVM, the LVM does not express *twi*. Evidence from other organisms suggests this may be due to the LVM-specific presence of *byn*. In mouse, *twi* is not expressed in the notochord, where *brachyury* expression is sustained. *twi* is, however, induced in notochord-derived cells in *brachyury*-knockdown mice, which possess an abnormally loose and mesenchymal morphology (Zhu, Kwan, & Mackem, 2016). It is possible, then, that *brachyury* plays a conserved role in vertebrates and invertebrates, downregulating *twi* expression to prevent loose, mesenchymal morphologies. Unlike in vertebrates, however, where *wnt* and *brachyury* expression often enhance one another through positive feedback loops, the putative interaction between *byn* and *wg* in the *Drosophila* LVM is antagonistic, with *byn* expression appearing to protect cells from the downstream effects of Wnt signaling.

The mode through which FGF expression directs LVM migration resembles that found in avian and mammalian mesodermal migration (Fig. 2C-D). Just as FGF8 and FGF4 activate FGFR in chick and mouse, the FGF ligands Pyramus (Pyr) and Thisbe (Ths) activate Htl in *Drosophila* (Stathopoulos, Tam, Ronshaugen, Frasch, & Levine, 2004). The spatiotemporal expression pattern of each ligand is distinct (Fig. 5C). *pyr* is expressed in the TVM from stage 10 through stage 11 and in the midgut throughout LVM migration (Reim, Hollfelder, Ismat, & Frasch, 2012). Conversely, *ths* is expressed in the TVM from stage 10 through stage 13, with expression in stages 11 and 12 being strongest in the posterior 2/3 of the TVM (Kadam, Ghosh, & Stathopoulos, 2012; Reim et al., 2012). *ths* is never expressed in the midgut (Reim et al., 2012). In a later section, we will discuss potential mechanisms by which FGF may be guiding migration in both cells of the *Drosophila* LVM and chick mesodermal precursors.

In summary, although *byn* expression and FGF signaling are not generally required for mesodermal induction in *Drosophila*, both are required for proper CVM development. Unlike in vertebrates, however, where *brachyury* expression is dependent on FGF activation, *byn* is expressed independently of FGF signaling in *Drosophila*, relying on the activation of Torso, a different RTK, instead. Thus, the roles of *byn* and FGF have been "remixed" in the CVM relative to their roles in the vertebrate mesoderm, making the CVM a useful alternative model for studying *byn* and FGF function in mesodermal specification and morphogenesis. Intriguingly, although *byn* expression and FGF activation are decoupled from one another in *Drosophila*, *byn* is still expressed downstream of Erk signaling at the embryonic poles, and FGF activation still triggers Erk activity in the migrating LVM (Mandal et al., 2004; San Martin & Bate, 2001). It is not yet known whether the Erk signaling dynamics triggered by Torso differ from those triggered by FGFR. Furthermore, it is not yet known how Erk signaling dynamics translate to *byn* expression dynamics, or how *byn* expression dynamics translate to cell specification and morphogenesis.

7. Moving mesoderm: Erk-dependent control of convergence and extension

The final sections of this review will be concerned with the processes by which newly-specified vertebrate mesoderm travels to its final location. While the details of mesodermal precursor travel vary among vertebrates, two of the most common morphogenetic modes employed are (1) convergence and extension and (2) epithelial-to-mesenchymal transition followed by directed cell migration. As in the *Drosophila* CVM, Erk signaling plays a role in orchestrating these movements. In this section, we cover how Erk signaling helps coordinate CE through the regulation of protrusions and cell-cell adhesion.

In both *Xenopus* and zebrafish, convergence and extension (CE) is required for lateral and paraxial mesoderm precursors to reach their final embryonic positions (Solnica-Krezel & Sepich, 2012). In these cells, CE occurs through a crawling mode in which the migrating tissue behaves as a loosely associated mesenchyme (Fig. 6A). Cells initiate movement by inserting protrusions between neighboring cells and establishing weak cell-cell adhesions, which allows cells to crawl between their neighbors. Tissue-level movement then proceeds through cellular rearrangements (Shindo, 2018).

Erk signaling plays a role in regulating both cell protrusion and cell-cell adhesion. Protrusions in the paraxial mesoderm are stabilized through noncanonical Wnt signaling, which is stimulated by the binding of noncanonical Wnt ligands to receptors (Davey & Moens, 2017; Heisenberg et al., 2000; Shindo, 2018; Tada & Smith, 2000; Ulrich et al., 2003; Wallingford et al., 2000). One such ligand, Wnt11, is induced and released in *Xenopus* and zebrafish embryos at the margin downstream of FGFR and Erk signaling (Conlon & Smith, 1999; Ku & Melton, 1993; Makita, Mizuno, Koshida, Kuroiwa, & Takeda, 1998; Saka, Tada, & Smith, 2000; Tada & Smith, 2000). While canonical Wnt signaling proceeds through localization of β-catenin to the nucleus, noncanonical Wnt signaling can initiate signaling

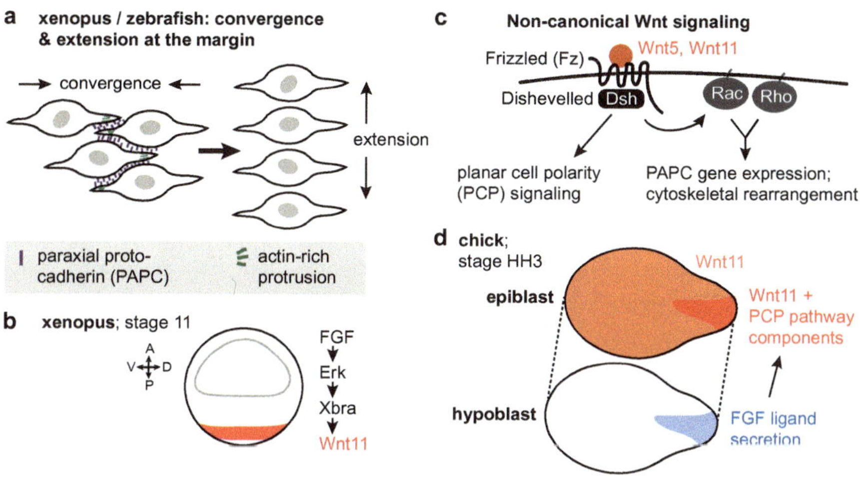

Fig. 6 Signaling control over cell rearrangements in convergent extension (CE) and primitive streak elongation. (A) Crawling processes coordinated by FGF signaling during CE in *Xenopus* and zebrafish. CE occur through a crawling mode in which cells insert actin-rich protrusions (green) between neighboring cells and establish new adhesions through PAPC (purple), a weak cell-cell adhesion molecule. The result of CE is convergence of cells along one axis and extension along an orthogonal axis. (B) Wnt11 expression (orange) downstream of FGF/Erk/Xbra at the *Xenopus* margin (stage 11). (C) Noncanonical Wnt signaling for cell rearrangement. Noncanonical Wnt ligands (orange), including Wnt5 and Wnt11, bind to Fz receptors. Noncanonical Wnt signaling through Dsh can trigger PCP signaling and activate Rho and Rac. Rho and Rac influence cytoskeletal rearrangement directly and indirectly through PAPC induction. (D) Primitive streak elongation through CE in chick. Localized noncanonical Wnt signaling is required for streak elongation. All cells of the epiblast express Wnt11 (light orange). FGF secreted by the hypoblast (blue) localizes downstream components of the Wnt/PCP pathway (dark orange) to the future streak.

through the planar cell polarity (PCP) pathway and through Rho and Rac activity downstream of Disheveled activation. Together, these pathways coordinate cytoskeletal rearrangements to generate protrusions (Shindo, 2018; Tada & Kai, 2009) (Fig. 6B).

In addition to regulating protrusions, Erk signaling also regulates cell-cell adhesion in the paraxial mesoderm through control of paraxial protocadherin (PAPC) expression. PAPC is a weakly adhesive molecule expressed in the paraxial mesoderm of *Xenopus* and zebrafish throughout gastrulation. It is required for paraxial mesoderm CE and is unique in its ability to promote cell movements while also providing cell-cell adhesion (Kim, Yamamoto, Bouwmeester, Agius, & Robertis, 1998; Yamamoto et al., 1998). This is likely due to PAPC's role in modulating Rho and JNK in a way that is non-redundant to Wnt/PCP signaling (Medina, Swain, Kuerner, & Steinbeisser, 2004; Unterseher et al., 2004). In zebrafish, PAPC expression is induced in an Erk-dependent manner downstream of the *ntl* target *spadetail* (Yamamoto et al., 1998). In *Xenopus*, noncanonical Wnt signaling induces PAPC, and FGF signaling induces a functional partner of PAPC, potentially through Erk signaling, which aids PAPC both in its adhesion and signaling roles (Chung, Yamamoto, & Ueno, 2007; Schambony & Wedlich, 2007).

Taken together, these observations indicate that Erk signaling helps coordinate CE through induction of Wnt/PCP signaling and control of PAPC activity (Fig. 6B). Although Wnt/PCP signaling and PAPC adhesion are conserved between *Xenopus* and zebrafish, it is interesting that the gene networks connecting Erk signaling to these processes differ, and that paraxial cell movement trajectories are distinct in the two organisms (Solnica-Krezel & Sepich, 2012), which raises the question: Does the timing of Wnt/PCP signaling relative to that of PAPC activation matter for dictating cell trajectories? Furthermore, does the transient Erk pulse that precedes these processes help to coordinate their relative timing?

Although CE is not used for mesodermal morphogenesis in chick and mouse, the chick embryo does employ CE in establishing its primitive streak (Fig. 6C). It is speculated that the chick primitive streak evolved from the ancestral blastopore through the acquisition of an additional cell-intercalation event (Chuai, Serrano-Najere, Serra, Mahadavan, & Weijer, 2021; Voiculescu, Bertocchini, Wolpert, Keller, & Stern, 2007). Like the CE movements initiated at the blastopores of *Xenopus* and zebrafish, lengthening of the primitive streak in chick depends upon FGF control of Wnt/PCP signaling through Wnt11. Unlike in *Xenopus* and zebrafish, *wnt11* is expressed globally throughout the chick epiblast. In order to localize PCP signaling to the future streak, it is thought that the hypoblast, which underlies the epiblast, releases FGF at positions corresponding to the future primitive streak. FGF signaling then induces expression of three PCP pathway components in the epiblast, allowing Wnt11 to locally activate PCP signaling in the region where the streak will form (Chuai et al., 2021; Voiculescu et al., 2007). It is likely that FGF induces these components via Erk signaling (Hardy et al., 2011).

In summary, CE is required for lateral and paraxial mesoderm morphogenesis in anamniotes and is regulated by Erk-dependent gene induction. Although different mechanisms are used for mesodermal gastrulation in amniotes than in anamniotes, CE and the mechanisms underlying it have been coopted by some amniotes for other morphogenetic events, like primitive streak extension. It remains to be seen whether the transient pulse of Erk signaling experienced by migratory mesodermal precursors in vertebrates coordinates the timing of protrusion and cell-cell adhesion events relative to one another. If this is the case, Erk coordination of mesodermal morphogenesis might prove a useful lens through which the evolution of mesodermal gastrulation in vertebrates can be examined.

8. Erk regulation of epithelial-to-mesenchymal transition in the primitive streak

In mouse and chick, mesodermal precursors that travel through the streak undergo an epithelial-to-mesenchymal transition (EMT) before migrating to their final embryonic positions (Nakaya & Sheng, 2009; Nakaya, Sukowati, Wu, & Sheng, 2008; Williams et al., 2012). EMT is a process through which cells organized as a closely adhered, two-dimensional sheet de-adhere from one another and lose their polarity to become migratory, loosely associated cells capable of navigating three-dimensional space. It is characterized by a change in cadherin expression, with mesodermal precursors expressing E-cadherin prior to transition and N-cadherin after (Gheldof & Berx, 2013). FGF signaling has been linked to both processes. In mouse, FGF signaling downregulates E-cadherin levels through the induction of Snail, a transcription factor that is thought to repress E-cadherin expression (Ciruna & Rossant, 2001). In chick, however, FGF signaling at the primitive streak affects neither Snail nor E-cadherin levels. Instead, FGF signaling initiates N-cadherin upregulation through induction of PDGFR at the streak (Hardy et al., 2011). In both cases, cadherin regulation is tied closely to the regulation of other mesoderm-related processes: Erk-dependent *brachyury* expression and canonical Wnt signaling in mouse and PI3K signaling in chick. In this section, we discuss how FGF-control of EMT can either be coupled to or decoupled from mesodermal induction through Erk-dependent mechanisms.

The primitive streak is formed through progressive EMT in mouse (Williams et al., 2012) (Fig. 7A). FGF signaling orchestrates this process through induction of *snail* (Ciruna & Rossant, 2001), an event that is Erk-dependent in the chick epiblast (Hardy et al., 2011) and many human cell lines, and is likely Erk-dependent in the mouse epiblast as well. FGF induction of *snail* subsequently downregulates E-cadherin in the mouse primitive streak (Ciruna & Rossant, 2001). A side effect of this E-cadherin

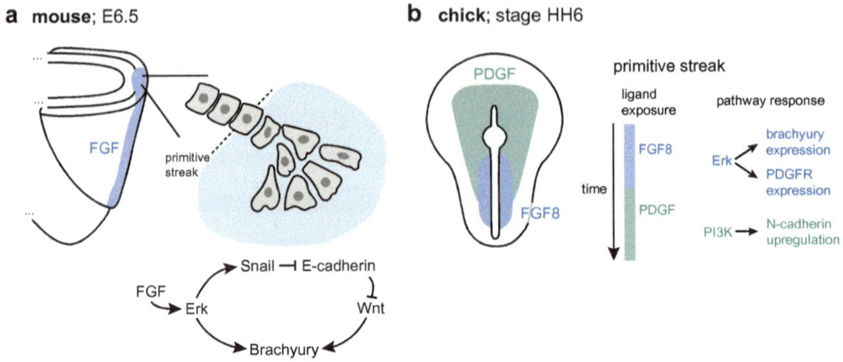

Fig. 7 FGF signaling for epithelial-to-mesenchymal transition (EMT). (A) FGF signaling downregulates E-cadherin in mouse mesodermal precursors, allowing for ingression via EMT through the primitive streak. FGF is expressed in the primitive streak (blue). Downregulation of E-cadherin, likely through FGF/Erk/Snail, upregulates Wnt signaling. FGF is also thought to induce *brachyury* directly through Erk phosphorylation of Ets family transcription factors. Thus, there are two Erk-dependent arms, one direct and one indirect, through which FGF signaling can upregulate *brachyury*. (B) FGF signaling promotes EMT through PDGFR expression. FGF8 (blue) is expressed by mesodermal precursors along the primitive streak, and PDGF (green) is expressed at the streak and more broadly. Cells from the streak that receive FGF8 activate Erk and express PDGFR. Binding of PDGF to PDGFR triggers PI3K signaling, which upregulates N-cadherin expression, triggering EMT. This gene circuit allows cell fate specification through *brachyury* to be decoupled from PDGF/PI3K-triggered EMT. An arrow representing time indicates the order in which signaling and gene expression events occur. FGF8 triggers the first set of events (blue) and PDGF triggers the second set (green).

downregulation is that canonical Wnt signaling is upregulated (Ciruna & Rossant, 2001); E-cadherin is potently capable of β-catenin sequestration, so high levels of E-cadherin sequester β-catenin at the membrane and limit its ability to induce transcription downstream of Wnt stimulation (Orsulic, Huber, Aberle, Arnold, & Kemler, 1999).

As discussed previously, Erk signaling induces *brachyury* in the mouse primitive streak (Yao et al., 2003), potentially via phosphorylation of Ets family transcription factors. It has been observed that mesodermal cells with high *brachyury* levels migrate further from the primitive streak than those with lower levels (Kispert & Herrmann, 1994; Wilson & Beddington, 1996; Yamaguchi et al., 1999). While this observation could be interpreted to mean that Brachyury promotes migration, a more likely scenario is that E-cadherin downregulation enhances *brachyury* expression through the upregulation of Wnt signaling. Thus, in the mouse migratory mesoderm, mesodermal specification and morphogenesis appear to be linked through

Erk signaling, which induces *brachyury* and downregulates E-cadherin simultaneously. Notably, this genetic architecture also provides two mechanisms through which Erk can modulate *brachyury* expression: a direct pathway and an indirect pathway through E-cadherin and Wnt signaling (Fig. 7A).

By contrast, the upregulation of N-cadherin in chick is decoupled from *brachyury* expression over the course of mesodermal gastrulation (Fig. 7B). Initially, Erk signaling in the primitive streak induces expression of a wide range of mesoderm-related genes including *brachyury*, noncanonical Wnt ligands, and PDGFR (Hardy et al., 2011). PDGFR is activated by PDGF ligands released by the surrounding epiblast. Like FGFR, PDGFR is an RTK capable of initiating both Erk and PI3K signaling. Unlike FGFR, in the context of mesodermal specification and morphogenesis, PDGFR preferentially stimulates the PI3K pathway. This PI3K signaling results in N-cadherin upregulation, which allows cells to migrate away from the streak (Yang, Chrisman, & Weijer, 2008). When FGF, Erk, PDGF, or PI3K signaling is inhibited, migration becomes impossible (Hardy et al., 2011; Yang et al., 2008).

Erk's ability to induce PDGFR expression in the primitive streak allows for the separation of mesoderm specification from mesodermal morphogenesis. While *brachyury* expression is Erk-dependent, N-cadherin upregulation is PI3K-dependent. By passing control of N-cadherin regulation to PDGFR, which only signals through PI3K and is activated by ligands that are expressed in a different spatiotemporal pattern than FGFs, EMT can be initiated in a spatiotemporally distinct manner from mesodermal specification. It remains unclear whether there is a functional significance to separating these events in chick. Similarly, it remains unclear whether coupling these events in mouse serves a function.

9. A potential role for Erk signaling in mesodermal directed cell migration

While CE is central to mesodermal morphogenesis in *Xenopus* and zebrafish, in mouse and chick, internalized cells do not undergo CE. Instead, mesodermal precursors that have traveled through the primitive streak migrate to their final positions using directions supplied by FGF signaling (Hardy et al., 2011; Sun et al., 1999; Yang et al., 2002). Mutant experiments in mouse have shown that mesodermal precursors are unable to migrate away from the primitive streak in the absence of FGF4 and

FGF8 (Sun et al., 1999). While it is unknown which pathways FGF signaling activates in migratory mesoderm, imaging experiments in mouse show that Erk activity increases in cells as they migrate away from the streak (Morgani et al., 2018), supporting a model where FGF signaling serves as a directional cue, potentially mediated by Erk.

In chick, tissue transplant experiments more directly support a chemotactic model of mesodermal migration. In wild type embryos FGF8 is released at the primitive streak, and FGF4 is released anterior to the primitive streak downstream of the FGF8 cue (Hardy et al., 2011) (Fig. 2C). Paraxial mesoderm precursors ingress through the anterior primitive streak, moving in an arc first anteriorly and away from the midline and then anteriorly and toward the midline (Fig. 1B). Experiments with FGF4 and FGF8-releasing beads show that middle streak cells are repelled by FGF8 release and attracted to FGF4 release, suggesting that the arc that paraxial mesoderm precursors travel is directed by the localized release of FGF8 and FGF4 (Yang et al., 2002). As in the mouse streak, mesodermal migration in chick also coincides with Erk activation. FGF8 triggers both Erk and PI3K activity, the latter of which induces FGF4 (Hardy et al., 2011). Erk levels decrease as mesodermal cells enter the streak and then increase again as cells emerge to migrate anteriorly (Lunn et al., 2007), presumably in response to FGF4. Thus, this mesodermal population experiences two waves of Erk signaling, the first prior to ingression through the streak and the second during anterior migration.

Although the mechanism by which different FGFs direct mesodermal migration in chick is still unclear, considerable progress has been made toward dissecting migration of the LVM in *Drosophila*, a superficially similar system. Normally, LVM cells migrate along the TVM in two parallel, synchronous lines. In the absence of the FGFR *htl*, however, cells from the two lines merge and lose their synchronicity (Kadam et al., 2012; Reim et al., 2012). They also lose Erk signaling (Mandal et al., 2004). In embryos lacking both FGFs (*pyr* and *ths*), ectopically providing either FGF is sufficient to cause LVM cells to migrate toward the ligand source. Conversely, when *pyr* and *ths* are expressed together ectopically, LVM cells stall (Kadam et al., 2012).

One interpretation of these results is that heterodimeric binding of *pyr* and *ths* to *htl* induces different cellular behaviors than the homodimeric binding of either (Kadam et al., 2012). Other experiments, however, have shown that high ectopic levels of either *pyr* or *ths* alone are sufficient to stall LVM migration, supporting the interpretation that it is the overall level of

FGF ligand present, regardless of identity, that determines whether cells migrate toward the ligand source or stall (Reim et al., 2012). This view supports a permissive model of FGF signaling, in which FGF signaling promotes association among mesodermal cells. In the presence of native levels of *pyr* and *ths*, cells of the LVM associate with one another but are primarily guided by signals emitted from the TVM. In the presence of excess *pyr* and *ths*, however, mesoderm-mesoderm contacts become too strong relative to mesoderm-substrate contacts and motility is hindered (Bae et al., 2012). Because Erk signaling is present in the migrating LVM and lost in the absence of *htl* (Mandal et al., 2004), it is tempting to speculate that Erk mediates this process.

It is unclear whether directed migration in chick is governed by a similar mechanism to that of the *Drosophila* CVM. In any case, intuition garnered from the *Drosophila* model can be used going forward to make testable predictions about directed migration of mesoderm in chick. For example, the chick paraxial mesoderm moves cohesively relative to the individually driven cell movements of the lateral mesoderm (Sweetman, Wagstaff, Cooper, Weijer, & Münsterberg, 2008). Furthermore, cells of the paraxial mesoderm are guided by FGF8 and FGF4 while cells of the lateral mesoderm appear to be guided by FGF8 alone (Yang et al., 2002). A model of migration based on what is seen in the *Drosophila* CVM would suggest that cells of the paraxial mesoderm experience higher levels of signaling in response to being exposed to more FGF than the lateral mesoderm, potentially leading to stronger cell-cell contacts within the paraxial mesoderm than the lateral mesoderm. It also suggests that Erk signaling dynamics in the paraxial mesoderm would differ from those in the lateral mesoderm due to differences in the timing and amount of FGF these two populations receive.

Alternatively, it is possible that a difference in timing between paraxial and lateral mesoderm induction is primarily responsible for the different cell migration patterns. Chick embryos gastrulate in an anterior-to-posterior manner, with the paraxial mesoderm migrating through the streak at an earlier stage than the lateral mesoderm (Frasch & Sink, 2006). As observed in *Xenopus*, where changes in negative regulator expression over time change the way in which FGFR activation is interpreted (Sivak et al., 2005), there may be a switch in regulation that occurs at some point in chick developmental time, causing the lateral mesoderm to interpret FGFR activation differently than the paraxial mesoderm. It is thus possible that Erk signaling mediates migration in one mesodermal population but not the other.

FGF may also be acting through Erk to activate the noncanonical Wnt and PCP pathways in the chick mesoderm. The noncanonical Wnt pathway is active in cells of the lateral mesoderm (Sweetman et al., 2008), and at least one of its ligands is induced by Erk signaling in the posterior streak (Hardy et al., 2011). Meanwhile, canonical Wnt induced in the anterior streak by the nearby neural plate represses noncanonical Wnt signaling in the paraxial mesoderm, resulting in a less migratory phenotype for the paraxial mesoderm than the lateral mesoderm (Sweetman et al., 2008). Thus, Erk signaling may attempt to trigger the noncanonical Wnt pathway in both the paraxial and lateral mesoderm but only succeeds in the lateral mesoderm due to the presence of repressors against noncanonical Wnt activity in the paraxial mesoderm.

In truth, it is likely that all three mechanisms—regulation of cell-cell adhesion, ingression timing, and noncanonical Wnt signaling—are employed to some extent in directing paraxial and lateral mesoderm migration in chick. More work, however, is required to determine exactly how directed migration is being achieved in each streak population and whether Erk signaling has more control over one mechanism than the others.

10. Open questions and new tools for Erk control over mesoderm specification and movement

Many open questions remain regarding the role of Erk signaling in the mesoderm. The most central question raised in this review revolves around the role that Erk dynamics play in mesoderm induction and morphogenesis. It has long been known that Erk signaling is involved in mesoderm induction in vertebrates, but in recent years, combined observations of signaling activity and cell movements have brought increased attention to the fact that Erk signaling is transient in mesodermal precursors. This transience poses interesting questions both from the perspective of germ layer induction as well as from that of morphogenetic programming.

In vertebrates, mesodermal and endodermal precursor cells arise from the same progenitor population (Grapin-Botton & Constam, 2007). As seen in the case of endoderm induction in zebrafish, neighboring cells at the margin can take on different germ layer fates depending on whether or not they activate Erk, despite the fact that both populations receive FGF ligand stimulation (van Boxtel et al., 2018). Perhaps the widespread observation of transient Erk activity during mesoderm specification is due to the high pre-existing expression level of negative regulators in mesendodermal precursors (e.g., DUSP6 in zebrafish) that is required to tune the Erk threshold such that FGF signaling is only able to trigger Erk activity in the presumptive mesoderm and not neighboring endodermal precursor cells. In this scenario, a DUSP6-expressing mesoderm precursor cell would be predicted to rapidly turn off Erk signaling upon exit from the margin, leading to a transient pathway response.

From the perspective of mesodermal morphogenesis, on the other hand, transient Erk signaling can be interpreted as playing a functional role in coordinating diverse morphogenetic processes. In organisms like *Xenopus* and zebrafish where convergence and extension drive morphogenesis, Erk signaling is upstream of processes that polarize cells and enable cell intercalation through protrusion generation and cadherin recycling. It is plausible that these events must occur in a precise order to properly position mesodermal tissues, necessitating an Erk signal that is present for a specific period in developmental time. Similarly, in organisms like chick and mouse where FGF ligands direct mesodermal cell migration, more than one sort of cell movement is required for cells to achieve their final position. A dynamic Erk signal would allow the level of active Erk in cells to rise and fall as cells' modes of migration evolve over developmental time.

10.1 New tools: Probing the function of signaling dynamics using optogenetic control

The recent development of optogenetic tools for controlling cell signaling pathways offers opportunities to further test these hypotheses. Optogenetic stimulation is especially powerful for interrogating developmental signaling because both the spatial range and dynamics of a light input can be easily controlled. In contrast, chemical stimuli typically diffuse over a broad range (limiting the ability to apply them to specific tissues) and often bind tightly, limiting dynamic control. Excellent light-based stimuli are now available for studying growth factor signaling through Erk, with optogenetic variants of growth factor receptors (Dine, Gil, Uribe, Brangwynne, & Toettcher, 2018; Grusch et al., 2014; Kim et al., 2014), Ras (Toettcher, Weiner, & Lim, 2013), Raf (Zhang et al., 2014), and MEK (Patel et al., 2019; Zhou, Fan, Li, Shen, & Lin, 2017). Thus, it is now possible to stimulate at virtually every node of the FGFR/Erk signaling cascade, controlling the timing and location of pathway activity, and measuring cells' resulting responses.

A roadmap for these sorts of experiments has also begun to emerge, primarily through studies in *Drosophila*. In the early fly embryo, activation of the Torso RTK produces a long-lived gradient of Erk at the anterior and posterior poles; later, during nuclear cycle 14, EGFR activation along two lateral stripes produces a 30 min pulse of Erk activity at these positions (Lim et al., 2015). These two RTK patterns are responsible for distinct cell fate outcomes, with Torso driving gut endoderm formation and EGFR producing neurogenic ectoderm. One may thus ask several questions with analogies to mesoderm specification. First, is Erk sufficient to recapitulate all functions downstream of RTK activation? Second, might varying the timing or duration of Erk signaling be sufficient to switch between ectoderm and endoderm cell fates?

Optogenetic control of Ras/Erk signaling proved an ideal technology for addressing these questions. Light-based membrane recruitment of a constitutively active Ras activator SOS (SOS^{cat}) (Gureasko et al., 2008) is sufficient to trigger rapid, reversible, and spatially localized Erk activation, producing an optogenetic tool termed the OptoSOS system (Johnson et al., 2017; Toettcher et al., 2013). We found that light-based OptoSOS stimulation could be used to test for a causal role for Erk signaling dynamics *in vivo*, revealing that increasing the duration of Erk signaling could produce neurogenic ectoderm after 30 min of illumination, or gut endoderm after 60 min of light (Johnson & Toettcher, 2019). Additionally, we found that light pattern at the termini can rescue complete genetic loss of the Torso RTK to rescue the entire fly life cycle, demonstrating that localized activation of the Ras/Erk pathway can indeed recapitulate all essential properties of a developmental RTK (Johnson, Djabrayan, Shvartsman, & Toettcher, 2020).

The above studies were successful in relating pathway activity to an eventual phenotypic response, but many details are missing from the intermediate layers of pathways, gene networks, and cell movements that link signaling to cell fate. These intermediate layers have begun to be filled in through a growing number of studies that combine precise light stimuli with quantitative biosensors of cell movement and gene expression (Guglielmi, Barry, Huber, & De Renzis, 2015; Huang, Amourda, Zhang, Tolwinski, & Saunders, 2017; Singh et al., 2021; Viswanathan et al., 2019). These approaches have already proven useful for dissecting how cells interpret incoming Erk signals to choose between mesodermal and endodermal fates in *Drosophila* mesendoderm (McFann et al., 2021). We found that mesendodermal cells are sensitive to two distinct Erk durations: a short duration is sufficient to inhibit mesoderm invagination along the ventral furrow through a dynamic gene network involving the terminal gap gene *hkb* and the EMT regulator *snail*, whereas longer Erk durations are needed to override the mesoderm fate and drive endoderm-associated morphogenetic movements. With the recent development of perturbation tools for vertebrate model systems (Čapek et al., 2019; LaBelle et al., 2021; Patel et al., 2019; Reade et al., 2017; Sako et al., 2016), as well as a rapidly growing arsenal of live biosensors for signaling and transcription, we will likely soon see similar approaches used to address questions of germ layer specification in other organisms.

10.2 New tools: ERK biosensors may reveal additional dynamics during mesoderm differentiation

Currently, Erk activity dynamics in the early mesoderm are inferred from the observation of a fixed region of high Erk activity through which mesoderm cells pass (e.g., the margin in lower vertebrates; the primitive streak in higher vertebrates). However, a key challenge for the field is the lack of direct visualization of early mesodermal single-cell Erk dynamics in any model system. This lack of direct observation raises the possibility that even more complex dynamics might be present but have simply gone unnoticed in the single snapshots afforded by fixed staining.

Supporting this possibility, live-cell Erk activity biosensors are now available in cultured cells and *in vivo* and have revealed that complex Erk dynamics—such as pulses, oscillations, or traveling waves—are widespread across cellular contexts. In general, these biosensors are based on engineering a synthetic, fluorescent Erk substrate whose phosphorylation can be read out in a change in fluorescence intensity or localization. For example, the EKAR family of biosensors takes advantage of a "clamshell" design, in which two fluorophores are appended to the N- and C-terminal ends of the substrate and are brought into proximity by a conformational change induced by Erk phosphorylation (Harvey et al., 2008; Komatsu et al., 2011). FRET-based biosensors have been used to generate transgenic mice in which Erk waves have been visualized in the skin, cochlea, and tumors (Hiratsuka et al., 2015; Ishii, Tateya, Matsuda, & Hirashima, 2021), and in zebrafish embryos with a particular focus on traveling Erk waves in a later mesodermal population, the presomitic mesoderm (Sari et al., 2018; Wong et al., 2018). An alternative to FRET-based designs, the "kinase translocation reporter" (KTR) was developed to encode Erk activity in a change in the synthetic substrate's localization, shifting from nuclear to cytosolic localization upon Erk activation (Regot, Hughey, Bajar, Carrasco, & Covert, 2014). The Erk KTR and subsequent variants have been used to visualize pulses and waves of Erk activity in mammalian cell lines (Hino et al., 2020; Regot et al., 2014), primary epidermal cells (Goglia et al., 2020), and *in vivo* in *C. elegans* (de la Cova, Townley, Regot, & Greenwald, 2017), the *Drosophila* pupal notum (Moreno, Valon, Levillayer, & Levayer, 2019), zebrafish scales (De Simone et al., 2021), and the early mouse embryo (Pokrass et al., 2020; Simon, Rahman, Raina, Schröter, & Hadjantonakis, 2020). Additionally, recent progress in biosensor multiplexing has made it possible to follow the dynamics of many signaling pathways at once downstream of RTK stimulation (Yang et al., 2021). We thus look forward to a detailed characterization of Erk signaling and other FGF-dependent signals across additional developmental transitions, including early mesodermal specification and migration during gastrulation.

10.3 New tools: CRISPR editing to define precise roles for components with overlapping function

A third class of questions relates to the identity and levels of different FGFs in controlling Erk activation across various mesodermal fates and movements. We have repeatedly seen examples of FGFs presented in distinct or overlapping regions, where genetic evidence suggests that they each play essential roles. Are all FGFs molecularly equivalent, but simply differ in their genetic control, thus leading to distinct expression domains and timing? Alternatively, are certain FGFs interpreted in specific ways, perhaps because of their affinity for cognate receptors, other ligands, or the extracellular matrix? Recent evidence from BMP ligands and receptors suggests a complex code, where even simple differences in binding affinity can lead to complex combinatorial logic (Antebi et al., 2017; Su et al., 2020). It remains to be seen whether similar logic holds *in vivo* or can be generalized to other ligand-receptor systems such as FGFs.

Excitingly, it is also now possible to envision addressing these types of questions experimentally. The recent advent of efficient CRISPR knock-in techniques (Gu, Posfai, & Rossant, 2018) opens the door to rapidly "swapping" coding sequences for different ligand or receptor isoforms. We can now envision replacing FGF8 with FGF4 in mouse (or *pyramus* and *thisbe* in *Drosophila*) and assessing phenotypic differences compared to wild-type embryos. Applying similar logic to sequential CRISPR knock-out of EGF family ligands has already shed light on the molecular requirements for Erk dynamics in epithelial cell lines (Lin et al., 2022), demonstrating that the future is bright for sophisticated ligand deletion and replacement studies.

Another class of questions revolves around how Erk-dependent processes translate to cell movement. While some mesodermal subpopulations move in a loosely associated manner, others move as a cohesive group. It is still unclear what morphogenetic processes are required for each mode of movement. FGF signaling through Erk has been implicated in cell polarization, protrusion generation, cadherin recycling, and induction of cadherins with varying adhesion strengths. An appealing hypothesis is that different modes of movement require different combinations of features, for example, the presence of different adhesion molecules or different levels of dynamism in adhesion molecule recycling. It has yet to be seen whether the level or duration of Erk signaling controls this balance of features, or whether additional, independent signaling cues are interpreted alongside Erk at different times and places to define specific migratory states. For example, both PLCγ and PI3K signaling play important roles during vertebrate morphogenesis around the same time that Erk signaling is also active (Böttcher & Niehrs, 2005; Carballada, Yasuo, & Lemaire, 2001; Geary & LaBonne, 2018; Hardy et al., 2011; Montero, Kilian, Chan, Bayliss, & Heisenberg, 2003; Sivak et al., 2005; Yang et al., 2008). Further investigation is needed to understand how specific pathways are triggered downstream of FGFR activation, as well as how these pathways work together to regulate cadherin turnover and other processes required for cytoskeletal rearrangement and morphogenesis.

In conclusion, Erk signaling is involved in mesoderm specification and morphogenesis across species, but the specific roles the pathway plays in gene induction, cell rearrangement, EMT, and migration remain unclear. In particular, the implications of Erk signaling being transient are still poorly understood. With modern tools for signaling and genetic perturbations, it is finally possible to manipulate the timing, level, and location of Erk signaling at nearly every node in the pathway, and to do the same for other signaling pathways downstream of FGFR. In the coming years, we hope to see these tools applied toward uncovering the Erk signaling requirements for the many processes involved in building a proper mesoderm, from genetic regulation for germ layer fate specification to cytoskeletal rearrangement and cadherin turnover.

Acknowledgments

The authors thank members of the Shvartsman and Toettcher laboratories for insightful discussions, as well as the broader developmental signaling community for its support. We apologize to the authors of many excellent studies that we were unable to discuss and cite for brevity. The writing of this review was supported by Hertz Fellowship and the NSF Graduate Research Fellowship Program (S.E.M.) and National Science Foundation CAREER Award 1750663 (to J.E.T.).

References

Amaya, E., Stein, P. A., Musci, T. J., & Kirschner, M. W. (1993). FGF signalling in the early specification of mesoderm in Xenopus. *Development, 118*, 477–487.

Antebi, Y. E., Linton, J. M., Klumpe, H., Bintu, B., Gong, M., Su, C., et al. (2017). Combinatorial signal perception in the BMP pathway. *Cell, 170*. https://doi.org/10.1016/j.cell.2017.08.015. 1184-1196.e24.

Aulehla, A., & Pourquié, O. (2010). Signaling gradients during paraxial mesoderm development. *Cold Spring Harbor Perspectives in Biology, 2*, a000869. https://doi.org/10.1101/cshperspect.a000869.

Bae, Y.-K., Trisnadi, N., Kadam, S., & Stathopoulos, A. (2012). The role of FGF signaling in guiding coordinate movement of cell groups: Guidance cue and cell adhesion regulator? *Cell Adhesion & Migration, 6*, 397–403. https://doi.org/10.4161/cam.21103.

Ben-Haim, N., Lu, C., Guzman-Ayala, M., Pescatore, L., Mesnard, D., Bischofberger, M., et al. (2006). The nodal precursor acting via activin receptors induces mesoderm by maintaining a source of its convertases and BMP4. *Developmental Cell, 11*, 313–323. https://doi.org/10.1016/j.devcel.2006.07.005.

Böttcher, R. T., & Niehrs, C. (2005). Fibroblast growth factor signaling during early vertebrate development. *Endocrine Reviews, 26*, 63–77. https://doi.org/10.1210/er.2003-0040.

Branney, P. A., Faas, L., Steane, S. E., Pownall, M. E., & Isaacs, H. V. (2009). Characterisation of the fibroblast growth factor dependent transcriptome in early development. *PLoS One, 4*, e4951. https://doi.org/10.1371/journal.pone.0004951.

Brewer, J. R., Mazot, P., & Soriano, P. (2016). Genetic insights into the mechanisms of Fgf signaling. *Genes & Development, 30*, 751–771. https://doi.org/10.1101/gad.277137.115.

Burdsal, C. A., Flannery, M. L., & Pedersen, R. A. (1998). FGF-2 alters the fate of mouse epiblast from ectoderm to mesoderm in vitro. *Developmental Biology, 198*, 231–244.

Čapek, D., Smutny, M., Tichy, A.-M., Morri, M., Janovjak, H., & Heisenberg, C.-P. (2019). Light-activated Frizzled7 reveals a permissive role of non-canonical wnt signaling in mesendoderm cell migration. *eLife, 8*. https://doi.org/10.7554/eLife.42093.

Carballada, R., Yasuo, H., & Lemaire, P. (2001). Phosphatidylinositol-3 kinase acts in parallel to the ERK MAP kinase in the FGF pathway during Xenopus mesoderm induction. *Development, 128*, 35–44.

Castanon, I., & Baylies, M. K. (2002). A Twist in fate: Evolutionary comparison of twist structure and function. *Gene, 287*, 11–22. https://doi.org/10.1016/s0378-1119(01)00893-9.

Caunt, C. J., & Keyse, S. M. (2013). Dual-specificity MAP kinase phosphatases (MKPs): Shaping the outcome of MAP kinase signalling. *The FEBS Journal, 280*, 489–504. https://doi.org/10.1111/j.1742-4658.2012.08716.x.

Chotteau-Lelievre, A., Dolle, P., Peronne, V., Coutte, L., de Launoit, Y., & Desbiens, X. (2001). Expression patterns of the Ets transcription factors from the PEA3 group during early stages of mouse development. *Mechanisms of Development, 108*, 191–195. https://doi.org/10.1016/s0925-4773(01)00480-4.

Christen, B., & Slack, J. M. (1997). FGF-8 is associated with anteroposterior patterning and limb regeneration in Xenopus. *Developmental Biology*, *192*, 455–466. https://doi.org/10.1006/dbio.1997.8732.

Christen, B., & Slack, J. M. (1999). Spatial response to fibroblast growth factor signalling in Xenopus embryos. *Development*, *126*, 119–125.

Chuai, M., Serrano-Najere, G., Serra, M., Mahadavan, L., & Weijer, C. J. (2021). Reconstruction of distinct vertebrate gastrulation modes via modulation of key cell behaviours in the chick embryo. *bioRxiv*.

Chung, H. A., Hyodo-Miura, J., Kitayama, A., Terasaka, C., Nagamune, T., & Ueno, N. (2004). Screening of FGF target genes in Xenopus by microarray: Temporal dissection of the signalling pathway using a chemical inhibitor. *Genes to Cells*, *9*, 749–761. https://doi.org/10.1111/j.1356-9597.2004.00761.x.

Chung, H. A., Yamamoto, T. S., & Ueno, N. (2007). ANR5, an FGF target gene product, regulates gastrulation in Xenopus. *Current Biology*, *17*, 932–939. https://doi.org/10.1016/j.cub.2007.04.034.

Ciruna, B., & Rossant, J. (2001). FGF signaling regulates mesoderm cell fate specification and morphogenetic movement at the primitive streak. *Developmental Cell*, *1*, 37–49. https://doi.org/10.1016/s1534-5807(01)00017-x.

Clark, J. F., & Soriano, P. (2022). Pulling back the curtain: The hidden functions of receptor tyrosine kinases in development. *Current Topics in Developmental Biology*, *149*, 123–152.

Conlon, F. L., & Smith, J. C. (1999). Interference with brachyury function inhibits convergent extension, causes apoptosis, and reveals separate requirements in the FGF and activin signalling pathways. *Developmental Biology*, *213*, 85–100. https://doi.org/10.1006/dbio.1999.9330.

Corson, L. B., Yamanaka, Y., Lai, K.-M. V., & Rossant, J. (2003). Spatial and temporal patterns of ERK signaling during mouse embryogenesis. *Development*, *130*, 4527–4537. https://doi.org/10.1242/dev.00669.

Dale, L., & Wardle, F. (2015). Xenopus embryo: Mesoderm induction. In *ELS*. John Wiley & Sons, Ltd: Chichester.

Davey, C. F., & Moens, C. B. (2017). Planar cell polarity in moving cells: Think globally, act locally. *Development*, *144*, 187–200. https://doi.org/10.1242/dev.122804.

de la Cova, C., Townley, R., Regot, S., & Greenwald, I. (2017). A real-time biosensor for ERK activity reveals signaling dynamics during C. elegans cell fate specification. *Developmental Cell*, *42*, 542–553.e4. https://doi.org/10.1016/j.devcel.2017.07.014.

De Simone, A., Evanitsky, M. N., Hayden, L., Cox, B. D., Wang, J., Tornini, V. A., et al. (2021). Control of osteoblast regeneration by a train of Erk activity waves. *Nature*, *590*, 129–133. https://doi.org/10.1038/s41586-020-03085-8.

Dine, E., Gil, A. A., Uribe, G., Brangwynne, C. P., & Toettcher, J. E. (2018). Protein phase separation provides long-term memory of transient spatial stimuli. *Cell Systems*, *6*, 655–663.e5. https://doi.org/10.1016/j.cels.2018.05.002.

Fletcher, R. B., & Harland, R. M. (2008). The role of FGF signaling in the establishment and maintenance of mesodermal gene expression in Xenopus. *Developmental Dynamics*, *237*, 1243–1254. https://doi.org/10.1002/dvdy.21517.

Frasch, M., & Sink, H. (2006). Development of the larval visceral musculature. In *Muscle development in Drosophila* (pp. 62–78). New York, New York, USA: Springer Science + Business Media.

Fukunaga, C. J. (2014). *True detective: The secret fate of all life*. HBO.

Fürthauer, M., Reifers, F., Brand, M., Thisse, B., & Thisse, C. (2001). sprouty4 acts in vivo as a feedback-induced antagonist of FGF signaling in zebrafish. *Development*, *128*, 2175–2186.

Fürthauer, M., Van Celst, J., Thisse, C., & Thisse, B. (2004). Fgf signalling controls the dorsoventral patterning of the zebrafish embryo. *Development*, *131*, 2853–2864. https://doi.org/10.1242/dev.01156.

Geary, L., & LaBonne, C. (2018). FGF mediated MAPK and PI3K/Akt signals make distinct contributions to pluripotency and the establishment of Neural Crest. *eLife*, 7. https://doi.org/10.7554/eLife.33845.

Gheldof, A., & Berx, G. (2013). Cadherins and epithelial-to-mesenchymal transition. *Progress in Molecular Biology and Translational Science*, *116*, 317–336. https://doi.org/10.1016/B978-0-12-394311-8.00014-5.

Gilbert, S. (2000). The early development of vertebrates: Fish, birds, and mammals. In *Developmental biology*. Sunderland (MA): Sinauer Associates.

Goglia, A. G., Wilson, M. Z., Jena, S. G., Silbert, J., Basta, L. P., Devenport, D., et al. (2020). A live-cell screen for altered Erk dynamics reveals principles of proliferative control. *Cell Systems*, *10*, 240–253.e6. https://doi.org/10.1016/j.cels.2020.02.005.

Golub, R., Adelman, Z., Clementi, J., Weiss, R., Bonasera, J., & Servetnick, M. (2000). Evolutionarily conserved and divergent expression of members of the FGF receptor family among vertebrate embryos, as revealed by FGFR expression patterns in Xenopus. *Development Genes and Evolution*, *210*, 345–357. https://doi.org/10.1007/s004270000076.

Gómez, A. R., López-Varea, A., Molnar, C., de la Calle-Mustienes, E., Ruiz-Gómez, M., Gómez-Skarmeta, J. L., et al. (2005). Conserved cross-interactions in Drosophila and Xenopus between Ras/MAPK signaling and the dual-specificity phosphatase MKP3. *Developmental Dynamics*, *232*, 695–708. https://doi.org/10.1002/dvdy.20227.

Gould, A. P., & White, R. A. (1992). Connectin, a target of homeotic gene control in Drosophila. *Development*, *116*, 1163–1174.

Grapin-Botton, A., & Constam, D. (2007). Evolution of the mechanisms and molecular control of endoderm formation. *Mechanisms of Development*, *124*, 253–278. https://doi.org/10.1016/j.mod.2007.01.001.

Griffin, K., Patient, R., & Holder, N. (1995). Analysis of FGF function in normal and no tail zebrafish embryos reveals separate mechanisms for formation of the trunk and the tail. *Development*, *121*, 2983–2994.

Grusch, M., Schelch, K., Riedler, R., Reichhart, E., Differ, C., Berger, W., et al. (2014). Spatio-temporally precise activation of engineered receptor tyrosine kinases by light. *The EMBO Journal*, *33*, 1713–1726. https://doi.org/10.15252/embj.201387695.

Gu, B., Posfai, E., & Rossant, J. (2018). Efficient generation of targeted large insertions by microinjection into two-cell-stage mouse embryos. *Nature Biotechnology*, *36*, 632–637. https://doi.org/10.1038/nbt.4166.

Guglielmi, G., Barry, J. D., Huber, W., & De Renzis, S. (2015). An optogenetic method to modulate cell contractility during tissue morphogenesis. *Developmental Cell*, *35*, 646–660. https://doi.org/10.1016/j.devcel.2015.10.020.

Gureasko, J., Galush, W. J., Boykevisch, S., Sondermann, H., Bar-Sagi, D., Groves, J. T., et al. (2008). Membrane-dependent signal integration by the Ras activator Son of sevenless. *Nature Structural & Molecular Biology*, *15*, 452–461. https://doi.org/10.1038/nsmb.1418.

Hanafusa, H., Matsumoto, K., & Nishida, E. (2009). Regulation of ERK activity duration by Sprouty contributes to dorsoventral patterning. *Nature Cell Biology*, *11*, 106–109. https://doi.org/10.1038/ncb1820.

Hardy, K. M., Yatskievych, T. A., Konieczka, J., Bobbs, A. S., & Antin, P. B. (2011). FGF signalling through RAS/MAPK and PI3K pathways regulates cell movement and gene expression in the chicken primitive streak without affecting E-cadherin expression. *BMC Developmental Biology*, *11*, 20. https://doi.org/10.1186/1471-213X-11-20.

Harvey, C. D., Ehrhardt, A. G., Cellurale, C., Zhong, H., Yasuda, R., Davis, R. J., et al. (2008). A genetically encoded fluorescent sensor of ERK activity. *Proceedings of the National Academy of Sciences of the United States of America*, *105*, 19264–19269. https://doi.org/10.1073/pnas.0804598105.

Heisenberg, C. P., Tada, M., Rauch, G. J., Saúde, L., Concha, M. L., Geisler, R., et al. (2000). Silberblick/Wnt11 mediates convergent extension movements during zebrafish gastrulation. *Nature*, *405*, 76–81. https://doi.org/10.1038/35011068.

Hino, N., Rossetti, L., Marín-Llauradó, A., Aoki, K., Trepat, X., Matsuda, M., et al. (2020). ERK-mediated mechanochemical waves direct collective cell polarization. *Developmental Cell*, *53*, 646–660.e8. https://doi.org/10.1016/j.devcel.2020.05.011.

Hiratsuka, T., Fujita, Y., Naoki, H., Aoki, K., Kamioka, Y., & Matsuda, M. (2015). Intercellular propagation of extracellular signal-regulated kinase activation revealed by in vivo imaging of mouse skin. *eLife*, *4*, e05178. https://doi.org/10.7554/eLife.05178.

Hoch, M., & Pankratz, M. J. (1996). Control of gut development by fork head and cell signaling molecules in Drosophila. *Mechanisms of Development*, *58*, 3–14. https://doi.org/10.1016/s0925-4773(96)00541-2.

Huang, A., Amourda, C., Zhang, S., Tolwinski, N. S., & Saunders, T. E. (2017). Decoding temporal interpretation of the morphogen Bicoid in the early Drosophila embryo. *eLife*, *6*. https://doi.org/10.7554/eLife.26258.

Isaacs, H. V., Pownall, M. E., & Slack, J. M. (1994). eFGF regulates Xbra expression during Xenopus gastrulation. *The EMBO Journal*, *13*, 4469–4481.

Isaacs, H. V., Pownall, M. E., & Slack, J. M. (1995). eFGF is expressed in the dorsal midline of Xenopus laevis. *The International Journal of Developmental Biology*, *39*, 575–579.

Ishii, M., Tateya, T., Matsuda, M., & Hirashima, T. (2021). Retrograde ERK activation waves drive base-to-apex multicellular flow in murine cochlear duct morphogenesis. *eLife*, *10*. https://doi.org/10.7554/eLife.61092.

Johnson, H. E., Djabrayan, N. J. V., Shvartsman, S. Y., & Toettcher, J. E. (2020). Optogenetic rescue of a patterning mutant. *Current Biology*, *30*, 3414–3424.e3. https://doi.org/10.1016/j.cub.2020.06.059.

Johnson, H. E., Goyal, Y., Pannucci, N. L., Schüpbach, T., Shvartsman, S. Y., & Toettcher, J. E. (2017). The spatiotemporal limits of developmental Erk signaling. *Developmental Cell*, *40*, 185–192. https://doi.org/10.1016/j.devcel.2016.12.002.

Johnson, H. E., & Toettcher, J. E. (2019). Signaling dynamics control cell fate in the early Drosophila embryo. *Developmental Cell*, *48*, 361–370.e3. https://doi.org/10.1016/j.devcel.2019.01.009.

Kadam, S., Ghosh, S., & Stathopoulos, A. (2012). Synchronous and symmetric migration of Drosophila caudal visceral mesoderm cells requires dual input by two FGF ligands. *Development*, *139*, 699–708. https://doi.org/10.1242/dev.068791.

Karabagli, H., Karabagli, P., Ladher, R. K., & Schoenwolf, G. C. (2002). Comparison of the expression patterns of several fibroblast growth factors during chick gastrulation and neurulation. *Anatomy and Embryology (Berlin)*, *205*, 365–370. https://doi.org/10.1007/s00429-002-0264-7.

Kawachi, K., Masuyama, N., & Nishida, E. (2003). Essential role of the transcription factor Ets-2 in Xenopus early development. *The Journal of Biological Chemistry*, *278*, 5473–5477. https://doi.org/10.1074/jbc.M211054200.

Kim, N., Kim, J. M., Lee, M., Kim, C. Y., Chang, K.-Y., & Heo, W. D. (2014). Spatiotemporal control of fibroblast growth factor receptor signals by blue light. *Chemistry & Biology*, *21*, 903–912. https://doi.org/10.1016/j.chembiol.2014.05.013.

Kim, S. H., Yamamoto, A., Bouwmeester, T., Agius, E., & Robertis, E. M. (1998). The role of paraxial protocadherin in selective adhesion and cell movements of the mesoderm during Xenopus gastrulation. *Development*, *125*, 4681–4690.

Kispert, A., & Herrmann, B. G. (1994). Immunohistochemical analysis of the Brachyury protein in wild-type and mutant mouse embryos. *Developmental Biology*, *161*, 179–193. https://doi.org/10.1006/dbio.1994.1019.

Kispert, A., Herrmann, B. G., Leptin, M., & Reuter, R. (1994). Homologs of the mouse Brachyury gene are involved in the specification of posterior terminal structures in Drosophila, Tribolium, and Locusta. *Genes & Development, 8*, 2137–2150. https://doi.org/10.1101/gad.8.18.2137.

Kispert, A., Ortner, H., Cooke, J., & Herrmann, B. G. (1995). The chick Brachyury gene: Developmental expression pattern and response to axial induction by localized activin. *Developmental Biology, 168*, 406–415. https://doi.org/10.1006/dbio.1995.1090.

Kjolby, R. A. S., Truchado-Garcia, M., Iruvanti, S., & Harland, R. M. (2019). Integration of Wnt and FGF signaling in the Xenopus gastrula at TCF and Ets binding sites shows the importance of short-range repression by TCF in patterning the marginal zone. *Development, 146*. https://doi.org/10.1242/dev.179580.

Komatsu, N., Aoki, K., Yamada, M., Yukinaga, H., Fujita, Y., Kamioka, Y., et al. (2011). Development of an optimized backbone of FRET biosensors for kinases and GTPases. *Molecular Biology of the Cell, 22*, 4647–4656. https://doi.org/10.1091/mbc.E11-01-0072.

Krens, S. F. G., Corredor-Adámez, M., He, S., Snaar-Jagalska, B. E., & Spaink, H. P. (2008). ERK1 and ERK2 MAPK are key regulators of distinct gene sets in zebrafish embryogenesis. *BMC Genomics, 9*, 196. https://doi.org/10.1186/1471-2164-9-196.

Krens, S. F. G., He, S., Lamers, G. E. M., Meijer, A. H., Bakkers, J., Schmidt, T., et al. (2008). Distinct functions for ERK1 and ERK2 in cell migration processes during zebrafish gastrulation. *Developmental Biology, 319*, 370–383. https://doi.org/10.1016/j.ydbio.2008.04.032.

Ku, M., & Melton, D. A. (1993). Xwnt-11: A maternally expressed Xenopus wnt gene. *Development, 119*, 1161–1173.

Kusch, T., & Reuter, R. (1999). Functions for Drosophila brachyenteron and forkhead in mesoderm specification and cell signalling. *Development, 126*, 3991–4003.

LaBelle, J., Ramos-Martinez, A., Shen, K., Motta-Mena, L. B., Gardner, K. H., Materna, S. C., et al. (2021). TAEL 2.0: An improved optogenetic expression system for zebrafish. *Zebrafish, 18*, 20–28. https://doi.org/10.1089/zeb.2020.1951.

LaBonne, C., & Whitman, M. (1994). Mesoderm induction by activin requires FGF-mediated intracellular signals. *Development, 120*, 463–472.

Lea, R., Papalopulu, N., Amaya, E., & Dorey, K. (2009). Temporal and spatial expression of FGF ligands and receptors during Xenopus development. *Developmental Dynamics, 238*, 1467–1479. https://doi.org/10.1002/dvdy.21913.

Lengyel, J. A., & Iwaki, D. D. (2002). It takes guts: The Drosophila hindgut as a model system for organogenesis. *Developmental Biology, 243*, 1–19. https://doi.org/10.1006/dbio.2002.0577.

Li, C., Scott, D. A., Hatch, E., Tian, X., & Mansour, S. L. (2007). Dusp6 (Mkp3) is a negative feedback regulator of FGF-stimulated ERK signaling during mouse development. *Development, 134*, 167–176. https://doi.org/10.1242/dev.02701.

Lim, B., Dsilva, C. J., Levario, T. J., Lu, H., Schüpbach, T., Kevrekidis, I. G., et al. (2015). Dynamics of inductive ERK signaling in the Drosophila embryo. *Current Biology, 25*, 1784–1790. https://doi.org/10.1016/j.cub.2015.05.039.

Lin, S., Hirayama, D., Maryu, G., Matsuda, K., Hino, N., Deguchi, E., et al. (2022). Redundant roles of EGFR ligands in the ERK activation waves during collective cell migration. *Life Science Alliance, 5*(1). https://doi.org/10.26508/lsa.202101206.

Lolas, M., Valenzuela, P. D. T., Tjian, R., & Liu, Z. (2014). Charting Brachyury-mediated developmental pathways during early mouse embryogenesis. *Proceedings of the National Academy of Sciences of the United States of America, 111*, 4478–4483. https://doi.org/10.1073/pnas.1402612111.

Lombardo, A., Isaacs, H. V., & Slack, J. M. (1998). Expression and functions of FGF-3 in Xenopus development. *The International Journal of Developmental Biology, 42*, 1101–1107.

Lunn, J. S., Fishwick, K. J., Halley, P. A., & Storey, K. G. (2007). A spatial and temporal map of FGF/Erk1/2 activity and response repertoires in the early chick embryo. *Developmental Biology, 302*, 536–552. https://doi.org/10.1016/j.ydbio.2006.10.014.

Maegawa, S., Varga, M., & Weinberg, E. S. (2006). FGF signaling is required for {beta}-catenin-mediated induction of the zebrafish organizer. *Development, 133*, 3265–3276. https://doi.org/10.1242/dev.02483.

Maillet, M., Purcell, N. H., Sargent, M. A., York, A. J., Bueno, O. F., & Molkentin, J. D. (2008). DUSP6 (MKP3) null mice show enhanced ERK1/2 phosphorylation at baseline and increased myocyte proliferation in the heart affecting disease susceptibility. *The Journal of Biological Chemistry, 283*, 31246–31255. https://doi.org/10.1074/jbc.M806085200.

Makita, R., Mizuno, T., Koshida, S., Kuroiwa, A., & Takeda, H. (1998). Zebrafish wnt11: Pattern and regulation of the expression by the yolk cell and No tail activity. *Mechanisms of Development, 71*, 165–176. https://doi.org/10.1016/s0925-4773(98)00013-6.

Mandal, L., Dumstrei, K., & Hartenstein, V. (2004). Role of FGFR signaling in the morphogenesis of the Drosophila visceral musculature. *Developmental Dynamics, 231*, 342–348. https://doi.org/10.1002/dvdy.20088.

Martin, B. L., & Kimelman, D. (2008). Regulation of canonical Wnt signaling by Brachyury is essential for posterior mesoderm formation. *Developmental Cell, 15*, 121–133. https://doi.org/10.1016/j.devcel.2008.04.013.

Maruoka, Y., Ohbayashi, N., Hoshikawa, M., Itoh, N., Hogan, B. L., & Furuta, Y. (1998). Comparison of the expression of three highly related genes, Fgf8, Fgf17 and Fgf18, in the mouse embryo. *Mechanisms of Development, 74*, 175–177. https://doi.org/10.1016/s0925-4773(98)00061-6.

Mathieu, J., Griffin, K., Herbomel, P., Dickmeis, T., Strähle, U., Kimelman, D., et al. (2004). Nodal and Fgf pathways interact through a positive regulatory loop and synergize to maintain mesodermal cell populations. *Development, 131*, 629–641. https://doi.org/10.1242/dev.00964.

McFann, S., Dutta, S., Toettcher, J. E., & Shvartsman, S. Y. (2021). Temporal integration of inductive cues on the way to gastrulation. *Proceedings of the National Academy of Sciences of the United States of America, 118*(23). https://doi.org/10.1073/pnas.2102691118.

McFaul, M., Hart, N., & Draper, B. (2020). Loss of Fgf-1 responsive Pea3 transcription factors results in ciliopathy-associated phenotypes during early zebrafish development. *bioRxiv*.

Medina, A., Swain, R. K., Kuerner, K.-M., & Steinbeisser, H. (2004). Xenopus paraxial protocadherin has signaling functions and is involved in tissue separation. *The EMBO Journal, 23*, 3249–3258. https://doi.org/10.1038/sj.emboj.7600329.

Molina, G. A., Watkins, S. C., & Tsang, M. (2007). Generation of FGF reporter transgenic zebrafish and their utility in chemical screens. *BMC Developmental Biology*, 7, 62. https://doi.org/10.1186/1471-213X-7-62.

Montero, J.-A., Kilian, B., Chan, J., Bayliss, P. E., & Heisenberg, C.-P. (2003). Phosphoinositide 3-kinase is required for process outgrowth and cell polarization of gastrulating mesendodermal cells. *Current Biology, 13*, 1279–1289. https://doi.org/10.1016/s0960-9822(03)00505-0.

Moreno, E., Valon, L., Levillayer, F., & Levayer, R. (2019). Competition for space induces cell elimination through compaction-driven ERK downregulation. *Current Biology, 29*, 23–34.e8. https://doi.org/10.1016/j.cub.2018.11.007.

Morgani, S. M., Saiz, N., Garg, V., Raina, D., Simon, C. S., Kang, M., et al. (2018). A Sprouty4 reporter to monitor FGF/ERK signaling activity in ESCs and mice. *Developmental Biology, 441*, 104–126. https://doi.org/10.1016/j.ydbio.2018.06.017.

Morley, R. H., Lachani, K., Keefe, D., Gilchrist, M. J., Flicek, P., Smith, J. C., et al. (2009). A gene regulatory network directed by zebrafish No tail accounts for its roles in mesoderm formation. *Proceedings of the National Academy of Sciences of the United States of America*, *106*, 3829–3834. https://doi.org/10.1073/pnas.0808382106.

Murphy, L. O., & Blenis, J. (2006). MAPK signal specificity: The right place at the right time. *Trends in Biochemical Sciences*, *31*, 268–275. https://doi.org/10.1016/j.tibs.2006.03.009.

Murphy, L. O., Smith, S., Chen, R.-H., Fingar, D. C., & Blenis, J. (2002). Molecular interpretation of ERK signal duration by immediate early gene products. *Nature Cell Biology*, *4*, 556–564. https://doi.org/10.1038/ncb822.

Nakamura, A., Goto, Y., Kondo, Y., & Aoki, K. (2021). Shedding light on developmental ERK signaling with genetically encoded biosensors. *Development*, *148*(18). https://doi.org/10.1242/dev.199767.

Nakaya, Y., & Sheng, G. (2009). An amicable separation: Chick's way of doing EMT. *Cell Adhesion & Migration*, *3*, 160–163. https://doi.org/10.4161/cam.3.2.7373.

Nakaya, Y., Sukowati, E. W., Wu, Y., & Sheng, G. (2008). RhoA and microtubule dynamics control cell-basement membrane interaction in EMT during gastrulation. *Nature Cell Biology*, *10*, 765–775. https://doi.org/10.1038/ncb1739.

Nose, A., Mahajan, V. B., & Goodman, C. S. (1992). Connectin: A homophilic cell adhesion molecule expressed on a subset of muscles and the motoneurons that innervate them in Drosophila. *Cell*, *70*, 553–567. https://doi.org/10.1016/0092-8674(92)90426-d.

Nutt, S. L., Dingwell, K. S., Holt, C. E., & Amaya, E. (2001). Xenopus Sprouty2 inhibits FGF-mediated gastrulation movements but does not affect mesoderm induction and patterning. *Genes & Development*, *15*, 1152–1166. https://doi.org/10.1101/gad.191301.

Ornitz, D. M., & Itoh, N. (2001). Fibroblast growth factors. *Genome Biology*, *2*. https://doi.org/10.1186/gb-2001-2-3-reviews3005. REVIEWS3005.

Orsulic, S., Huber, O., Aberle, H., Arnold, S., & Kemler, R. (1999). E-cadherin binding prevents beta-catenin nuclear localization and beta-catenin/LEF-1-mediated transactivation. *Journal of Cell Science*, *112*(Pt. 8), 1237–1245.

Ota, S., Tonou-Fujimori, N., Tonou-Fujimori, N., Nakayama, Y., Ito, Y., Kawamura, A., et al. (2010). FGF receptor gene expression and its regulation by FGF signaling during early zebrafish development. *Genesis*, *48*, 707–716. https://doi.org/10.1002/dvg.20682.

Patel, A. L., Yeung, E., McGuire, S. E., Wu, A. Y., Toettcher, J. E., Burdine, R. D., et al. (2019). Optimizing photoswitchable MEK. *Proceedings of the National Academy of Sciences of the United States of America*, *116*, 25756–25763. https://doi.org/10.1073/pnas.1912320116.

Pokrass, M. J., Ryan, K. A., Xin, T., Pielstick, B., Timp, W., Greco, V., et al. (2020). Cell-cycle-dependent ERK signaling dynamics direct fate specification in the mammalian preimplantation embryo. *Developmental Cell*, *55*, 328–340.e5. https://doi.org/10.1016/j.devcel.2020.09.013.

Ray, A. T., Mazot, P., Brewer, J. R., Catela, C., Dinsmore, C. J., & Soriano, P. (2020). FGF signaling regulates development by processes beyond canonical pathways. *Genes & Development*, *34*, 1735–1752. https://doi.org/10.1101/gad.342956.120.

Reade, A., Motta-Mena, L. B., Gardner, K. H., Stainier, D. Y., Weiner, O. D., & Woo, S. (2017). TAEL: A zebrafish-optimized optogenetic gene expression system with fine spatial and temporal control. *Development*, *144*, 345–355. https://doi.org/10.1242/dev.139238.

Regot, S., Hughey, J. J., Bajar, B. T., Carrasco, S., & Covert, M. W. (2014). High-sensitivity measurements of multiple kinase activities in live single cells. *Cell*, *157*, 1724–1734. https://doi.org/10.1016/j.cell.2014.04.039.

Reifers, F., Böhli, H., Walsh, E. C., Crossley, P. H., Stainier, D. Y., & Brand, M. (1998). Fgf8 is mutated in zebrafish acerebellar (ace) mutants and is required for maintenance of midbrain-hindbrain boundary development and somitogenesis. *Development, 125*, 2381–2395.

Reim, I., Hollfelder, D., Ismat, A., & Frasch, M. (2012). The FGF8-related signals Pyramus and Thisbe promote pathfinding, substrate adhesion, and survival of migrating longitudinal gut muscle founder cells. *Developmental Biology, 368*, 28–43. https://doi.org/10.1016/j.ydbio.2012.05.010.

Rodaway, A., Takeda, H., Koshida, S., Broadbent, J., Price, B., Smith, J. C., et al. (1999). Induction of the mesendoderm in the zebrafish germ ring by yolk cell-derived TGF-beta family signals and discrimination of mesoderm and endoderm by FGF. *Development, 126*, 3067–3078.

Saka, Y., Tada, M., & Smith, J. C. (2000). A screen for targets of the Xenopus T-box gene Xbra. *Mechanisms of Development, 93*, 27–39. https://doi.org/10.1016/s0925-4773(00)00260-4.

Sako, K., Pradhan, S. J., Barone, V., Inglés-Prieto, Á., Müller, P., Ruprecht, V., et al. (2016). Optogenetic control of nodal signaling reveals a temporal pattern of nodal signaling regulating cell fate specification during gastrulation. *Cell Reports, 16*, 866–877. https://doi.org/10.1016/j.celrep.2016.06.036.

San Martin, B., & Bate, M. (2001). Hindgut visceral mesoderm requires an ectodermal template for normal development in Drosophila. *Development, 128*, 233–242.

Sari, D. W. K., Akiyama, R., Naoki, H., Ishijima, H., Bessho, Y., & Matsui, T. (2018). Time-lapse observation of stepwise regression of Erk activity in zebrafish presomitic mesoderm. *Scientific Reports, 8*, 4335. https://doi.org/10.1038/s41598-018-22619-9.

Schambony, A., & Wedlich, D. (2007). Wnt-5A/Ror2 regulate expression of XPAPC through an alternative noncanonical signaling pathway. *Developmental Cell, 12*, 779–792. https://doi.org/10.1016/j.devcel.2007.02.016.

Schulte-Merker, S., & Smith, J. C. (1995). Mesoderm formation in response to Brachyury requires FGF signalling. *Current Biology, 5*, 62–67. https://doi.org/10.1016/s0960-9822(95)00017-0.

Shindo, A. (2018). Models of convergent extension during morphogenesis. *Wiley Interdisciplinary Reviews: Developmental Biology*, 7(1). https://doi.org/10.1002/wdev.293.

Simon, C. S., Rahman, S., Raina, D., Schröter, C., & Hadjantonakis, A.-K. (2020). Live visualization of ERK activity in the mouse blastocyst reveals lineage-specific signaling dynamics. *Developmental Cell, 55*, 341–353.e5. https://doi.org/10.1016/j.devcel.2020.09.030.

Singh, A. P., Wu, P., Ryabichko, S., Raimundo, J., Swan, M., Wieschaus, E., et al. (2021). Optogenetic control of the Bicoid morphogen reveals fast and slow modes of gap gene regulation. *bioRxiv*.

Sivak, J. M., Petersen, L. F., & Amaya, E. (2005). FGF signal interpretation is directed by Sprouty and Spred proteins during mesoderm formation. *Developmental Cell, 8*, 689–701. https://doi.org/10.1016/j.devcel.2005.02.011.

Smith, J. C. (1989). Mesoderm induction and mesoderm-inducing factors in early amphibian development. *Development, 105*, 665–677.

Smith, J. C., Price, B. M., Green, J. B., Weigel, D., & Herrmann, B. G. (1991). Expression of a Xenopus homolog of Brachyury (T) is an immediate-early response to mesoderm induction. *Cell, 67*, 79–87. https://doi.org/10.1016/0092-8674(91)90573-h.

Solnica-Krezel, L., & Sepich, D. S. (2012). Gastrulation: Making and shaping germ layers. *Annual Review of Cell and Developmental Biology, 28*, 687–717. https://doi.org/10.1146/annurev-cellbio-092910-154043.

Stathopoulos, A., Tam, B., Ronshaugen, M., Frasch, M., & Levine, M. (2004). pyramus and thisbe: FGF genes that pattern the mesoderm of Drosophila embryos. *Genes & Development, 18*, 687–699. https://doi.org/10.1101/gad.1166404.

Su, C. J., Murugan, A., Linton, J. M., Yeluri, A., Bois, J., Klumpe, H., et al. (2020). Ligand-receptor promiscuity enables cellular addressing. *bioRxiv*.

Sun, J., Macabenta, F., Akos, Z., & Stathopoulos, A. (2020). Collective migrations of Drosophila embryonic trunk and caudal mesoderm-derived muscle precursor cells. *Genetics, 215*, 297–322. https://doi.org/10.1534/genetics.120.303258.

Sun, X., Meyers, E. N., Lewandoski, M., & Martin, G. R. (1999). Targeted disruption of Fgf8 causes failure of cell migration in the gastrulating mouse embryo. *Genes & Development, 13*, 1834–1846. https://doi.org/10.1101/gad.13.14.1834.

Sweetman, D., Wagstaff, L., Cooper, O., Weijer, C., & Münsterberg, A. (2008). The migration of paraxial and lateral plate mesoderm cells emerging from the late primitive streak is controlled by different Wnt signals. *BMC Developmental Biology, 8*, 63. https://doi.org/10.1186/1471-213X-8-63.

Tada, M., & Kai, M. (2009). Noncanonical Wnt/PCP signaling during vertebrate gastrulation. *Zebrafish, 6*, 29–40. https://doi.org/10.1089/zeb.2008.0566.

Tada, M., & Smith, J. C. (2000). Xwnt11 is a target of Xenopus Brachyury: Regulation of gastrulation movements via Dishevelled, but not through the canonical Wnt pathway. *Development, 127*, 2227–2238.

Takada, S., Stark, K. L., Shea, M. J., Vassileva, G., McMahon, J. A., & McMahon, A. P. (1994). Wnt-3a regulates somite and tailbud formation in the mouse embryo. *Genes & Development, 8*, 174–189. https://doi.org/10.1101/gad.8.2.174.

Theodosiou, A., & Ashworth, A. (2002). MAP kinase phosphatases. *Genome Biology, 3*. https://doi.org/10.1186/gb-2002-3-7-reviews3009. REVIEWS3009.

Toettcher, J. E., Weiner, O. D., & Lim, W. A. (2013). Using optogenetics to interrogate the dynamic control of signal transmission by the Ras/Erk module. *Cell, 155*, 1422–1434. https://doi.org/10.1016/j.cell.2013.11.004.

Tortelote, G. G., Hernández-Hernández, J. M., Quaresma, A. J. C., Nickerson, J. A., Imbalzano, A. N., & Rivera-Pérez, J. A. (2013). Wnt3 function in the epiblast is required for the maintenance but not the initiation of gastrulation in mice. *Developmental Biology, 374*, 164–173. https://doi.org/10.1016/j.ydbio.2012.10.013.

Tsang, M., Maegawa, S., Kiang, A., Habas, R., Weinberg, E., & Dawid, I. B. (2004). A role for MKP3 in axial patterning of the zebrafish embryo. *Development, 131*, 2769–2779. https://doi.org/10.1242/dev.01157.

Ulrich, F., Concha, M. L., Heid, P. J., Voss, E., Witzel, S., Roehl, H., et al. (2003). Slb/Wnt11 controls hypoblast cell migration and morphogenesis at the onset of zebrafish gastrulation. *Development, 130*, 5375–5384. https://doi.org/10.1242/dev.00758.

Umbhauer, M., Marshall, C. J., Mason, C. S., Old, R. W., & Smith, J. C. (1995). Mesoderm induction in Xenopus caused by activation of MAP kinase. *Nature, 376*, 58–62. https://doi.org/10.1038/376058a0.

Unterseher, F., Hefele, J. A., Giehl, K., De Robertis, E. M., Wedlich, D., & Schambony, A. (2004). Paraxial protocadherin coordinates cell polarity during convergent extension via Rho A and JNK. *The EMBO Journal, 23*, 3259–3269. https://doi.org/10.1038/sj.emboj.7600332.

van Boxtel, A. L., Economou, A. D., Heliot, C., & Hill, C. S. (2018). Long-range signaling activation and local inhibition separate the mesoderm and endoderm lineages. *Developmental Cell, 44*, 179–191.e5. https://doi.org/10.1016/j.devcel.2017.11.021.

van Noort, M., & Clevers, H. (2002). TCF transcription factors, mediators of Wnt-signaling in development and cancer. *Developmental Biology, 244*, 1–8. https://doi.org/10.1006/dbio.2001.0566.

Viswanathan, R., Necakov, A., Trylinski, M., Harish, R. K., Krueger, D., Esposito, E., et al. (2019). Optogenetic inhibition of Delta reveals digital Notch signalling output during tissue differentiation. *EMBO Reports*, *20*, e47999. https://doi.org/10.15252/embr.201947999.

Voiculescu, O., Bertocchini, F., Wolpert, L., Keller, R. E., & Stern, C. D. (2007). The amniote primitive streak is defined by epithelial cell intercalation before gastrulation. *Nature*, *449*, 1049–1052. https://doi.org/10.1038/nature06211.

Vonica, A., & Gumbiner, B. M. (2002). Zygotic Wnt activity is required for Brachyury expression in the early Xenopus laevis embryo. *Developmental Biology*, *250*, 112–127. https://doi.org/10.1006/dbio.2002.0786.

Wallingford, J. B., Rowning, B. A., Vogeli, K. M., Rothbächer, U., Fraser, S. E., & Harland, R. M. (2000). Dishevelled controls cell polarity during Xenopus gastrulation. *Nature*, *405*, 81–85. https://doi.org/10.1038/35011077.

Warga, R. M., & Nüsslein-Volhard, C. (1999). Origin and development of the zebrafish endoderm. *Development*, *126*, 827–838.

Williams, M., Burdsal, C., Periasamy, A., Lewandoski, M., & Sutherland, A. (2012). Mouse primitive streak forms in situ by initiation of epithelial to mesenchymal transition without migration of a cell population. *Developmental Dynamics*, *241*, 270–283. https://doi.org/10.1002/dvdy.23711.

Wilson, V., & Beddington, R. S. (1996). Cell fate and morphogenetic movement in the late mouse primitive streak. *Mechanisms of Development*, *55*, 79–89. https://doi.org/10.1016/0925-4773(95)00493-9.

Wong, K.-L., Akiyama, R., Bessho, Y., & Matsui, T. (2018). ERK activity dynamics during zebrafish embryonic development. *International Journal of Molecular Sciences*, *20*(1). https://doi.org/10.3390/ijms20010109.

Wu, L. H., & Lengyel, J. A. (1998). Role of caudal in hindgut specification and gastrulation suggests homology between Drosophila amnioproctodeal invagination and vertebrate blastopore. *Development*, *125*, 2433–2442.

Yamaguchi, T. P., Conlon, R. A., & Rossant, J. (1992). Expression of the fibroblast growth factor receptor FGFR-1/flg during gastrulation and segmentation in the mouse embryo. *Developmental Biology*, *152*, 75–88. https://doi.org/10.1016/0012-1606(92)90157-c.

Yamaguchi, T. P., Harpal, K., Henkemeyer, M., & Rossant, J. (1994). fgfr-1 is required for embryonic growth and mesodermal patterning during mouse gastrulation. *Genes & Development*, *8*, 3032–3044. https://doi.org/10.1101/gad.8.24.3032.

Yamaguchi, T. P., Takada, S., Yoshikawa, Y., Wu, N., & McMahon, A. P. (1999). T (Brachyury) is a direct target of Wnt3a during paraxial mesoderm specification. *Genes & Development*, *13*, 3185–3190. https://doi.org/10.1101/gad.13.24.3185.

Yamamoto, A., Amacher, S. L., Kim, S. H., Geissert, D., Kimmel, C. B., & De Robertis, E. M. (1998). Zebrafish paraxial protocadherin is a downstream target of spadetail involved in morphogenesis of gastrula mesoderm. *Development*, *125*, 3389–3397.

Yang, J.-M., Chi, W.-Y., Liang, J., Takayanagi, S., Iglesias, P. A., & Huang, C.-H. (2021). Deciphering cell signaling networks with massively multiplexed biosensor barcoding. *Cell*, *184*(25), 6193–6206. https://doi.org/10.1016/j.cell.2021.11.005.

Yang, X., Chrisman, H., & Weijer, C. J. (2008). PDGF signalling controls the migration of mesoderm cells during chick gastrulation by regulating N-cadherin expression. *Development*, *135*, 3521–3530. https://doi.org/10.1242/dev.023416.

Yang, X., Dormann, D., Münsterberg, A. E., & Weijer, C. J. (2002). Cell movement patterns during gastrulation in the chick are controlled by positive and negative chemotaxis mediated by FGF4 and FGF8. *Developmental Cell*, *3*, 425–437. https://doi.org/10.1016/s1534-5807(02)00256-3.

Yao, Y., Li, W., Wu, J., Germann, U. A., Su, M. S. S., Kuida, K., et al. (2003). Extracellular signal-regulated kinase 2 is necessary for mesoderm differentiation. *Proceedings of the National Academy of Sciences of the United States of America*, *100*, 12759–12764. https://doi.org/10.1073/pnas.2134254100.

Zhang, K., Duan, L., Ong, Q., Lin, Z., Varman, P. M., Sung, K., et al. (2014). Light-mediated kinetic control reveals the temporal effect of the Raf/MEK/ERK pathway in PC12 cell neurite outgrowth. *PLoS One*, *9*, e92917. https://doi.org/10.1371/journal.pone.0092917.

Zhou, X. X., Fan, L. Z., Li, P., Shen, K., & Lin, M. Z. (2017). Optical control of cell signaling by single-chain photoswitchable kinases. *Science*, *355*, 836–842. https://doi.org/10.1126/science.aah3605.

Zhu, J., Kwan, K. M., & Mackem, S. (2016). Putative oncogene Brachyury (T) is essential to specify cell fate but dispensable for notochord progenitor proliferation and EMT. *Proceedings of the National Academy of Sciences of the United States of America*, *113*, 3820–3825. https://doi.org/10.1073/pnas.1601252113.

Znosko, W. A., Yu, S., Thomas, K., Molina, G. A., Li, C., Tsang, W., et al. (2010). Overlapping functions of Pea3 ETS transcription factors in FGF signaling during zebrafish development. *Developmental Biology*, *342*, 11–25. https://doi.org/10.1016/j.ydbio.2010.03.011.

CHAPTER EIGHT

Establishment and interpretation of NODAL and BMP signaling gradients in early vertebrate development

Caroline S. Hill*
Developmental Signalling Laboratory, The Francis Crick Institute, London, United Kingdom
*Corresponding author: e-mail address: caroline.hill@crick.ac.uk

Contents

Abstract

Transforming growth factor β (TGF-β) family ligands play crucial roles in orchestrating early embryonic development. Most significantly, two family members, NODAL and BMP form signaling gradients and indeed in fish, frogs and sea urchins these two opposing gradients are sufficient to organize a complete embryonic axis. This review focuses on how these gradients are established and interpreted during early vertebrate development. The review highlights key principles that are emerging, in particular the importance of signaling duration as well as ligand concentration in both gradient generation and their interpretation. Feedforward and feedback loops involving other signaling pathways are also essential for providing spatial and temporal information downstream of the NODAL and BMP signaling pathways. Finally, new data suggest the existence of buffering mechanisms, whereby early signaling defects can be readily corrected downstream later in development, suggesting that signaling gradients do not have to be as precise as previously thought.

Current Topics in Developmental Biology, Volume 149
ISSN 0070-2153
https://doi.org/10.1016/bs.ctdb.2021.12.002

1. Introduction

The transforming growth factor β (TGF-β) family comprises 33 members, of which 30 are ligands and include the TGF-βs themselves, activins, NODAL, bone morphogenetic proteins (BMPs), growth and differentiation factors (GDFs), and anti-Mullerian hormone (AMH). The other three members are the antagonists, LEFTY1/2 and Inhibin α (Morikawa, Derynck, & Miyazono, 2016). Several TGF-β family ligands play essential roles in germ layer specification and tissue patterning in early embryonic development and in tissue homeostasis in adult organisms. Many of these functions, as well as the sequences of the ligands and pathway components have been evolutionarily conserved in organisms ranging from simple placozoa like *Trichoplax* to humans (Huminiecki et al., 2009). Mutations or deletions in key pathway components deregulate signaling and are responsible for many human diseases and syndromes (Miller & Hill, 2016).

The ligands act as disulphide-linked dimers that can be both homodimers or heterodimers. They are all synthesized as long precursors, comprising an N-terminal prodomain and a C-terminal mature domain, separated by a Furin cleavage site (Hinck, Mueller, & Springer, 2016), with the mature domain constituting the active ligand. Generally, the ligands are cleaved during secretion by Furin-like proteases, although in the case of NODAL it has been reported that cleavage can occur extracellularly (Ben-Haim et al., 2006).

All TGF-β family members require two types of plasma membrane-localized serine/threonine kinase receptors to signal—so-called type I and type II receptors (Fig. 1). A ligand dimer binds to a receptor heterotetramer comprising two type Is and two type IIs. Most of the ligands bind one receptor with higher affinity than the other, suggesting a mechanism of sequential receptor assembly. TGF-βs, activins, and some GDFs bind their type II receptors with high affinity and their type I receptors with low affinity. For the BMPs and other GDFs, it is the other way round (Martinez-Hackert, Sundan, & Holien, 2021). In addition, some ligands also require co-receptors. For example, NODAL requires GPI-linked co-receptors of the EGF-CFC family, TDGF1 or CFC1, and some BMPs require members of the RGM family (Hill, 2018; Nickel, Ten Dijke, & Mueller, 2018; Siebold, Yamashita, Monnier, Mueller, & Pasterkamp, 2017). The function of the co-receptors is not entirely resolved. In the case of TDGF1/CFC1, they are thought to promote binding of the NODAL ligand to the receptors,

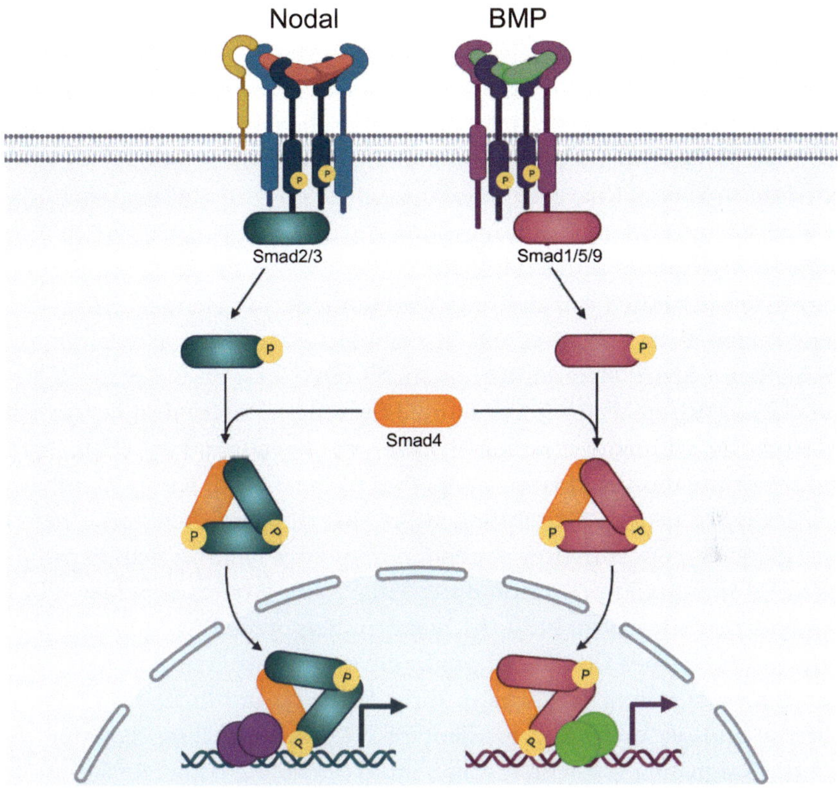

Fig. 1 A schematic of the Nodal and BMP signaling pathways. Ligands signal though a heterotetrameric complex of type I and type II receptors, which for Nodal also requires the GPI-linked co-receptor Tdgf1 (called Oep in fish, colored in yellow). Upon ligand binding constitutively active type II receptors phosphorylate and activate type I receptors, resulting in the phosphorylation of R-Smads: Smad2/3 for Nodal signaling, and Smad1/5/9 for BMP signaling. Phosphorylated Smads form trimers with Smad4 which accumulate in the nucleus where they regulate transcription in conjunction with other non-Smad transcription factors. The figure was prepared using BioRender.

but it has also been suggested that TDGF1 could be involved in both promoting NODAL processing and for internalization of the NODAL–receptor complexes to the endosomes where signaling occurs (Constam, 2009; Hill, 2018). In the receptor complex, the type II receptor phosphorylates the type I receptor on serines and threonines in a juxta membrane region called the GS domain (for glycine- and serine-rich domain). This phosphorylation activates the type I receptor kinase and the phosphoserines and phosphothreonines also contribute to a binding site for the receptor-regulated SMADs (R-SMADs) which are the major signal transducers of the pathway

(Derynck & Budi, 2019). In the case of TGF-β, activin, NODAL and a subset of the GDFs, the relevant R-SMADS are SMAD2 and SMAD3. For the BMPs and other GDFs, the activated R-SMADs are SMAD1, SMAD5 and SMAD9 (Miller & Hill, 2016). Once phosphorylated, these R-SMADs form complexes either amongst themselves or with the common mediator SMAD, SMAD4. These complexes accumulate in the nucleus where they act directly as transcriptional regulators in conjunction with other non-SMAD transcription factors (Hill, 2016) (Fig. 1). The first such transcription factor to be identified was the Forkhead transcription factor FOXH1 (formerly called Fast1) that recruits a complex comprising phosphorylated SMAD2 and SMAD4 to enhancers of NODAL-responsive genes (Chen, Rubock, & Whitman, 1996). A number of other transcription factors with different DNA-binding specificities have since been identified (Hill, 2016).

One of the most fascinating properties of the TGF-β family members is their ability to behave as morphogens, in that they can act across a field of cells, inducing concentration-dependent responses in target cells to provide positional information (Economou & Hill, 2020; Rogers & Schier, 2011). This is particularly true of NODAL and members of the BMP subfamily, and these are the ligands that I shall predominantly concentrate on here. Both ligands form signaling gradients in early embryos, and indeed, in *Xenopus*, zebrafish and sea urchins, a gradient of NODAL signaling, counteracted by a BMP signaling gradient, is sufficient to organize a complete embryonic axis (De Robertis, 2009; Lapraz, Haillot, & Lepage, 2015; Xu, Houssin, Ferri-Lagneau, Thisse, & Thisse, 2014). In developing vertebrate embryos NODAL is required for the induction of mesoderm (prospective muscle, blood, bone, cartilage, heart, spleen, kidney and adipose tissue) and endoderm (gut and other internal organs such as liver and pancreas), for the convergence and extension (C/E) movements of gastrulation, as well as in controlling left–right asymmetry (Grimes & Burdine, 2017; Hill, 2018; Schier, 2003). BMP signaling, in contrast, has a conserved role in many Metazoa in dorsal–ventral (D/V) patterning and later in development in patterning the neural tube and in digit patterning (Andrews, Kong, Novitch, & Butler, 2019; Benazet & Zeller, 2009; Zinski, Tajer, & Mullins, 2018).

Although the gradients of NODAL and BMP signaling have been extensively studied, and the functional roles of these ligands are well understood, there has been much discussion about the mechanism whereby these gradients are established and shaped, and how exactly they are interpreted to give the observed outputs. Substantial progress has been made in these areas in the last 5 years, with a glut of papers published in the last year. I will therefore concentrate on these aspects of NODAL and BMP signaling here, predominantly focusing on work in zebrafish embryos, but bringing in insights from other models like mouse embryos, human and mouse embryonic stem cells (ESCs) and *Drosophila* as appropriate.

2. NODAL signaling

2.1 Establishment of the NODAL signaling gradient

2.1.1 The Nodal and Lefty signaling system satisfies a Turing-like reaction–diffusion system

There are three Nodal ligands in the zebrafish genome, Ndr1 (also known as Squint), Ndr2 (also known as Cyclops) and Southpaw (Feldman et al., 1998; Long, Ahmad, & Rebagliati, 2003; Rebagliati, Toyama, Haffter, & Dawid, 1998). Of these, Ndr1 and Ndr2 are the first expressed and are required for mesoderm and endoderm specification; Southpaw is required later in development for left–right asymmetry (Grimes & Burdine, 2017; Schier, 2009). Beginning at blastula stages (around 4 h post fertilization (hpf)) Ndr1 and Ndr2 establish a signaling gradient at the margin of the embryo. At this stage, the embryo consists of a ball of cells sitting on top of a large yolk cell. Maternally-supplied Ndr1 is localized to the dorsal side of the embryo where it initiates signaling in the dorsal-most embryonic cells. Zygotic Ndr1/2 are expressed in an extra-embryonic tissue called the yolk syncytial layer (YSL) which lies directly under the blastoderm. Their expression is regulated by the transcription factors Nanog and Mxtx2 (itself controlled by Wnt signaling), and they initiate signaling around the entire embryonic margin (Erter, Solnica-Krezel, & Wright, 1998; Fan et al., 2007; Feldman et al., 1998; Gore et al., 2005; Hong, Jang, Brown, McBride, & Feldman, 2011; van Boxtel et al., 2015; Xu et al., 2012). During the subsequent 90 min or so, while the embryo is undergoing epiboly where it thins and spreads out over the yolk cell, a gradient of Nodal signaling is established around the embryonic margin. During gastrulation, these cells involute and migrate anteriorly and dorsally (Schier & Talbot, 2005).

The original model of how the Nodal signaling gradient is established was inspired by the observation that the Nodal ligands and their antagonists, Lefty1 and Lefty2, obey the requirements for a Turing-type reaction–diffusion system (Chen & Schier, 2002; Meinhardt & Gierer, 2000; Saijoh et al., 2000). These requirements state that the activator should induce its own expression and that of the antagonist. In turn, the antagonist(s) should function at a longer

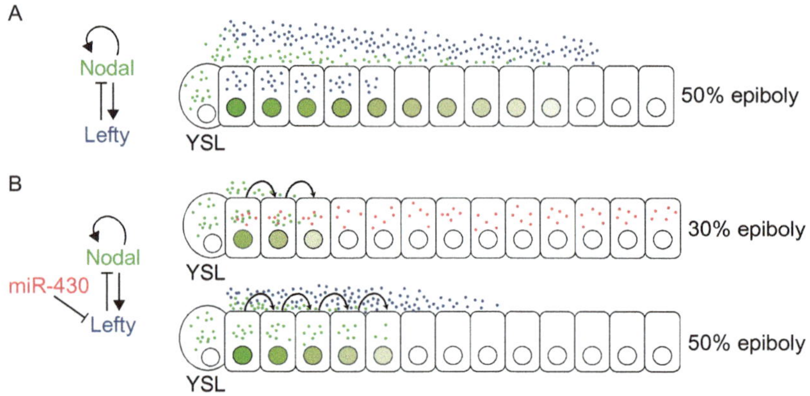

Fig. 2 Schematics showing different models of Nodal signaling gradient formation. (A) The Turing reaction–diffusion model whereby Nodal and Lefty are diffusible, but Lefty diffuses further than Nodal. This results in a gradient over around 10 cell tiers, indicated by the green nuclei which represent levels of pSmad2. The Nodal ligands are initially produced in the YSL (the left-most cell) and signal to the cells of the blastoderm (shown to the right of the YSL). The schematic showing how Nodal and Lefty regulate one another is shown on the far left. (B) An alternative model of Nodal signaling gradient establishment. Lefty translation has been shown to be repressed by the microRNA, miR-430 (pink dots). This keeps Lefty levels low at up to around 30% epiboly. Nodal acts at short range and is thought to spread predominantly via a relay (shown by black arrows). By 50% epiboly, the miR-430 no longer represses Lefty translation and Lefty is produced that prevents the gradient spreading further than around 5 cell tiers. The signaling gradient is depicted as in A. The schematic showing how Nodal, Lefty and miR-430 regulate one another is shown on the far left. In both parts, Nodal is shown as green dots and Lefty as blue dots.

range than the activator(s) (Schier & Talbot, 2005). Ndr1 and Ndr2 are certainly known to directly induce their own expression, as well as that of Lefty1 and Lefty2 (Chen & Schier, 2002 and references therein) (Fig. 2A). Furthermore, early evidence indicated that the Lefty1/2 functioned at a greater distance than Ndr1 and 2 (Chen & Schier, 2001, 2002) (Fig. 2A). Direct measurements of the signaling ranges of these ligands and antagonists using overexpressed fluorescently-tagged versions in fluorescence recovery assays confirmed these observations and showed that Ndr2 has the shortest range, followed by Ndr1, which has an intermediate range, with Lefty1/2 acting as long range inhibitors (Muller et al., 2012). In the absence at that time of a direct readout of Nodal signaling, the Schier lab took advantage of the expression patterns of Nodal target genes. In particular, they focused on *goosecoid* (*gsc*), which is thought to be a low sensitivity target gene requiring high levels of Ndr1 for its induction and high sensitivity targets such as *tbxta* (also known as *no-tail, ta* or *brachyury*), *noto* (also known as *floating head*) and *bhikhari* (*bik*) which require low levels of Ndr1 for their induction (Chen & Schier, 2001, 2002). Given that the high sensitivity targets are expressed up to 10 cell tiers from the YSL (Schier, 2003), the authors concluded that a Nodal signaling gradient is established via a Turing-type reaction–diffusion system where the Nodal ligands diffuse from their source to activate signaling in up to 10 cell tiers, but their activity is inhibited beyond that due to the long range activity of the Lefty proteins (Chen & Schier, 2001, 2002; Muller et al., 2012; Schier, 2003) (Fig. 2A).

2.1.2 Nodal acts as a short-range ligand

In 2015, the establishment of the Nodal gradient was revisited following the development of new methods to directly read out the Nodal signaling gradient (van Boxtel et al., 2015). A transgenic zebrafish line was developed, stably expressing a Nodal reporter comprising *eGFP* under the control of three Foxh1 and Smad binding sites, termed activin response elements (AREs) (Germain, Howell, Esslemont, & Hill, 2000). This reporter showed that the Nodal signaling gradient did not extend as far as previously thought but was confined to the first 5–6 cell tiers from the margin and did not extend beyond the domain of Ndr1/2 ligand expression (van Boxtel et al., 2015). This was confirmed by immunostaining for activated phosphorylated Smad2 (pSmad2) (van Boxtel et al., 2015). This restricted activity of Nodal has also been observed when clones of Ndr1 expression are induced in the animal pole of zebrafish embryos, or when cells expressing fluorescent Ndr1 are transplanted in the animal pole (Soh, Pomreinke, & Muller, 2020; van Boxtel, Economou, Heliot, & Hill, 2018).

The finding that the Nodal gradient is restricted raised the question of what was responsible for the expression of Nodal target genes in a broader domain up to 10 cell tiers from the margin. van Boxtel et al. showed that expression of *tbxta* and *noto* beyond the Nodal expression domain is driven by Fgf signaling, resulting from the expression of Fgfs in the margin under the control of Nodal signaling (Bennett et al., 2007; Mathieu et al., 2004; Rodaway et al., 1999; van Boxtel et al., 2015). Thus, the range of Nodal target gene expression is expanded because of a relay between Nodal and Fgf signaling (Fig. 3A, B). Moreover, the observation that Nodal signaling did not extend beyond the domain of ligand expression, suggested that in the zebrafish embryo Nodal acts at short range and spreads by means of a relay whereby Nodal induces its own expression in adjacent cells (van Boxtel et al., 2015). Furthermore, the length of this gradient is determined by a

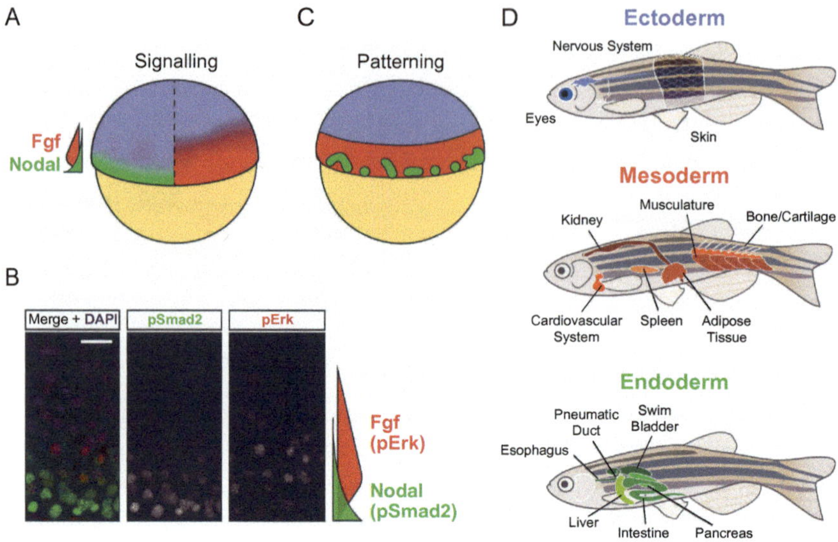

Fig. 3 Specification of mesoderm and endoderm require Nodal and Fgf signaling. (A) A schematic showing the short-range Nodal signaling gradient (read out by pSmad2) and longer range Fgf gradient (read out by pErk) at 50% epiboly. (B) Examples of immunostaining for pSmad2 and pErk in the margin of a 50% epiboly embryo. Nuclei are labeled with DAPI (blue). Scale bar corresponds to 25 μm. (C) The combination of Nodal and Fgf signaling lead to formation of mesoderm progenitors (red) in the first 10 cell tiers from the YSL and endoderm progenitors (green) in the first 2 cell tiers from the margin. Prospective ectoderm is shown in blue. (D) Schematics of adult embryos with the tissues derived from the different germ layers indicated.

delay in the translation of *lefty* mRNA, which is regulated by the miRNA, *miR-430*. Thus, although *lefty1/2* mRNA is synthesized downstream of Nodal, it is not initially translated, and thus the domain of Nodal signaling can grow. At early gastrulation, around 50% epiboly (5.3 hpf), the ability of *miR-430* to prevent Lefty1/2 translation is repressed, functional Lefty1/2 are produced and the spread of Nodal between cell tiers is terminated (van Boxtel et al., 2015) (Fig. 2B). This mechanism explains how a window of signaling activity is transformed into both a spatial and temporal gradient, the latter being evident from the observation that the cells closest to the YSL that have the highest levels of pSmad2, have been signaling the longest.

The idea that the Nodal signaling gradient spreads by a relay has been further investigated recently in both zebrafish embryos and in micropatterned human ESC gastruloids. The Warmflash lab knocked in a fluorescent tag into the endogenous *NODAL* locus in human ESCs, to enable the visualization of endogenous NODAL protein for the first time in living cells

(Liu et al., 2022). Using donor cells expressing the labeled NODAL and receiver cells that can respond to NODAL, but cannot produce it, they showed that NODAL protein is restricted to the immediate environment of the source cells, and spreads by a relay mechanism, consistent with the conclusions drawn from the analyses described above in zebrafish embryos (Liu et al., 2022; van Boxtel et al., 2015). The Warmflash study also showed that although LEFTY is capable of long-range activity, in the human gastruloid system it acts locally to control the timing of NODAL spread (Liu et al., 2022). Recent work from the Schier lab, using transplants into zebrafish embryos of sensor cells that, as in the experiment above, can respond to Nodal, but cannot produce it, has now shed more light on the mechanism underlying the short-range signaling properties of Nodal (Lord, Carte, Abitua, & Schier, 2021). These authors showed that the co-receptor Oep (One-eyed pinhead, the zebrafish orthologue of human TDGF1) acts to restrict the diffusability of Nodal by ligand capture. In the absence of Oep, Nodal can diffuse substantially beyond 100 μm, whereas when Oep is present, it is restricted to the observed signaling gradient of 5–6 cells tiers (~90 μm) (Lord et al., 2021). Moreover, these authors also showed that as well as restricting Nodal diffusion, Oep sensitizes cells to Nodal induction. Furthermore, constant replenishment of Oep at the plasma membrane is necessary for this function. The authors proved this as they observed in a zygotic *oep* mutant, which has only a limited supply of maternal Oep, the Nodal signal is restricted to the margin at dome stage (~4.3 hpf) but propagates outward to form a broad band of signaling by shield stage (~6 hpf) (Lord et al., 2021).

2.1.3 Nodal functions as a heterodimer with GDF1/3

It had generally been assumed that TGF-β family ligands act as homodimers, but in recent years, it is becoming clear that many act as heterodimers (Tajer & Mullins, 2017). This was shown to be the case for the Nodal ligands Ndr1 and Ndr2 in zebrafish embryos by three labs in 2017, when they characterized mutants of another TGF-β family member, Gdf3 (also known as Vg1) (Bisgrove, Su, & Yost, 2017; Montague & Schier, 2017; Pelliccia, Jindal, & Burdine, 2017). These studies indicated that although a zygotic *gdf3* mutant did not have a phenotype, maternal and maternal–zygotic mutants showed strong phenotypes resembling the loss of Nodal signaling seen in MZ*oep* mutants or $ndr1^{-/-};ndr2^{-/-}$ double mutants. These authors went on to show that maternal Gdf3 is required for Nodal function, and that the graded domain of Nodal signaling at the embryonic margin is due to the activity of heterodimers of Ndr1–Gdf3 and Ndr2–Gdf3, with the expression of Ndr1 and Ndr2 being spatially restricted as discussed above, while the expression of Gdf3 is ubiquitous. Probing the underlying mechanism further, Montague and Schier demonstrated that although Ndr1/2 can be processed and secreted without Gdf3, they require Gdf3 for their endogenous activity. Gdf3 is unprocessed and resides in the endoplasmic reticulum without Ndr1/2, and is only processed, secreted, and active in the presence of Ndr1/2 (Montague & Schier, 2017). They suggested that this mechanism enables a rapid response to low levels of ligand in vivo. Exactly how processing and secretion of Gdf3 is regulated by Ndr1/2 has not yet been determined, but similar observations were previously shown for mammalian NODAL, GDF1 and GDF3 (Andersson, Bertolino, & Ibanez, 2007). In addition, work in mammalian cells has shown that heterodimers of NODAL and GDF1 are more potent than NODAL homodimers (Fuerer, Nostro, & Constam, 2014), and in mouse embryos GDF1 directly interacts with NODAL which greatly increases its specific activity (Tanaka, Sakuma, Nakamura, Hamada, & Saijoh, 2007).

2.1.4 Lefty provides robustness in Nodal signaling gradient establishment

Inherent in the models for establishment of the Nodal signaling gradient both in zebrafish embryos and in human gastruloids is the role for the Lefty proteins in determining the range of the Nodal signaling gradient. Characterization of *lft1*$^{-/-}$;*lft2*$^{-/-}$ double zebrafish mutants which lack both Lefty proteins, demonstrated that a gradient of Nodal signaling can be achieved in the absence of Lefty proteins, but the negative feedbacks via Lefty confer robustness to Nodal function (Rogers et al., 2017). In the absence of Lefty1 and Lefty2, the Nodal signaling gradient is expanded. However, the *lft1*$^{-/-}$;*lft2*$^{-/-}$ embryos could be rescued by reducing the dose of Nodal signaling via introducing *ndr1*$^{-/-}$ or *ndr2*$^{-/-}$ mutations, or by treating the embryos with a low dose of the Nodal receptor inhibitor SB-505124 (DaCosta Byfield, Major, Laping, & Roberts, 2004). These embryos could also be rescued with a Lefty-producing clone situated at the animal pole, demonstrating that Lefty production does not need to be coupled to Nodal signaling and that Lefty clearly is capable of substantial diffusion (Rogers et al., 2017). Thus, it appears that the main role of Lefty is to provide negative feedback and buffer any fluctuations in Nodal signaling. This function of Lefty1/2 was also revealed in the recent study of a MZ*smad4a* zebrafish mutant (Guglielmi et al., 2021). Here, the lack of maternal and zygotic Smad4 completely abolished BMP signaling, but surprisingly Nodal signaling was maintained in the absence of Smad4, although lacked robustness. The reason for the fragility of the Nodal signaling in these mutants was because of an initial delay in *lefty1/2* expression (Guglielmi et al., 2021).

2.2 Nodal gradient interpretation

2.2.1 The duration of signaling is an important determinant of gradient interpretation

The traditional view of interpretation of morphogen gradients has been greatly influenced by Lewis Wolpert's French Flag model, which proposed that a graded morphogen acts as a source of positional information across a tissue, with the ligand concentration encoding a positional value (Wolpert, 2011). It has become clear though that although ligand concentration may play a role, duration of ligand exposure is also important (Economou & Hill, 2020; Huang & Saunders, 2020; Pages & Kerridge, 2000). This was first illustrated for Nodal signaling in the experiments of Hagos and Dougan who performed timed inhibition of Nodal signaling using a well characterized small molecule receptor inhibitor and showed that cell fate specification depends on the duration of exposure to Nodal ligands (Hagos & Dougan, 2007). They showed that during the mid-to-late blastula period, Nodal sequentially specifies the somites, notochord, blood, Kupffer's vesicle, hatching gland, heart, and endoderm. They concluded that the total Nodal dose that cells are exposed to is a function of both the length of time a cell is exposed to Nodal and the distance of the cell from the Nodal source (the YSL). Thus, cells at the margin which are closest to the YSL receive a high Nodal dose for an extended period and adopt cell fates such as prechordal plate or definitive endoderm. Conversely, cells further away from the source, which receive lower doses for a shorter time become mesodermal derivatives like somites (Hagos & Dougan, 2007).

This idea that time of ligand exposure is an important determinant of gene expression was explored further in the zebrafish embryo by the Schier lab (Dubrulle et al., 2015). They showed that transcriptional output did not correlate with absolute levels of signaling, thus arguing against the French Flag model. Instead, they found that genes expressed at long range like *tbxta* required just 1 h exposure to Nodal, whereas a short-range gene like *gsc* required 2 h of signaling. They argued that these differences could be explained by differences in the transcription rates of *tbxta* and *gsc* and concluded that morphogen interpretation may be shaped by the kinetics of target gene induction, such that the higher the rate of transcription and the earlier the onset of induction, the greater the spatial range of expression.

The work described above investigating the role of ligand exposure in determining gene expression patterns downstream of Nodal relied on either turning off the pathway at specific time points (Hagos & Dougan, 2007), or transplanting Nodal sensor cells into the animal pole of an embryo that makes Nodal, but cannot respond to it (Dubrulle et al., 2015). The development of optogenetics has more recently made it possible to control the duration of signaling more precisely. The Heisenberg lab made photo-activatable Nodal receptors by fusing the kinase domains of Acvr1b and Acvr2b to light-oxygen-voltage domains which dimerize upon exposure to blue light, thereby activating Nodal signaling (Sako et al., 2016). In these experiments, the length of the light pulse can be used to dictate the duration of Nodal signaling. In the context of MZ*oep* embryos, which are entirely deficient in endogenous Nodal signaling, these authors found that altering Nodal signaling duration within the dorsal organizer had differential effects on mesoderm and endoderm induction. They showed that extended Nodal signaling promoted prechordal plate specification and suppressed endoderm differentiation (Sako et al., 2016). The underlying mechanism involves the up-regulation of Gsc, a marker of prechordal plate and transcriptional repressor, which then acts to repress Nodal-induced Sox17 expression which is required for endoderm differentiation. This mechanism may regulate the balance between prechordal plate and endoderm precursors in the embryonic shield, which is on the dorsal side and experiences the longest duration of Nodal signaling.

2.2.2 The output of Nodal signaling requires feedforward and feedback loops involving multiple signaling pathways

The importance of duration of Nodal signaling in inducing sequential transcriptional events has also been clearly demonstrated in other studies focused on how the Nodal gradient separates the endoderm and mesoderm lineages (van Boxtel et al., 2015, 2018). The observation that Nodal signaling leads to the upregulation of the Fgf ligands Fgf8a and Fgf3 readily explained how Fgf target genes, *tbxta* and *noto* were expressed beyond the cells expressing or responding to Nodal ligands. However, this observation also created a potential paradox, as it had long been known that Fgf signaling inhibits endoderm induction, which occurs in the first two cell tiers from the YSL that clearly express Fgf ligands (Mizoguchi, Izawa, Kuroiwa, & Kikuchi, 2006; Poulain, Furthauer, Thisse, Thisse, & Lepage, 2006; van

Boxtel et al., 2015, 2018) (Fig. 3). Using specific inhibitors of pathways downstream of Fgf, van Boxtel et al. showed that it is Erk MAP kinase signaling that is responsible for the ability of Fgf to inhibit endoderm specification. They went on to show that in the cells closest to the margin, Nodal induces the production of Dusp4, which is a phosphatase that dephosphorylates phosphorylated Erk (pErk). Dusp4 thus suppresses the pErk activity downstream of Fgf specifically in the first two cell tiers, allowing a subset of them to be specified as endoderm (van Boxtel et al., 2018) (Fig. 3A, B). They also identified the importance of the timing of Fgf inhibition and the underlying mechanism. Dusp4 is a direct Nodal target gene but requires the synthesis of the transcription factor Mixl1 for its expression (Nelson et al., 2017), which is itself a target of Nodal signaling (Nelson et al., 2014). This introduces a delay in the expression of Dusp4, compared with other Nodal targets, and confines it to the first two cell tiers. This requirement for the successive expression of *mixl1*, then *dusp4* to repress Fgf signaling before the master regulator of endodermal cell fate *sox32* can be expressed, may at least partly explain the requirement of an extended duration of Nodal signaling for endoderm versus mesoderm (van Boxtel et al., 2018). This work has also highlighted the necessity of feedforward and feedback interactions between multiple signaling pathways for providing cells with temporal and positional information (Fig. 3). Why only a subset of the cells in the first two cell tiers of the zebrafish embryo are fated to become endoderm, even though they apparently receive the same signals as neighboring cells that become mesoderm remains to be determined.

Nodal signaling is not just required for cell fate specification during blastula and gastrula stages but is also required for the C/E movements of gastrulation that are required to extend the anteroposterior (AP) axis of the embryo. To study this, the Solnica-Krezel lab uncoupled the C/E movements from Nodal-induced patterning by investigating the behavior of neuroectoderm of Nodal-deficient zebrafish gastrulae (Williams & Solnica-Krezel, 2020). They found defective C/E movements in these embryos, which were not fully explained by a lack of planar cell polarity (PCP) signaling downstream of Nodal. They also used blastula explants and showed that Nodal induced a robust PCP-dependent cell polarization and C/E movements. Treatment of such explants with a Nodal inhibitor revealed a continuous requirement for Nodal signaling for this extension even after mesoderm was specified and also in the absence of mesoderm, demonstrating a mesoderm-independent role for Nodal in regulating C/E. These authors thus concluded that Nodal acts both upstream of and in parallel with PCP signaling to regulate embryonic axis extension cooperatively (Williams & Solnica-Krezel, 2020).

3. BMP signaling

3.1 Establishment of the BMP signaling gradient

In early zebrafish embryos BMP signaling is required for D/V patterning and does so via a ventral to dorsal BMP activity gradient, which is first evident at around 40% epiboly (~5 hpf) (Ramel & Hill, 2013; Tucker, Mintzer, & Mullins, 2008). This role for BMPs is evident from the dorsalized phenotypes of mutants in receptors (Alk8 (*lost-a-fin*)), ligands (Bmp2b and Bmp7 (*swirl* and *snailhouse* respectively) and pathway components (Smad5 (*somitabun*)) (Mullins et al., 1996), and from the ventralized phenotypes of BMP antagonist mutants, for example Chordin mutants (Chrd (*chordino*)) (Schulte-Merker, Lee, McMahon, & Hammerschmidt, 1997). The mechanism whereby this D/V BMP activity gradient is formed has been the subject of much study in the last decade or so, with some recent papers, which I shall focus on here, making substantial progress.

3.1.1 The BMP D/V gradient in zebrafish is not diffusion based

In some developmental systems, BMP gradients are known to evolve through dispersion from a spatially restricted source. A good example of this is the *Drosophila* wing disc, where a hindered extracellular diffusion model is thought to operate in which the *Drosophila* BMP ligand Dpp spreads by diffusion which is inhibited by binding to extracellular matrix components such as heparin sulphate proteoglycans (Muller, Rogers, Yu, Brand, & Schier, 2013; Schwank et al., 2011). Note that other models, such as one relying on transcytosis have also been proposed (Entchev, Schwabedissen, & Gonzalez-Gaitan, 2000). Strikingly, a recent elegant study using an artificial engineered GFP gradient in the *Drosophila* wing disc together with BMP receptors modified to respond to GFP instead of Dpp, has indicated that a diffusion-based gradient was sufficient to rescue growth and patterning defects in a Dpp-null wing disc (Stapornwongkul, de Gennes, Cocconi, Salbreux, & Vincent, 2020). For optimal rescue, expression of GPI-anchored nonsignaling receptors was also required, which may mimic the activity of heparin sulphate proteoglycans in the wild-type system, and thus giving more weight to the hindered diffusion model.

In contrast, the mechanism is different in the early zebrafish embryo. The gradient is not formed by secretion of ligands from a localized source, but arises by remodeling initially ubiquitous expression of BMP ligands (Ramel & Hill, 2013; Tucker et al., 2008). Visualization of pSmad1/5 activity throughout blastula/gastrula stages indicated a uniform low level of signaling throughout the early blastula embryo (~4 hpf), consistent with the uniform ligand expression. This is progressively cleared dorsally and intensifies in the most ventral regions, then persisting through gastrulation (Tucker et al., 2008). This graded spatial expression pattern of pSmad1/5 is also seen for the transcription of the BMP ligands *bmp2b, bmp7* and *bmp4* (Ramel & Hill, 2013). Initial ubiquitous expression of these ligands is remodeled as a result of first the dorsal expression of the transcription repressor Dharma (Koos & Ho, 1999), followed by inhibitory dorsal Fgf signaling (Furthauer, Van Celst, Thisse, & Thisse, 2004), and finally by the dorsal expression of the BMP antagonist Chrd (Ramel & Hill, 2013; Schulte-Merker et al., 1997). The observation that the activity gradient faithfully mapped onto the transcriptional expression gradient of the ligands, and the absence of BMP signaling activity beyond the ligand-expressing cells, led to the idea that the BMPs do not substantially diffuse, and the activity gradient is dependent on the spatial regulation of ligand transcription, which autoregulate their own expression after the 40% epiboly stage (Ramel & Hill, 2013).

Following on from this work, two groups used a combination of careful quantitation of the BMP activity gradient using pSmad1/5 immunostaining in wild-type and mutant embryos with modeling approaches to address in more detail how the activity gradient is established (Pomreinke et al., 2017; Zinski et al., 2017). Zinski and colleagues considered four possible models for gradient formation: a transcriptional gradient of ligand expression, a shuttling model, a counter gradient model, and a source–sink model (Zinski et al., 2017). The transcriptional gradient model is as described above. The shuttling model is one shown to operate during *Drosophila* D/V patterning. In this model, the BMP ligands form complexes with the antagonist Chrd which allows the BMPs to diffuse towards the BMP source midpoint and away from the site of highest Chrd concentration. There, a metalloprotease, Tolloid cleaves the Chrd, and releases the BMP ligands. This mechanism thus generates an activity gradient across a field of cells uniformly expressing BMPs (Eldar et al., 2002; Sawala, Sutcliffe, & Ashe, 2012; Shimmi, Umulis, Othmer, & O'Connor, 2005). In the counter gradient model, diffusion of Chrd from its expression domain at the dorsal side counteracts the activity of BMPs expressed in the ventral half of the zebrafish embryo (Blitz, Shimmi, Wunnenberg-Stapleton,

O'Connor, & Cho, 2000; Little & Mullins, 2006). Lastly, in the source–sink model, dorsally-expressed Chrd depletes BMP ligands, thus establishing a BMP diffusion gradient from ventral to dorsal (Zinski et al., 2017). These authors showed that most of their data fitted best to the source–sink or counter gradient models, and not the shuttling or transcriptional models. They then distinguished between the source–sink and counter gradient models by measuring BMP diffusivity. The source–sink model predicts high BMP diffusivity, while the counter gradient model predicts low diffusivity. The BMP diffusion rates measured by fluorescence recovery after photobleaching (FRAP) were consistent with a source–sink model (Zinski et al., 2017). Note that the graded BMP production resulting from the transcriptional gradient of *bmp2b, bmp7* and *bmp4* described above is built into all these models, indicating that this transcriptional *bmp* gradient also plays a key role in shaping the pSmad1/5 gradient, in addition to the source–sink mechanism.

In a similar study, Pomreinke and colleagues also considered the source–sink, transcriptional and shuttling models for establishment of a BMP signaling gradient, and additionally an alternative long-range accumulation and feedback model (Inomata, Shibata, Haraguchi, & Sasai, 2013; Pomreinke et al., 2017). Again, by quantifying the evolution of the BMP activity gradient in both wild-type and Chrd mutants and measuring the stability and diffusion of fluorescently-tagged BMP and Chrd, these authors argued that their measured properties were most consistent with a source–sink model (Pomreinke et al., 2017).

More recent work from the Mullins lab on the metalloproteases Tolloid and Bmp1a that degrade the BMP antagonist Chrd has shed more light on the regulation of Chrd activity during the establishment of the D/V BMP activity gradient (Tuazon, Wang, Andrade, Umulis, & Mullins, 2020). By combining rigorous analysis of *tolloid* and *bmp1a* loss-of-function mutants with quantitative immunofluorescence for pSmad1/5 to measure the BMP activity gradient, these authors showed that these proteases play a crucial role in restricting the activity of Chrd to the dorsal side of the embryo. They showed that maternally deposited Bmp1a plays a crucial non-redundant role in establishing the D/V BMP activity gradient, which is then maintained by zygotic Tolloid and Bmp1a, which act redundantly. In the absence of these proteases, Chrd can freely diffuse all the way to the ventral side, leading to a strongly dorsalized phenotype (Tuazon et al., 2020). To corroborate their view that the role of these proteases is therefore to restrict Chrd to the dorsal half of the embryo, the authors made a membrane-tethered non-diffusible version of Chrd by fusing the membrane protein CD2 to the C-terminus of Chrd. They showed that it could rescue an M-*bmp1a*$^{-/-}$;*chrd*$^{-/-}$ mutant if expressed in the dorsal half of the embryo, although not if restricted to the dorsal-most cells (Tuazon et al., 2020). This suggests that embryo-wide Chrd diffusion that would be predicted from a counter-gradient model is not required for BMP gradient formation and supports the idea that Chrd restricted to the dorsal half of the embryo, generates a sink that drives BMP gradient formation (Tuazon et al., 2020) (Fig. 4).

During this study, the authors uncovered a fascinating example of how embryos can recover from early patterning defects. They found that the M-*bmp1a*$^{-/-}$ mutants exhibited a wild-type phenotype at 1 dpf, despite the fact that measurement of their pSmad1/5 gradient in early gastrula embryos revealed that it was significantly diminished compared with wild type, with associated defects in expression of BMP target genes. They went on to show that the gradient had remarkably recovered by mid-gastrulation (Tuazon et al., 2020). The underlying mechanism remains to be determined, but this may be an example of a more general phenomenon of buffering whereby early patterning processes do not have to be as precise as previously

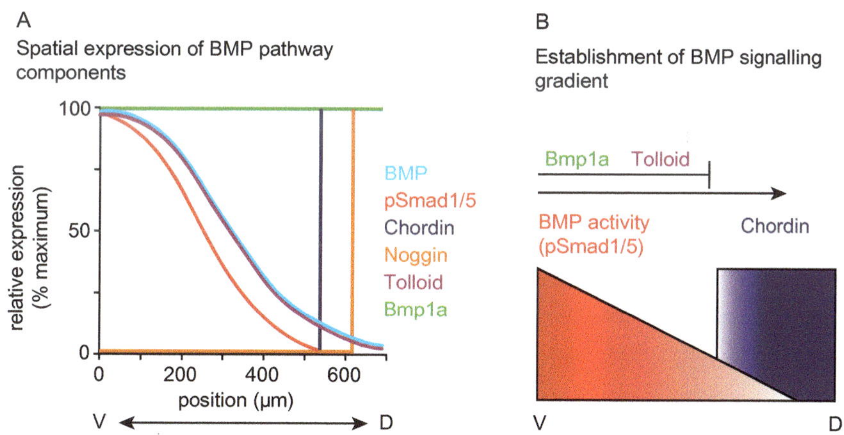

Fig. 4 Establishment of the ventral to dorsal BMP signaling gradient. (A) Expression patterns of components of the BMP signaling pathway in gastrula stage zebrafish embryos. (B) Source-sink model for BMP signaling gradient formation with the activity of the metalloproteases, Tolloid and Bmp1a acting to restrict the spread of Chordin to the dorsal half of the embryo. V, ventral; D, dorsal. *Figure adapted from Tuazon, F. B., Wang, X., Andrade, J. L., Umulis, D., & Mullins, M. C. (2020). Proteolytic restriction of chordin range underlies BMP gradient formation.* Cell Reports, 32, *108039.*

thought, as correction mechanisms exist downstream. Indeed, similar evidence of buffering was made in embryos transiently experiencing ectopic activation of BMP signaling, as these embryos were shown to lack gross morphological defects (Rogers, ElGamacy, Jordan, & Muller, 2020). Rogers et al. suggested that an important feature of BMP signaling is the fact that several BMP target genes are themselves regulators of BMP signaling, and thus embryos may be able to recover because of negative and/or positive feedback. Cell sorting strategies to sharpen gene expression boundaries may also contribute to the observed recovery from BMP signaling manipulation (Rogers et al., 2020).

3.1.2 The importance of heterodimers in BMP signaling

It has been known for some time that in the early zebrafish embryo, Bmp2b and Bmp7a act as obligate heterodimers (Little & Mullins, 2009). The first evidence for this came from the observation that *bmp2b*$^{-/-}$ and *bmp7*$^{-/-}$ mutants have identical phenotypes, indicating that they are genetically equivalent and function non-redundantly. Furthermore, it was demonstrated that only Bmp2b–Bmp7 heterodimers were active; homodimers could form but did not signal. Two classes of type I receptor were required for signaling by the heterodimers, Alk8 (the equivalent of ALK2 or ACVR1 in mammals) and Alk3 (BMPR1A in mammals) (Little & Mullins, 2009). A recent paper from the same group has delved into the mechanism in more detail (Tajer, Dutko, Little, & Mullins, 2021). They showed that if overexpressed, Bmp2 homodimers could signal, but unexpectedly, still required the two different type I receptors, Alk8 (Acvr1) and Alk3 (Bmpr1). Furthermore, the kinase domain of Acvr1 was required, but not that of Bmpr1. Interestingly, Acvr1 binds Bmp2 very poorly, whilst Bmpr1 is a high affinity receptor. The authors therefore concluded that the reason that the heterodimers are much better at signaling than homodimers is because they preferentially recruit both Acvr1 and Bmpr1 into the same signaling complex. In this complex, the type II receptor phosphorylates and activates Acvr1 for downstream signaling, while Bmpr1 acts to promote ligand recruitment. They suggested that the presence of a low affinity receptor in the complex might make the complex more sensitive to fluctuations in BMP concentration, and that too many high affinity receptors at surface might interfere with BMP diffusion (Tajer et al., 2021). Heterodimers of BMPs are not unique to zebrafish embryos. Recently the Christian lab generated an allele of *Bmp7* in mice (Bmp7$^{\text{R-GFlag}}$) which cannot be cleaved to produce the active mature ligand. This mutation abolishes the function of BMP7 homodimers and all other BMPs that normally heterodimerize with BMP7. Analysis of the phenotype of the $Bmp7^{R\text{-}GFlag}$ homozygotes indicated that BMP7 preferentially functions as a heterodimer with BMP2 or BMP4 during mouse embryogenesis (Kim, Neugebauer, McKnite, Tilak, & Christian, 2019).

3.1.3 BMP signaling gradients: Insights from the human blastula

Confluent hESCs form a pseudostratified, apicobasal polarized epithelium which provides a good model for the organization of the embryonic epiblast, where tight junctions separate the apical and the basolateral cell membranes. The Brivanlou and Siggia labs have used this in vitro system both in 2D using micropatterns, and also in 3D to investigate how BMP signaling initiates all the events downstream that lead to gastrulation (Etoc et al., 2016; Simunovic et al., 2019). These studies revealed that BMP4 reception in the polarized epithelium is restricted to the basolateral cell domain, because the BMP receptors are localized basolaterally, and are absent from the apical side (Etoc et al., 2016). A similar phenomenon is also seen in the mouse embryo (Zhang, Zwick, Loew, Grimley, & Ramanathan, 2019) and in other tissue culture systems (Nallet-Staub et al., 2015). Furthermore, uniquely in the human system, BMP4 activates expression of the BMP antagonist, NOGGIN (Etoc et al., 2016). In a recent study, the Brivanlou lab have investigated this further, asking how this agonist–antagonist pair behave to regulate spatial BMP signaling (Phan-Everson et al., 2021). Surprisingly, they showed that BMP4 and NOGGIN are secreted from the opposite sides of the polarized epithelium, thereby segregating the ligand and its inhibitor into distinct extracellular spaces, and making it hard to understand how NOGGIN could regulate BMP signaling activity. The finding that NOGGIN applied apically would inhibit BMP4 applied basally suggested a mechanism involving trans-epithelial transport. By tracing the transcytosis route of apical NOGGIN through the endosomal system, they showed that NOGGIN can diffuse apically and then is internalized in a dynamin-dependent manner, transcytosed, and trafficked to the basolateral surface close to where BMP receptors are located (Phan-Everson et al., 2021). This system may have evolved to prevent immediate binding and inactivation of both ligand and antagonist and thus allow a window of time to initiate BMP signaling before it is subject to reshaping by NOGGIN (Phan-Everson et al., 2021). There are interesting parallels between this system and that described above for the Nodal–Lefty pair, where in that case a microRNA prevents translation of the antagonist Lefty, thus building in a delay in the system to allow the signaling gradient time to get established (van Boxtel et al., 2015). It will be important in the future to determine whether the secretion of agonists and antagonists are on opposite sides of the epithelium for ligands other than BMPs both in the human embryo and in other organisms. Indeed, Activin/NODAL receptors have also been shown to be localized basolaterally in human gastruloids grown on micropatterns, as well as when cultured on filters, although nothing is yet known about which side of the epithelium LEFTY1/2 or the NODAL antagonist Cerberus are secreted (Etoc et al., 2016).

3.2 BMP gradient interpretation

Despite intense study of the mechanism underlying the evolution of the D/V BMP gradient in multiple organisms, much less is known about how this gradient is interpreted to induce different cell fates across the D/V axis. In broad terms, low or no BMP signaling is associated with the development of neurectoderm, an intermediate level with the neural crest and high levels of BMP signaling are associated with specification of the epidermis (Sauka-Spengler & Bronner-Fraser, 2008). Moreover, timing of BMP signaling is important. Using heat shock-inducible transgenes to turn BMP signaling on or off rapidly at different times, the Mullins lab demonstrated that BMP patterns anterior D/V fates at the beginning of gastrulation, while more posterior cell fates require BMP signaling later during gastrulation. For the latter, it appears that a late temporal window of signaling is required, rather than a long duration of signaling (Tucker et al., 2008). The authors suggested that this is a mechanism whereby D/V and AP identities can be acquired simultaneously. Another nice example of the importance of time in BMP signaling gradient interpretation is in neural patterning, in the decision between eye fate and telencephalon fate. The Houart group showed that a well-defined period of BMP signaling in pre and early gastrula embryos is required to protect the telencephalon from adopting a retinal identity through repression of expression of the transcription factor Rx3 (Bielen & Houart, 2012).

Two recent papers from the Mullins and Müller labs have now investigated in more detail how exactly the initial BMP D/V gradient is read out during blastula stages (Greenfeld, Lin, & Mullins, 2021; Rogers et al., 2020). Both labs first identified a cohort of direct BMP target genes. The Müller lab then investigated how well the patterns of expression of these genes mirrored the BMP signaling gradient. They found no clear correlation with the very diverse expression patterns observed for the target genes (Rogers et al., 2020). Since the BMP signaling gradient is formed from an initial uniform signaling domain that is cleared from the dorsal side, the gradient has a temporal as well as a spatial component. These authors therefore used optogenetic BMP receptors, similar in design to those previously used to induce Nodal signaling (Sako et al., 2016), to investigate whether the amplitude or duration of BMP signaling were key determinants of the spatial expression pattern of the target genes. However, they found that neither could account fully for the spatiotemporal expression patterns observed (Rogers et al., 2020). Additionally, they found negligible correlation between spatiotemporal expression and transcription kinetics for the majority of the target genes. Instead, they observed that most of the spatial differences between the BMP target genes disappeared when Fgf and Nodal signaling were inhibited. Fgf and Nodal were shown to affect BMP target gene expression in two ways: they restrict BMP signaling on the dorsal side by activating expression BMP antagonists, and they exclude expression of some BMP targets at the margin. Thus, they concluded that combinatorial signaling is a major driver of spatial expression of BMP target genes (Rogers et al., 2020). This conclusion is very similar to that reached to explain the spatial expression patterns of Nodal target genes at the margin, where again the spatial patterns could not be explained by the activity of a single morphogen gradient but required interplay between Nodal and Fgf signaling (van Boxtel et al., 2018).

The work described above showed that many of the major differences in the spatial patterns of BMP target gene expression were due to combinatorial signaling, but these authors did not investigate how the gradient is read out per se. The Mullins lab deliberately focused on BMP target genes that they surmised were not influenced by other signaling pathways and used a combination of modeling approaches and quantitative imaging to investigate how the BMP signaling gradient is interpreted (Greenfeld et al., 2021). They considered three possible mechanisms: the cells read out the slope of the gradient, or the duration of signaling, or a threshold of signaling activation. By observing the behavior of three specific target genes (*sizzled*, *foxi1* and *bambia*) expressed at different ranges within the BMP signaling gradient and observing how their expression pattern changed in a *chrd*$^{-/-}$ mutant, and in *bmp7*$^{-/-}$ mutants rescued with different amounts of recombinant BMP2/7, they ruled out gradient slope and signaling duration as determinants of BMP gene spatial expression. Instead, they showed that cells respond to three distinct pSmad5 threshold levels to activate and position target gene expression along the D/V axis (Greenfeld et al., 2021). Further work is now needed to determine how the enhancers of these target genes are so sensitive to levels of nuclear pSmad1/5.

4. Interactions between NODAL and BMP signaling

The Nodal and BMP signaling pathways do not operate in isolation in the developing embryo, but both synergise and antagonize each other, depending on the context. Several mechanisms have been described that function at different levels within the signaling pathways. Concerning synergistic interactions, these have been well documented in mouse and human systems. In embryonic day 5.5 mouse embryos BMP signaling has been shown to act upstream of NODAL signaling. Uncleaved NODAL precursor produced in the epiblast is both activated through prodomain cleavage by the secreted proprotein-convertases at the interface of the extraembryonic ectoderm and epiblast. It also upregulates BMP4 expression in the extraembryonic ectoderm, which in turn signals back to the epiblast to enhance WNT3 expression, which then activates NODAL expression (Arnold & Robertson, 2009; Ben-Haim et al., 2006). This feedforward motif between BMPs and NODAL is also evident in human ESCs, where BMP signaling is sufficient to induce all three germ layers in self organizing gastruloids, through its ability to induce NODAL expression (Etoc et al., 2016; Warmflash, Sorre, Etoc, Siggia, & Brivanlou, 2014). This feedforward motif has not been found in zebrafish embryos, where the expression of these ligands is independent of one another (Schier & Talbot, 2005).

In other contexts, these pathways antagonize each other, and this can occur via different mechanisms. One obvious mechanism is via competition for SMAD4, which is shared between the two pathways. This is thought to occur during left–right axis determination in mouse embryos where BMP signaling attenuates NODAL signaling (Furtado et al., 2008). BMP signaling through SMAD1 sets a threshold for NODAL signaling in the left lateral plate mesoderm (LPM) by limiting the availability of SMAD4, thus determining the level of NODAL-dependent *Nodal* activation in the LPM (Furtado et al., 2008). Additional feedbacks between NODAL and BMP may also occur during left–right patterning through the activity of the BMP antagonists, CHRD and NOGGIN (Mine, Anderson, & Klingensmith, 2008). Intriguingly, antagonism between NODAL and BMP signaling via competition for SMAD4 is not seen in early zebrafish embryos, as inhibition of BMP signaling with the small molecule type I receptor inhibitor, DMH1 or via an *MZalk8* mutation does not result in promoting Nodal signaling (Guglielmi et al., 2021; Mintzer et al., 2001) These pathways may also inhibit one another through the formation of so-called mixed R-SMAD complexes comprising activated SMAD2 or SMAD3 with activated SMAD1 or SMAD5. This mechanism has been demonstrated in tissue culture systems (Gronroos et al., 2012), and is also thought to be responsible for the observed antagonism of NODAL signaling by BMP–SMAD5 signaling that prevents primitive streak formation in the amnion of normal mouse embryos (Pereira et al., 2012). These authors observed the formation of an ectopic primitive streak-like structure accompanied by ectopic *Nodal* expression in the amnion of the *smad5*$^{-/-}$ mutant mouse. Antagonism between these pathways can also occur downstream, as in the induction of expression of BMP antagonists like Chrd and Noggin by Nodal in the dorsal organizer of fish and frog embryos (Langdon & Mullins, 2011; Rogers et al., 2020; Tuazon & Mullins, 2015).

Prompted by the observation that opposing sources of BMP and Nodal ligands are sufficient to induce the formation of a full axis in zebrafish embryos the Müller lab has explored how these two major TGF-β family signaling pathways interact with one another to determine cell fate decisions (Soh et al., 2020). Having demonstrated that adjacent transplants of clones of fluorescently labeled BMP and Nodal in the sphere stage embryos resulted in zebrafish embryos that generated secondary axes with high efficiency, they mimicked Nodal and BMP signaling using constitutively activate versions of the R-Smads, Smad2 and Smad5, in which the serines that are normally phosphorylated by the appropriate type I receptors, are replaced by phosphoserine mimicking aspartates (Souchelnytskyi et al., 1997). They found that specific ratios of active Smad2 and Smad5 were able to generate a variety of embryonic structures, suggesting the importance of different ratios of active Smad2 and Smad5 in cell fate decision making (Soh et al., 2020). They found that Smad2 and Smad5 selectively antagonize each other for certain cell fates, like prechordal plate marked by *gsc,* which requires high activated Smad2 levels and low activated Smad5 levels, and epidermis, marked by *foxi1* that conversely requires high activated Smad5 levels and low activated Smad2 levels. For other fates, the pathways act synergistically as shown for the ventral mesoderm marker *eve1* which is expressed in the ventral margin and requires inputs from both activated Smad2 and activated Smad5 (Soh et al., 2020). Further work is now needed to fully understand in this system to what extent these interactions are direct at the enhancers of these target genes, or whether they are mediated by intermediary factors or signaling molecules.

5. Outlook

The NODAL and BMP signaling pathways act as morphogens to specify the germ layers and pattern them in early vertebrate embryos. Recent work, particularly in the context of zebrafish embryos and human gastruloids, has shed significant light on how these morphogen gradients are established, shaped, and interpreted. Several key principles are emerging from these studies. First, generally these ligands seem not to generally function as homodimers, but rather as heterodimers with other TGF-β family members. In the case of the BMP heterodimers, this reflects the requirement for two distinct type I receptors for signaling, but the molecular basis for the Nodal–GDF1/3 heterodimers is still not clear. Second, these pathways do not function alone, but rather as a network with other signaling pathways including Fgf and Wnt that are activated downstream. The patterns of gene expression activated in response to BMP or Nodal signaling are therefore modulated by inputs from these other pathways. The result of this is that it is not only the strength of BMP or Nodal signaling that is important, but also the duration, because sequential rounds of gene expression are required to activate the downstream pathways and to remodel them. Finally, work discussed in this review has suggested that these gradients may not be read out with the precision that was previously thought. Several recent studies have shown for BMP signaling that either insufficient or excessive BMP signaling early on in development is compensated for later and therefore does not result in the D/V patterning defects that would be expected. This suggests that robust buffering mechanisms must exist to correct imprecise signaling gradients. Deciphering what these mechanisms are will be important for the future, as well as establishing whether the same is true of Nodal signaling.

Acknowledgments

I would like to thank James Briscoe, Andrew Economou, Luca Guglielmi and Scott Wilcockson for very helpful comments on the manuscript. I thank Scott Wilcockson for Fig. 3. Work in the Hill lab is funded by the Francis Crick Institute, which receives its core funding from Cancer Research UK (FC001095), the UK Medical Research Council (FC001095), and the Wellcome Trust (FC001095) and the EPSRC.

References

Andersson, O., Bertolino, P., & Ibanez, C. F. (2007). Distinct and cooperative roles of mammalian Vg1 homologs GDF1 and GDF3 during early embryonic development. *Developmental Biology*, *311*, 500–511.

Andrews, M. G., Kong, J., Novitch, B. G., & Butler, S. J. (2019). New perspectives on the mechanisms establishing the dorsal-ventral axis of the spinal cord. *Current Topics in Developmental Biology*, *132*, 417–450.

Arnold, S. J., & Robertson, E. J. (2009). Making a commitment: Cell lineage allocation and axis patterning in the early mouse embryo. *Nature Reviews. Molecular Cell Biology*, *10*, 91–103.

Benazet, J. D., & Zeller, R. (2009). Vertebrate limb development: Moving from classical morphogen gradients to an integrated 4-dimensional patterning system. *Cold Spring Harbor Perspectives in Biology*, *1*, a001339.

Ben-Haim, N., Lu, C., Guzman-Ayala, M., Pescatore, L., Mesnard, D., Bischofberger, M., et al. (2006). The nodal precursor acting via activin receptors induces mesoderm by maintaining a source of its convertases and BMP4. *Developmental Cell*, *11*, 313–323.

Bennett, J. T., Joubin, K., Cheng, S., Aanstad, P., Herwig, R., Clark, M., et al. (2007). Nodal signaling activates differentiation genes during zebrafish gastrulation. *Developmental Biology*, *304*, 525–540.

Bielen, H., & Houart, C. (2012). BMP signaling protects telencephalic fate by repressing eye identity and its Cxcr4-dependent morphogenesis. *Developmental Cell*, *23*, 812–822.

Bisgrove, B. W., Su, Y. C., & Yost, H. J. (2017). Maternal Gdf3 is an obligatory cofactor in nodal signaling for embryonic axis formation in zebrafish. *eLife*, *6*, e28534.

Blitz, I. L., Shimmi, O., Wunnenberg-Stapleton, K., O'Connor, M. B., & Cho, K. W. (2000). Is chordin a long-range- or short-range-acting factor? Roles for BMP1-related metalloproteases in chordin and BMP4 autofeedback loop regulation. *Developmental Biology*, *223*, 120–138.

Chen, X., Rubock, M. J., & Whitman, M. (1996). A transcriptional partner for MAD proteins in TGF-β signalling. *Nature*, *383*, 691–696.

Chen, Y., & Schier, A. F. (2001). The zebrafish nodal signal squint functions as a morphogen. *Nature*, *411*, 607–610.

Chen, Y., & Schier, A. F. (2002). Lefty proteins are long-range inhibitors of squint-mediated nodal signaling. *Current Biology*, *12*, 2124–2128.

Constam, D. B. (2009). Riding shotgun: A dual role for the epidermal growth factor-Cripto/FRL-1/cryptic protein Cripto in nodal trafficking. *Traffic*, *10*, 783–791.

DaCosta Byfield, S., Major, C., Laping, N. J., & Roberts, A. B. (2004). SB-505124 is a selective inhibitor of transforming growth factor-β type I receptors ALK4, ALK5, and ALK7. *Molecular Pharmacology*, *65*, 744–752.

De Robertis, E. M. (2009). Spemann's organizer and the self-regulation of embryonic fields. *Mechanisms of Development*, *126*, 925–941.

Derynck, R., & Budi, E. H. (2019). Specificity, versatility, and control of TGF-β family signaling. *Science Signaling*, *12*, eaav5183.

Dubrulle, J., Jordan, B. M., Akhmetova, L., Farrell, J. A., Kim, S. H., Solnica-Krezel, L., et al. (2015). Response to nodal morphogen gradient is determined by the kinetics of target gene induction. *eLife*, *4*, e05042.

Economou, A. D., & Hill, C. S. (2020). Temporal dynamics in the formation and interpretation of nodal and BMP morphogen gradients. *Current Topics in Developmental Biology*, *137*, 363–389.

Eldar, A., Dorfman, R., Weiss, D., Ashe, H., Shilo, B. Z., & Barkai, N. (2002). Robustness of the BMP morphogen gradient in Drosophila embryonic patterning. *Nature*, *419*, 304–308.

Entchev, E. V., Schwabedissen, A., & Gonzalez-Gaitan, M. (2000). Gradient formation of the TGF-β homolog Dpp. *Cell*, *103*, 981–991.

Erter, C. E., Solnica-Krezel, L., & Wright, C. V. (1998). Zebrafish nodal-related 2 encodes an early mesendodermal inducer signaling from the extraembryonic yolk syncytial layer. *Developmental Biology*, *204*, 361–372.

Etoc, F., Metzger, J., Ruzo, A., Kirst, C., Yoney, A., Ozair, M. Z., et al. (2016). A balance between secreted inhibitors and edge sensing controls gastruloid self-organization. *Developmental Cell, 39*, 302–315.

Fan, X., Hagos, E. G., Xu, B., Sias, C., Kawakami, K., Burdine, R. D., et al. (2007). Nodal signals mediate interactions between the extra-embryonic and embryonic tissues in zebrafish. *Developmental Biology, 310*, 363–378.

Feldman, B., Gates, M. A., Egan, E. S., Dougan, S. T., Rennebeck, G., Sirotkin, H. I., et al. (1998). Zebrafish organizer development and germ-layer formation require nodal-related signals. *Nature, 395*, 181–185.

Fuerer, C., Nostro, M. C., & Constam, D. B. (2014). Nodal.Gdf1 heterodimers with bound prodomains enable serum-independent nodal signaling and endoderm differentiation. *The Journal of Biological Chemistry, 289*, 17854–17871.

Furtado, M. B., Solloway, M. J., Jones, V. J., Costa, M. W., Biben, C., Wolstein, O., et al. (2008). BMP/SMAD1 signaling sets a threshold for the left/right pathway in lateral plate mesoderm and limits availability of SMAD4. *Genes & Development, 22*, 3037–3049.

Furthauer, M., Van Celst, J., Thisse, C., & Thisse, B. (2004). Fgf signalling controls the dorsoventral patterning of the zebrafish embryo. *Development, 131*, 2853–2864.

Germain, S., Howell, M., Esslemont, G. M., & Hill, C. S. (2000). Homeodomain and winged-helix transcription factors recruit activated Smads to distinct promoter elements via a common Smad interaction motif. *Genes & Development, 14*, 435–451.

Gore, A. V., Maegawa, S., Cheong, A., Gilligan, P. C., Weinberg, E. S., & Sampath, K. (2005). The zebrafish dorsal axis is apparent at the four-cell stage. *Nature, 438*, 1030–1035.

Greenfeld, H., Lin, J., & Mullins, M. C. (2021). The BMP signaling gradient is interpreted through concentration thresholds in dorsal-ventral axial patterning. *PLoS Biology, 19*, e3001059.

Grimes, D. T., & Burdine, R. D. (2017). Left-right patterning: Breaking symmetry to asymmetric morphogenesis. *Trends in Genetics, 33*, 616–628.

Gronroos, E., Kingston, I. J., Ramachandran, A., Randall, R. A., Vizan, P., & Hill, C. S. (2012). Transforming growth factor β inhibits bone morphogenetic protein-induced transcription through novel phosphorylated Smad1/5-Smad3 complexes. *Molecular and Cellular Biology, 32*, 2904–2916.

Guglielmi, L., Heliot, C., Kumar, S., Alexandrov, Y., Gori, I., Papaleonidopoulou, F., et al. (2021). Smad4 controls signaling robustness and morphogenesis by differentially contributing to the nodal and BMP pathways. *Nature Communications, 12*, 6374.

Hagos, E. G., & Dougan, S. T. (2007). Time-dependent patterning of the mesoderm and endoderm by nodal signals in zebrafish. *BMC Developmental Biology, 7*, 22.

Hill, C. S. (2016). Transcriptional control by the SMADs. *Cold Spring Harbor Perspectives in Biology, 8*, a022079.

Hill, C. S. (2018). Spatial and temporal control of NODAL signaling. *Current Opinion in Cell Biology, 51*, 50–57.

Hinck, A. P., Mueller, T. D., & Springer, T. A. (2016). Structural biology and evolution of the TGF-β family. *Cold Spring Harbor Perspectives in Biology, 8*, a022103.

Hong, S. K., Jang, M. K., Brown, J. L., McBride, A. A., & Feldman, B. (2011). Embryonic mesoderm and endoderm induction requires the actions of non-embryonic nodal-related ligands and Mxtx2. *Development, 138*, 787–795.

Huang, A., & Saunders, T. E. (2020). A matter of time: Formation and interpretation of the Bicoid morphogen gradient. *Current Topics in Developmental Biology, 137*, 79–117.

Huminiecki, L., Goldovsky, L., Freilich, S., Moustakas, A., Ouzounis, C., & Heldin, C. H. (2009). Emergence, development and diversification of the TGF-β signalling pathway within the animal kingdom. *BMC Evolutionary Biology, 9*, 28.

Inomata, H., Shibata, T., Haraguchi, T., & Sasai, Y. (2013). Scaling of dorsal-ventral patterning by embryo size-dependent degradation of Spemann's organizer signals. *Cell*, *153*, 1296–1311.
Kim, H. S., Neugebauer, J., McKnite, A., Tilak, A., & Christian, J. L. (2019). BMP7 functions predominantly as a heterodimer with BMP2 or BMP4 during mammalian embryogenesis. *eLife*, *8*, e48872.
Koos, D. S., & Ho, R. K. (1999). The nieuwkoid/dharma homeobox gene is essential for bmp2b repression in the zebrafish pregastrula. *Developmental Biology*, *215*, 190–207.
Langdon, Y. G., & Mullins, M. C. (2011). Maternal and zygotic control of zebrafish dorsoventral axial patterning. *Annual Review of Genetics*, *45*, 357–377.
Lapraz, F., Haillot, E., & Lepage, T. (2015). A deuterostome origin of the Spemann organiser suggested by nodal and ADMPs functions in echinoderms. *Nature Communications*, *6*, 8434.
Little, S. C., & Mullins, M. C. (2006). Extracellular modulation of BMP activity in patterning the dorsoventral axis. *Birth Defects Research. Part C, Embryo Today*, *78*, 224–242.
Little, S. C., & Mullins, M. C. (2009). Bone morphogenetic protein heterodimers assemble heteromeric type I receptor complexes to pattern the dorsoventral axis. *Nature Cell Biology*, *11*, 637–643.
Liu, L, Nemashkalo, A, Rezende, L, Jung, J. Y., Chhabra, S, Guerra, M. C., & Warmflash, A. (2022). Nodal is a short-range morphogen with activity that spreads through a relay mechanism in human gastruloids. *Nature Communications*, *13*, 497. https://doi.org/10.1038/s41467-022-28149-3.
Long, S., Ahmad, N., & Rebagliati, M. (2003). The zebrafish nodal-related gene southpaw is required for visceral and diencephalic left-right asymmetry. *Development*, *130*, 2303–2316.
Lord, N. D., Carte, A. N., Abitua, P. B., & Schier, A. F. (2021). The pattern of nodal morphogen signaling is shaped by co-receptor expression. *eLife*, *10*, e54894.
Martinez-Hackert, E., Sundan, A., & Holien, T. (2021). Receptor binding competition: A paradigm for regulating TGF-β family action. *Cytokine & Growth Factor Reviews*, *57*, 39–54.
Mathieu, J., Griffin, K., Herbomel, P., Dickmeis, T., Strahle, U., Kimelman, D., et al. (2004). Nodal and Fgf pathways interact through a positive regulatory loop and synergize to maintain mesodermal cell populations. *Development*, *131*, 629–641.
Meinhardt, H., & Gierer, A. (2000). Pattern formation by local self-activation and lateral inhibition. *BioEssays*, *22*, 753–760.
Miller, D. S. J., & Hill, C. S. (2016). TGF-β superfamily signalling. In R. A. Bradshaw, & P. D. Stahl (Eds.), *Encyclopedia of cell biology* (pp. 37–50). Elsevier.
Mine, N., Anderson, R. M., & Klingensmith, J. (2008). BMP antagonism is required in both the node and lateral plate mesoderm for mammalian left-right axis establishment. *Development*, *135*, 2425–2434.
Mintzer, K. A., Lee, M. A., Runke, G., Trout, J., Whitman, M., & Mullins, M. C. (2001). Lost-a-fin encodes a type I BMP receptor, Alk8, acting maternally and zygotically in dorsoventral pattern formation. *Development*, *128*, 859–869.
Mizoguchi, T., Izawa, T., Kuroiwa, A., & Kikuchi, Y. (2006). Fgf signaling negatively regulates nodal-dependent endoderm induction in zebrafish. *Developmental Biology*, *300*, 612–622.
Montague, T. G., & Schier, A. F. (2017). Vg1-nodal heterodimers are the endogenous inducers of mesendoderm. *eLife*, *6*, e28183.
Morikawa, M., Derynck, R., & Miyazono, K. (2016). TGF-β and the TGF-β family: Context-dependent roles in cell and tissue physiology. *Cold Spring Harbor Perspectives in Biology*, *8*.

Muller, P., Rogers, K. W., Jordan, B. M., Lee, J. S., Robson, D., Ramanathan, S., et al. (2012). Differential diffusivity of nodal and lefty underlies a reaction-diffusion patterning system. *Science, 336*, 721–724.

Muller, P., Rogers, K. W., Yu, S. R., Brand, M., & Schier, A. F. (2013). Morphogen transport. *Development, 140*, 1621–1638.

Mullins, M. C., Hammerschmidt, M., Kane, D. A., Odenthal, J., Brand, M., van Eeden, F. J., et al. (1996). Genes establishing dorsoventral pattern formation in the zebrafish embryo: The ventral specifying genes. *Development, 123*, 81–93.

Nallet-Staub, F., Yin, X., Gilbert, C., Marsaud, V., Ben Mimoun, S., Javelaud, D., et al. (2015). Cell density sensing alters TGF-β signaling in a cell-type-specific manner, independent from hippo pathway activation. *Developmental Cell, 32*, 640–651.

Nelson, A. C., Cutty, S. J., Gasiunas, S. N., Deplae, I., Stemple, D. L., & Wardle, F. C. (2017). In vivo regulation of the zebrafish endoderm progenitor niche by T-box transcription factors. *Cell Reports, 19*, 2782–2795.

Nelson, A. C., Cutty, S. J., Niini, M., Stemple, D. L., Flicek, P., Houart, C., et al. (2014). Global identification of Smad2 and Eomesodermin targets in zebrafish identifies a conserved transcriptional network in mesendoderm and a novel role for Eomesodermin in repression of ectodermal gene expression. *BMC Biology, 12*, 81.

Nickel, J., Ten Dijke, P., & Mueller, T. D. (2018). TGF-β family co-receptor function and signaling. *Acta Biochimica et Biophysica Sinica (Shanghai), 50*, 12–36.

Pages, F., & Kerridge, S. (2000). Morphogen gradients. A question of time or concentration? *Trends in Genetics, 16*, 40–44.

Pelliccia, J. L., Jindal, G. A., & Burdine, R. D. (2017). Gdf3 is required for robust nodal signaling during germ layer formation and left-right patterning. *eLife, 6*, e28635.

Pereira, P. N., Dobreva, M. P., Maas, E., Cornelis, F. M., Moya, I. M., Umans, L., et al. (2012). Antagonism of nodal signaling by BMP/Smad5 prevents ectopic primitive streak formation in the mouse amnion. *Development, 139*, 3343–3354.

Phan-Everson, T., Etoc, F., Li, S., Khodursky, S., Yoney, A., Brivanlou, A. H., et al. (2021). Differential compartmentalization of BMP4/NOGGIN requires NOGGIN trans-epithelial transport. *Developmental Cell, 56*(1930–1944), e1935.

Pomreinke, A. P., Soh, G. H., Rogers, K. W., Bergmann, J. K., Blassle, A. J., & Muller, P. (2017). Dynamics of BMP signaling and distribution during zebrafish dorsal-ventral patterning. *eLife, 6*, e25861.

Poulain, M., Furthauer, M., Thisse, B., Thisse, C., & Lepage, T. (2006). Zebrafish endoderm formation is regulated by combinatorial nodal, FGF and BMP signalling. *Development, 133*, 2189–2200.

Ramel, M. C., & Hill, C. S. (2013). The ventral to dorsal BMP activity gradient in the early zebrafish embryo is determined by graded expression of BMP ligands. *Developmental Biology, 378*, 170–182.

Rebagliati, M. R., Toyama, R., Haffter, P., & Dawid, I. B. (1998). Cyclops encodes a nodal-related factor involved in midline signaling. *Proceedings of the National Academy of Sciences of the United States of America, 95*, 9932–9937.

Rodaway, A., Takeda, H., Koshida, S., Broadbent, J., Price, B., Smith, J. C., et al. (1999). Induction of the mesendoderm in the zebrafish germ ring by yolk cell-derived TGF-β family signals and discrimination of mesoderm and endoderm by FGF. *Development, 126*, 3067–3078.

Rogers, K. W., ElGamacy, M., Jordan, B. M., & Muller, P. (2020). Optogenetic investigation of BMP target gene expression diversity. *eLife, 9*, e58641.

Rogers, K. W., Lord, N. D., Gagnon, J. A., Pauli, A., Zimmerman, S., Aksel, D. C., et al. (2017). Nodal patterning without lefty inhibitory feedback is functional but fragile. *eLife, 6*, e28785.

Rogers, K. W., & Schier, A. F. (2011). Morphogen gradients: From generation to interpretation. *Annual Review of Cell and Developmental Biology*, *27*, 377–407.
Saijoh, Y., Adachi, H., Sakuma, R., Yeo, C. Y., Yashiro, K., Watanabe, M., et al. (2000). Left-right asymmetric expression of lefty2 and nodal is induced by a signaling pathway that includes the transcription factor FAST2. *Molecular Cell*, *5*, 35–47.
Sako, K., Pradhan, S. J., Barone, V., Ingles-Prieto, A., Muller, P., Ruprecht, V., et al. (2016). Optogenetic control of nodal Signaling reveals a temporal pattern of nodal signaling regulating cell fate specification during gastrulation. *Cell Reports*, *16*, 866–877.
Sauka-Spengler, T., & Bronner-Fraser, M. (2008). A gene regulatory network orchestrates neural crest formation. *Nature Reviews. Molecular Cell Biology*, *9*, 557–568.
Sawala, A., Sutcliffe, C., & Ashe, H. L. (2012). Multistep molecular mechanism for bone morphogenetic protein extracellular transport in the Drosophila embryo. *Proceedings of the National Academy of Sciences of the United States of America*, *109*, 11222–11227.
Schier, A. F. (2003). Nodal signaling in vertebrate development. *Annual Review of Cell and Developmental Biology*, *19*, 589–621.
Schier, A. F. (2009). Nodal morphogens. *Cold Spring Harbor Perspectives in Biology*, *1*, a003459.
Schier, A. F., & Talbot, W. S. (2005). Molecular genetics of axis formation in zebrafish. *Annual Review of Genetics*, *39*, 561–613.
Schulte-Merker, S., Lee, K. J., McMahon, A. P., & Hammerschmidt, M. (1997). The zebrafish organizer requires chordino. *Nature*, *387*, 862–863.
Schwank, G., Dalessi, S., Yang, S. F., Yagi, R., de Lachapelle, A. M., Affolter, M., et al. (2011). Formation of the long range Dpp morphogen gradient. *PLoS Biology*, *9*, e1001111.
Shimmi, O., Umulis, D., Othmer, H., & O'Connor, M. B. (2005). Facilitated transport of a Dpp/Scw heterodimer by Sog/Tsg leads to robust patterning of the drosophila blastoderm embryo. *Cell*, *120*, 873–886.
Siebold, C., Yamashita, T., Monnier, P. P., Mueller, B. K., & Pasterkamp, R. J. (2017). RGMs: Structural insights, molecular regulation, and downstream Signaling. *Trends in Cell Biology*, *27*, 365–378.
Simunovic, M., Metzger, J. J., Etoc, F., Yoney, A., Ruzo, A., Martyn, I., et al. (2019). A 3D model of a human epiblast reveals BMP4-driven symmetry breaking. *Nature Cell Biology*, *21*, 900–910.
Soh, G. H., Pomreinke, A. P., & Muller, P. (2020). Integration of nodal and BMP signaling by mutual Signaling effector antagonism. *Cell Reports*, *31*, 107487.
Souchelnytskyi, S., Tamaki, K., Engstrom, U., Wernstedt, C., ten Dijke, P., & Heldin, C. H. (1997). Phosphorylation of Ser465 and Ser467 in the C terminus of Smad2 mediates interaction with Smad4 and is required for transforming growth factor-β signaling. *The Journal of Biological Chemistry*, *272*, 28107–28115.
Stapornwongkul, K. S., de Gennes, M., Cocconi, L., Salbreux, G., & Vincent, J. P. (2020). Patterning and growth control in vivo by an engineered GFP gradient. *Science*, *370*, 321–327.
Tajer, B., Dutko, J. A., Little, S. C., & Mullins, M. C. (2021). BMP heterodimers signal via distinct type I receptor class functions. *Proceedings of the National Academy of Sciences of the United States of America*, *118*, e2017952118.
Tajer, B., & Mullins, M. C. (2017). Heterodimers reign in the embryo. *Elife*, *6*, e33682.
Tanaka, C., Sakuma, R., Nakamura, T., Hamada, H., & Saijoh, Y. (2007). Long-range action of nodal requires interaction with GDF1. *Genes & Development*, *21*, 3272–3282.
Tuazon, F. B., & Mullins, M. C. (2015). Temporally coordinated signals progressively pattern the anteroposterior and dorsoventral body axes. *Seminars in Cell & Developmental Biology*, *42*, 118–133.
Tuazon, F. B., Wang, X., Andrade, J. L., Umulis, D., & Mullins, M. C. (2020). Proteolytic restriction of chordin range underlies BMP gradient formation. *Cell Reports*, *32*, 108039.

Tucker, J. A., Mintzer, K. A., & Mullins, M. C. (2008). The BMP signaling gradient patterns dorsoventral tissues in a temporally progressive manner along the anteroposterior axis. *Developmental Cell*, *14*, 108–119.

van Boxtel, A. L., Chesebro, J. E., Heliot, C., Ramel, M. C., Stone, R. K., & Hill, C. S. (2015). A temporal window for signal activation dictates the dimensions of a nodal signaling domain. *Developmental Cell*, *35*, 175–185.

van Boxtel, A. L., Economou, A. D., Heliot, C., & Hill, C. S. (2018). Long-range Signaling activation and local inhibition separate the mesoderm and endoderm lineages. *Developmental Cell*, *44*(179–191), e175.

Warmflash, A., Sorre, B., Etoc, F., Siggia, E. D., & Brivanlou, A. H. (2014). A method to recapitulate early embryonic spatial patterning in human embryonic stem cells. *Nature Methods*, *11*, 847–854.

Williams, M. L., & Solnica-Krezel, L. (2020). Nodal and planar cell polarity signaling cooperate to regulate zebrafish convergence and extension gastrulation movements. *eLife*, *9*, e54445.

Wolpert, L. (2011). Positional information and patterning revisited. *Journal of Theoretical Biology*, *269*, 359–365.

Xu, C., Fan, Z. P., Muller, P., Fogley, R., DiBiase, A., Trompouki, E., et al. (2012). Nanog-like regulates endoderm formation through the Mxtx2-nodal pathway. *Developmental Cell*, *22*, 625–638.

Xu, P. F., Houssin, N., Ferri-Lagneau, K. F., Thisse, B., & Thisse, C. (2014). Construction of a vertebrate embryo from two opposing morphogen gradients. *Science*, *344*, 87–89.

Zhang, Z., Zwick, S., Loew, E., Grimley, J. S., & Ramanathan, S. (2019). Mouse embryo geometry drives formation of robust signaling gradients through receptor localization. *Nature Communications*, *10*, 4516.

Zinski, J., Bu, Y., Wang, X., Dou, W., Umulis, D., & Mullins, M. C. (2017). Systems biology derived source-sink mechanism of BMP gradient formation. *eLife*, *6*, e22199.

Zinski, J., Tajer, B., & Mullins, M. C. (2018). TGF-β family signaling in early vertebrate development. *Cold Spring Harbor Perspectives in Biology*, *10*, a033274.

CHAPTER NINE

Signaling oscillations in embryonic development

Sabine L. Bosman and Katharina F. Sonnen*
Hubrecht Institute—KNAW (Royal Netherlands Academy of Arts and Sciences) and University Medical Center Utrecht, Utrecht, The Netherlands
*Corresponding author: e-mail address: k.sonnen@hubrecht.eu

Contents

Abstract

Tight spatiotemporal control of cellular behavior and cell fate decisions is paramount to the formation of multicellular organisms during embryonic development. Intercellular communication via signaling pathways mediates this control. Interestingly, these signaling pathways are not static, but dynamic and change in activity over time. Signaling oscillations as a specific type of dynamics are found in various signaling pathways and model systems. Functions of oscillations include the regulation of periodic events or the transmission of information by encoding signals in the dynamic properties of a signaling pathway. For instance, signaling oscillations in neural or pancreatic progenitor cells modulate their proliferation and differentiation. Oscillations between neighboring cells can also be synchronized, leading to the emergence of waves traveling through the tissue. Such population-wide signaling oscillations regulate for example the consecutive segmentation of vertebrate embryos, a process called somitogenesis. Here, we outline our current understanding of signaling oscillations in embryonic development, how signaling oscillations are generated, how they are studied and how they contribute to the regulation of embryonic development.

Current Topics in Developmental Biology, Volume 149
ISSN 0070-2153
https://doi.org/10.1016/bs.ctdb.2022.02.011

1. Introduction

Since the beginning of the twentieth century, the regulation of embryonic development has been explained by the morphogen gradient theory (Neumann & Cohen, 1997; Turing, 1952). According to this theory, cell fates are specified by the establishment of concentration gradients of diffusing morphogens or signaling molecules. In the 1960s, Lewis Wolpert proposed the French flag problem, which states that morphogen gradients provide positional information in the embryo (Wolpert, 1969). Cells would then differentiate into specific cell types based on locations where the morphogen concentration reached a threshold for a specific response, thereby patterning the embryo (Wolpert, 1989). The first morphogens were identified in screens for developmentally relevant genes in *Drosophila* by Nüsslein-Volhard and Wieschaus (Nüsslein-Volhard & Wieschaus, 1980; Wieschaus & Nüsslein-Volhard, 2016). These have subsequently been shown to induce cellular signaling cascades that modulate intracellular processes, such as changes in gene expression or mechanical properties, which lie at the bases of the progression through development.

Research over the last decades has shown that the transmission of information through signaling pathways is more complex than previously thought. Cells do not only respond to absolute morphogen concentrations, but also to the dynamics of a signal. Thus, a signal cannot only be decoded based on the absolute intensity, but also the duration, fold change and frequency of a signal (Behar & Hoffmann, 2010; Sonnen & Aulehla, 2014; Sonnen & Janda, 2021). For example, variations in the duration of Sonic hedgehog (Shh) signaling have been found to determine the cell fate of neural progenitor cells during development of the neural tube (Dessaud et al., 2007). By now, dynamics have been observed in many signaling pathways and their functions are starting to be revealed (Purvis & Lahav, 2013; Sonnen & Aulehla, 2014). A specific type of dynamics is signaling oscillations, the regular periodic activation of signaling activity, which is found in NOTCH, WNT (Wingless and Int-1) and FGF (fibroblast growth factor) signaling, but also other pathways such as NFκB and P53. Signaling oscillations can occur in single cells within a tissue, not being synchronized to their neighbors. In contrast, neighboring cells can be synchronized in their oscillations, leading to population-wide oscillation dynamics to coordinate embryonic development.

In this chapter, we describe the principles of signaling oscillations and elaborate on the mechanisms and cellular processes in which signaling has been found to be oscillatory. We particularly focus on NOTCH, WNT and EGF (epidermal growth factor)/FGF signaling oscillations and their impact on embryonic development in vertebrates. By comparing similarities and differences between the function of signaling oscillations during various processes, we will finally discuss general roles for signaling oscillations in biological information transmission.

1.1 Signaling oscillations by delayed negative feedback

Upon pathway stimulation, an intracellular signal transduction cascade is initiated, which leads to the activation of effector proteins. These effector proteins can exert different functions in cells, such as changing activity of enzymes or modulating the cytoskeleton. Often, effector proteins function as transcription factors and directly or indirectly regulate gene expression. Targets of every signaling pathway include negative regulators of the pathway itself, inducing negative feedback loops to reduce pathway activation (Alon, 2007). Depending on the amount of activation and inactivation and the timing in-between both, periodic activation and inactivation can generate oscillations (Fig. 1A). Relative to pathway initiation, activation of the negative regulator should be delayed allowing the increase in signal activity prior to downregulation, thereby generating oscillations. The delay depends for instance on the timing of transcription, splicing and translation of the regulator (Kageyama, Shimojo, & Isomura, 2018; Takashima, Ohtsuka, González, Miyachi, & Kageyama, 2011). Speeding up gene expression of the negative regulator in the segmenting mouse embryo by removal of introns leads to damped oscillations with a higher period (Harima, Takashima, Ueda, Ohtsuka, & Kageyama, 2013) (Fig. 1B). This exemplifies that stable oscillations are only generated in a window of optimal combinations of biochemical parameters of the signaling network. Delayed negative feedback mechanisms inducing oscillatory activity are found in various signaling pathways. For example, NFκB signaling induces expression of its inhibitor IκB, which results in oscillations in nucleo-cytoplasmic shuttling of NFκB itself (Ashall et al., 2009; Nelson, 2004). The NOTCH signaling pathway induces expression of several negative regulators including members of the HES/HER basic helix–loop–helix (bHLH) family of transcriptional repressors. NOTCH signaling is oscillatory in various systems, such as neural

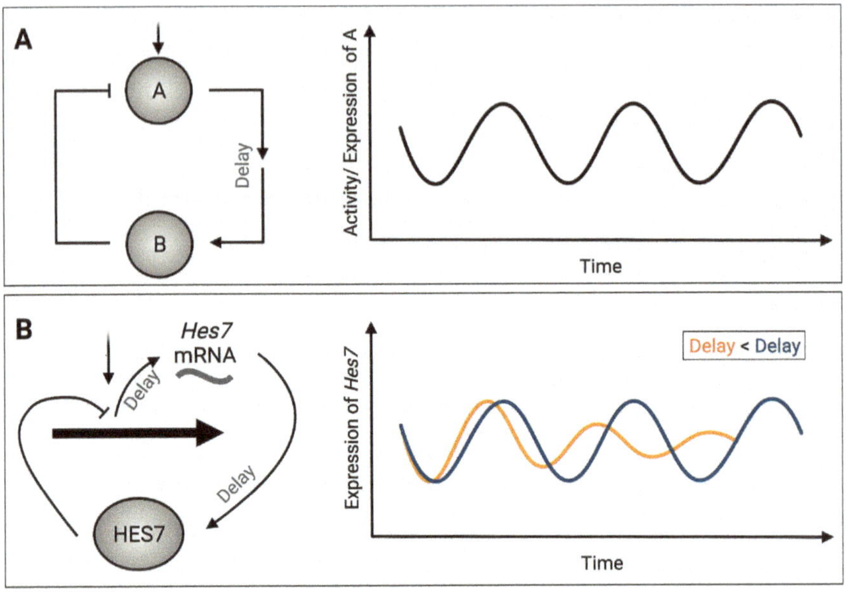

Fig. 1 Signaling oscillations can be generated by delayed negative feedback. (A) Signaling activates protein A, which induces the negative regulator B with a delay. B subsequently inhibits A. Under optimal conditions, stable oscillations are generated. (B) In the context of the segmenting embryo, Hes7 expression oscillates upon signaling stimulation. HES7 then functions as negative regulator by repressing its own expression and expression of other genes. When changing the timing of Hes7 expression by removing introns from the Hes7 gene, the period of HES7 oscillations decreases and oscillations are dampened.

stem cells, pancreas or presomitic mesoderm of the segmenting vertebrate embryo (Guo et al., 2020; Kageyama et al., 2018; Palmeirim, Henrique, Ish-Horowicz, & Pourquié, 1997; Shimojo et al., 2016).

1.2 Functions of signaling oscillations

In biological systems, signaling oscillations have diverse functions (Fig. 2): Several biological processes occur periodically, such as cell division or the synchronization with the day-night cycle via the circadian clock. To ensure proper progression and keep the timing of simultaneous or subsequent steps, such processes are controlled by oscillatory activities (Fig. 2A). In the context of the developing vertebrate embryo, periodic segmentation is coordinated with tissue growth by signaling oscillations (Hubaud & Pourquie, 2014).

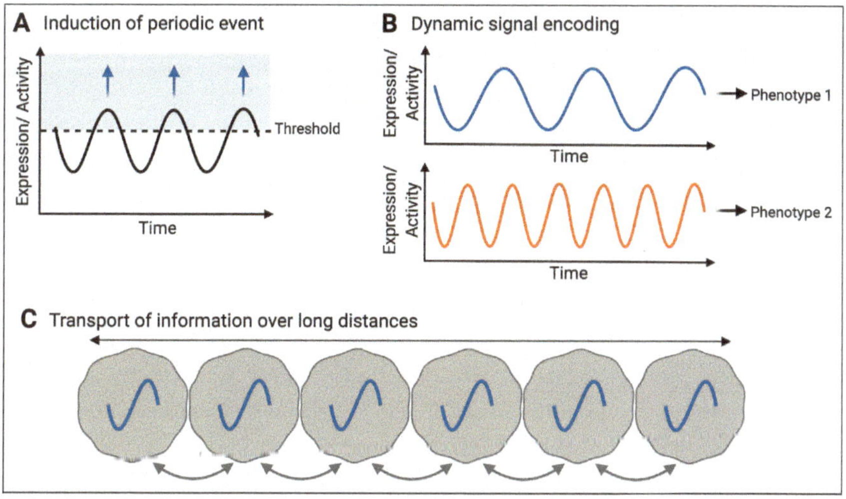

Fig. 2 Signaling oscillations regulate cellular behavior. (A) Signaling oscillations can induce periodic events by periodically crossing the threshold of activity required for induction. (B) Cells use signaling oscillations to transmit biological information precisely and efficiently. Information can be encoded in the absolute signal, the frequency, or the number of oscillations. (C) By synchronization, oscillations between neighboring cells can be coordinated. This allows to synchronize periodic events or transmit biological information across tissues.

Besides periodically initiating a cellular process, cells use signaling oscillations to encode information in the dynamics (also see chapter "Putting in the Erk: Growth factor signaling and mesoderm morphogenesis" by McFann et al. in this volume). How a limited number of signaling pathways regulates diverse processes in developing organisms remains largely unknown. Conserved signaling pathways have been found to transmit different types of information through variations in their dynamics, thereby allowing cells to have various fates employing the same signaling molecules (Fig. 2B) (Purvis & Lahav, 2013; Sonnen & Aulehla, 2014; Zinner, Lukonin, & Liberali, 2020). Moreover, dynamic signal encoding can make cells less sensitive to noise (Gonze et al., 2018). In an oscillatory signal, information can be encoded in the absolute signal, but also in the oscillation amplitude, the frequency, or the number of oscillations. Whether information is encoded in the amplitude or frequency of a signal, is dependent on the topology of the signaling network (Alon, 2007). For transmitting information in the dynamic properties of a signaling pathway, an absolute signal (e.g., ligand type or concentration) is encoded in signaling dynamics, which then have to be decoded into an absolute response (e.g., expression of certain target genes).

The mechanism of this encoding and decoding process is often not understood in detail. One example for a pathway which uses frequency modulation to encode information is the NFκB pathway, which is induced by stimuli such as pathogens, cytokines or radiation to regulate the immune response (Ashall et al., 2009). The transcription factor NFκB shuttles between the nucleus and the cytosol upon stimulation with cytokines. The resulting cellular response and induced gene expression are then dependent on the oscillation frequency (Ashall et al., 2009; Kellogg & Tay, 2015; Lane et al., 2017).

In multicellular systems, all cells within a tissue have to be coordinated to allow development and maintenance of the organism. Besides morphogen gradients, tissues are organized by population-wide oscillation dynamics. This is achieved by synchronization of neighboring cells via oscillating signaling pathways. In somitogenesis, neighboring cells are synchronized by NOTCH signaling (Riedel-Kruse, Müller, & Oates, 2007; Tsiairis & Aulehla, 2016). When blocking this intercellular coupling by inhibiting NOTCH signaling, cells gradually shift out of synchrony (Riedel-Kruse et al., 2007). Such a mechanism allows for coordinated alterations in gene expression along neighboring cells, thereby transferring spatiotemporal information across tissues and providing a mechanism for cells to jointly differentiate into specific cell types with spatiotemporal precision (Fig. 2C).

1.3 Functional investigation of signaling oscillations

The period of signaling oscillations in mammalian systems ranges from several minutes to a few hours. Signaling reporters have been established to visualize either pathway activity itself or downstream target gene expression. Pathway activation can be visualized by labelling components of the pathway that change subcellular localization upon activation. This way nucleocytoplasmic shuttling of, for example, NFκB or ERK (extracellular signal-regulated kinase) has been detected in cultured cells (Burack & Shaw, 2005; Fuseler, Merrill, Rogers, Grisham, & Wolf, 2006). In addition, several signaling pathways contain enzymatic reactions in their signaling cascade, which can be employed in research to develop signaling reporters. For instance, activity of the kinase ERK has been visualized by reporters that—upon phosphorylation—change subcellular localization or show a shift in fluorescence resonance energy transfer (FRET) (de la Cova, Townley, Regot, & Greenwald, 2017; Regot, Hughey, Bajar, Carrasco, & Covert, 2014; Zhang et al., 2018). In the somitogenesis field, oscillations have first been detected by variations in staining patterns of cyclic genes that were later identified as downstream targets of signaling pathways (Aulehla & Johnson, 1999; Dequéant et al., 2006; Palmeirim et al., 1997). For real-time imaging of the segmentation clock, such target genes were the basis to establish signaling reporters by using their promoter region to drive expression of destabilized Luciferase or fluorescent proteins (Aulehla et al., 2008; Masamizu et al., 2006; Yoshioka-Kobayashi et al., 2020). With advances in reporter systems and microscopy techniques, the quality of real-time imaging has increased over the last years to even allow single-cell tracking and quantification of oscillations within tissues (Delaune, Francois, Shih, & Amacher, 2012).

The function of signaling pathways has traditionally been studied by continuous perturbation of the pathway using genetic or pharmacological means. This, however, does not reveal the role of the oscillations themselves. To address this, tools had to be developed that allow the specific modulation of oscillations without affecting overall signaling levels. Two broadly applicable technologies that allow temporally controlled perturbations of signaling pathways are microfluidics and optogenetics. In microfluidic systems, cells and tissues are cultured in a confined environment and microfluidic pumps are used to externally control and change culture conditions. This allows for instance to modulate signaling oscillations by periodic application of pathway modulators (reviewed in Sonnen & Merten, 2019). Using this approach, the decoding of NFκB signaling oscillations has been investigated by synchronizing a large number of cells to pulses of the cytokine TNFα (Ashall et al., 2009; Tay et al., 2010). Another method for the dynamic perturbation of signaling oscillations is optogenetics. Optogenetic systems make use of light-sensitive proteins that change conformation upon light stimulation (Johnson & Toettcher, 2018). Various optogenetic systems for activation of signaling pathways have been developed that are based on receptor dimerization or the recruitment of pathway components to the plasma membrane (Kim et al., 2014; Toettcher, Weiner, & Lim, 2013; Viswanathan et al., 2019). Moreover, systems for the light-controlled induction of gene expression can also model the periodic expression of cyclic genes (Imayoshi et al., 2013; Sueda, Imayoshi, Harima, & Kageyama, 2019). The combination of visualization and quantification of signaling dynamics and their functional manipulation now allow an in-depth analysis of signaling oscillations in multicellular systems (Weterings, van Oostrom, & Sonnen, 2021).

2. Single-cell oscillation dynamics

Signaling oscillations occur in various cell and tissue types. Especially for the NOTCH and ERK signaling pathways, signaling reporters are available that have allowed the detection of signaling oscillations in single cells within single-cell cultures or in the context of complex tissues. There, oscillations often correlate with a progenitor and/or proliferative cell state, however, the functional significance and the mechanism of action are often not known in detail.

2.1 NOTCH signaling oscillations

The NOTCH pathway is a conserved signaling pathway found in metazoans, which regulates cell fate decisions in several tissues during embryonic development (Artavanis-Tsakonas, 1999). In canonical NOTCH signaling, the transmembrane protein NOTCH is activated by Delta-SERRATE/JAGGED-LAG2 (DSL) ligands that are expressed by neighboring cells (Fig. 3A) (Bray, 2006). This leads to the proteolytic cleavage of NOTCH (S2 cleavage), thereby releasing a membrane-bound NOTCH domain (Baron, 2003). This domain is subsequently cleaved by γ-SECRETASE at the S3 cleavage site, causing the formation of a NOTCH intracellular domain (NICD) that is released from the membrane and travels to the nucleus (Fig. 2) (Baron, 2003). In the nucleus, NICD acts as transcription factor and—together with the DNA-binding protein CSL (CBF1, Su(H) and LAG1)—regulates the expression of NOTCH target genes, such as *bHLH* genes from the *Hes* family (Bray, 2006; Ohtsuka et al., 1999).

NOTCH signaling is juxtacrine, i.e., communication occurs through interaction between neighboring cells (Fig. 3A). Active NOTCH signaling in one cell can either inhibit or stimulate NOTCH signaling in neighboring cells, which is termed lateral inhibition or lateral induction, respectively (Sjöqvist & Andersson, 2019). Through this mechanism, NOTCH can instruct neighboring cells to obtain the same or different cell fates during development (Henrique & Schweisguth, 2019). Integral to NOTCH signaling is the induction of several negative regulators upon pathway activation, such as members of the bHLH family of transcription factors, which repress their own expression and that of other NOTCH signaling components (Harima, Imayoshi, Shimojo, Kobayashi, & Kageyama, 2014). For instance, the bHLH protein HES1 is able to bind to its own promoter to decrease its expression and both *Hes1* mRNA and HES1 protein have a short half-life, which results in oscillations with a period of approximately 2h (Fig. 3B) (Hirata et al., 2002). NOTCH signaling has been found to be oscillatory in stem cells in multiple developmental processes and in cultured embryonic

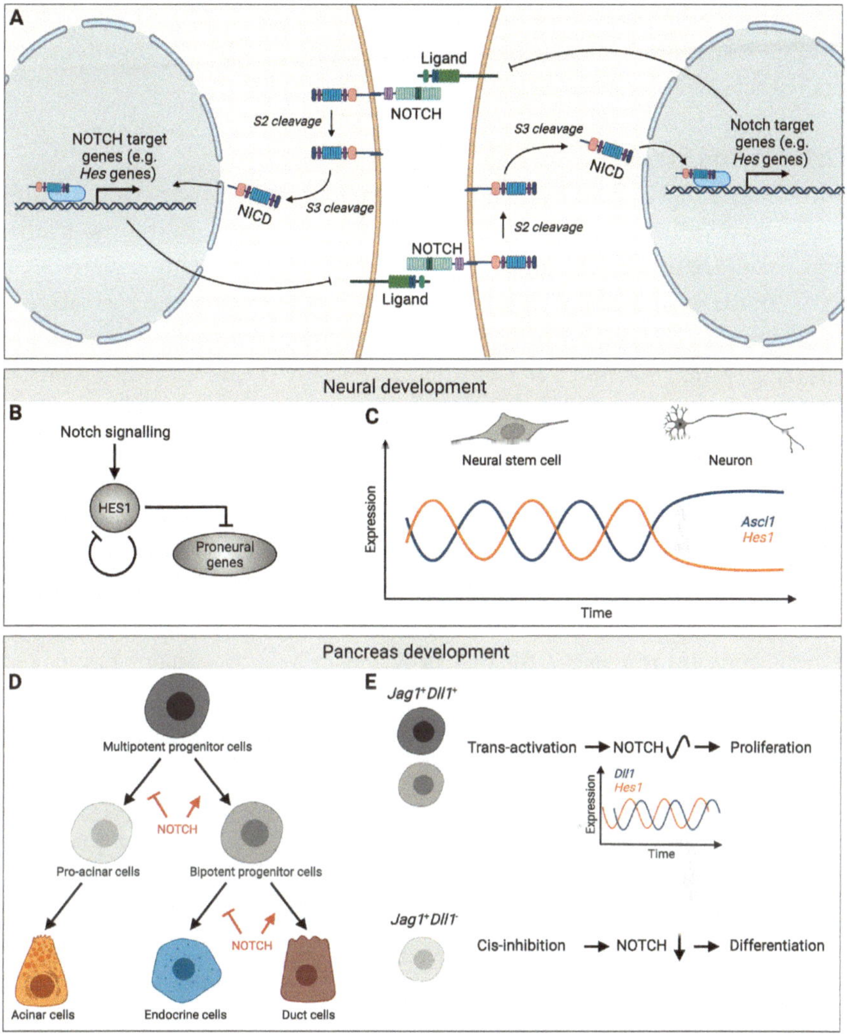

Fig. 3 NOTCH signaling oscillations modulate proliferation and differentiation of progenitor cells. (A) NOTCH signaling functions in a juxtacrine manner. Upon interaction of the ligand and the NOTCH receptor, NOTCH is cleaved and the intracellular domain NICD travels to the nucleus to regulate gene expression. (B) NOTCH signaling induces expression of the transcription factor Hes1, which represses its own expression and that of other genes, including proneural genes. (C) In neural progenitor cells, expression of Hes1 and several proneural genes such as Ascl1 oscillates. Expression is stabilized when cells differentiate. (D) In pancreas development, NOTCH signaling regulates the differentiation of multipotent progenitors into acinar, endocrine and duct cells. (E) HES1 and DLL1 oscillate in multipotent and bipotent progenitor cells, which promotes proliferation. When NOTCH signaling decreases in pro-acinar cells, cells differentiate. NOTCH signaling dynamics depend on the expression of the ligands Dll1 and Jag1.

stem cells (Kobayashi et al., 2009). Here, we will focus on the role of NOTCH signaling oscillations in stem cells during neurogenesis and pancreatic development.

2.1.1 Neurogenesis

In the beginning of vertebrate neurogenesis, the neural tube contains symmetrically dividing neural stem cells (NSCs) that span from the basal to the apical side (Sharma & Goffinet, 2012). Cell differentiation is initiated once NSCs divide asymmetrically, which subsequently leads to the formation of the two main cell types of the central nervous system, the neurons and glial cells (Kageyama, Shimojo, & Ohtsuka, 2019). NOTCH signaling is critical in regulating neural differentiation during development of the nervous system (Kageyama et al., 2018; Sharma & Goffinet, 2012).

Proneural genes are upregulated in cells that undergo differentiation, while active NOTCH signaling represses expression of those genes in cells that maintain stemness. Lateral inhibition was proposed to ensure that adjacent cells either have high levels of Hes1 or high levels of proneural genes, which ensures the maintenance of NSCs throughout the process of neurogenesis (Shimojo, Ohtsuka, & Kageyama, 2008). Via negative feedback mechanisms, NOTCH signaling is oscillatory in NSCs. These oscillations modulate the differentiation of progenitor cells to neuronal and glial cells through the oscillatory expression of bHLH transcription factors Hes1 and Hes5 (Imayoshi et al., 2013; Kageyama et al., 2019; Manning et al., 2019; Ohtsuka et al., 1999). While *Hes1* and *Hes5* are expressed in different parts of the developing nervous system, the brain and the spinal cord, respectively, *Hes1* can be upregulated if *Hes5* is knocked out and *vice versa* (Hatakeyama et al., 2004). Neurogenesis is only (severely) disrupted if both *Hes1* and *Hes5* are knocked out, indicating the redundancy of both HES proteins (Hatakeyama et al., 2004). Sustained or high *Hes1* expression led to downregulation of the proneural gene *Neurogenin 2* (*Ngn2*) and the NOTCH ligand *Dll1* in mice, while they were upregulated upon decreased *Hes1* expression (Shimojo et al., 2008). Expression of proneural genes and NOTCH ligands oscillates during neurogenesis, presumably due to the periodic inhibition by HES1 (Fig. 3B) (Shimojo et al., 2008, 2016). When introducing a mutation into mice to induce stable expression of *Dll1*, HES1 oscillations were dampened, which caused more NSCs to differentiate into neural cells and ultimately led to microcephaly (Shimojo et al., 2016).

How *Hes* oscillations regulate the differentiation of NSCs into specific neural cell types has been a matter of intense investigation. The three proneural genes *Ngn2*, *Ascl1* and *Olig2* participate in cell fate determination during development of the murine central nervous system (Nieto, Schuurmans, Britz, & Guillemot, 2001; Sun et al., 2001; Zhou & Anderson, 2002). When quantifying expression levels using luciferase-based reporters, these transcription factors showed oscillatory expression patterns in NSCs during early neurogenesis (Imayoshi et al., 2013). While *Hes1* and *Ascl1* had a period of 2–3 h, the period of *Olig2* expression was approximately 6 h (Imayoshi et al., 2013; Shimojo et al., 2008). Even though uncommitted NSCs displays oscillatory expression of *Hes1*, *Ngn2*, *Ascl1* and *Olig2*, these genes are eventually only expressed in specific neural cell types (Imayoshi et al., 2013; Imayoshi & Kageyama, 2014). Whereas *Ascl1* oscillations contribute to the proliferation of NSCs before they commit to a specific cell fate, sustained *Ascl1* expression later becomes restricted to neurons (Fig. 3C) (Imayoshi & Kageyama, 2014). During the differentiation of astrocytes, *Hes1* levels increase, but remain oscillatory (Imayoshi et al., 2013). Lastly, the levels of *Olig2* increase in a subset of cells and this induces differentiation to oligodendrocytes (Lu et al., 2002; Zhou & Anderson, 2002). In quiescent NSCs, HES1 oscillates at a higher level, or its expression is sustained, which prevents expression of *Ascl1* and thus prevents differentiation (Sueda et al., 2019). While Hes1 oscillations are not required to enter or maintain quiescence, they are essential for exit from quiescence (Marinopoulou et al., 2021). To determine how *Ascl1*, *Hes1* and *Olig2* oscillations can regulate proliferation and differentiation in neural development, an optogenetic system was established, which allowed the induction of *Ascl1* expression by light pulses (Imayoshi et al., 2013). When externally inducing pulses of Ascl1, proliferation was promoted. In contrast, sustained activation induced differentiation. These different responses seemed to be independent of absolute expression levels, but rather depend on the differences in expression dynamics (Imayoshi et al., 2013).

Thus, dynamic NOTCH signaling not only regulates NSC maintenance during neurogenesis but is also essential for the differentiation toward different neural cell types. It remains to be elucidated how the oscillatory expression of specific genes are stabilized, while others are downregulated, and how this subsequently causes NSCs to differentiate into specific cell types.

2.1.2 Pancreas development

The pancreas consists of three main cell types, which are the ductal, acinar, and endocrine cells. The ductal and acinar cells together form the exocrine compartment, which regulates the production and transportation of digestive enzymes (Jennings, Berry, Strutt, Gerrard, & Hanley, 2015). In contrast, the endocrine compartment produces hormones such as insulin and glucagon. Pancreatic development starts when the foregut endoderm forms pancreatic buds, which contain multipotent pancreatic progenitor cells (MPPCs). Next, the pancreatic buds form tubular structures that move into the surrounding mesenchyme, which is a process called branching (Shih, Wang, & Sander, 2013). At this stage, the fate of the MPPCs is specified (Fig. 3D). Cells located in the tip domain of the branches form acinar cells, whereas cells in the trunk domain differentiate into ductal or endocrine cells (Pan & Wright, 2011; Shih et al., 2013). The adult pancreas consists of a tubular network containing ductal and acinar cells, which is surrounded by the islets of endocrine cells.

During pancreatic development, NOTCH signaling modulates the specification of MPPCs into the three pancreatic cell types, as well as MPPC proliferation. Like during neural development, MPPCs are maintained in a proliferative state by NOTCH signaling (Bastidas-Ponce, Scheibner, Lickert, & Bakhti, 2017; Jensen et al., 2000). NOTCH signaling is also involved in the patterning of the tip and trunk compartments during the process of pancreatic branching (Afelik et al., 2012): It promotes the formation of bipotent trunk cells in the branches, while its absence promotes the formation of acinar tip cells (Fig. 3D). Lastly, NOTCH signaling regulates the cell fate choice of bipotent trunk progenitor cells into duct or endocrine cells. Bipotent progenitor cells differentiate into endocrine cells when *Neurogenin 3* (*Ngn3*) is expressed (Shih et al., 2013). Like the NOTCH-mediated inhibition of Ngn2 in the developing nervous system, Hes1 negatively regulates the expression of *Ngn3* in the pancreas (Jensen et al., 2000; Qu et al., 2013). Conversely, *Ngn3* expression induces the expression of *Dll1*, thereby activating NOTCH signaling in adjacent cells and thus inhibiting the endocrine cell fate (Jensen et al., 2000; Qu et al., 2013).

While the role of NOTCH signaling in pancreatic development has been studied extensively, only recently has it been demonstrated that NOTCH signaling is dynamic in this process. Seymour et al. showed that both the ligand *Dll1* and the downstream target *Hes1* had an oscillatory expression pattern with a period of 90 min during the early stages of pancreas development, while the ligand Jag1 was stably expressed (Fig. 3E) (Seymour et al., 2020). Like the functioning of NOTCH signaling in neural progenitors, NOTCH signaling oscillations in the pancreas were implicated in regulating the proliferation of MPPCs, since the stable expression of Dll1 caused a decrease in the size of pancreatic buds. Dll1 expression became restricted to bipotent progenitor cells, while the NOTCH ligand Jag1 had a uniform expression pattern that became restricted to acinar cells (Seymour et al., 2020). Moreover, using small molecule inhibitors to modulate NOTCH signaling oscillations, it was found that the parameters of signaling oscillations, namely the oscillation period and amplitude, seemed to influence cell fate determination during pancreas development (Seymour et al., 2020). Decreasing the amplitude of NOTCH oscillations resulted in the formation of more acinar cells, while increasing the period of the oscillations resulted in the formation of more bipotent progenitor cells. Further studies including precise microfluidic and optogenetic modulation of these oscillations can reveal the mechanism of how NOTCH signaling oscillations contribute to the Jag- and Dll1-mediated control of pancreas development.

2.2 ERK signaling oscillations

The ERK signaling pathway regulates various biological processes in metazoan, including embryonic development and cancer (Lavoie, Gagnon, & Therrien, 2020). ERK is activated through a signaling cascade that is initiated by the binding of growth factors to receptor tyrosine kinases (RTKs) at the cell membrane (Peyssonnaux & Eychène, 2001). Upon stimulation, RTKs activate the G protein RAS, which results in the activation of the kinase RAF (termed *mitogen activated protein kinase kinase kinase* or MAPKKK). RAF phosphorylates and activates MEK (MAPKK), which subsequently results in the phosphorylation and activation of ERK (MAPK). ERK phosphorylates and modulates the activity of diverse proteins within cells; it also travels to the nucleus, where it activates transcription factors to regulate gene expression. ERK target genes, such as *Myc* or *Cyclins*, are involved in processes such as cell proliferation, differentiation, survival, and motility (Guo et al., 2020; Shaul & Seger, 2007). In addition, ERK signaling induces expression of multiple negative regulators, such as *Sprouty* or phosphatases of the DUSP (dual specificity phosphatase) family (Brondello, Brunet, Pouysségur, & McKenzie, 1997; Hacohen, Kramer, Sutherland, Hiromi, & Krasnow, 1998). Besides induction of gene expression of negative regulators, phosphatases, already present in cells, can directly inactivate the MAPK cascade by dephosphorylation. Such negative feedback loops can result in various types of dynamics. ERK signaling dynamics were first identified in the PC12 cell line, which is a murine cell model that can be used to study neuronal differentiation *in vitro* (Marshall, 1995). Interestingly, the cell fate of these PC12 cells was found to be affected by the duration of ERK activity. While transient ERK activation induced by EGF causes PC12 cells to proliferate, sustained ERK activity induced by nerve growth factor (NGF) results in the differentiation of PC12 cells into neural cells (Marshall, 1995). At single-cell level, PC12 cells have a more heterogeneous response to EGF and NGF, as both growth factors induce transient or sustained ERK activity, albeit at different ratios (Ryu et al., 2016).

ERK signaling can also transmit information through oscillations (Shankaran & Wiley, 2010). ERK signaling oscillations were first demonstrated using an ERK-GFP fusion protein reporter in a human mammary epithelial cell line, in which oscillations of nuclear localization of ERK were induced by EGF and had a period of 15 min (Shankaran et al., 2009). Interestingly, in mammary epithelial cells ligand concentration seems to be encoded in the frequency of ERK activity pulses, as increasing the ligand concentration resulted in faster pulses (Albeck, Mills, & Brugge, 2013). By now, ERK signaling oscillations have been observed in several cell types. Stimulation of the mouse cell lines C3H 10T1/2 and NIH 3T3 with FGF resulted in both RAS and ERK oscillations (Nakayama, Satoh, Igari, Kageyama, & Nishida, 2008). Also in mouse embryonic stem cells, ERK signaling oscillations with a period of only 7 min have been observed (Raina et al., 2021). These oscillations were shown to be induced by FGF4, and the number of cells oscillating increases with the concentration of FGF4. Finally, ERK signaling oscillations occur in skin and intestinal cells, in which ERK activity oscillations have been implicated in the control of cell proliferation and survival (Gagliardi et al., 2021; Hiratsuka, Bordeu, Pruessner, & Watt, 2020; Muta et al., 2018).

The functions of ERK signaling oscillations in the context of developing organisms are less well understood. With advances in fluorescent reporters and optogenetic tools to modulate ERK signaling, the significance of ERK signaling oscillations is starting to be elucidated (Gagliardi et al., 2021; Johnson & Toettcher, 2019; Toettcher et al., 2013). As the frequency of ERK signaling oscillations differs per cell type and ERK stimulant, the mechanism controlling ERK activity oscillations is likely not the same in the different tissue types. In the future, the mechanisms of how information is encoded in ERK signaling oscillations and how these oscillations are decoded to induce a certain cellular effect has to be revealed.

3. Tissue-wide oscillation dynamics

In the developing embryo, oscillations coordinated between cells of a tissue can orchestrate developmental processes throughout a tissue or even an entire organism. Such tissue-wide dynamics mediate for instance the synchronized cycles of mitoses and spatial relocalization of nuclei in the early fly embryo (Deneke et al., 2019). These collective movements are regulated by oscillations in biochemical and mechanical activities traveling through the organism. Oscillations in signaling activity spanning entire tissues ensure the proper spatiotemporal progression of development. While such coordinated signaling activity is starting to be observed in other contexts, such as ERK signaling waves in epithelial tissue (Gagliardi et al., 2021; Muta et al., 2018; Valon et al., 2021), the best understood model of tissue-wide signaling dynamics is the periodic segmentation of vertebrate embryos. Somitogenesis is regulated by an oscillatory network of signaling pathways that ensures that new segments are formed with spatial and temporal precision (Hubaud & Pourquie, 2014).

3.1 Somitogenesis

During somitogenesis, the growing presomitic mesoderm (PSM) is sequentially subdivided into blocks of cells, termed somites (Fig. 4A). Somites are temporary structures during development, which differentiate further into sclerotome, myotome and dermatome giving rise to axial skeleton, muscles, and skin, respectively. This regular periodic segmentation has fascinated biologists for decades. A theoretical model was presented by Cooke and Zeeman in 1976, which predicted that the formation of "repeated structures during animal morphogenesis" might be mediated by a system consisting of a clock and a wavefront (Cooke & Zeeman, 1976). This clock-and-wavefront model entails a regressing wavefront, which makes the tissue that has been passed by the wavefront (or determination front) competent to differentiate, and a clock, which periodically induces segmentation of the competent tissue. This way, the combination of clock and wavefront determines when and where a segment forms and thereby the size of newly forming segments. Interestingly, in the decades following this study, cyclic genes as downstream targets of signaling pathways have indeed been identified in the PSM of chicken and mouse embryos, whose oscillation period correlated with the formation of new somites, which hinted at the existence of a molecular

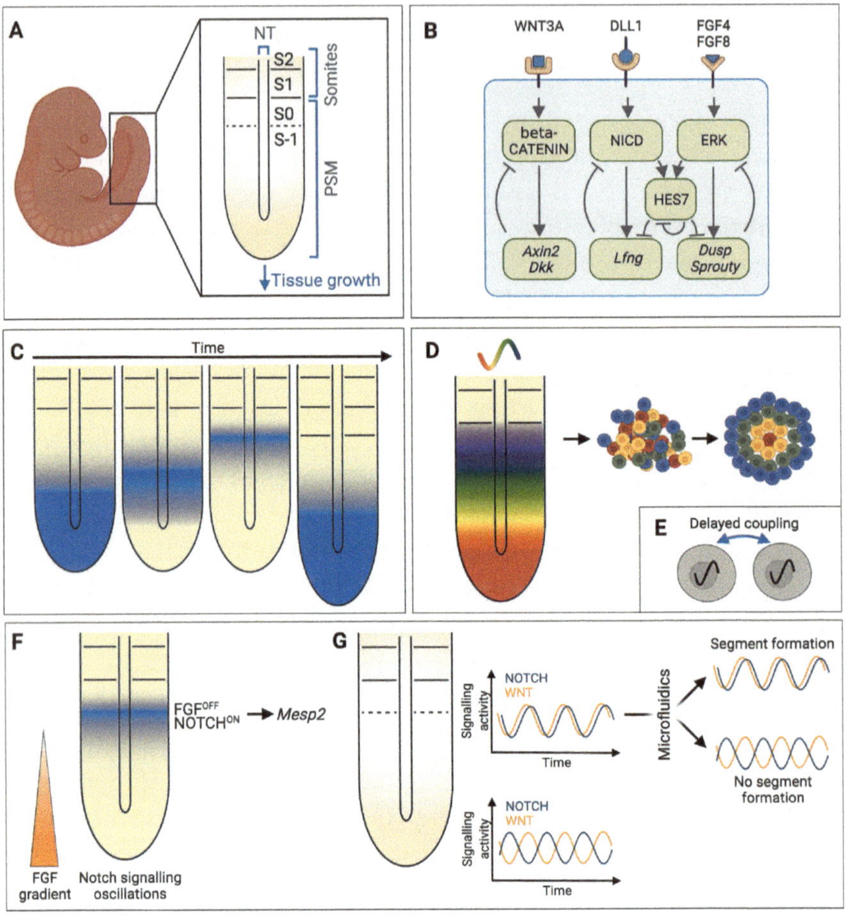

Fig. 4 Somitogenesis is regulated by a network of oscillatory signaling pathways. (A) As the tissue grows, somites form sequentially at the anterior sides of the PSM (NT = neural tube). (B) The segmentation clock consists of oscillations of WNT, NOTCH and FGF signaling in mice. (C) Signaling waves travel through the PSM from posterior toward the anterior side, where a pair of somites forms. Subsequently, the next wave starts. (D) When separating PSM cells into single cells and allowing them to re-aggregate, cells synchronize and re-establish signaling waves and gradients over time. (E) Delayed coupling between neighboring cells results in the formation of traveling waves. (F) When FGF signaling is turned off in anterior PSM and NOTCH signaling is turned on, expression of the differentiation factor Mesp2 is induced. (G) WNT and NOTCH signaling oscillate out-of-phase in posterior and in-phase in anterior PSM. When changing this phase-relationship using microfluidics to out-of-phase oscillations, segmentation is impaired.

segmentation clock (Fig. 4B) (Aulehla & Johnson, 1999; Palmeirim et al., 1997). In addition, gradients of signaling activity were found to determine the site of segmentation and have been suggested to represent the wavefront (Dubrulle, McGrew, & Pourquié, 2001). While the exact mechanism of somitogenesis is still debated and other models have been suggested, such as a reaction–diffusion system or a two-oscillator model (Beaupeux & François, 2016; Cotterell, Robert-Moreno, & Sharpe, 2015), it is generally accepted nowadays that a combination of signaling oscillations and signaling gradients regulates somite formation.

3.1.1 The segmentation clock

NOTCH signaling was the first pathway identified as a component of the segmentation clock with downstream targets being found to oscillate in chicken, mouse and zebrafish embryos (Aulehla & Johnson, 1999; Barrantes et al., 1999; Jouve et al., 2000; Oates & Ho, 2002; Palmeirim et al., 1997). Besides NOTCH signaling, FGF and WNT signaling pathways are oscillatory in mouse somitogenesis (Fig. 4B) (Aulehla et al., 2003; Dequéant et al., 2006; Niwa et al., 2007, 2011). A key pathway downstream of FGF stimulation is ERK signaling and the ERK target gene *Dusp4* has indeed been identified as cyclic gene in mouse somitogenesis (Dequéant et al., 2006; Niwa et al., 2007). WNT signaling is a conserved pathway that upon activation prevents degradation of the transcription factor β-CATENIN, which leads to its accumulation in the cytoplasm (Niehrs, 2012). In somitogenesis, target genes of WNT signaling include the negative regulators *Axin2*, *Dickkopf* or *Dact-1*, which form feedback loops that might drive the WNT signaling oscillations (Aulehla et al., 2003; Aulehla & Pourquié, 2008; Dequéant et al., 2006). *In vitro* differentiation of human induced pluripotent stem cells (hiPSCs) into PSM-like cells has recently revealed that the same pathways oscillate in human somitogenesis (Diaz-Cuadros et al., 2020; Matsuda, Hayashi, et al., 2020; Matsuda, Yamanaka, et al., 2020). In fact, broad gene expression analyses have indicated that oscillatory expression is much more complex and encompasses components of various other signaling pathways, such as HIPPO and TGFβ signaling (Matsuda, Yamanaka, et al., 2020). Whether these oscillations are merely indirect results of the segmentation clock or active players in somite formation has to be investigated in the future.

While the exact mechanism is not known, signaling oscillations are thought to be regulated by a complex network of delayed negative feedback mechanisms. Prominent among these are members of the *Hes*/*Her* family of bHLH transcription factors, HES7 in mice and Her7 in zebrafish, that can induce oscillations as negative regulators of their own expression as well as that of other genes (Fig. 4B) (Bessho, Hirata, Masamizu, & Kageyama, 2003). Alterations in the stability of the *Hes7* gene induces changes in the oscillation period in mice, indeed implying that Hes7 is part of the clock mechanism of somitogenesis (Harima et al., 2013; Hirata et al., 2004; Takashima et al., 2011). Moreover, inhibition of NOTCH activity by the downstream target *Lunatic fringe* (*Lfng*) and oscillations in expression of the ligand *DLL1* contributes to the negative feedback mechanism driving oscillations (Dale et al., 2003; Hirata et al., 2004). DLL1 activity is inhibited by HES7, causing NOTCH activity in the adjacent cell to decrease (Shimojo et al., 2016). Mutations in *Dll1* have been found to disrupt somitogenesis, since alterations in the frequency of DLL1 oscillations through shortening or lengthening of the *Dll1* gene led to dampened HES7 oscillations and downregulation of *Lfng* (Shimojo et al., 2016). The period of the segmentation clock is dependent on the species with 30 min in zebrafish, 90 min in chicken, 2 h in mice and 4–5 h in humans (Gomez et al., 2008). A detailed comparison between the mouse and human segmentation clock suggested that differences in the parameters of biochemical reactions in mouse versus human cells could account for these differences (Matsuda, Hayashi, et al., 2020; Rayon et al., 2020).

3.1.2 Cell–cell coupling leads to the formation of signaling waves

While single PSM cells are thought to be independent oscillators (Masamizu et al., 2006; Webb et al., 2016), neighboring cells oscillate in synchrony. At tissue level, cells do not oscillate in phase, but are slightly phase-shifted in their oscillations, which leads to the generation of signaling waves traveling through the PSM from the posterior to the anterior side (Fig. 4C) (Masamizu et al., 2006; Palmeirim et al., 1997). Whereas posterior PSM cells show NOTCH signaling oscillations that correlate with the pace of somite formation, oscillations slow down toward the anterior PSM and are halted in the anterior, where segment formation occurs (Masamizu et al., 2006; Palmeirim et al., 1997). Neighboring cells are coordinated by synchronization via NOTCH signaling (Riedel-Kruse et al., 2007; Tsiairis & Aulehla, 2016). Without a functional NOTCH pathway, the PSM no longer shows coordinated expression levels of NOTCH target genes including the NOTCH ligand Delta (Jiang et al., 2000; Riedel-Kruse et al., 2007). Instead, the distribution of Delta and NOTCH target gene expression becomes variable between neighboring cells. When PSM tissue is separated into single cells and then re-aggregated, cells re-synchronize (Hubaud, Regev, Mahadevan, & Pourquié, 2017; Tsiairis & Aulehla, 2016). While all cells within this culture initially synchronize to in-phase oscillations, traveling waves and signaling gradients re-establish over time in a center-to-periphery direction with cells differentiating in the periphery (Fig. 4D), which highlights the robustness of the system (Tsiairis & Aulehla, 2016).

A delayed coupling mechanism has been suggested to induce a phase-shift in the oscillations between neighboring cells and thereby lead to the formation of traveling waves along the PSM (Fig. 4E) (Ares, Morelli, Jörg, Oates, & Jülicher, 2012; Morelli et al., 2009). Induction of oscillatory expression of *Lfng* in adjacent cells, which as glycosyltransferase negatively modulates NOTCH activity, was shown to be required for the synchronization of the segmentation clock and delayed coupling in mice (Jiang et al., 2000; Okubo et al., 2012; Yoshioka-Kobayashi et al., 2020). In *Lfng*-knockout PSM, the oscillatory expression of HES7 showed a decreased amplitude and loss of synchronization between cells, while HES7 oscillations were maintained in single cells from both mutant and wild-type PSM (Yoshioka-Kobayashi et al., 2020). Thus, a delayed coupling mechanism mediated by NOTCH signaling seems to induce traveling waves along the PSM. With improved microscopes, fluorescent reporters and image analysis pipelines, researchers have started to quantify the segmentation clock at single-cell resolution, which will reveal the mechanistic details of the oscillations guiding somitogenesis (Delaune et al., 2012; Rohde et al., 2021; Yoshioka-Kobayashi et al., 2020).

3.1.3 Signaling gradients and signaling oscillations regulate somite formation

In addition to being oscillatory, FGF and WNT signaling are graded along the PSM with highest levels in the posterior. The FGF signaling gradient is established by mRNA decay with fgf8 only being expressed in posterior PSM (Dubrulle et al., 2001; Dubrulle & Pourquié, 2004). Similarly, WNT3a is expressed in a graded fashion along the tissue (Aulehla et al., 2003). When FGF or WNT levels are decreased, the region of somite differentiation is shifted posteriorly (Dubrulle et al., 2001). In contrast, increasing the levels of WNT or FGF signaling shifts somite formation anteriorly (Aulehla et al., 2008; Dubrulle et al., 2001). The interplay between these signaling gradients and signaling oscillations then controls somite formation. The NOTCH effector protein NICD and the transcription factor Tbx6 were shown to be necessary to induce the expression of *Mesp2*, a transcription factor that is required for segment boundary formation (Oginuma, Niwa, Chapman, & Saga, 2008). Conversely, FGF signaling prevents the expression of *Mesp2* (Niwa et al., 2011). Since both the activity of NOTCH and FGF signaling are oscillatory, *Mesp2* expression is induced by NICD and TBX6, once FGF signaling activity has ceased. Therefore, the combination of the step-wisely regressing gradients and signaling oscillations determines which cells become somites (Fig. 4F) (Niwa et al., 2011).

How signaling oscillations are generated, if and how oscillations and gradients are linked and how they coordinate tissues to control somitogenesis are some of the questions that are currently being addressed. To do this, methods to selectively change specific components of the regulatory network are necessary. This is particularly challenging for FGF and WNT signaling, which are both oscillatory and graded. Genetic and chemical manipulation and perturbations with growth factor-soaked beads have been used to change signaling gradients, which revealed their impact on the spatial regulation of somite formation (Aulehla et al., 2008; Dubrulle et al., 2001). Modulation of signaling oscillations requires perturbations with temporal precision. In the first study that showed the external control of the segmentation clock, Soza-Ried, Ozturk, Ish-Horowicz, and Lewis (2014) used heat shock-inducible expression of the ligand DeltaC in zebrafish to apply periodic pulses of NOTCH activation. This allowed the authors to vary the segmentation period, which correlated with the size of forming segments. To control the segmentation clock in mice, a microfluidic system was established, which allows the periodic application of small molecule-inhibitors and growth factors with high spatiotemporal precision (Sonnen et al., 2018; van Oostrom, Meijer, & Sonnen, 2021). This technique was applied to study the interaction between the oscillatory pathways.

Whether there is a core oscillator driving the segmentation clock is a long-standing question, since delayed negative feedback loops, which could account for the generation of oscillations, can be found in each pathway. Constitutive inhibition or activation has given indications on hierarchies but remained controversial (Aulehla et al., 2003; Niwa et al., 2007, 2011; Wahl, Deng, Lewandoski, & Pourquié, 2007). The HES7 feedback loop was for instance suggested as a core oscillator, since knock-out of *Hes7* in mice caused NOTCH and FGF oscillations to cease; however, the WNT pathway still showed oscillatory activity (Feller, Schneider, Schuster-Gossler, & Gossler, 2008; Hirata et al., 2004; Niwa et al., 2007). This indicates that either the identity of the core oscillator remains unknown, or the different types of oscillations might be generated separately.

However, constitutive perturbations are only partially qualified to address the interactions between oscillations, since this way direct and indirect effects cannot be distinguished. Using dynamic perturbations by microfluidics, oscillations of the different signaling pathways were found to be reciprocally coupled: When modulating NOTCH signaling oscillations by applying pulses of NOTCH signaling inhibitor, WNT signaling oscillations were changed accordingly and *vice versa* (Sonnen et al., 2018). Interestingly, WNT and NOTCH signaling oscillated out-of-phase in the posterior and in-phase in the anterior PSM (Fig. 4G) (Aulehla et al., 2003; Dequéant et al., 2006; Sonnen et al., 2018). Using microfluidics, the phase relationship between WNT and NOTCH signaling was changed in the anterior, which interfered with somite formation suggesting that somitogenesis is dependent on the phase-shift between the two oscillatory pathways (Fig. 4G) (Niwa et al., 2011; Sonnen et al., 2018).

Such findings illustrate the complexity of somitogenesis that we are only starting to reveal with new technologies to analyze and perturb signaling oscillations in segmenting embryos. Diversity between different vertebrates adds to the complexity in somitogenesis. While WNT, FGF and NOTCH signaling oscillate in mice, in zebrafish somitogenesis only NOTCH signaling oscillations have been found (Oates, 2020). The mechanism of how signaling oscillations and signaling gradients control somite formation and how these mechanisms differ between organisms, has to be addressed in the future.

3.2 Coordinated oscillations in other tissues

Oscillations in single cells within tissues can only be revealed by live imaging, since salt-and-pepper expression between neighboring cells in fixed samples can merely hint at dynamics (Shimojo et al., 2008). Coordinated tissue-wide oscillation dynamics could already be visualized 3 decades ago by staining of fixed samples and comparing stainings for cyclic genes in multiple embryos of one litter or using half cultures, in which one half of a symmetrical tissue was fixed immediately and the other half was cultured for a certain amount of time (Palmeirim et al., 1997). Similar techniques were applied to identify dynamics in other tissue types. For instance, limb development is regulated by a complex network involving multiple signaling pathways. The formation of different limb elements is dependent on various morphogen gradients, including the WNT, FGF and SHH and

BMP pathways that determine the axes of the forming limb (Petit, Sears, & Ahituv, 2017). In addition, NOTCH signaling has been found to inhibit the differentiation of mesenchymal progenitor cells into chondrocytes during limb development in mice (Dong et al., 2010). Analyzing the expression of the NOTCH target genes *Hairy2* and *Hes1* in limbs of chicken embryos and *in vitro* cultures of limb buds, respectively, indicated that NOTCH signaling oscillates with a period of 6h during limb development in chicken (Bhat, Glimm, Linde-Medina, Cui, & Newman, 2019; Pascoal et al., 2007; Sheeba, Andrade, & Palmeirim, 2012). Only real-time imaging of dynamic signaling reporters can ultimately confirm these oscillation dynamics in limb development. In addition, using ERK activity reporters, ERK signaling waves have been observed in epithelia such as mouse small intestinal organoids or the regenerating scale of zebrafish, where they regulate proliferation and survival of cells and tissue growth (De Simone et al., 2021; Gagliardi et al., 2021; Muta et al., 2018).

4. Perspective

Cells constantly send and receive signals. This communication orchestrates the formation of a complex organism from a single cell and ensures the subsequent homeostasis. Over the years, it has become apparent that intercellular signaling is more complex than previously known. One aspect is the fact that signaling does not work as an on–off switch but is dynamic. These dynamics are used by cells to transmit information properly. A specific type of dynamics is signaling oscillations, which have been observed in various tissue types, developmental stages, and species. Examples include neural and pancreas development as well as the periodic segmentation of vertebrates. Within tissues, oscillations mostly correlate with a proliferative or stem cell-like stage, while constitutive activation or downregulation coincides with differentiation. A functional link between oscillations and cellular behavior has been shown in neural stem cells, in which optogenetic modulation of oscillatory transcription factors allowed the external control of cellular behavior dependent on the applied dynamics (Imayoshi et al., 2013). Likewise, external control of signaling oscillations in segmenting vertebrates affects somitogenesis (Sonnen et al., 2018; Soza-Ried et al., 2014). For other systems such a functional link still has to be shown, but a general view emerges that signaling oscillations keep cells undifferentiated and at the same time capable to quickly respond to differentiation cues.

In many instances, cells within a tissue show signaling oscillations apparently independent of their neighbors, such as in the developing pancreas (Seymour et al., 2020). This is in part due to the heterogeneity in cell types and spatial organization in the respective tissue with progenitor cells being intermingled with niche cells. In such systems, signaling oscillations are therefore rather used to regulate cellular behavior and cell fate decisions at the single-cell level. When taking a tissue- or organism-wide perspective, additional signaling components such as morphogen gradients are required to set up the spatial arrangement of cells. How tissue-wide information is channeled to signaling oscillations within single cells to determine when and where signaling oscillations should maintain the stem cell identity of a neural or pancreatic progenitor cell or oscillations should cease to induce differentiation, has to be addressed in the future. In contrast to such single-cell oscillations, in other systems oscillations between neighboring—often more uniform—cells are synchronized, which often leads to the appearance of kinematic waves traveling through the tissue. This way, information can be transported through an entire tissue. In somitogenesis, signaling waves coordinate growth of the tissue with consecutive rounds of segmentation. In epithelia, ERK waves can promote survival of cells surrounding a dying cell to maintain epithelial integrity (Gagliardi et al., 2021; Valon et al., 2021). While in the case of somitogenesis NOTCH signaling seems to couple neighboring cells to each other, for ERK waves the identity of the coupling mechanism is not fully understood, but might be based on mechanical coupling between cells (Hino et al., 2020).

The oscillations of a signaling pathway are not isolated, discrete phenomena, but are integrated within the complexities of the cell including inputs from other signaling pathways, metabolism, circadian clock, or mechanics. The crosstalk between these different processes determines the final cellular outcome. For instance, signaling oscillations in mouse somitogenesis are influenced by metabolic activity in the PSM (Bulusu et al., 2017; Oginuma et al., 2017, 2020). Moreover, multiple oscillatory pathways can work together to regulate development, as the relationship between WNT, FGF and NOTCH signaling oscillations controls somite formation (Niwa et al., 2011; Sonnen et al., 2018). In the anterior region, where the tissue segments, FGF signaling has to be off and NOTCH signaling on to induce expression of the differentiation marker Mesp2 (Niwa et al., 2011). At the same time, WNT and NOTCH signaling have to oscillate in phase to allow the formation of somites (Sonnen et al., 2018). How the relationship between the three pathways is regulated along the tissue and how somite formation is ultimately induced, has to be investigated in the future. With technologies to quantify and manipulate each process specifically, together with theoretical modeling, a comprehensive picture of development can be achieved (Negrete Jr. & Oates, 2021).

To date, our understanding of signaling oscillations is restricted to a few signaling pathways and species due to the limited availability of appropriate tools for visualization and analysis of signaling dynamics within organisms and tissues. The combination of new dynamic signaling reporter systems, based on e.g. fast-maturing fluorescent proteins, FRET or the reversible interaction of nanobodies with fluorescent proteins (Bothma, Norstad, Alamos, & Garcia, 2018; Maryu et al., 2018; Yoshioka-Kobayashi et al., 2020), and more sensitive, less phototoxic microscopy systems (Krzic, Gunther, Saunders, Streichan, & Hufnagel, 2012; Serra et al., 2019) enable researchers to study dynamics of different signaling pathways in more and more tissues and model systems during development. Open questions are for instance how NFκB oscillations or calcium oscillations function in developing embryos (Dupont, Combettes, Bird, & Putney, 2011; Nelson, 2004). With the establishment of stem cell-based embryo-like culture systems such as gastruloids or blastoids the study and manipulation of embryonic development becomes more amenable and even allows the study of human development at mechanistic detail in the future (Moris et al., 2020; Rivron et al., 2018; van den Brink et al., 2020). While this chapter mainly focused on signaling oscillations in vertebrates, information transmission through oscillations has been observed in other species. Reminiscent of vertebrate somitogenesis, in the invertebrate *Tribolium castaneum*, a molecular clock with a period of 95 min was discovered to coordinate this beetle's segmentation (Sarrazin, Peel, & Averof, 2012). Oscillations are also found in plants, where tissue-wide oscillations of Auxin signaling were found throughout the process of lateral root formation and control the positioning of lateral roots along the primary root in *Arabidopsis thaliana* (Moreno-Risueno et al., 2010). Expanding the study of oscillations to other signaling pathways and organisms might therefore unveil an even larger role for signaling oscillations during embryonic development than is currently appreciated.

Thus, signaling oscillations are an essential parameter of information transmission in embryonic development. Revealing their function and mechanisms will contribute to our understanding of how development occurs in a controlled and robust way and how mutations in signaling components induce developmental defects.

Acknowledgments

We thank Hidenobu Miyazawa, Sonja Weterings, Marek van Oostrom and Pascal Schulthess for feedback on the manuscript. Figures were partially generated with biorender.com.

References

Afelik, S., Qu, X., Hasrouni, E., Bukys, M. A., Deering, T., Nieuwoudt, S., et al. (2012). Notch-mediated patterning and cell fate allocation of pancreatic progenitor cells. *Development, 139*, 1744–1753.

Albeck, J. G., Mills, G. B., & Brugge, J. S. (2013). Frequency-modulated pulses of ERK activity transmit quantitative proliferation signals. *Molecular Cell, 49*, 249–261.

Alon, U. (2007). *An introduction to systems biology: Design principles of biological circuits*. Boca Raton, FL: Chapman & Hall/CRC.

Ares, S., Morelli, L. G., Jörg, D. J., Oates, A. C., & Jülicher, F. (2012). Collective modes of coupled phase oscillators with delayed coupling. *Physical Review Letters, 108*, 204101.

Artavanis-Tsakonas, S. (1999). Notch signaling: Cell fate control and signal integration in development. *Science, 284*, 770–776.

Ashall, L., Horton, C. A., Nelson, D. E., Paszek, P., Harper, C. V., Sillitoe, K., et al. (2009). Pulsatile stimulation determines timing and specificity of NF-kappaB-dependent transcription. *Science, 324*, 242–246.

Aulehla, A., & Johnson, R. L. (1999). Dynamic expression of lunatic fringe suggests a link between notch signaling and an autonomous cellular oscillator driving somite segmentation. *Developmental Biology, 207*, 49–61.

Aulehla, A., & Pourquié, O. (2008). Oscillating signaling pathways during embryonic development. *Current Opinion in Cell Biology, 20*, 632–637.

Aulehla, A., Wehrle, C., Brand-Saberi, B., Kemler, R., Gossler, A., Kanzler, B., et al. (2003). Wnt3a plays a major role in the segmentation clock controlling somitogenesis. *Developmental Cell, 4*, 395–406.

Aulehla, A., Wiegraebe, W., Baubet, V., Wahl, M. B., Deng, C., Taketo, M., et al. (2008). A beta-catenin gradient links the clock and wavefront systems in mouse embryo segmentation. *Nature Cell Biology, 10*, 186–193.

Baron, M. (2003). An overview of the notch signalling pathway. *Seminars in Cell & Developmental Biology, 14*, 113–119.

Barrantes, I. B., Elia, A. J., Wünsch, K., Hrabe de Angelis, M. H., Mak, T. W., Rossant, J., et al. (1999). Interaction between notch signalling and lunatic fringe during somite boundary formation in the mouse. *Current Biology, 9*, 470–480.

Bastidas-Ponce, A., Scheibner, K., Lickert, H., & Bakhti, M. (2017). Cellular and molecular mechanisms coordinating pancreas development. *Development, 144*, 2873–2888.

Beaupeux, M., & François, P. (2016). Positional information from oscillatory phase shifts: Insights fromin silicoevolution. *Physical Biology, 13*, 1–14.

Behar, M., & Hoffmann, A. (2010). Understanding the temporal codes of intra-cellular signals. *Current Opinion in Genetics & Development, 20*, 684–693.

Bessho, Y., Hirata, H., Masamizu, Y., & Kageyama, R. (2003). Periodic repression by the bHLH factor Hes7 is an essential mechanism for the somite segmentation clock. *Genes & Development, 17*, 1451–1456.

Bhat, R., Glimm, T., Linde-Medina, M., Cui, C., & Newman, S. A. (2019). Synchronization of Hes1 oscillations coordinates and refines condensation formation and patterning of the avian limb skeleton. *Mechanisms of Development, 156*, 41–54.

Bothma, J. P., Norstad, M. R., Alamos, S., & Garcia, H. G. (2018). LlamaTags: A versatile tool to image transcription factor dynamics in live embryos. *Cell, 173*, 1810–1822.

Bray, S. J. (2006). Notch signalling: A simple pathway becomes complex. *Nature Reviews. Molecular Cell Biology*, *7*, 678–689.

Brondello, J.-M., Brunet, A., Pouysségur, J., & McKenzie, F. R. (1997). The dual specificity mitogen-activated protein kinase phosphatase-1 and -2 are induced by the p42/p44MAPK cascade. *Journal of Biological Chemistry*, *272*, 1368–1376.

Bulusu, V., Prior, N., Snaebjornsson, M. T., Kuehne, A., Sonnen, K. F., Kress, J., et al. (2017). Spatiotemporal analysis of a glycolytic activity gradient linked to mouse embryo mesoderm development. *Developmental Cell*, *40*, 331–341. e334.

Burack, W. R., & Shaw, A. S. (2005). Live cell imaging of ERK and MEK: Simple binding equilibrium explains the regulated nucleocytoplasmic distribution of ERK. *The Journal of Biological Chemistry*, *280*, 3832–3837.

Cooke, J., & Zeeman, E. C. (1976). A clock and wavefront model for control of the number of repeated structures during animal morphogenesis. *Journal of Theoretical Biology*, *58*, 455–476.

Cotterell, J., Robert-Moreno, A., & Sharpe, J. (2015). A local, self-organizing reaction-diffusion model can explain somite patterning in embryos. *Cell Systems*, *1*, 257–269.

Dale, J. K., Maroto, M., Dequeant, M. L., Malapert, P., McGrew, M., & Pourquié, O. (2003). Periodic notch inhibition by lunatic fringe underlies the chick segmentation clock. *Nature*, *421*, 275–278.

de la Cova, C., Townley, R., Regot, S., & Greenwald, I. (2017). A real-time biosensor for ERK activity reveals signaling dynamics during C. elegans cell fate specification. *Developmental Cell*, *42*, 542–553. e544.

De Simone, A., Evanitsky, M. N., Hayden, L., Cox, B. D., Wang, J., Tornini, V. A., et al. (2021). Control of osteoblast regeneration by a train of Erk activity waves. *Nature*, *590*, 129–133.

Delaune, E. A., Francois, P., Shih, N. P., & Amacher, S. L. (2012). Single-cell-resolution imaging of the impact of notch signaling and mitosis on segmentation clock dynamics. *Developmental Cell*, *23*, 995–1005.

Deneke, V. E., Puliafito, A., Krueger, D., Narla, A. V., De Simone, A., Primo, L., et al. (2019). Self-organized nuclear positioning synchronizes the cell cycle in Drosophila embryos. *Cell*, *177*, 925–941.e917.

Dequéant, M. L., Glynn, E., Gaudenz, K., Wahl, M., Chen, J., Mushegian, A., et al. (2006). A complex oscillating network of signaling genes underlies the mouse segmentation clock. *Science*, *314*, 1595–1598.

Dessaud, E., Yang, L. L., Hill, K., Cox, B., Ulloa, F., Ribeiro, A., et al. (2007). Interpretation of the sonic hedgehog morphogen gradient by a temporal adaptation mechanism. *Nature*, *450*, 717–720.

Diaz-Cuadros, M., Wagner, D. E., Budjan, C., Hubaud, A., Tarazona, O. A., Donelly, S., et al. (2020). In vitro characterization of the human segmentation clock. *Nature*, *580*, 113–118.

Dong, Y., Jesse, A. M., Kohn, A., Gunnell, L. M., Honjo, T., Zuscik, M. J., et al. (2010). RBPjκ-dependent notch signaling regulates mesenchymal progenitor cell proliferation and differentiation during skeletal development. *Development*, *137*, 1461–1471.

Dubrulle, J., McGrew, M. J., & Pourquié, O. (2001). FGF signaling controls somite boundary position and regulates segmentation clock control of spatiotemporal Hox gene activation. *Cell*, *106*, 219–232.

Dubrulle, J., & Pourquié, O. (2004). fgf8 mRNA decay establishes a gradient that couples axial elongation to patterning in the vertebrate embryo. *Nature*, *427*, 419–422.

Dupont, G., Combettes, L., Bird, G. S., & Putney, J. W. (2011). Calcium oscillations. *Cold Spring Harbor Perspectives in Biology*, *3*, 1–18.

Feller, J., Schneider, A., Schuster-Gossler, K., & Gossler, A. (2008). Noncyclic notch activity in the presomitic mesoderm demonstrates uncoupling of somite compartmentalization and boundary formation. *Genes & Development*, *22*, 2166–2171.

Fuseler, J. W., Merrill, D. M., Rogers, J. A., Grisham, M. B., & Wolf, R. E. (2006). Analysis and quantitation of NF-kappaB nuclear translocation in tumor necrosis factor alpha (TNF-alpha) activated vascular endothelial cells. *Microscopy and Microanalysis, 12*, 269–276.

Gagliardi, P. A., Dobrzyński, M., Jacques, M.-A., Dessauges, C., Ender, P., Blum, Y., et al. (2021). Collective ERK/Akt activity waves orchestrate epithelial homeostasis by driving apoptosis-induced survival. *Developmental Cell, 56*, 1712–1726.e1716.

Gomez, C., Özbudak, E. M., Wunderlich, J., Baumann, D., Lewis, J., & Pourquié, O. (2008). Control of segment number in vertebrate embryos. *Nature, 454*, 335–339.

Gonze, D., Gérard, C., Wacquier, B., Woller, A., Tosenberger, A., Goldbeter, A., et al. (2018). Modeling-based investigation of the effect of noise in cellular systems. *Frontiers in Molecular Biosciences, 5*, 1–12.

Guo, Y. J., Pan, W. W., Liu, S. B., Shen, Z. F., Xu, Y., & Hu, L. L. (2020). ERK/MAPK signalling pathway and tumorigenesis. *Experimental and Therapeutic Medicine, 19*, 1997–2007.

Hacohen, N., Kramer, S., Sutherland, D., Hiromi, Y., & Krasnow, M. A. (1998). Sprouty encodes a novel antagonist of FGF signaling that patterns apical branching of the Drosophila airways. *Cell, 92*, 253–263.

Harima, Y., Imayoshi, I., Shimojo, H., Kobayashi, T., & Kageyama, R. (2014). The roles and mechanism of ultradian oscillatory expression of the mouse Hes genes. *Seminars in Cell & Developmental Biology, 34*, 85–90.

Harima, Y., Takashima, Y., Ueda, Y., Ohtsuka, T., & Kageyama, R. (2013). Accelerating the tempo of the segmentation clock by reducing the number of introns in the Hes7 gene. *Cell Reports, 3*, 1–7.

Hatakeyama, J., Bessho, Y., Katoh, K., Ookawara, S., Fujioka, M., Guillemot, F., et al. (2004). Hes genes regulate size, shape and histogenesis of the nervous system by control of the timing of neural stem cell differentiation. *Development, 131*, 5539–5550.

Henrique, D., & Schweisguth, F. (2019). Mechanisms of notch signaling: A simple logic deployed in time and space. *Development, 146*, 1–11.

Hino, N., Rossetti, L., Marín-Llauradó, A., Aoki, K., Trepat, X., Matsuda, M., et al. (2020). ERK-mediated mechanochemical waves direct collective cell polarization. *Developmental Cell, 53*, 646–660.e648.

Hirata, H., Bessho, Y., Kokubu, H., Masamizu, Y., Yamada, S., Lewis, J., et al. (2004). Instability of Hes7 protein is crucial for the somite segmentation clock. *Nature Genetics, 36*, 750–754.

Hirata, H., Yoshiura, S., Ohtsuka, T., Bessho, Y., Harada, T., Yoshikawa, K., et al. (2002). Oscillatory expression of the bHLH factor Hes1 regulated by a negative feedback loop. *Science, 298*, 840–843.

Hiratsuka, T., Bordeu, I., Pruessner, G., & Watt, F. M. (2020). Regulation of ERK basal and pulsatile activity control proliferation and exit from the stem cell compartment in mammalian epidermis. *PNAS, 117*, 17796–17807.

Hubaud, A., & Pourquie, O. (2014). Signalling dynamics in vertebrate segmentation. *Nature Reviews. Molecular Cell Biology, 15*, 709–721.

Hubaud, A., Regev, I., Mahadevan, L., & Pourquié, O. (2017). Excitable dynamics and yap-dependent mechanical cues drive the segmentation clock. *Cell, 171*, 668–682.e611.

Imayoshi, I., Isomura, A., Harima, Y., Kawaguchi, K., Kori, H., Miyachi, H., et al. (2013). Oscillatory control of factors determining multipotency and fate in mouse neural progenitors. *Science, 342*, 1203–1208.

Imayoshi, I., & Kageyama, R. (2014). bHLH factors in self-renewal, multipotency, and fate choice of neural progenitor cells. *Neuron, 82*, 9–23.

Jennings, R. E., Berry, A. A., Strutt, J. P., Gerrard, D. T., & Hanley, N. A. (2015). Human pancreas development. *Development, 142*, 3126–3137.

Jensen, J., Pedersen, E. E., Galante, P., Hald, J., Heller, R. S., Ishibashi, M., et al. (2000). Control of endodermal endocrine development by Hes-1. *Nature Genetics*, *24*, 36–44.

Jiang, Y.-J., Aerne, B. L., Smithers, L., Haddon, C., Ish-Horowicz, D., & Lewis, J. (2000). Notch signalling and the synchronization of the somite segmentation clock. *Nature*, *408*, 475–479.

Johnson, H. E., & Toettcher, J. E. (2018). Illuminating developmental biology with cellular optogenetics. *Current Opinion in Biotechnology*, *52*, 42–48.

Johnson, H. E., & Toettcher, J. E. (2019). Signaling dynamics control cell fate in the early Drosophila embryo. *Developmental Cell*, *48*, 361–370.e363.

Jouve, C., Palmeirim, I., Henrique, D., Beckers, J., Gossler, A., Ish-Horowicz, D., et al. (2000). Notch signalling is required for cyclic expression of the hairy-like gene HES1 in the presomitic mesoderm. *Development*, *127*, 1421–1429.

Kageyama, R., Shimojo, H., & Isomura, A. (2018). Oscillatory control of notch signaling in development. *Adv. Exp. Med. Biol.*, *1066*, 265–277.

Kageyama, R., Shimojo, H., & Ohtsuka, T. (2019). Dynamic control of neural stem cells by bHLH factors. *Neuroscience Research*, *138*, 12–18.

Kellogg, R. A., & Tay, S. (2015). Noise facilitates transcriptional control under dynamic inputs. *Cell*, *160*, 381–392.

Kim, N., Kim, J. M., Lee, M., Kim, C. Y., Chang, K.-Y., & Heo, W. D. (2014). Spatiotemporal control of fibroblast growth factor receptor signals by blue light. *Chemistry & Biology*, *21*, 903–912.

Kobayashi, T., Mizuno, H., Imayoshi, I., Furusawa, C., Shirahige, K., & Kageyama, R. (2009). The cyclic gene Hes1 contributes to diverse differentiation responses of embryonic stem cells. *Genes & Development*, *23*, 1870–1875.

Krzic, U., Gunther, S., Saunders, T. E., Streichan, S. J., & Hufnagel, L. (2012). Multiview light-sheet microscope for rapid in toto imaging. *Nature Methods*, *9*, 730–733.

Lane, K., Van Valen, D., DeFelice, M. M., Macklin, D. N., Kudo, T., Jaimovich, A., et al. (2017). Measuring signaling and RNA-Seq in the same cell links gene expression to dynamic patterns of NF-κB activation. *Cell Systems*, *4*, 458–469.e455.

Lavoie, H., Gagnon, J., & Therrien, M. (2020). ERK signalling: A master regulator of cell behaviour, life and fate. *Nature Reviews. Molecular Cell Biology*, *21*, 607–632.

Lu, Q. R., Sun, T., Zhu, Z., Ma, N., Garcia, M., Stiles, C. D., et al. (2002). Common developmental requirement for Olig function indicates a motor neuron/oligodendrocyte connection. *Cell*, *109*, 75–86.

Manning, C. S., Biga, V., Boyd, J., Kursawe, J., Ymisson, B., Spiller, D. G., et al. (2019). Quantitative single-cell live imaging links HES5 dynamics with cell-state and fate in murine neurogenesis. *Nature Communications*, *10*, 1–19.

Marinopoulou, E., Biga, V., Sabherwal, N., Miller, A., Desai, J., Adamson, A. D., et al. (2021). HES1 protein oscillations are necessary for neural stem cells to exit from quiescence. *iScience*, *24*, 103198.

Marshall, C. J. (1995). Specificity of receptor tyrosine kinase signaling: Transient versus sustained extracellular signal-regulated kinase activation. *Cell*, *80*, 179–185.

Maryu, G., Miura, H., Uda, Y., Komatsubara, A. T., Matsuda, M., & Aoki, K. (2018). Live-cell imaging with genetically encoded protein kinase activity reporters. *Cell Structure and Function*, *43*, 61–74.

Masamizu, Y., Ohtsuka, T., Takashima, Y., Nagahara, H., Takenaka, Y., Yoshikawa, K., et al. (2006). Real-time imaging of the somite segmentation clock: Revelation of unstable oscillators in the individual presomitic mesoderm cells. *Proceedings of the National Academy of Sciences of the United States of America*, *103*, 1313–1318.

Matsuda, M., Hayashi, H., Garcia-Ojalvo, J., Yoshioka-Kobayashi, K., Kageyama, R., Yamanaka, Y., et al. (2020). Species-specific segmentation clock periods are due to differential biochemical reaction speeds. *Science*, *369*, 1450–1455.

Matsuda, M., Yamanaka, Y., Uemura, M., Osawa, M., Saito, M. K., Nagahashi, A., et al. (2020). Recapitulating the human segmentation clock with pluripotent stem cells. *Nature, 580*, 124–129.

Morelli, L. G., Ares, S., Herrgen, L., Schröter, C., Jülicher, F., & Oates, A. C. (2009). Delayed coupling theory of vertebrate segmentation. *Human Frontier Science Program, 3*, 55–66.

Moreno-Risueno, M. A., Van Norman, J. M., Moreno, A., Zhang, J., Ahnert, S. E., & Benfey, P. N. (2010). Oscillating gene expression determines competence for periodic Arabidopsis root branching. *Science, 329*, 1306–1311.

Moris, N., Anlas, K., van den Brink, S. C., Alemany, A., Schröder, J., Ghimire, S., et al. (2020). An in vitro model of early anteroposterior organization during human development. *Nature, 582*, 410–415.

Muta, Y., Fujita, Y., Sumiyama, K., Sakurai, A., Taketo, M. M., Chiba, T., et al. (2018). Composite regulation of ERK activity dynamics underlying tumour-specific traits in the intestine. *Nature Communications, 9*, 1–16.

Nakayama, K., Satoh, T., Igari, A., Kageyama, R., & Nishida, E. (2008). FGF induces oscillations of Hes1 expression and Ras/ERK activation. *Current Biology, 18*, R332–R334.

Negrete, J., Jr., & Oates, A. C. (2021). Towards a physical understanding of developmental patterning. *Nature Reviews. Genetics, 22*, 518–531.

Nelson, D. E. (2004). Oscillations in NF-kB signaling control the dynamics of gene expression. *Science, 306*, 704–708.

Neumann, C., & Cohen, S. (1997). Morphogens and pattern formation. *Bioessays, 19*, 721–729.

Niehrs, C. (2012). The complex world of WNT receptor signalling. *Nature Reviews. Molecular Cell Biology, 13*, 767–779.

Nieto, M., Schuurmans, C., Britz, O., & Guillemot, F. (2001). Neural bHLH genes control the neuronal versus glial fate decision in cortical progenitors. *Neuron, 29*, 401–413.

Niwa, Y., Masamizu, Y., Liu, T., Nakayama, R., Deng, C.-X., & Kageyama, R. (2007). The initiation and propagation of Hes7 oscillation are cooperatively regulated by Fgf and notch signaling in the somite segmentation clock. *Developmental Cell, 13*, 298–304.

Niwa, Y., Shimojo, H., Isomura, A., González, A., Miyachi, H., & Kageyama, R. (2011). Different types of oscillations in notch and Fgf signaling regulate the spatiotemporal periodicity of somitogenesis. *Genes & Development, 25*, 1115–1120.

Nüsslein-Volhard, C., & Wieschaus, E. F. (1980). Mutations affecting segment number and polarity in Drosophila. *Nature, 287*, 795–801.

Oates, A. C. (2020). Waiting on the fringe: Cell autonomy and signaling delays in segmentation clocks. *Current Opinion in Genetics & Development, 63*, 61–70.

Oates, A. C., & Ho, R. K. (2002). Hairy/E(spl)-related (her) genes are central components of the segmentation oscillator and display redundancy with the Delta/Notch signaling pathway in the formation of anterior segmental boundaries in the zebrafish. *Development, 129*, 2929–2946.

Oginuma, M., Harima, Y., Tarazona, O. A., Diaz-Cuadros, M., Michaut, A., Ishitani, T., et al. (2020). Intracellular pH controls WNT downstream of glycolysis in amniote embryos. *Nature, 584*, 98–101.

Oginuma, M., Moncuquet, P., Xiong, F., Karoly, E., Chal, J., Guevorkian, K., et al. (2017). A gradient of glycolytic activity coordinates FGF and Wnt signaling during elongation of the body axis in amniote embryos. *Developmental Cell, 40*, 342–353. e310.

Oginuma, M., Niwa, Y., Chapman, D. L., & Saga, Y. (2008). Mesp2 and Tbx6 cooperatively create periodic patterns coupled with the clock machinery during mouse somitogenesis. *Development, 135*, 2555–2562.

Ohtsuka, T., Ishibashi, M., Gradwohl, G., Nakanishi, S., Guillemot, F., & Kageyama, R. (1999). Hes1 and Hes5 as notch effectors in mammalian neuronal differentiation. *The EMBO Journal*, *18*, 2196–2207.

Okubo, Y., Sugawara, T., Abe-Koduka, N., Kanno, J., Kimura, A., & Saga, Y. (2012). Lfng regulates the synchronized oscillation of the mouse segmentation clock via trans-repression of notch signalling. *Nature Communications*, *3*, 1–9.

Palmeirim, I., Henrique, D., Ish-Horowicz, D., & Pourquié, O. (1997). Avian hairy gene expression identifies a molecular clock linked to vertebrate segmentation and somitogenesis. *Cell*, *91*, 639–648.

Pan, F. C., & Wright, C. (2011). Pancreas organogenesis: From bud to plexus to gland. *Developmental Dynamics*, *240*, 530–565.

Pascoal, S., Carvalho, C. R., Rodriguez-León, J., Delfini, M.-C., Duprez, D., Thorsteinsdóttir, S., et al. (2007). A molecular clock operates during chick autopod proximal-distal outgrowth. *Journal of Molecular Biology*, *368*, 303–309.

Petit, F., Sears, K. E., & Ahituv, N. (2017). Limb development: A paradigm of gene regulation. *Nature Reviews. Genetics*, *18*, 245–258.

Peyssonnaux, C., & Eychène, A. (2001). The Raf/MEK/ERK pathway: New concepts of activation. *Biology of the Cell*, *93*, 53–62.

Purvis, J. E., & Lahav, G. (2013). Encoding and decoding cellular information through signaling dynamics. *Cell*, *152*, 945–956.

Qu, X., Afelik, S., Jensen, J. N., Bukys, M. A., Kobberup, S., Schmerr, M., et al. (2013). Notch-mediated post-translational control of Ngn3 protein stability regulates pancreatic patterning and cell fate commitment. *Developmental Biology*, *376*, 1–12.

Raina, D., Bahadori, A., Stanoev, A., Protzek, M., Koseska, A., & Schroter, C. (2021). Cell-cell communication through FGF4 generates and maintains robust proportions of differentiated cell types in embryonic stem cells. *Development*, *148*.

Rayon, T., Stamataki, D., Perez-Carrasco, R., Garcia-Perez, L., Barrington, C., Melchionda, M., et al. (2020). Species-specific pace of development is associated with differences in protein stability. *Science*, *369*, 1–13.

Regot, S., Hughey, J. J., Bajar, B. T., Carrasco, S., & Covert, M. W. (2014). High-sensitivity measurements of multiple kinase activities in live single cells. *Cell*, *157*, 1724–1734.

Riedel-Kruse, I. H., Müller, C., & Oates, A. C. (2007). Synchrony dynamics during initiation, failure, and rescue of the segmentation clock. *Science*, *317*, 1911–1915.

Rivron, N. C., Frias-Aldeguer, J., Vrij, E. J., Boisset, J.-C., Korving, J., Vivié, J., et al. (2018). Blastocyst-like structures generated solely from stem cells. *Nature*, *557*, 106–111.

Rohde, L. A., Bercowsky-Rama, A., Negrete, J., Valentin, G., Naganathan, S. R., Desai, R. A., et al. (2021). Cell-autonomous generation of the wave pattern within the vertebrate segmentation clock. *BioRxiv*, 1–15.

Ryu, H., Chung, M., Dobrzyński, M., Fey, D., Blum, Y., Lee, S. S., et al. (2016). Frequency modulation of ERK activation dynamics rewires cell fate. *Molecular Systems Biology*, *12*, 1–14.

Sarrazin, A. F., Peel, A. D., & Averof, M. (2012). A segmentation clock with two-segment periodicity in insects. *Science*, *336*, 338–341.

Serra, D., Mayr, U., Boni, A., Lukonin, I., Rempfler, M., Challet Meylan, L., et al. (2019). Self-organization and symmetry breaking in intestinal organoid development. *Nature*, *569*, 66–72.

Seymour, P. A., Collin, C. A., Egeskov-Madsen, A. R., Jorgensen, M. C., Shimojo, H., Imayoshi, I., et al. (2020). Jag1 modulates an oscillatory Dll1-notch-Hes1 signaling module to coordinate growth and fate of pancreatic progenitors. *Developmental Cell*, *52*, 731–747. e738.

Shankaran, H., Ippolito, D. L., Chrisler, W. B., Resat, H., Bollinger, N., Opresko, L. K., et al. (2009). Rapid and sustained nuclear-cytoplasmic ERK oscillations induced by epidermal growth factor. *Molecular Systems Biology*, *5*, 1–13.

Shankaran, H., & Wiley, H. S. (2010). Oscillatory dynamics of the extracellular signal-regulated kinase pathway. *Current Opinion in Genetics & Development, 20*, 650–655.

Sharma, S. C., & Goffinet, A. M. (2012). *Development of the central nervous system in vertebrates*. Springer Science & Business Media.

Shaul, Y. D., & Seger, R. (2007). The MEK/ERK cascade: From signaling specificity to diverse functions. *Biochimica et Biophysica Acta, 1773*, 1213–1226.

Sheeba, C. J., Andrade, R. P., & Palmeirim, I. (2012). Joint interpretation of AER/FGF and ZPA/SHH over time and space underlies hairy2 expression in the chick limb. *Biology Open, 1*, 1102–1110.

Shih, H. P., Wang, A., & Sander, M. (2013). Pancreas organogenesis: From lineage determination to morphogenesis. *Annual Review of Cell and Developmental Biology, 29*, 81–105.

Shimojo, H., Isomura, A., Ohtsuka, T., Kori, H., Miyachi, H., & Kageyama, R. (2016). Oscillatory control of Delta-like1 in cell interactions regulates dynamic gene expression and tissue morphogenesis. *Genes & Development, 30*, 102–116.

Shimojo, H., Ohtsuka, T., & Kageyama, R. (2008). Oscillations in notch signaling regulate maintenance of neural progenitors. *Neuron, 58*, 52–64.

Sjöqvist, M., & Andersson, E. R. (2019). Do as I say, Not(ch) as I do: Lateral control of cell fate. *Developmental Biology, 447*, 58–70.

Sonnen, K. F., & Aulehla, A. (2014). Dynamic signal encoding—From cells to organisms. *Seminars in Cell & Developmental Biology, 34*, 91–98.

Sonnen, K. F., & Janda, C. Y. (2021). Signalling dynamics in embryonic development. *Biochemical Journal, 478*, 4045–4070.

Sonnen, K. F., Lauschke, V. M., Uraji, J., Falk, H. J., Petersen, Y., Funk, M. C., et al. (2018). Modulation of phase shift between Wnt and Notch signaling oscillations controls mesoderm segmentation. *Cell, 172*, 1079–1090. e1012.

Sonnen, K. F., & Merten, C. A. (2019). Microfluidics as an emerging precision tool in developmental biology. *Developmental Cell, 48*, 293–311.

Soza-Ried, C., Ozturk, E., Ish-Horowicz, D., & Lewis, J. (2014). Pulses of notch activation synchronise oscillating somite cells and entrain the zebrafish segmentation clock. *Development, 141*, 1780–1788.

Sueda, R., Imayoshi, I., Harima, Y., & Kageyama, R. (2019). High Hes1 expression and resultant Ascl1 suppression regulate quiescent vs. active neural stem cells in the adult mouse brain. *Genes & Development, 33*, 511–523.

Sun, Y., Nadal-Vicens, M., Misono, S., Lin, M. Z., Zubiaga, A., Hua, X., et al. (2001). Neurogenin promotes neurogenesis and inhibits glial differentiation by independent mechanisms. *Cell, 104*, 365–376.

Takashima, Y., Ohtsuka, T., González, A., Miyachi, H., & Kageyama, R. (2011). Intronic delay is essential for oscillatory expression in the segmentation clock. *Proceedings of the National Academy of Sciences of the United States of America, 108*, 3300–3305.

Tay, S., Hughey, J. J., Lee, T. K., Lipniacki, T., Quake, S. R., & Covert, M. W. (2010). Single-cell NF-κB dynamics reveal digital activation and analogue information processing. *Nature, 466*, 267–271.

Toettcher, J. E., Weiner, O. D., & Lim, W. A. (2013). Using optogenetics to interrogate the dynamic control of signal transmission by the ras/erk module. *Cell, 155*, 1422–1434.

Tsiairis, C. D., & Aulehla, A. (2016). Self-organization of embryonic genetic oscillators into spatiotemporal wave patterns. *Cell, 164*, 656–667.

Turing, A. M. (1952). The chemical basis of morphogenesis. *Philosophical Transactions of the Royal Society, 237*, 37–72.

Valon, L., Davidović, A., Levillayer, F., Villars, A., Chouly, M., Cerqueira-Campos, F., et al. (2021). Robustness of epithelial sealing is an emerging property of local ERK feedback driven by cell elimination. *Developmental Cell, 56*, 1700–1711. e1708.

van den Brink, S. C., Alemany, A., van Batenburg, V., Moris, N., Blotenburg, M., Vivie, J., et al. (2020). Single-cell and spatial transcriptomics reveal somitogenesis in gastruloids. *Nature*, *582*, 405–409.

van Oostrom, M. J., Meijer, W. H. M., & Sonnen, K. F. (2021). A microfluidics approach for the functional investigation of signaling oscillations governing somitogenesis. *Journal of Visualized Experiments*, *169*, 1–17.

Viswanathan, R., Necakov, A., Trylinski, M., Harish, R. K., Krueger, D., Esposito, E., et al. (2019). Optogenetic inhibition of Delta reveals digital notch signalling output during tissue differentiation. *EMBO Reports*, *20*, 1–13.

Wahl, M. B., Deng, C., Lewandoski, M., & Pourquié, O. (2007). FGF signaling acts upstream of the NOTCH and WNT signaling pathways to control segmentation clock oscillations in mouse somitogenesis. *Development*, *134*, 4033–4041.

Webb, A. B., Lengyel, I. M., Jörg, D. J., Valentin, G., Jülicher, F., Morelli, L. G., et al. (2016). Persistence, period and precision of autonomous cellular oscillators from the zebrafish segmentation clock. *eLife*, *5*, 1–17.

Weterings, S. D. C., van Oostrom, M. J., & Sonnen, K. F. (2021). Building bridges between fields: Bringing together development and homeostasis. *Development*, *148*, 1–11.

Wieschaus, E., & Nüsslein-Volhard, C. (2016). The Heidelberg screen for pattern mutants of Drosophila: A personal account. *Annual Review of Cell and Developmental Biology*, *32*, 1–46.

Wolpert, L. (1969). Positional information and the spatial pattern of cellular differentiation. *Journal of Theoretical Biology*, *25*, 1–47.

Wolpert, L. (1989). Positional information revisited. *Development*, *107*(Suppl), 3–12.

Yoshioka-Kobayashi, K., Matsumiya, M., Niino, Y., Isomura, A., Kori, H., Miyawaki, A., et al. (2020). Coupling delay controls synchronized oscillation in the segmentation clock. *Nature*, *580*, 119–123.

Zhang, Q., Huang, H., Zhang, L., Wu, R., Chung, C.-I., Zhang, S.-Q., et al. (2018). Visualizing dynamics of cell signaling in vivo with a phase separation-based kinase reporter. *Molecular Cell*, *69*, 334–346.

Zhou, Q., & Anderson, D. J. (2002). The bHLH transcription factors OLIG2 and OLIG1 couple neuronal and glial subtype specification. *Cell*, *109*, 61–73.

Zinner, M., Lukonin, I., & Liberali, P. (2020). Design principles of tissue organisation: How single cells coordinate across scales. *Current Opinion in Cell Biology*, *67*, 37–45.

CHAPTER TEN

Parallels in signaling between development and regeneration in ectodermal organs

Neha Pincha[a], Pauline Marangoni[a], Ameera Haque[a], and Ophir D. Klein[a,b,*]
[a]Program in Craniofacial Biology and Department of Orofacial Sciences, University of California, San Francisco, CA, United States
[b]Department of Pediatrics and Institute for Human Genetics, University of California, San Francisco, CA, United States
*Corresponding author: e-mail address: ophir.klein@ucsf.edu

Contents

Current Topics in Developmental Biology, Volume 149
ISSN 0070-2153
https://doi.org/10.1016/bs.ctdb.2022.02.006

Abstract

Ectodermal organs originate from the outermost germ layer of the developing embryo and include the skin, hair, tooth, nails, and exocrine glands. These organs develop through tightly regulated, sequential and reciprocal epithelial-mesenchymal crosstalk, and they eventually assume various morphologies and functions while retaining the ability to regenerate. As with many other tissues in the body, the development and morphogenesis of these organs are regulated by a set of common signaling pathways, such as Shh, Wnt, Bmp, Notch, Tgf-β, and Eda. However, subtle differences in the temporal activation, the multiple possible combinations of ligand-receptor activation, the various cofactors, as well as the underlying epigenetic modulation determine how each organ develops into its adult form. Although each organ has been studied separately in considerable detail, the mechanisms underlying the parallels and differences in signaling that regulate their development have rarely been investigated. First, we will use the tooth, the hair follicle, and the mammary gland as representative ectodermal organs to explore how the development of signaling centers and establishment of stem cell populations influence overall growth and morphogenesis. Then we will compare how some of the major signaling pathways (Shh, Wnt, Notch and Yap/Taz) differentially regulate developmental events. Finally, we will discuss how signaling regulates regenerative processes in all three.

1. Introduction

Ectodermal organs comprise a wide array of morphologically distinct appendages localized in various positions in the body and fulfilling highly specialized functions. Some—like the tooth, hair follicle, nails, and scales—develop mineralized structures as they grow, while their dermal counterparts—the sweat, sebaceous and mammary glands—have a secretory role. Despite these pronounced differences, ectodermal organs share striking similarities in their developmental strategies. Indeed, they all form through tightly regulated crosstalk between an epithelium and its underlying mesenchyme.

In this chapter, we focus on three ectodermal appendages: the tooth, the hair follicle and the mammary gland. These organs encompass a great diversity of morphology, localization, and function, and they differ in their cycling properties and regenerative capacities. The tooth is specialized for chewing/grinding and is the only organ in the body covered with enamel, the hardest mineralized substance. The hair consists of keratin and lies embedded in a follicle, allowing it to perform the function of insulation and, to some extent, sensory perception. The mammary gland is an accessory reproductive organ, restricted to females and specialized to produce milk during lactation. Despite their vastly different functions and morphologies, these organs share a common initiation stage involving the formation of a thickened epithelial placode in response to mesenchymal signals. They start building morphological differences as the epithelial structures invaginate into the underlying mesenchyme, and these differences continue to accumulate as each organ grows and reaches maturity.

In humans, the development of deciduous teeth is followed by a second developmental wave leading to the formation of the permanent adult dentition. However, rodent teeth, which are the most common models used to study mammalian odontogenesis, are somewhat different. After their initial development, the incisors maintain dental stem cells, allowing for the continuous growth of this organ throughout the life of the animal. Mineralized tissues located at the distal tip of the incisor are abraded, while new cells and mineralized dental matrices are produced on the proximal end, where dental stem and progenitor cells reside (Harada, Toyono, Toyoshima, & Ohuchi, 2002; Kuang-Hsien Hu, Mushegyan, & Klein, 2014). Recently, we have shown that a population of actively cycling epithelial cells enables the dynamic and uninterrupted renewal of all epithelial lineages, and that quiescent cells remain in the epithelial niche without contributing to adult homeostasis (Sharir et al., 2019). In the mesenchyme, the relationship between stem cells and actively cycling progenitors is somewhat less clear, but these cell types support the maintenance of dental pulp and dentin-producing odontoblasts. This continuous equilibrium allows for the replacement of the entire incisor approximately every 4 weeks, making it one of the most rapidly renewing ectodermal organs in mammals (Sharir et al., 2019). The continuous growth also allows the incisor to rapidly regenerate after injuries by efficiently replenishing the damaged tissues (Chen et al., 2020; Sharir et al., 2019; Zhao et al., 2017).

The hair follicle develops through three stages: anagen (growth), catagen (regression) and telogen (resting). At any given time, the hair on our bodies is divided among these three stages, with most of it being in the anagen phase (Müller-Röver et al., 2001; Zhang, Cheong, Ciapurin, McDermitt, & Tumbar, 2009). The duration of anagen correlates with the length of hair, lasting between 2 and 6 years on the scalp and only 30–45 days on our arms and legs (Willems & Sinclair, 2021). In mice, the first two waves of hair growth are synchronous from the anterior to the posterior regions of the body. After that, the follicles on the skin are present as a mosaic of different phases (Nguyen & Aragona, 2021). The growth of the follicle is regulated by two pools of stem cells—the hair germ and bulge stem cells—which in turn are under the control of signals arising from the dermal papilla. In contrast to the incisor, the hair follicle stem cells undergo successive periods of quiescence and activation to regulate the formation of the hair shaft and its subsequent shedding at the end of telogen. The ability of the hair follicle stem cells to differentiate into all the cell types in the skin makes them the major replenishing source in the skin after injuries.

The mammary gland grows in bursts during distinct phases throughout the life of females. Humans possess one pair of mammary glands known as the breasts, while mice have five pairs lying in between the forelimb and hindlimb buds. The gland's development can be broken down into three stages: embryonic, pubertal, and adult (including pregnancy and involution). Embryonic development produces a rudimentary ductal structure at predetermined locations. Further branching morphogenesis occurs throughout puberty (Slepicka, Somasundara, & dos Santos, 2021), when the progression of branching is regulated by the female reproductive hormones and leads to the formation of the mammary gland luminal branches and alveoli (Macias & Hinck, 2012). The adult mammary gland remains very plastic as the organ undergoes important remodeling upon pregnancy to support lactation, and it then reverts to a pre-pregnancy state upon diminishing demand for milk, leading to alternating cycles of growth and involution (Denson et al., 2018; Macias & Hinck, 2012). Mammary epithelial cells have substantial regenerative potential, possessing the ability to repopulate cleared fat pads, and the identity and nature of mammary stem cells (MaSCs) have been of intense interest over the past decade (Fu, Nolan, Lindeman, & Visvader, 2020).

In all of these organs, signaling networks regulate morphogenesis as well as trigger the self-renewal and regeneration phases. Several classical recombination experiments have established that the inductive signal at the early stages of tooth development is produced in the epithelium, whereas the underlying mesenchyme produces regulatory cues that direct subsequent organ development. For example, recombination of the enzymatically separated embryonic day (E) 9–11 mandibular arch epithelium with E8 neural crest cells that need an activation signal to form the tooth, was sufficient to induce their differentiation into the dental mesenchyme (Lumsden, 1988). Moreover, recombining the same stage mandibular epithelium with 2nd branchial arch mesenchymal cells (that normally do not participate in tooth formation) also resulted in tooth formation (Mina & Kollar, 1987). These experiments helped demonstrate the inductive role of the epithelium in morphogenesis. Combining chick pharyngeal arch epithelium with E16–18 mouse mandibular molar mesenchyme led to the formation of enamel-secreting ameloblasts (Kollar & Fisher, 1980). This showed that not only is the regulation of the later stages of tooth development limited to the odontogenic mesenchyme, but also that teeth can be induced to form in tissues such as the chick epithelium which evolutionarily no longer normally do so. A similar mechanism exists in the mammary gland and the hair follicle. Combining the mammary gland epithelium with the salivary gland mesenchyme resulted in the development of an organ very similar to the salivary gland (Kratochwil, 1969). Interestingly, the salivary mesenchyme has a stronger morphogenetic influence than the mammary mesenchyme, regardless of the positional and temporal identity it is combined with. Experiments involving the removal or transplantation of the dermal papillae also showed that the potential for hair follicle development after the initial stages resides within the mesenchyme (Oliver, 1966). It is the lower third of the follicle that gives the initiation signal and leads to the formation of the dermal papilla during subsequent cycling (Sengel, 1976). Taken together, these experiments not only demonstrate that organ growth, in the case of the tooth, hair follicle and mammary gland, is regulated by the mesenchyme at later stages, but also that it can be induced in different regions if the necessary cues are provided. The epithelium or the mesenchyme from different ectodermal organs retain the capacity to influence the development of uncommitted mesenchymal or epithelial tissues, respectively, at particular stages of development. Recombination experiments involving the epithelium and mesenchyme from different species were also performed to further test this hypothesis. Presumptive chick corneal epithelium, when combined with either the scale-forming dermis or the feather-forming dermis from the inner ear region, developed into scales and feathers, respectively (Coulombre & Coulombre, 1971). Additional intra- and inter-species epidermal-mesenchymal recombination experiments further demonstrated that the initiation signal for scale, hair and feather formation arises from the epidermis, while their next developmental phase is guided by signals originating in the dermis (Coulombre & Coulombre, 1971; Dhouailly et al., 2019). Interestingly, upon recombination of tissues from two distinct species, the developmental sequence invariably halts after reaching a certain step, suggesting that certain intrinsic signals are missing to complete morphogenesis.

The ability of the mesenchyme to regulate the growth of ectodermal appendages led biologists to probe the plasticity of the epithelium to form different organs when provided with specific sets of cues. More recently, transplants of portions of E8 mouse neural tube, which still contains neural crest cells, into chick embryos led to the migration of the murine cells into the host's oral epithelium. The authors demonstrated that cells expressing *Msx1*, *Pax9* and *Barx1* could induce localized expression of *Bmp4* and *Shh*, ultimately resulting in the formation of mineralized tooth buds in the chick (Mitsiadis, Chéraud, Sharpe, & Fontaine-Pérus, 2003). This study introduced some important points for consideration: first, that neural crest cells possess the ability to interpret generic epithelial signals to promote tooth initiation, and second that, unlike previous assumptions, it might be the chick neural crest cells and not the epithelium that have lost the ability to form the tooth. A more recent study further highlights the idea of specific signaling cues possessing the ability to drive the development of specific ectodermal organs. The authors showed that the inhibition of BMP signaling in cichlids leads to the loss of *Shh* expression, resulting in shorter and misshapen replacement teeth that express the taste bud marker *Calb2* (Bloomquist et al., 2019). In addition, they showed that reciprocally, in mutants for the BMP antagonist *Follistatin*, the tongue epithelium becomes enriched for odontogenic markers such as *Bmp7*, *Shh*, *Pitx2*, *Pax9*, *Barx1*, and *Msx2*. The importance of signaling pathways in determining the identity of an ectodermal organ was further highlighted by work showing that inhibition of BMP or activation of *β-CATENIN* can lead to the formation of hair follicles instead of sweat glands (Närhi et al., 2008; Plikus et al., 2009) or instead of the mammary gland nipple (Mayer, Foley, De La Cruz, Chuong, & Widelitz, 2008). These results raise the possibility that such a capacity would require the epithelial cells of the various organs to be somewhat plastic in nature. Indeed, recent studies show that dental epithelial stem cells from postnatal day (P) 3 mouse incisors are able to form branched ductal structures when transplanted into a murine mammary fat pad, and when these dental stem cells are intermixed with mammary epithelial cells, additional signaling cues they receive enable their differentiation into all the epithelial cell types of the mammary gland (Jimenez-Rojo, Pagella, Harada, & Mitsiadis, 2019). Other reports have demonstrated that dental cells exhibit a high degree of fate plasticity. Deletion of the key transcription factor *Msx2* triggered conversion to an epidermal fate and the expression of hair-specific keratin in the incisor outer enamel epithelium (Nakatomi et al., 2018), while removing *Med1*, *Sox21* and *Fam38h* led to actual hair growth (but no follicle formation) alongside the incisor (Saito et al., 2020; Yoshizaki, Fukumoto, Bikle, & Oda, 2020). Collectively, we now understand much more about not only the shared signaling cues and cellular machinery at play during ectodermal appendage early development, but also the fine-tuning of signaling pathways that are responsible for the diversity observed during organ morphogenesis.

In this chapter, we will briefly discuss how earlier literature brought to light the parallels and divergences between these appendages during the various stages of growth. Our focus will be on how more recent findings change our understanding of these ectodermal organs. We will start with an outline of the changes in signaling regulation during development and morphogenesis of each organ at the macroscopic level and then highlight four major signaling pathways—Shh, Wnt, Notch and Yap/Taz—with a concluding discussion on their regenerative capacities.

2. From signaling centers during development to stem cell regulated morphogenesis in the adult

2.1 Tooth

Like with other ectodermal appendages, signaling through Wnt, Shh, Fgfs, Bmps and other well-established pathways regulates the initial development of the tooth. Placode formation (epithelial thickening) can be considered the first signaling center, as it produces the initiation signals that trigger tooth development. These signals are provided by the reciprocal interactions between *Fgf8* and *Bmp4* that demarcate the regions where the dental lamina will form (Ahn, 2015; Mikkola & Millar, 2006; Pispa & Thesleff, 2003). This is followed by the epithelial expression of *Wnt* genes and *Shh* in the regions where tooth initiation takes place. In the absence of inhibitory signals from the mesenchyme, *Wnt* and *Bmp* overexpression have been shown to mediate the growth of supernumerary incisors. Similarly, the overexpression of *Eda* or *Edar* leads to the development of supernumerary teeth, but both genes are dispensable for placode initiation. The dental placode is also known to express *Msx2*, *Lef1* and *p21*, and develops as the result of extensive epithelial-mesenchymal crosstalk (Ahn, 2015) (Fig. 1).

As the dental epithelium further invaginates into the mesenchyme, within the mass of proliferating cells, two more signaling centers called the Initiation Knot (IK) and the Enamel Knot (EK) develop at the bud stage and the cap stage, respectively (Ahtiainen, Uski, Thesleff, & Mikkola, 2016; Jernvall, Keranen, & Thesleff, 2002; Laurikkala et al., 2001; Sarkar & Sharpe, 1999). These signaling centers are comprised of non-proliferating cells expressing *p21* (cell cycle inhibitor), *Shh*, *Eda*, members of the *Wnt*, *Fgf*,

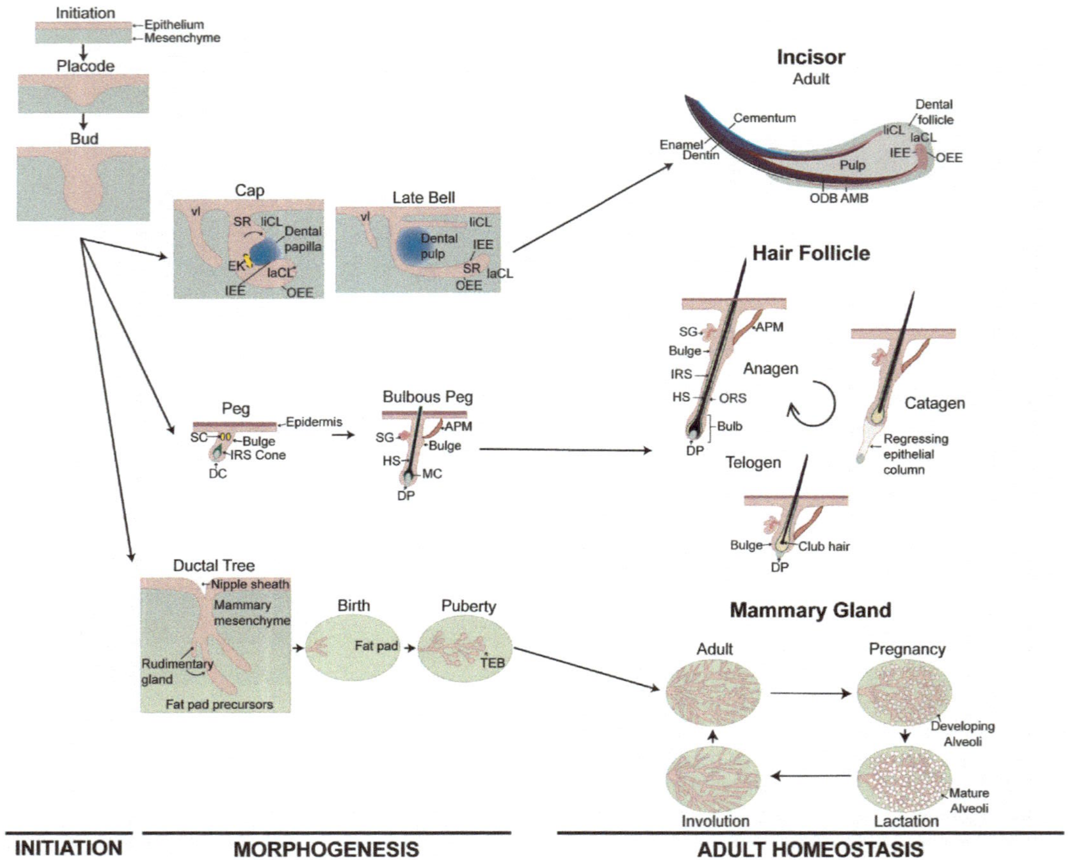

Fig. 1 See figure legend on opposite page.

Bmp families as well as several other factors. The IK and EK regulate the invagination of the tooth bud and epithelial folding, thus marking the bud-to-cap transition and subsequent progression into the bell stage. Throughout the successive stages of early tooth development, distinct pools of cells express markers typical of stem cells such as *Lef1* or *Sox2*, while they remain highly proliferative in nature. It is only with the development of the labial cervical loop, the future stem cell niche, that a pool of adult stem cells forms in the rodent incisor. The expression of *Sox2* becomes progressively restricted to the labial cervical loop with the growth of the incisor (Juuri, Isaksson, Jussila, Heikinheimo, & Thesleff, 2013).

Once the cervical loop is formed, embryonic signaling centers dissolve and pools of stem cells appear that regulate the continuous growth of the

Fig. 1 Schematic showing initiation, morphogenesis, and adult homeostasis stages of incisor, hair follicle and mammary gland development. The incisor, hair follicle, and mammary gland share the initial stages of development—all three begin from an epithelial thickening called the placode that invaginates into the underlying mesenchyme to form the bud (the last shared stage). The **incisor** bud forms two symmetrical outgrowths that grow parallel to the jawline, and this phase is regulated by two signaling centers—the initiation knot (not shown) and the enamel knot. As the incisor grows, an actively proliferating pool of stem cells forms in the labial cervical loop (laCL) which is maintained throughout the life of the animal and helps replenish the mineralized tissue as it gets abraded in a continuous process. The **hair follicle** (HF) bud grows and cycles through three stages: anagen (growth), catagen (regression) and telogen (resting) throughout the life of the animal. The mesenchymal cells below the bud condense to form the dermal papilla (DP), which activates and coordinates the functions of the two pools of stem cells—the hair germ and bulge stem cells (Bu-SCs) thereby regulating the regeneration cycle of the follicle. HFSCs undergo successive periods of quiescence and activation to regulate the formation of the hair shaft and its subsequent shedding at the end of telogen. The **mammary gland** bud descends into the underlying mesenchyme and grows into a rudimentary structure consisting of a few small ducts, with a nipple sheath and an open lumen. It remains in this form until birth and develops minimally till puberty. At puberty, with the onset of hormonal signals, the gland undergoes ductal morphogenesis and fills the fat pad with the epithelial mammary tree, which persists in the mature virgin. The regenerative phase of the mammary gland occurs during pregnancy, when extensive alveologenesis occurs for milk production/lactation followed by involution back to its original adult state once the demand for milk is gone. All these stages are regulated by lineage restricted stem cells distributed throughout the ducts and TEBs. Abbreviations: vl = vestibular lamina, SR = stellate reticulum, liCL = lingual cervical loop, laCL = labial cervical loop, EK = enamel knot, IEE = inner enamel epithelium, OEE = outer enamel epithelium, ODB = odontoblasts, AMB = ameloblasts, SC = sebocytes, IRS = inner root sheath, ORS = outer root sheath, DC = dermal condensate, SG = sebaceous gland, HS = hair shaft, AP = arrector pili muscle, MC = melanocytes, DP = dermal papilla, TEB = terminal end bud.

tooth and replacement of the worn-out enamel and dentin (Harada & Ohshima, 2004; Yu & Klein, 2020). Stem cells in the rodent incisor were initially thought to reside in a quiescent, label-retaining region in the proximal labial cervical loop (Harada et al., 2002). In vitro, these cells were thought to possess the ability to differentiate into fast cycling Transit Amplifying (TA) cells that further differentiate into all mature epithelial cell types including ameloblasts in the adult incisor. A conveyor belt model was proposed, in which the quiescent stem cells occasionally get activated, move away from their niche and differentiate into TA cells, which in turn differentiate into the pre-ameloblasts that further mature as they move distally (Juuri et al., 2012). Recently, this theory has been challenged and a new model has been proposed, in which actively cycling epithelial progenitors (previously called TA cells) but not quiescent cells replenish the epithelium (Sharir et al., 2019). This study further demonstrated that these progenitors not only generate the enamel-producing ameloblasts but also the adjacent non-ameloblast cells. These findings were further corroborated by another group, which also demonstrated that the incisor mesenchymal compartment is maintained through similar dynamics (Krivanek et al., 2020). Markers for the epithelial progenitor cells have been identified through various assays and include *Sox2*, *Gli1*, *Lgr5*, *Bmi1* and *Lrig1* (Biehs et al., 2013; Juuri et al., 2013; Sanz-Navarro et al., 2018; Seidel et al., 2017). The adjacent pool of stem cells in the mesenchyme are labeled by *Gli1* (Zhao et al., 2014). The expression of *Wnt*, *Tgf-β*, *Bmp*, *Fgf*, *Shh* factors have been implicated in the maintenance of continuous growth in the adult incisor.

2.2 Hair follicle

Unlike the incisors, of which the mouse has only 4, the hair follicles in the mouse are much more numerous. Hair develops in four successive, defined waves during embryogenesis, and interestingly the embryonic skin generates new follicles as it grows to compensate for the decrease in local follicle density (Cheng et al., 2014; Gong, Xu, & Li, 2018). Like the incisor, hair follicle development also starts with the formation of the placode as the initial signaling center. *Wnt* is important for placode initiation as well as subsequent invagination into the mesenchyme (Ahn, 2015; Mikkola & Millar, 2006; Pispa & Thesleff, 2003). The local induction of WNT in the epidermis and subsequently in the dermis causes the epidermal cells expressing high levels of WNT to cluster and form the hair placodes (Rognoni & Walko, 2019). An interesting discovery is that, at the initiation of placodogenesis, cells enter and remain in the G1 stage of the cell cycle (Ahtiainen et al., 2016). Placode invagination and entry into the germ stage is facilitated by *Shh* and *Edar* expression (Ahn, 2015; Mikkola & Millar, 2006; Pispa & Thesleff, 2003). The expression of *Shh* at the leading edge of the hair bud (although not formally described as a signaling center) is necessary to drive epithelial cell proliferation (Mill et al., 2003). Also, similar to the tooth, the invaginating cells of the placode express *Lef1* (Gong et al., 2018) and other adult stem cell specific markers such as *Sox9*, *Lrig1* and *Lgr6* (Rognoni & Watt, 2018). On the other hand, unlike the tooth bud, Bmp signaling suppresses the formation of hair placodes, and the inhibition of *Sostdc1* (a Bmp antagonist) leads to the formation of supernumerary hair follicles. This reciprocal interaction between the *Eda* and *Bmp* signaling pathways is also thought to be responsible for regulating hair follicle patterning (Ahn, 2015; Mikkola & Millar, 2006; Pispa & Thesleff, 2003). As the follicle matures further, the stem cell markers start segregating into the distinct populations seen in the adult including the hair germ, bulge, junctional zone and isthmus (Rognoni & Watt, 2018) (Fig. 1).

After morphogenesis is completed, the hair follicle undergoes periodic regeneration throughout the life of the animal. Although continuous like the tooth, the regeneration of the hair follicle occurs in three long and distinct stages and involves two stem cell compartments as well as a signaling center. A group of specialized mesenchymal cells, called the dermal papilla, functions as the signaling center and regulates the entire regeneration cycle of the follicle postnatally (Hsu, Li, & Fuchs, 2014; Millar, 2002). At the beginning of each regeneration cycle, or anagen, activating *Wnt* and inhibitory *Bmp* signals from the dermal papilla first lead to activation of the "primed" stem cell population (i.e., the hair germ) (Greco et al., 2009; Millar, 2002; Suen, Li, & Yang, 2020). Once activated, these cells proliferate and generate the transit amplifying cells, also called the matrix of the follicle. The transit amplifying cells are the active population of the follicle that differentiate and form the cells of the inner root sheath and the hair shaft. They have two other tasks: first, generation of the activation signal (i.e., the secretion of SHH) for the other stem cell population known as the bulge stem cells, and at the same time regulating the surrounding niche, sustaining the dermal papilla signaling and thus maintaining their own population in a feedback loop (Hsu et al., 2014). The quiescent bulge stem cells are activated by the matrix after a couple of days of anagen entry and are responsible for differentiating into the cells of the outer root sheath. The outer root sheath encases the hair shaft and is responsible for the downward growth and increasing length of the follicle. This downward growth provides the break in the cycle; as the distance between the matrix (transit amplifying cells) and the bulge increases, this results in the loss of the bulge activation signals from the matrix, and the stem cells become quiescent again. The hair then enters the catagen phase where the matrix and most of the outer root sheath cells undergo apoptosis. The remaining outer root sheath cells form a new bulge and hair germ adjacent to the old one, which serves as an anchor for the hair generated in the previous cycle (Hsu et al., 2014; Hsu, Pasolli, & Fuchs, 2011). Both stem cell populations in the follicle, although activated at different times and performing different functions, share many molecular features such as the presence of *Cd34* or *Krt15* expressing cells, and both also remain quiescent during the resting phase or telogen (Gay et al., 2020; Prochazkova et al., 2015). Thus, the crosstalk between the dermal papilla (signaling center), the hair germ and bulge stem cells and the transit amplifying cells regulates progress through the different stages of the hair follicle regeneration cycle.

2.3 Mammary gland

In rodents, the development of the mammary gland initiates with the designation of the mammary line. *Wnt*, *Shh* and *Fgf10* expression determine the location of the mammary line, but *Shh* then needs to be suppressed before placode formation (Ahn, 2015; Mikkola & Millar, 2006; Pispa & Thesleff, 2003). As in the tooth and hair follicle, *Wnt/β-catenin* signaling is important for placode formation and *Shh-Gli3* expression is necessary for later epithelial and mesenchymal morphogenesis and proliferation. The overexpression of *Eda* results in the development of supernumerary mammary placodes as well. However, mammary gland development has more similarities with the hair placode. Indeed, *Bmp* inhibition results in placode initiation, and *Sostdc1* (Bmp antagonist) expression plays an important role in fine-tuning correct patterning. Also, the mammary placode cells retain their lower mitotic index (when compared to the adjacent epidermal cells) and do not show BrdU incorporation until the late bud stage (Balinsky, 1950; Lee et al., 2011). Formation of the mammary placodes also requires the expression of *Fgf* ligands and receptors, and of the transcription factors *Lef1* and *Tbx3* (Fig. 1).

Another interesting difference in mammary gland development is that, although gland growth occurs throughout life, it is limited to specific time periods: embryonic development, puberty, and pregnancy. Also, a major regulatory factor in mammary gland regeneration is production of the female reproductive hormones—estrogen and progesterone. Although we will not be focusing on these in this chapter, it is important to note that the glands secreting these hormones, mainly the ovaries, act as the distant signaling centers in all phases of mammary gland growth (Brisken & O'Malley, 2010). In brief, the maternal hormones provide the first stimulus to the developing mammary gland for the initiation of duct development (Slepicka et al., 2021). These signals cease after birth and the gland stays in the rudimentary state until puberty. At the onset of puberty, the female reproductive hormones produced every ovarian cycle trigger increased tertiary branching (and alveogenesis during puberty), thus enabling mammary gland expansion along with the formation of the ductal network. At the end of the lactation cycle, the mammary epithelial cells are cleared through apoptosis, lobule cell shedding and tissue resident macrophages (Dawson et al., 2020). During both puberty and pregnancy, hormone production in part regulates the proliferation and differentiation of the mammary epithelial cells, resulting in the branched morphology (globular in human), while pro-apoptotic factors help clear out the newly formed ducts to allow for lumen formation.

The origin and nature of MaSCs has been a subject of great interest. Similar to the tooth and hair follicle, the embryonic mammary gland is rich in bipotent epithelial progenitor cells that are proliferative in nature. Approximately 10% of these primordial cells possess the signature of quiescent adult stem cells (Boras-granic, Dann, & Wysolmerski, 2014; Fu et al., 2020). Lineage tracing experiments using long-term label retention techniques have shown that these embryonic cells switch to quiescence postnatally and thus seed the adult stem cell pool. Postnatally and in adults, during puberty, MaSCs become progressively more unipotent (lineage restricted) and are present in the ducts, where they drive the normal homeostasis of the adult gland. Recent studies using a genetic tracing technique that labels a small population of proliferating cells, followed by a long-term chase of the randomly labeled progeny, showed that the labeled cells are randomly distributed at different densities in multiple ducts (Davis et al., 2016). These data are consistent with the observations that, during puberty, proliferating lineage restricted stem cells are present in the terminal end buds and are lost when these buds regress at the termination of puberty. Similar observations were made with a single cell RNA sequencing study that mapped the mammary cell dynamics across different developmental stages (Bach et al., 2017). Although these authors did not find evidence of cells expressing both the basal and luminal lineage markers, they proposed the existence of a metastable state that could represent the transit amplifying cells with the potential to differentiate into either hormone sensing or secretory cells. Using a marker-free algorithm to define a stem cell expression module in the adult human mammary gland, bipotent cells expressing *Ybx1* and *Eno1* (modulators of basal breast cancer) were largely represented in the basal cell population (Chen et al., 2019). These cells map to the periphery of the basal and immature luminal clusters and the lower potency cells lie in the luminal cluster. Interestingly, there are suggestions that under the influence of pregnancy hormones, bipotent MaSCs can generate luminal progenitor cells required for faster alveoli development during pregnancy. Together, the data suggests that MaSCs are proliferating and unipotent in a physiological context and bipotent MaSCs may exist but are induced upon transplantation (Wang et al., 2015). Furthermore, the MaSCs may show a lineage bias depending on the physiological requirements and hormonal status of the gland (Fu et al., 2020). Another difference of the adult mammary gland from the tooth and hair follicle lies in its branched morphology. Epithelial-mesenchymal transition regulators such as *Ovol2* drive branching morphogenesis, while the expression of *Tgf-β*, along with mechanical cues, is responsible for the periodic pauses during adult morphogenesis of the gland.

In conclusion, the tooth, hair follicle and mammary gland share many molecular signaling pathways during placode development. Inhibition of the Wnt/β-catenin pathway prevents the formation of all three organs in mice (Dhouailly, 2009). Slight but significant differences arise that change the way these organs develop further. The contribution of cell proliferation toward organ development begins at progressively later stages in each organ, from the placode in the tooth, the hair germ in the hair follicle, and spout formation in the mammary gland. Upon the onset of adult morphogenesis, there is a loss of the signaling center in the tooth, whereas this is retained in the hair follicle (in close proximity to the organ) and the mammary gland (at a considerable distance from the organ). Moreover, the proliferating multipotent stem cell pool in the incisor is replaced by two multipotent pools in the hair that alternate between states of activity and quiescence. In the mammary gland, the stem cells seem to become lineage restricted and distributed in the multiple ducts across the epithelium.

3. Shh

The Hedgehog (Hh) signaling pathway plays a central role in ectodermal appendage morphogenesis and regeneration in vertebrates. Sonic hedgehog (SHH), Indian hedgehog (IHH), and Desert hedgehog (DHH) play many regulatory roles in both embryonic development and regeneration (Zheng et al., 2019). In this section, we will summarize the recent studies on the role of Hh signaling in the tooth, hair follicle and mammary gland (Fig. 2).

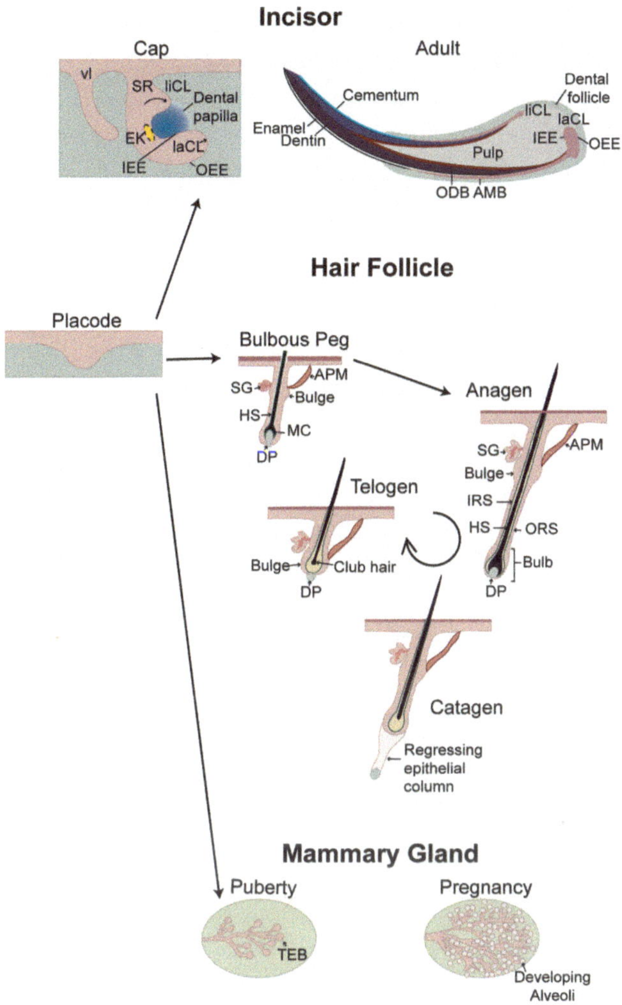

	Incisor	Hair Follicle	Mammary Gland
Placode	**Epi:** Shh, Ptc1, Smo, Gli1, Wnt3/4/6/7b/10b, Notch1/2/3, YAP (c)	**Epi:** Shh, Wnt10a/b, Lef1, β-cat, Tcf3/4	**Epi:** Wnt1/2/3/5a/6/7b/10a/b/11, Notch1
	Mes: Ptc1, Smo, Gli1, Wnt5a, Srfp2/3, Jagged 1/2	**Mes:** Ptc1, Gli1, Wnt5a, Dkk1	**Mes:** Gli3
Morphogenesis	**EK:** Shh, Wnt3/6/10a/b, YAP (c) **Epi:** Ptc1, Smo, Gli1, Notch1/2/3, liCL/laCL: YAP (n)	**ORS:** Tcf3/4, YAP (n), Terminally Differentiated: Wnt3/11/4, Dvl29 **IRS:** Wnt10a/b, YAP (c)	**TEB:** Ptch1, Wnt 2/4/5a/b/6/7b **Ductal:** Wnt4/5b/6, Notch3
	Mes: Ptc1, Smo, Gli1/2/3, Wnt5a/b, Jagged 1/2	**Mes:** Ptc1, Smo, Gli1/2, Wnt5a	**Stromal:** Gli2, Wnt2, Jagged
Adult	**Epi:** Shh, Gli1, Ptc2, Wnt4/6/11, Notch1/2/3, Jagged2, Delta 1, YAP (c) **AMB:** Wnt3a/4/6/7a/10a, Jagged2, Delta1 **ODB:** Wnt5a/6/10a	**Matrix:** Shh, Lef1, Jagged1, Notch1/3, RBPjκ **Bulge:** (HFSC): Gli1 Refractory Telogen: Tcf3/4, YAP (c) Anagen: Wnt10a/b, β-cat, Lef1, YAP (n)	**Epi:** Ihh, Wnt 2/4/5a/b/6/7b/10b **Alveolar Secretory Cells:** Gli2, YAP
	Mes: Ptc1, Wnt4/5a/b/7b/10a/11, Dlk1	**Mes:** Wnt10a/b	
	Mandibular Nerve: Dhh		

Fig. 2 See figure legend on next page.

3.1 Tooth

In early tooth development, *Shh* expression becomes restricted to the odontogenic epithelium, in the region where individual incisors will form, in both the maxilla and mandible (Cobourne, Hardcastle, & Sharpe, 2001; Keränen, Kettunen, Åberg, Thesleff, & Jernvall, 1999). At later stages, *Shh* is exclusively expressed in the signaling center called the enamel knot, and the receptors and downstream effectors PTCH, SMO, and GLI are expressed throughout the epithelium and mesenchyme. Pharmacological inhibition of Shh signaling resulted in the arrest of tooth development at the initiation stage, while epithelial *Shh* deletion (driven by the *Krt14* promoter) resulted in misshapen teeth (Cobourne et al., 2001; Dassule, Lewis, Bei, Maas, & McMahon, 2000). Additionally, in $Gli2^{-/-};Gli3^{+/-}$ double knockout mice, epithelial thickenings were observed in the oral epithelium, but these did not proceed to develop into the enamel organ (Hardcastle, Mo, Hui, & Sharpe, 1998). Also, implantation of SHH-soaked beads into non-odontogenic oral epithelium induced ectopic epithelial invaginations, indicating the importance of SHH in initiating odontogenesis (Hosoya, Shalehin, Takebe, Shimo, & Irie, 2020). In more detailed recent studies

Fig. 2 Signaling pathways in ectodermal appendage initiation and morphogenesis. The murine incisor, mammary gland, and hair follicle development is regulated by several signaling pathways. This schematic summarizes the expression patterns of the components of the Shh, Wnt, Notch, and YAP/TAZ pathways at the three major stages of development (placode, morphogenesis, and adult) for all three ectodermal appendages. In the **incisor**, Shh and Wnt regulate the formation of the placode while Yap/Taz signaling plays an essential role in the formation of the enamel knot. In the adult organ, Shh, Wnt and Yap/Taz all regulate the proliferation and maintenance of the stem cells, thereby controlling organ regeneration, while Notch plays an essential role in stem cell differentiation and amelogenesis. **Hair follicle** placode development is largely driven by Wnt. Shh signaling from the DP activates both the hair germ and Bu-SCs sequentially and drives hair follicle cycling in combination with Wnt. Notch acts largely during anagen where it regulates stem cell maturation and shaft development. Yap/Taz mainly helps maintain stem cell proliferation and identity. **Mammary gland** initiation is regulated by the inhibition of Shh and activation of Wnt. Both Shh and Wnt actively regulate ductal morphogenesis and stem cell maintenance in the adult gland during puberty as well as pregnancy. Notch plays a role in the differentiation of the MaSCs and specification of a luminal cell fate in the adult gland. TAZ regulates branching morphogenesis while YAP signaling is more important during pregnancy and is involved in the terminal differentiation of the secretory alveolar cells. YAP is also essential for mammary epithelial cell proliferation and survival. Abbreviations: vl = vestibular lamina, SR = stellate reticulum, liCL = lingual cervical loop, laCL = labial cervical loop, EK = enamel knot, IEE = inner enamel epithelium, OEE = outer enamel epithelium, ODB = odontoblasts, AMB = ameloblasts, TEB = terminal end bud, SC = sebocytes, IRS = inner root sheath, ORS = outer root sheath, DC = dermal condensate, SG = sebaceous gland, HS = hair shaft, AP = arrector pili muscle, MC = melanocytes, DP = dermal papilla.

of murine molar development, it was found that SHH is not required for cell proliferation nor for tooth bud growth during early development, but instead plays an important role in determining the shape of the tooth placode (Li, Chatzeli, Panousopoulou, Tucker, & Green, 2016). Molar explants treated with the SHH signaling inhibitor cyclopamine at various developmental stages resulted in shallower and wider tooth buds similar to those of the epithelial *Shh* knockout mutants (Dassule et al., 2000). Furthermore, in these explant studies the authors showed that Shh signaling complements the role of FGF and is necessary for invagination of the tooth bud. Suprabasal cell morphology at the transition from placode to bud stage suggests SHH-dependent convergent migration and intercalation, which likely drives the post-stratification placode invagination leading to the appropriate bud morphology, as evidenced by the loss of this phenotype when Shh signaling is perturbed (Li, Chatzeli, et al., 2016).

During the later stages of tooth crown formation, the cells in the inner enamel epithelium (IEE), which differentiate into enamel-secreting ameloblasts, express both *Shh* and its receptor *Ptch1* (Hosoya et al., 2020). Inhibition of Shh signaling at this stage results in reduced proliferation of the epithelial cells and suppresses differentiation of these cells into ameloblasts. In the continuously growing incisor, Shh signaling also simultaneously regulates *Sox2*-expressing stem cells (Binder et al., 2019). The loss of Shh signaling in ameloblast lineage cells resulted in unpolarized ameloblast differentiation, enamel hypoplasia, and abnormal tooth morphology (Hosoya et al., 2020). Together, these data suggest that in the adult incisor, Shh plays an important role in amelogenesis. A recent study found that PTCH1 and PTCH2 have functionally distinct roles in the murine incisor as well as receptor-dependent specificity in response to the Hh ligand (Binder et al., 2019). These results indicate that the *Sox2*$^{+}$ stem cell population is maintained by Shh signaling transduced through PTCH1, whereas *P-cadherin* expression is negatively regulated through *Dhh-Ptch2* signaling. Additionally, inhibiting Shh signaling with cyclopamine resulted in a decrease in the number of *Sox2*$^{+}$ stem cells as a direct consequence of increased cell death in this lineage. Similarly, inhibiting Shh signaling with anti-SHH antibodies (5E1) induced an increase in the number of apoptotic cells and a decrease in *Sox2*$^{+}$ cells in the apical bud (Ishikawa, Ida-Yonemochi, Saito, Nakatomi, & Ohshima, 2021). Interestingly, apoptotic cells appeared mainly in the SR, while *Shh* expression and cell proliferation in the IEE remained normal and the number of *Sox2*$^{+}$ cells in the OEE was not affected. These findings suggest that Shh signaling regulates transitions in the progenitor zone, resulting in fewer *Sox2*$^{+}$ cells in the IEE.

3.2 Hair follicle

Hedgehog signaling is active in both the epithelium and the mesenchyme during hair follicle development, and although it is not required to initiate hair follicle development, it is essential for the downgrowth of epithelial cells as the follicle grows after the induction stage (Zheng et al., 2019). Mice lacking Shh signaling were able to produce hair placodes and dermal condensates, but development was arrested at the hair bud stage. At E15.5, PTC1, SMO, GLI1 and GLI2 are detected in the dermal components of the hair follicle. Loss of Shh resulted in smaller than normal dermal papilla (DP) size, while SMO inhibition by RNAi resulted in the loss of DP precursor cells altogether, indicating that Shh signaling is required for DP maintenance.

Adult hair follicles cycle through growth (anagen), regression (catagen), and resting (telogen) phases. Bulge stem cells (Bu-SCs), which participate in hair follicle regeneration, are known to remain quiescent until proliferation is triggered by SHH secreted by hair follicle transit amplifying cells (HF-TACs), also called the matrix cells. During telogen, *Gli1* expression is limited to two populations: $GLI^{+}/LGR5^{+}$ stem cells (in contact with hair germ) and $GLI^{+}/SOX9^{+}$ stem cells (upper fringe of the bulge) (Brownell, Guevara, Bai, Loomis, & Joyner, 2011). Shh signaling in the DP regulates the GLI+/LGR5+ stem cells as these differentiate into new hair follicle structures during the hair follicle growth cycle. Meanwhile, upper bulge cells that express *Gli1* are induced by SHH released from the neurons associated with the bulge (Zheng et al., 2019). The upper bulge cells are also thought to be a long-term stem cell bank, which can contribute to the repair of the epidermis between follicles when triggered by SHH. As the HF progresses from early to late anagen, HF-TACs migrate away from Bu-SCs (due to downgrowth of the follicle) and by late anagen, the distance between the HF-TACs and Bu-SCs has increased beyond the signaling range of SHH, allowing Bu-SCs to resume quiescence.

Shh signaling has also been recently shown to play a major role in hair follicle neogenesis (HFN). In mice, large wounds typically undergo regenerative healing and produce functional tissue capable of HFN, while small wounds normally form scar tissue (Lim et al., 2018). However, the relationship between scarring and hair follicle regeneration has remained elusive. During healing, *Gli1* expression, indicative of hair placode formation, was detected in large wounds but was conspicuously absent from small ones. It was found that activating the Shh pathway during wound healing enables the development of a regenerative DP, which is required and sufficient to induce HFN. Both epidermal overexpression of Shh and constitutive dermal activation of Smoothened, specifically in dermal myofibroblasts, resulted in extensive HFN in wounds that otherwise would form scar tissue. The resulting hair follicles exhibited normal morphogenesis. A recent study showed de novo induction of HFs in unwounded skin by experimentally elevating Hh signaling (Sun et al., 2020). Utilizing a *Gli1*^{CreER} driver to conditionally knockout the repressive *Ptch1* receptor in epithelial cells and the adjacent stroma, the authors were able to induce HFN in adult skin, as well as in hairless paw skin.

3.3 Mammary gland

Components of the Shh signaling pathway are expressed in both the epithelium and associated mesenchyme of the mammary bud, but Shh signaling does not seem to be necessary for bud growth (Hatsell & Frost, 2007; Hui et al., 2013; Zheng et al., 2019). The inhibition of *Shh* or *Ihh* individually does not affect mammary bud development, indicating that either they are not required for mammary gland growth or their function is redundant (Michno, Boras-Granic, Mill, Hui, & Hamel, 2003). In addition, promoter activity for both *Ptch* and *Gli* has been shown to be absent in the developing mammary bud (Hatsell & Cowin, 2006). Studies using epithelial overexpression mutants of *Shh* or lacking a functional GLI3 (Shh repressor) fail to develop certain mammary buds, suggesting that the active inhibition of *Shh* via GLI3 drives mammary bud initiation (Gritli-Linde et al., 2007; Hatsell & Cowin, 2006).

At puberty, several studies demonstrated a role for active Shh signaling in mammary gland morphogenesis. Epithelial *Shh* regulates mammary gland ductal morphogenesis via the activation of a non-canonical signaling cascade involving c-SRC and ERα (Estrogen receptor α) (Okolowsky, Furth, & Hamel, 2014). An epithelium-specific knockout of *Ptch1* leads to hyperproliferation and increased branching of the mammary gland, and this role is independent of the canonical pathway mediated via *Smo* (Monkkonen, Landua, Visbal, & Lewis, 2017). On the other hand, homozygous stromal *Ptch1* ablation leads to decreased proliferation and branching, with stunted ducts enriched in luminal cells with altered hormonal receptor expression. The function of the stromal *Ptch1* is mediated via the canonical Smo pathway. Together, these data suggest that non-canonical Shh signaling in the epithelium and the canonical pathway in the mesenchyme regulate ductal morphogenesis in the virgin pubescent gland. This function continues in the adult gland, where the stromal deletion of *Gli2* mediated by Fsp1Cre results in delayed ductal development, decreased number of stromal cells supporting the mammary gland, abnormal ductal extension, and ultimately affects ductal regeneration (Zhao et al., 2017). In fact, single-cell RNA-sequencing analysis highlighted that Igf, Wnt, Fgf and Hgf signaling, all active in epithelial stem cell renewal, ductal outgrowth, and morphogenesis, along with several hormone receptor signaling pathways, act in a GLI2-dependent manner. Together, these findings suggest that Shh signaling, through *Gli2*, specifies a stromal niche that regulates MaSC activity. During pregnancy, the overexpression of epithelial *Gli1* disrupts the alveolar network, resulting in an inability to lactate and ultimately the tissue becoming cancerous (Hui et al., 2013). Hh activity also induces mammosphere formation in vitro and an expansion of the progenitor population via differential *Tp63* promoter selection in vivo. This property is utilized for mammary gland expansion during pregnancy.

In conclusion, Shh regulation is mediated through different mechanisms at the initial developmental stages of all three organs. Shh is important for tooth bud initiation as well as the later invagination and shape determination of the growing bud. In the hair follicle, it is not important for initiation but comes into play during subsequent stages and regulates follicle downgrowth, as well as dermal papilla maintenance. However, elevated Shh expression is sufficient to induce hair follicle neogenesis in large wound beds as well as unwounded skin in adults. In contrast, in the mammary gland, it is the inhibition of Shh signaling that induces bud growth. Indeed, inhibition of *Shh* in ventral skin epithelium via the ablation of *Smo* leads to a loss of hair and instead causes an induction of glandular structures from the defunct hair follicles (Gritli-Linde et al., 2007). In all three adult organs, a combination of epithelial and mesenchymal Shh signaling participates in the maintenance and proliferation of the stem cell populations as well as overall growth of the organ specific structures. In the incisor, *Shh* is expressed in the epithelium and regulates progenitor cell proliferation and differentiation as well as maintenance of the quiescent cells. On the other hand, in the hair follicle, *Shh* is expressed in the dermal papilla but performs the same function of regulating the proliferation and differentiation of the HFSCs in the hair germ and bulge. Finally, Shh signaling, non-canonical in the epithelium and canonical in the mesenchyme, has been implicated in adult virgin mammary gland ductal morphogenesis, and stromal *Shh* has been shown to regulate MaSC activity.

4. Wnt

Wnt genes regulate many aspects of embryonic development, adult tissue homeostasis, and tissue repair and regeneration (Amerongen, Bowman, & Nusse, 2012; Steinhart & Angers, 2018; Whyte, Smith, & Helms, 2012). Most Wnt genes signal through the canonical pathway, where they interact with Fzd and Lrp5/6 and regulate gene transcription via β-catenin. However, certain Wnts such as Wnt5a and Wnt11, function via non-canonical β-catenin independent mechanisms. These signal through a variety of receptors, including Fzd (involving the planar cell polarity pathway), Ror and Ryk receptor tyrosine kinases, and signaling can be mediated through small GTPases and Calcium fluxes (Anastas & Moon, 2013; Komiya & Habas, 2008). Wnt signaling has also been implicated in the maintenance of the stem cell niche as well as self-renewal in multiple organs in mammalian models (Nusse, 2008; Nusse & Clevers, 2017). In this section, we highlight the recent additions to the literature focusing on the tooth, the hair follicle, and the mammary gland (Fig. 2).

4.1 Tooth

The expression of Wnt molecules during early tooth development is critical for the regulation of cell proliferation, cell migration, and the epithelial-mesenchymal interactions regulating incisor morphogenesis (Lee, Kim, Otsu, Harada, & Jung, 2016; Lu, Yang, Zhao, & Liu, 2019). Inhibition of epithelial Wnt signaling following either the ablation of *β-catenin* or the overexpression of the Wnt antagonist *Dkk1* arrests tooth bud development at the bud stage, while the bud to cap transition is mediated via mesenchymal Wnt signaling (Ahn, 2015). In fact, stabilization of mesenchymal β-catenin signaling resulted in the formation of tooth-placode like structures (Chen, Lan, & Baek, 2009), and early expression of a dominant-active β-catenin protein perturbs tooth formation, producing a large domed evagination at early stages and supernumerary teeth later on (Kim et al., 2021). A recent study suggested that Wnt-mediated regulation of cell migration and subsequent tooth morphogenesis could also be regulated via *Sox2* expression in the murine molar (Lee et al., 2016). It was also shown that epithelial *Gpr177* (Wnt receptor) regulates Wnt ligand secretion and canonical Wnt signaling in the enamel knot while interfering with cell proliferation (Zhu et al., 2013).

In the adult incisor, Wnt performs several distinct functions. It is an important regulator of enamel maturation during the later stages of amelogenesis (Fan et al., 2018). The constitutive activation of β-catenin in ameloblasts disrupts the expression of several proteinases, thus inhibiting further mineralization. After the secretory stage, Wnt-mediated secretion of proteinases, such as MMP20 and KLK4, enables enamel matrix digestion, which is a necessary step for enamel crystallite growth. Wnt signaling is active in the proliferating IEE, where its expression is controlled by the epigenetic regulator PRC1 (An et al., 2018). Like in other adult stem cell populations, the authors show that the PRC1 component Ring1b regulates Wnt signaling via *Zic1/2* transcription factors in the TACs. *Ring1b* expression is also necessary for cell proliferation, and its depletion results in both the loss of proliferating cells and, surprisingly, in apoptosis in the cervical loop. Thus, Wnt regulates the maintenance of the proliferating incisor epithelial cell pool. A similar feed forward loop regulates the mesenchymal stem cells in the incisor. Using *Gli1* and *Axin2* to label the mesenchymal stem cells (MSCs) and transit amplifying cells (TACs), respectively, canonical IGF-Wnt signaling in the MSCs and non-canonical Wnt5a/Ror2 signaling in the TACs have been proven to maintain a dynamic reciprocal interaction between the two cell populations (Jing et al., 2021). Therefore, unlike the epithelium, Wnt signaling plays a direct role in MSC maintenance in the tooth.

4.2 Hair follicle

Wnt genes play a central role throughout hair follicle embryonic development and morphogenesis. Canonical Wnt/β-catenin signaling in both the epithelium as well as the dermis induce placode formation and the initiation of the hair follicle synthesis (Lim, Tan, Yu, Lim, & Nusse, 2016; Tripurani et al., 2018). Genetic ablation of β-catenin in the embryonic epidermis results in the loss of the ability to form the hair placode, inhibits dermal condensation, and ultimately impedes hair follicle morphogenesis. At the same time, stabilization of activated epidermal β-catenin results in the de novo formation of premature hair follicles, as well as in defects in follicle down growth into the dermis. Apart from being indispensable during placode initiation, broad dermal Wnt signaling also mediates the formation of the dermal papilla. The specific ablation of canonical Wnt signaling in the dermal papilla after hair follicle induction, i.e., after the formation of a *Sox2*$^{+}$ dermal condensate, results in the subsequent loss of both guard hairs (formed during the initial morphogenesis wave) and whiskers. This suggests that Wnt signaling promotes hair follicle initiation as well as its initial morphogenetic steps.

Wnt continues to play an important role during the adult follicle growth cycle. Indeed, specific Wnts are expressed in the various hair follicle compartments and regulate different aspects of the growth cycle (Lim et al., 2016). Like in embryos, the overexpression of β-catenin in the adult epidermis results in the de novo formation of hair follicles (Lim et al., 2016;

Tripurani et al., 2018). In contrast, the epidermal ablation of β-catenin or the overexpression of *Dkk1* causes the follicular stem cells and matrix cells to differentiate. Additional studies have further demonstrated that the levels of β-catenin signaling determine the differentiation lineage of the follicular cells (Baker, Verstuyf, Jensen, & Watt, 2010). It is widely accepted that canonical Wnt/β-catenin signaling (overexpression of *Wnt10b* in the follicle or the transient activation of β-catenin) in the epithelial stem cells enables stem cell activation and proliferation, and therefore functions as the anagen stage inducer (Lim et al., 2016; Tripurani et al., 2018). However, it has been speculated that, although necessary for stem cell proliferation, β-catenin signaling does not help with stem cell maintenance (Choi et al., 2013). The authors showed that the expression of certain stem cell markers such as CD34 and KRT15 persists in the β-catenin-depleted bulge. However, the hair follicle growth stalled in the telogen phase, a feature seen in other previous and later studies (Lim et al., 2016; Tripurani et al., 2018). Moreover, additional data suggest that the expression of β-catenin persists in the bulge stem cells through all stages of the hair cycle, arguing in favor of a role in stem cell maintenance. Use of an *Axin2* driver showed that autocrine Wnt signaling promotes the maintenance of stemness in the bulge, while Wnt inhibitors promote cell differentiation in a paracrine manner in the inner bulge region. Other recent work demonstrates that, in the $Sox9^+$ bulge stem cells, ablation of EGFR results in upregulation of Wnt signaling, thereby leading to unrestrained proliferation and finally exhaustion of these cells. The authors show that EGFR-mediated repression of the transcription of several Wnt signaling genes in the bulge stem cells regulates their proliferation and enables maintenance of the quiescent state. Therefore, it is possible that EGFR mediated inhibition of Wnt signaling plays a key role in the transition from anagen to catagen stage. Similarly, a BMP-mediated inhibition of Wnt activity in the bulge helps in the persistence of quiescence in the stem cells during telogen, and a feedback loop between BMP and Wnt pathways regulates the telogen-anagen transition (Wu et al., 2019). Together, these data indicate that a balance of epithelial Wnt signaling activation and inhibition regulates hair follicle maturation and cycling through the different stages.

Wnt signaling plays an equally important role in the hair follicle mesenchyme. The expression of Wnt agonists such as *Rspo3* in the dermal papilla maintains anagen, and the expression of Wnt antagonists such as *Dkk2* and *Notum* help initiate catagen (Harshuk-Shabso, Dressler, Niehrs, Aamar, & Enshell-Seijffers, 2020). Similarly, the activation of β-catenin in the dermal sheath cells led to the reprogramming of these cells into aberrant fibroblasts, disruption of existing hair follicles' homeostasis, and the induction of ectopic hair follicle growth (Tao et al., 2019).

4.3 Mammary gland

In the mammary gland, Wnt signaling is involved in both the initiation as well as the growth of the placode (Ahn, 2015; Yu, Verheyen, & Zeng, 2016). Genetic deletion of epithelial Wnt genes in mice resulted in defects in placode development, while their overexpression/activation prompted the formation of ectopic mammary placodes. Once placode initiation has happened, Wnt-Lef1 signaling in the mesenchyme is necessary for the maintenance of mammary epithelial fate and placode invagination. Wnt activation in the pre-pubertal gland in the early postnatal stages suggests that it regulates early gland morphogenesis in a hormone independent manner (Rajaram et al., 2015). *Wnt4*-ablated embryonic mammary tissue loses the ability to regenerate upon serial transplantation, suggesting that Wnt4 activity regulates mammary stem cell maintenance. β-Catenin activity in the mesenchyme has also been shown to regulate stem cell activation and branching in the embryonic gland (Rangel et al., 2016).

Both canonical and non-canonical Wnt signaling continue to function in the development of the pubertal mammary gland (Fu et al., 2020). Wnt/β-catenin signaling promotes ductal morphogenesis in basal epithelial cells in a paracrine manner and alveologenesis in the hormone receptor negative luminal cells (Rangel et al., 2016). A reciprocal interaction between canonical Wnt signaling driven by *Wnt4a* expression and non-canonical Wnt signaling driven by *Wnt5a* expression regulates optimal branching and ductal expansion (Kessenbrock et al., 2017; Yu et al., 2016). The stimulatory function of *Wnt4* continues in the pregnant gland, where it promotes side branching from the luminal progenitors and thus lobulo-alveolar development (Brisken et al., 2000; Rangel et al., 2016). Wnt signaling has been shown to enhance mammosphere formation and MaSC-mediated regeneration in vitro and is suspected to help in MaSC maintenance and proliferation in the adult gland (Amerongen et al., 2012; Zeng & Nusse, 2010). In response to hormonal signaling, *Wnt4* and RSPO1 function as mediators of MaSC expansion (Joshi et al., 2015; Rajaram et al., 2015). The basal cells in the terminal end bud region, which represent stem cell populations, have been shown to express *Axin2* and to be responsive to Wnt signals in vitro (Amerongen et al., 2012; Yu et al., 2016). *Wnt-Rspo* expression in the luminal cells regulates the regenerative capacity of the basal stem cells (Alexander, 2018). Nevertheless, further analyses need to be conducted to successfully establish the role of Wnt in the regulation of the MaSC biology.

In conclusion, it is evident that epithelial and mesenchymal Wnt/β-catenin signaling play a significant role in placode development and invagination in all three organs. In the tooth bud, epithelial Wnt plays an important role in bud growth while mesenchymal Wnt mediates further development. A combination of epithelial and mesenchymal Wnt signaling induces hair placode and dermal papilla formation. In fact, constitutively active epidermal β-catenin results in the de novo formation of premature hair follicles. Similarly, in the mammary gland, both epithelial and mesenchymal Wnt signaling are involved in the initiation as well as growth of the placode. Wnt signaling also regulates MaSC maintenance and gland morphogenesis in the early postnatal stages. In all three adult glands, an interplay of canonical and non-canonical signaling plays a paramount role in morphogenesis. In addition, different Wnt genes function through similar targets to regulate homeostasis. In the incisor, Wnt is important for maintenance of both epithelial and mesenchymal progenitor pools. It also regulates epithelial stem cell differentiation into ameloblasts. In the hair follicle, Wnt signaling regulates the maintenance and expansion of the progenitor cells but does not seem to play a role in cell differentiation. Like in the tooth, in the mammary gland, a combination of canonical and non-canonical Wnt signaling drives MaSC maintenance and proliferation as well as ductal morphogenesis. Together, the data suggest that, along with other factors, the type of active Wnt may play an important role in generating the morphological and functional differences in the various ectodermal organs.

5. Notch

The Notch signaling pathway has been implicated in development as well as disease, with contexts ranging from stem cell maintenance, cytodifferentiation, tissue growth to development of cancers (Bray, 2016). Binding of the ligands (Delta, Jagged) to the transmembrane Notch receptors induces a series of events that results in the transcriptional regulation of multiple genes. We will be highlighting the versatile role of this seemingly cis-restricted signaling pathway in regulating tooth, hair follicle and mammary gland development (Fig. 2).

5.1 Tooth

Notch pathway genes are expressed in the incisor tooth germ from the bud stage, where they may play a role in regulating the epithelial-mesenchymal interaction (Cai, Gong, Huang, & Lin, 2011; Mitsiadis, Regaudiat, & Gridley, 2005). In the E18.5 embryo, Notch signaling genes (Notch receptors 1, 2 and 3 and the Jagged and Delta ligands) are expressed widely in the enamel organ, and as the incisor gets close to eruption, their expression becomes restricted to the stratum intermedium (SI) region, while the ligands *Delta* and *Jagged 1* and *2* are expressed in the SI derived pre-ameloblasts and ameloblasts. The ablation of *Jag2* in mouse embryos leads to defects in ameloblast and odontoblast differentiation, eventually disturbing enamel and dentin matrix deposition (Mitsiadis, Graf, Luder, Gridley, & Bluteau, 2010). Recent work has helped to elucidate additional mechanistic details in Notch-mediated regulation of enamel matrix deposition. The cleaved Notch-MED1 complex directly activates the transcription of *Alpl* in the enamel organ, especially in the SI (Yoshizaki et al., 2017). The expression of *Alpl* in the SI region is necessary for enamel matrix mineralization during incisor development.

Earlier work established that the major role of *Notch* in the adult incisor is to regulate the differentiation of the precursor cells in the SI into ameloblasts. For example, knocking out *Jag2* leads to a decrease in the expression of *Tbx1* in the ameloblast precursors, which in turn causes defects in cytodifferentiation as well as enamel formation (Mitsiadis et al., 2010). Notch-Jagged or Notch-Perp signaling regulates the formation and/or maintenance of desmosomes between the SI cells and the differentiating ameloblasts (Jheon et al., 2016). Disruption of the Notch pathway using monoclonal antibodies against NOTCH or JAGGED disrupts the SI-ameloblast interface as well as the adhesion between the SI cells themselves, ultimately resulting in defective enamel formation. The function of *Notch* extends into the incisor mesenchyme, where the *Notch* ligand *Dlk1* regulates the transition of the cervical loop associated MSCs to the TACs and the eventual lineage differentiation of the TACs (Walker et al., 2019). Disruption of *Notch* signaling in the MSCs results in premature TAC differentiation and abnormal dentin formation, thus interfering with incisor regeneration during homeostasis. In addition, the murine incisor repairs itself from 5-FU mediated injuries or mechanical severing via increased proliferation and differentiation of *Notch1*$^{+}$ SI cells into ameloblasts (Sharir et al., 2019). A similar pathway exists in the molar mesenchyme, where *Notch2*$^{+}$ dental pulp stem cells have been shown to proliferate and migrate into injured dental tissue, where they presumably help with recovery after injury (Mitsiadis, Catón, Pagella, Orsini, & Jimenez-Rojo, 2017).

5.2 Hair follicle

Notch signaling components are expressed in the hair follicle epithelium as well as the mesenchyme, starting from the placode and throughout initial stages of morphogenesis (Aubin-Houzelstein, 2012). However, loss of function or gain of function mutants for the various Notch ligands and receptors demonstrate that this signaling does not affect placode formation or hair follicle patterning. Notch becomes functionally essential in hair follicle development during late embryogenesis, where it plays a role in the differentiation of the hair follicle stem cells (Aubin-Houzelstein, 2012; Rishikaysh et al., 2014). Canonical Notch-Rbpj signaling restricts the bulge stem cells to the follicular fate during differentiation. In addition, in *Notch-* or *Jagged*-deficient hair follicles, the stem cells seem to adopt an epidermal differentiation fate. Loss of *Notch* ($K14\text{-}Cre/+;Notch1^{flox/flox}$) or *Jagged* ($K5\text{-}Cre/+;\ Jag1^{\ flox/flox}$) also results in a depletion of hair follicles and eventual baldness in adult mice. Notch-Hes1 signaling also facilitates the responsiveness of the lower bulge stem cells and the hair germ to *Shh* signaling, thereby facilitating the initiation of anagen as well as maintenance of the HFSCs (Suen et al., 2020). The ablation of *Notch* in the hair follicle also results in the formation of smaller bulbs and a premature entry into catagen, suggesting that it regulates hair follicle cycling too. Mutations in the gene *Tspear* (that result in premature truncation of protein translation) are associated with ectodermal dysplasias characterized by oligodontia and alopecia. They have been found to alter the expression of *Notch* signaling genes involved in both tooth and hair follicle development (Peled et al., 2016). Interestingly, Notch has also been shown to function in a non-cell autonomous manner to regulate hair follicle growth and cycling. Indeed, the overexpression of constitutively active *Notch* in the hair cortex results in a wavy shiny first coat followed by excessive hair shedding in every hair cycle during the catagen-telogen transition (Lin, Leimeister, Gessler, & Kopan, 2000). Similarly, overexpression of *Notch* in the suprabasal epidermal layer results in a complete hair loss every cycle and the new hair growing at each anagen is also wavy (Uyttendaele, Panteleyev, De Berker, Tobin, & Christiano, 2004). In both of these studies, the defects in hair follicle morphogenesis seem to arise from the disruption of the differentiation program regulated by Notch. The expression of the ligand *JAG1* on T regulatory cells (Tregs) has also been shown to facilitate hair follicle cycling and regeneration from telogen to anagen through the modulation of HFSC proliferation and differentiation by the activation of Notch signaling in these cells (Ali et al., 2017).

5.3 Mammary gland

Notch receptor and ligand expression have been observed in the mammary gland during all stages of development (Chen et al., 2021). Different Notch pathway genes are expressed at a basal level throughout development and adulthood, and peak during and around pregnancy as shown by whole gland qPCRs (Raafat et al., 2012). In the embryo, *Notch1* is a marker for unipotent stem cells as early as E12.5, and *Notch1* signaling determines fate specification of the embryonic mammary cells into luminal populations (Lilja et al., 2018). This feature persists in adults, as Notch overexpression is sufficient to cause committed basal progenitors to switch fate into luminal cell types. Similarly, conditional induction of the Notch intracellular domain in the luminal compartment led to defects in ductal organization and lactation, as well as to a redistribution of the luminal cell subtypes within the gland (Phoon et al., 2020). The expression of Notch inhibitors *Numb/Numbl* helps retain the identity of mammary myoepithelial cells and thus regulate the balance between myoepithelial and luminal lineages (Zhang et al., 2016). Deletion or ectopic expression of various components of Notch signaling in vivo have also shown that it plays a role in branching and ductal growth during adulthood and also secretory lobule formation during pregnancy (Chen et al., 2021; Edwards & Brennan, 2021). The ablation of *Numb/Numbl* results in the loss of the contractile myoepithelial layer leading to lactation failure during pregnancy. Thus, Notch functions in the regulation of luminal cell fate specification throughout development and morphogenesis of the mammary gland. The expansion in MaSC numbers upon the downregulation of Notch signaling suggests that Notch helps maintain the size of the MaSC population.

The paracrine regulation of Notch signaling in the mammary epithelium also plays an important role in gland homeostasis. A recent study shows that in the basal compartment, ROBO1 controls the expression of the Notch ligand *Jag1* by regulating β-catenin, which in turn regulates the activation of Notch signaling in the adjacent luminal cells (Cazares et al., 2021). In a disease context, PTEN expression in the mammary gland stromal fibroblasts regulates Jagged expression in these cells, which in turn signals to NOTCH3 in the basal myoepithelium (Sizemore et al., 2017). The loss of PTEN disrupts this paracrine signaling, leading to an expansion in the MaSC population at the expense of mammary ductal elongation and disrupting the alveolar side branching. The expression of the Notch ligand DLL1 is enriched in MaSCs, and these cells can generate both basal as well as luminal populations (Chakrabarti et al., 2021). This activity is the result of a feedback loop from the niche resident macrophages. *Dll1*$^{+}$ MaSCs bind to NOTCH2 and NOTCH3 on the macrophages, thereby sustaining their numbers and inducing them to produce Wnt ligands. As a result, these ligands mediate the maintenance of the MaSCs and ductal morphogenesis.

In conclusion, although expressed, Notch signaling components seem dispensable in the initiation stages of all three organs but become functionally important from late embryogenesis onwards, when they start dictating progenitor cell maintenance as well as cell fate decisions. In the late embryonic stage incisor, Notch regulates the differentiation and maturation of the pre-ameloblasts and ameloblasts, and therefore enamel deposition. This function of Notch continues in the adult incisor and plays an important role in injury repair. Similarly, in the hair follicle, Notch regulates the differentiation of the HFSCs into follicular cells. In addition, since hair follicles go through a cycle of growth, regression and rest, Notch mediates the cyclical transitions through the regulation of the differentiation program. In the mammary gland, Notch expression promotes cell differentiation to a luminal cell fate and thus ultimately affects ductal organization and lactation. Thus, Notch seems to regulate similar functions irrespective of its source of expression. Of note, in both the hair follicle and the mammary gland, Notch signaling within the epithelium and the maintenance of stemness is also regulated by non-cell autonomous mechanisms - the Treg cells in case of the hair follicle and macrophages in case of the mammary gland. Such a crosstalk has not yet been explored for the rodent incisor.

6. Yap/Taz

Hippo-YAP/TAZ signaling has emerged as a key player in the regulation of tissue/organ development as well as morphogenesis during the past 2 decades. The balance between the cytoplasmic and nuclear localization of these proteins controls organ size via the regulation of progenitor cell proliferation in various tissues (Varelas, 2014). YAP/TAZ expression also helps with establishing and maintaining the undifferentiated stem cell state in different organs (Panciera et al., 2016). We will describe the recent work exploring YAP/TAZ signaling in the tooth, hair follicle and mammary gland in this section (Fig. 2).

6.1 Tooth

The relevance and importance of YAP/TAZ signaling has only begun to be explored in the murine incisor. It was recently established that YAP/TAZ nuclear localization correlates with cell proliferation in the developing incisor tooth bud (Li, Hu, et al., 2016). The expression of both *Yap* and *Taz* genes is detected in the cytoplasm of enamel knot cells, which are strictly non-proliferative cells. Deletion of *Ctnna1* caused a defect in incisor morphogenesis, with perturbed mesenchymal condensation and absence of an enamel knot. Interestingly, knocking out *Yap* and *Taz* in the *Ctnna1* deletion mutants was sufficient to restore enamel knot formation and tooth morphogenesis. This paper led to the discovery that *Ctnna1* acts as a mechanical signal transducer leading to the cytoplasmic sequestering of YAP/TAZ proteins in the prospective enamel knot region, thereby regulating its formation and further progression of tooth morphogenesis.

As the tooth matures, the function of YAP/TAZ switches from the regulation of signaling center development during initiation to the maintenance and regulation of progenitor cells in the adult organ (Berthet & Varelas, 2017; Hu et al., 2017). It has been shown that YAP and TAZ have functionally redundant roles that enable adequate proliferation of the progenitor cells. Localized induction of the integrin signaling pathway leads to YAP activation, in turn resulting in cell proliferation and differentiation through multiple downstream effectors such as *Rheb* and mTOR. YAP signaling can be functionally compensated by TAZ, ensuring that there is a robust system to maintain adult tissue homeostasis, supporting continuous growth of the incisor. Another significant discovery highlighted in this study is the fact that the ablation of YAP and TAZ disrupts the maintenance of the labial cervical loop. Thus, cytoplasmic YAP/TAZ regulate not only the maintenance of progenitor cells but also their differentiation, hence impacting the formation of the outer enamel epithelium and the labial cervical loop.

6.2 Hair follicle

The mechanistic regulation as well as the contribution of YAP/TAZ signaling to skin morphogenesis, in particular to hair follicle development, has been scrutinized more extensively than in the tooth. Initial observations demonstrated that YAP localization is nuclear in the proliferating cell population (i.e., the outer root sheath and the transit amplifying cells), while it remains cytoplasmic in the differentiating lineages of the inner root sheath and the hair shaft, suggesting that YAP/TAZ regulates embryonic hair morphogenesis (Rognoni & Walko, 2019; Zhang, Pasolli, & Fuchs, 2011). Induction of a mutant version of *Yap* (*Yap-S127A*) that results in enhanced nuclear localization of the protein in basal keratinocytes showed that at the initial stages of hair follicle morphogenesis, the epidermis became hyper-thickened and a burst of dermal papillae (signaling centers) formed (Zhang et al., 2011). Instead of growing downwards, the signaling crosstalk with the epidermis appeared disrupted, which led to the upward movement of the papillae, and consequently to evaginated hair follicles. Moreover, these follicles remained immature and display no expression of Notch signaling effectors nor any other marker characteristic of terminally differentiated hair follicle cells. In contrast, the deletion of both *Yap* and *Taz* completely blocks hair growth in neonates (Elbediwy et al., 2016). Together, these data show that *Yap* helps in the maintenance of the undifferentiated state in the follicle progenitor cells, a property essential for continuing hair growth. Mice harboring the ablation of *WW45* (an upstream regulator of the Hippo pathway) have an elevation of YAP activity and thus show a similar epidermal cell hyperproliferation and a lack of terminal differentiation (Lee et al., 2008). Importantly, they also show a complete absence of hair follicle morphogenesis, further supporting that Hippo exerts a regulatory influence and acts upstream of hair follicle induction and maturation. The complete blockage of hair follicle initiation in the *WW45* mutants, compared to the formation of immature follicles in active mutant YAP, indicates that in addition to YAP there are other regulatory factors (signaling pathways or a temporal effect) that regulate hair follicle induction.

The function of YAP/TAZ in maintaining the progenitor state of the proliferating cells in the embryonic and neonatal follicle persists during adult follicle cycling. YAP localization becomes increasingly nuclear in the actively proliferating hair follicle stem cells with the initiation of every anagen stage, and remains prominent but completely cytoplasmic during the quiescent period or telogen (Rognoni & Walko, 2019). Similar to the neonates, the inducible YAP-*S127A* mutant protein leads to an expansion of the basal progenitor cell population, subsequent epidermal thickening and an inhibition of differentiation marker expression in adult mice too (Schlegelmilch et al., 2011; Totaro et al., 2017). To observe the longer-term effects of YAP-S127A, since the mutants do not survive adulthood, skin grafts were made from the transgenics onto nude mice. These revealed that hyperkeratosis was promoted in the grafted skin and resulted in stunted hair growth (Schlegelmilch et al., 2011). Interestingly, the transgenic skin lacked the expression of the hair follicle stem cell markers *Krt15* and *Sox9*. It can be speculated that this could be explained if these markers are expressed in the quiescent stem cells but it requires further exploration. The authors also demonstrate that the mechanical signal transducer *Ctnna1* inhibits YAP activity by modulating its phosphorylation status, thus suppressing unnecessary expansion of the epidermis. Recent observations also show that nuclear YAP/TAZ promotes the stem cell state via inhibition of Notch signaling in epidermal stem cells (Berthet & Varelas, 2017; Totaro et al., 2017), although these mechanisms have yet to be addressed in the hair follicle. Also, similar to the neonates, ablation of *Yap* and *Taz* in the basal layer of adult skin results in the upregulation of Notch targets and the induction of differentiation markers (Totaro et al., 2017). Independent observations on the same mice have shown a dramatic loss of hair in patches within a few weeks of *Yap* and *Taz* deletion (Elbediwy et al., 2016).

6.3 Mammary gland

Although both YAP/TAZ are redundantly expressed, studies in the mammary gland show that only TAZ functions in virgin glands. Pre-pubescent virgin mammary glands appear to be normal upon deletion of both *Yap* and *Taz* (Chen et al., 2014; Denson et al., 2018; Skibinski et al., 2014). However, a tissue wide expression of TAZ regulates the epithelial side branching morphogenesis in adult mammary glands (Skibinski et al., 2014). The deletion of *Taz* reduces the number and complexity of the branches. *Taz* null glands also show a decrease in the proportion of basal cells suggesting that TAZ activity is necessary for basal cell maintenance. Interestingly, during puberty, Hippo signaling is active in the luminal cells by restricting TAZ to the cytoplasm and thus preventing its activity. These data led to the notion that TAZ activity in the basal cells regulates the branching during gland development, a conclusion further supported by the display of increased branching and advanced mammary gland development in the earlier stages of puberty with the overexpression of TAZ induces (Denson et al., 2018). Additionally, the activation of TAZ also results in an increase in the fat mass in the gland. In addition to influencing the basal and luminal population ratios in the adult gland, TAZ can control the differentiation state of the mammary epithelial cells (Skibinski et al., 2014). Unlike *Taz*, *Yap* depletion does not result in any observable changes in cellular morphology, indicating that YAP and TAZ do not exhibit functional redundancy in the normal mammary gland. Moreover, the loss of *Yap* in the virgin mammary gland has been proposed to suppress tumor development while its hyperactivation when combined with other oncogenes induces epithelial tumor growth (Chen et al., 2014).

While dispensable in the virgin gland, WW45 and YAP are very important during the pregnancy phase of mammary gland cycle (Chen et al., 2014; Denson et al., 2018). Deletion of *WW45* or the overexpression of YAP inhibited the terminal differentiation of the secretory alveolar cells during pregnancy. Unlike other tissues (tooth or skin), the hyperactivation of YAP does not induce increased proliferation in the mammary gland. Instead, YAP is essential for mammary epithelial cell survival, since the loss of YAP function induces apoptosis in these cells (Chen et al., 2014). Changes in *Yap* expression during pregnancy are a result of the suppression of Hippo activity and regulate the expansion and differentiation of the mammary gland. The ability of YAP to induce a highly proliferative, undifferentiated state has been further explored in the ability to dedifferentiate epithelial tissues. The transient expression of YAP/TAZ in the differentiated mammary epithelial cells induces stem like properties allowing them to freely proliferate (Panciera et al., 2016). The expression of these factors is also required to preserve the self-renewal potential of MaSCs (Panciera et al., 2016; Skibinski et al., 2014).

In conclusion, all three organs—the tooth, hair follicle and mammary gland—require YAP/TAZ function for the maintenance of their stem/progenitor cell populations at adult stages of development. This function of YAP/TAZ is essential for the maintenance of the labial cervical loop in the incisor, whereas in the hair follicle it regulates complete hair growth, since the presence of a pool of proliferating progenitors is essential for the eventual differentiation and shaft formation. Similar to the hair follicle, in mammary glands, TAZ also regulates branching morphogenesis. However, in the mammary gland, YAP is not only essential for cell proliferation, but also cell survival. Interestingly, the regulatory features of YAP/TAZ signaling differ both functionally and temporally between them. The *Ctnna1*-mediated regulation of YAP/TAZ function in the embryonic tooth leads to development of the signaling center, whereas in the adult epidermis this pathway regulates homeostatic growth and promotes basal cell maintenance and proliferation. Hippo functions as a regulator of YAP signaling during the anagen or growth phase of hair follicle cycling, at a tissue wide level. In contrast, in the mammary gland, Hippo activity is restricted to the luminal cells where TAZ is inhibited in the pubertal and adult gland, thereby allowing YAP activation. Importantly, while YAP/TAZ function in the tooth and hair follicle seems to be redundant and compensatory, in the mammary gland each factor acts in a specific manner at defined stages of development.

Although we have described all these pathways separately, it is important to note that these and others such as Bmp, Eda, Fgf, etc. all act in concert with each other. The morphogenic and functional specialization of the ectodermal appendages arise due to the differential regulation of these signaling pathways. Given the intricacy of the signaling events driving both development and morphogenesis of ectodermal organs, a new area of focus over the past decade has been to investigate the epigenetic regulation of such events. Indeed, chromatin organization plays a key role in enabling the establishment of various transcriptional networks and is thus evidently a potential master-regulator of organogenesis. Whether investigating post-translational modification of the core histones, DNase I hypersensitivity of open accessible chromatin or DNA methylation (transcriptional repression), researchers have started to assemble strong evidence that epigenetic regulation is a key factor of tooth, hair follicle and mammary gland development.

7. Conclusion—The regenerative properties of the ectoderm

The presence of hair and mammary glands are two major features that distinguish mammals from other vertebrates. This is probably why both organs retain their function and developmental aspects across multiple species. The tooth, on the other hand, is present in most vertebrates, and although there is enormous variation seen in the morphological characteristics of the organ even within different mammalian species, the developmental mechanisms are largely conserved. Despite the differences, all three organs possess stem/progenitor populations that can regenerate, giving rise to the whole structure.

The possibility of using the property of regeneration of these organs for injury repair has prompted much research into the mechanisms underlying these processes. In the rodent incisor, injuries caused mechanically, by clipping the tooth, or chemically, by treating mice with 5-Fluorouracil (a potent chemotherapeutic drug), both led to regenerative phases characterized by a massive increase of proliferation in the SI and surrounding regions and a marked decrease in cell cycle length of the *Notch1*$^{+}$ progenitor cells, culminating in rapid repair of the tissue (Sharir et al., 2019). A point to note is that it is the stem/progenitor cells of the organ itself which help with renewal of lost tissue. Very few studies have been conducted to characterize the cellular and molecular response of recovery from injury in the mammary gland, but it has been shown that doxycycline induced apoptosis and ablation of the luminal cells, induces a hybrid state of multipotency in the basal cells and ultimately replenishes the adjacent luminal cells (Centonze et al., 2020). This increased basal cell proliferation and differentiation is regulated by a combination of EGFR, Wnt and Notch pathways. Studies on chemotherapy induced hair loss/alopecia have shown that chemotherapeutic agents induce P53-mediated apoptosis in proliferating HFSCs resulting in hair cycling defects (dystrophic anagen or catagen), but a permanent loss of hair happens only upon a complete loss of stem cells (Kim et al., 2019; Paus, Haslam, Sharov, & Botchkarev, 2013). Activation of Fgf signaling as well as of the DNA damage response pathway have both been shown to have beneficial effects on hair restoration. However, unlike the tooth and mammary gland, the hair follicle possesses the ability to newly form naturally in vivo. In a large enough wound ($> 1\ cm^2$), neogenesis of the follicle occurs from the epidermal stem cells at the center around day 15–21 of healing (Gong et al., 2018; Ito et al., 2007). The newly developed hair follicle reconstitutes its stem cell population through a cascade of molecular events governed by Wnt signaling. Interestingly, the epidermal basal cells rather than the follicle stem cells undergo a fate change to regenerate the new follicles.

The ability of stem cells to regenerate organ systems ex vivo has also been extensively explored. Ex vivo organ germ culture experiments, which involve culturing mesenchymal cells with epithelial cells layered on top using a gel matrix, have shown that the epithelial and mesenchymal cells from embryonic tooth germs can be used to grow functional teeth (Oshima, Ogawa, & Tsuji, 2017) in serum-supplemented media. The tooth mesenchymal stem cell populations can regenerate dental pulp and periodontal tissues (Huang et al., 2021; Li, Tang, Wang, & Chen, 2019; Volponi, Zaugg, Neves, Liu, & Sharpe, 2018). Fgf, Tnfα, Bmp, Wnt, along with various additional growth factors, are involved in the various aspects of dentin regeneration, odontogenic mesenchymal stem cell migration and differentiation. A specific set of niche factors consisting of a Rock pathway inhibitor, FGF2 and VEGFa allows hair follicle stem cells to retain their multipotency in 3D culture (Chacón-Martínez, Klose, Niemann, Glauche, & Wickström, 2017). The bidirectional interconversion state between the stem cells and their progeny obtained in this system is maintained via the interaction of Bmp and Shh signaling. Going a step further, it was recently shown using the organ germ method that bioengineered hair germs grown in media containing Noggin (Bmp inhibitor), FGFs, SHH-agonist and EGF, become enriched in $CD34^{+}/CD49f^{+}/ITG\beta5^{+}$ cells that endowed the ability to sustain long-term cycling of the hair follicles (Takeo, Asakawa, Toyoshima, Ogawa, & Tong, 2021). Serial transplantation experiments done with $CD24^{+}/CD29^{+}$ mammary epithelial progenitors bearing Wnt responsive genes such as *Lrp5/6* and *Lgr5* have demonstrated their ability to reconstitute the branched mammary epithelium in pre-cleared fat pads (Inman, Robertson, Mott, & Bissell, 2015). Although this property has been demonstrated multiple times, it needs to be kept in mind that transplant and reconstitution significantly alter the microenvironment. Thus, these properties of the MaSCs might be applicable only in non-physiological contexts.

Using data from the injury models and the ex vivo transplantation assays, a large body of work has demonstrated the importance of targeting specific pathways to promote organ regeneration. Abrogating the inhibition of Bmp via USAG-1 neutralizing antibodies enables the rescue of potential tooth germs and thus development of supernumerary or extra teeth (Murashima-Suginami et al., 2021). This suggests a possible use of this method to treat congenital tooth agenesis caused by genetic mutations such as in the Eda pathway. Small molecules that facilitate the growth of hair via targeting specific metabolic or molecular pathways and promoting the induction and maintenance of stemness have been proposed and some have even been established by clinical trials (Feldman et al., 2021; Son et al., 2018). Although the ability of small molecules to influence mammary gland regeneration has been largely understudied, the possibility of being able to enhance milk production in cattle holds some interest in the field (Boutinaud et al., 2017).

In conclusion, how the interplay of signaling factors generates the vast variety of ectodermal organs is a fascinating field of study. The small but significant differences during their initiation/morphogenesis will provide tremendous insights into how the morphological and functional differences in the adult organs arise. It is promising that someday a cocktail of factors may be developed to direct the desired type of ectodermal development, thereby enhancing the potential applications of regenerative medicine.

Acknowledgments

Members of the Klein laboratory were supported by NIDCR grants R35-DE026602, R01-DE028496 and R01-DE027620. We thank Dr. Oscar Cazares for helpful discussions and feedback on mammary gland signaling.

References

Ahn, Y. (2015). Signaling in tooth, hair, and mammary placodes. In *Current topics in developmental biology* (1st ed.). Elsevier Inc. https://doi.org/10.1016/bs.ctdb.2014.11.013.

Ahtiainen, L., Uski, I., Thesleff, I., & Mikkola, M. L. (2016). Early epithelial signaling center governs tooth budding morphogenesis. *The Journal of Cell Biology*, *214*, 753–767. https://doi.org/10.1083/jcb.201512074.

Alexander, C. M. (2018). *The Wnt signaling landscape of mammary stem cells and breast tumors, Progress in molecular biology and translational science*. Elsevier Inc. https://doi.org/10.1016/bs.pmbts.2017.11.020.

Ali, N., Zirak, B., Rodriguez, R. S., Pauli, M. L., Truong, H.-A., Lai, K., et al. (2017). Regulatory T cells in skin facilitate epithelial stem cell differentiation. *Cell*, *169*, 1119–1129. e11. https://doi.org/10.1016/j.cell.2017.05.002.

Amerongen, V., Bowman, A.N., Nusse, R., 2012. Article Developmental Stage and Time Dictate the Fate of Wnt / b -Catenin-Responsive Stem Cells in the Mammary Gland 387–400. https://doi.org/10.1016/j.stem.2012.05.023.

An, Z., Akily, B., Sabalic, M., Zong, G., Chai, Y., & Sharpe, P. T. (2018). Regulation of mesenchymal stem to transit-amplifying cell transition in the continuously growing mouse incisor. *Cell Reports*, *23*, 3102–3111. https://doi.org/10.1016/j.celrep.2018.05.001.

Anastas, J. N., & Moon, R. T. (2013). WNT signalling pathways as therapeutic targets in cancer. *Nature Reviews. Cancer*, *13*, 11–26. https://doi.org/10.1038/nrc3419.

Aubin-Houzelstein, G. (2012). Notch signaling and the developing hair follicle. *Adv. Exp. Med. Biol.*, 142–160.

Bach, K., Pensa, S., Grzelak, M., Hadfield, J., Adams, D. J., Marioni, J. C., et al. (2017). Differentiation dynamics of mammary epithelial cells revealed by single-cell RNA sequencing. *Nature Communications*, *8*. https://doi.org/10.1038/s41467-017-02001-5.

Baker, C. M., Verstuyf, A., Jensen, K. B., & Watt, F. M. (2010). Differential sensitivity of epidermal cell subpopulations to beta-catenin-induced ectopic hair follicle formation. *Developmental Biology*, *343*, 40–50. https://doi.org/10.1016/J.YDBIO.2010.04.005.

Balinsky, B. I. (1950). On the prenatal growth of the mammary gland rudiment in the mouse. *Journal of Anatomy*, *84*, 227.

Berthet, J., & Varelas, X. (2017). Integrin-FAK-CDC42-PP1A signaling gnaws at YAP/TAZ activity to control incisor stem cells. *BioEssays*, *39*. https://doi.org/https://doi.org/10.1002/bies.201700116.Integrin-FAK-CDC42-PP1A.

Biehs, B., Hu, J. K.-H., Strauli, N. B., Sangiorgi, E., Jung, H., Heber, R.-P., et al. (2013). BMI1 represses Ink4a/Arf and hox genes to regulate stem cells in the rodent incisor. *Nature Cell Biology*, *15*, 846–852. https://doi.org/10.1038/ncb2766.

Binder, M., Chmielarz, P., Mckinnon, P. J., Biggs, L. C., Thesleff, I., & Balic, A. (2019). Functionally distinctive Ptch receptors establish multimodal hedgehog signaling in the tooth epithelial stem cell niche. *Stem Cells*, *37*, 1238–1248. https://doi.org/10.1002/stem.3042.

Bloomquist, R. F., Fowler, T. E., An, Z., Yu, T. Y., Abdilleh, K., Fraser, G. J., et al. (2019). Developmental plasticity of epithelial stem cells in tooth and taste bud renewal. *Proceedings of the National Academy of Sciences of the United States of America*, *116*, 17858–17866. https://doi.org/10.1073/pnas.1821202116.

Boras-granic, K., Dann, P., & Wysolmerski, J. J. (2014). Embryonic cells contribute directly to the quiescent stem cell population in the adult mouse mammary gland. *Breast Cancer Research*, *16*(6), 487. https://doi.org/10.1186/s13058-014-0487-6.

Boutinaud, M., Isaka, N., Gandemer, E., Lamberton, P., Taranilla, A. I. D. P., Sordillo, L. M., et al. (2017). Inhibiting prolactin by cabergoline accelerates mammary gland remodeling during the early dry period in dairy cows to cite this version : HAL id : HAL-01613619 remodeling during the early dry period in dairy cows. *Journal of Dairy Science*, *100*(12), 9787–9798. https://doi.org/10.3168/jds.2017-12783.

Bray, S. J. (2016). Notch signalling in context. *Nature Reviews. Molecular Cell Biology*, *17*, 722–735. https://doi.org/10.1038/nrm.2016.94.

Brisken, C., Heineman, A., Chavarria, T., Elenbaas, B., Tan, J., Dey, S. K., et al. (2000). Essential function of Wnt-4 in mammary gland development downstream of progesterone signaling. *Genes & Development*, *14*(6), 650–654.

Brisken, C., & O'Malley, B. (2010). Hormone action in the mammary gland. *Cold Spring Harbor Perspectives in Biology*, *2*. https://doi.org/10.1101/CSHPERSPECT.A003178.

Brownell, I., Guevara, E., Bai, C. B., Loomis, C. A., & Joyner, A. L. (2011). Nerve-derived sonic hedgehog defines a niche for hair follicle stem cells capable of becoming epidermal stem cells. *Cell Stem Cell*, *8*, 552–565. https://doi.org/10.1016/j.stem.2011.02.021.

Cai, X., Gong, P., Huang, Y., & Lin, Y. (2011). Notch signalling pathway in tooth development and adult dental cells. *Cell Proliferation*, *44*, 495–507. https://doi.org/10.1111/j.1365-2184.2011.00780.x.

Cazares, O., Chatterjee, S., Lee, P., Strietzel, C., Bubolz, J. W., Harburg, G., et al. (2021). Alveolar progenitor differentiation and lactation depends on paracrine inhibition of notch via ROBO1/CTNNB1/JAG1. *Development*, *148*(21). https://doi.org/10.1242/dev.199940.

Centonze, A., Lin, S., Tika, E., Sifrim, A., Fioramonti, M., Malfait, M., et al. (2020). Heterotypic cell–cell communication regulates glandular stem cell multipotency. *Nature*, *584*, 608–613. https://doi.org/10.1038/s41586-020-2632-y.

Chacón-Martínez, C. A., Klose, M., Niemann, C., Glauche, I., & Wickström, S. A. (2017). Hair follicle stem cell cultures reveal self-organizing plasticity of stem cells and their progeny. *The EMBO Journal*, *36*, 151–164. https://doi.org/10.15252/embj.201694902.

Chakrabarti, R., Celià-terrassa, T., Kumar, S., Hang, X., Wei, Y., Gao, J., et al. (2021). Dll1-mediated macrophageal niche for mammary gland stem cells. *Science*, *360*. https://doi.org/10.1126/science.aan4153.Dll1-mediated.

Chen, S., Jing, J., Yuan, Y., Feng, J., Han, X., Wen, Q., et al. (2020). Runx2+ niche cells maintain incisor mesenchymal tissue homeostasis through IGF signaling. *Cell Reports*, *32*, 108007. https://doi.org/10.1016/j.celrep.2020.108007.

Chen, J., Lan, Y., Baek, J.-A., Gao, Y., & Jiang, R. (2009). Wnt/beta-catenin signaling plays an essential role in activation of odontogenic mesenchyme during early tooth development. *Developmental Biology*, *334*, 174–185. https://doi.org/10.1016/j.ydbio.2009.07.015.Wnt/beta-catenin.

Chen, W., Morabito, S. J., Kessenbrock, K., Enver, T., Meyer, K. B., & Teschendorff, A. E. (2019). Single-cell landscape in mammary epithelium reveals bipotent-like cells associated with breast cancer risk and outcome. *Communications Biology*, *2*. https://doi.org/10.1038/s42003-019-0554-8.

Chen, W., Wei, W., Yu, L., Ye, Z., Huang, F., Zhang, L., et al. (2021). Mammary development and breast cancer: A notch perspective. *Journal of Mammary Gland Biology and Neoplasia*, *26*, 309–320. https://doi.org/10.1007/s10911-021-09496-1.

Chen, Q., Zhang, N., Gray, R. S., Li, H., Ewald, A. J., Zahnow, C. A., et al. (2014). A temporal requirement for hippo signaling in mammary gland differentiation, growth, and tumorigenesis. *Genes & Development*, *28*, 432–437. https://doi.org/10.1101/gad.233676.113.

Cheng, C. W., Niu, B., Warren, M., Pevny, L. H., Lovell-Badge, R., Hwa, T., et al. (2014). Predicting the spatiotemporal dynamics of hair follicle patterns in the developing mouse. *Proceedings of the National Academy of Sciences of the United States of America, 111*, 2596–2601. https://doi.org/10.1073/pnas.1313083111.

Choi, Y. S., Zhang, Y., Xu, M., Yang, Y., Ito, M., Peng, T., et al. (2013). Distinct functions for Wnt/β-catenin in hair follicle stem cell proliferation and survival and interfollicular epidermal homeostasis. *Cell Stem Cell, 13*, 720–733. https://doi.org/10.1016/j.stem.2013.10.003.Distinct.

Cobourne, M. T., Hardcastle, Z., & Sharpe, P. T. (2001). Sonic hedgehog regulates epithelial proliferation and cell survival in the developing tooth germ. *Journal of Dental Research, 80*, 1974–1979. https://doi.org/10.1177/00220345010800110501.

Coulombre, J. L., & Coulombre, A. J. (1971). Metaplastic induction of scales and feathers in the corneal anterior epithelium of the chick embryo. *Developmental Biology, 25*, 464–478. https://doi.org/10.1016/0012-1606(71)90042-X.

Dassule, H. R., Lewis, P., Bei, M., Maas, R., & McMahon, A. P. (2000). Sonic hedgehog regulates growth and morphogenesis of the tooth. *Development, 127*, 4775–4785. https://doi.org/10.1242/DEV.127.22.4775.

Davis, F. M., Lloyd-Lewis, B., Harris, O. B., Kozar, S., Winton, D. J., Muresan, L., et al. (2016). Single-cell lineage tracing in the mammary gland reveals stochastic clonal dispersion of stem/progenitor cell progeny. *Nature Communications*, 7. https://doi.org/10.1038/ncomms13053.

Dawson, C. A., Pal, B., Vaillant, F., Gandolfo, L. C., Liu, Z., Bleriot, C., et al. (2020). Tissue-resident ductal macrophages survey the mammary epithelium and facilitate tissue remodelling. *Nature Cell Biology, 22*, 546–558. https://doi.org/10.1038/s41556-020-0505-0.

Denson, K. E., Mussell, A. L., Shen, H., Truskinovsky, A., Yang, N., Parashurama, N., et al. (2018). The hippo signaling transducer TAZ regulates mammary gland morphogenesis and carcinogen-induced mammary tumorigenesis. *Scientific Reports, 8*, 1–10. https://doi.org/10.1038/s41598-018-24712-5.

Dhouailly, D. (2009). A new scenario for the evolutionary origin of hair, feather, and avian scales. *Journal of Anatomy, 214*, 587–606. https://doi.org/10.1111/j.1469-7580.2008.01041.x.

Dhouailly, D., Godefroit, P., Martin, T., Nonchev, S., Caraguel, F., & Oftedal, O. (2019). Getting to the root of scales, feather and hair: As deep as odontodes? *Experimental Dermatology, 28*, 503–508. https://doi.org/10.1111/exd.13391.

Edwards, A., & Brennan, K. (2021). Notch Signalling in breast development and cancer. *Frontiers in Cell and Development Biology, 9*. https://doi.org/10.3389/fcell.2021.692173.

Elbediwy, A., Vincent-Mistiaen, Z. I., Spencer-Dene, B., Stone, R. K., Boeing, S., Wculek, S. K., et al. (2016). Integrin signalling regulates YAP and TAZ to control skin homeostasis. *Development, 143*, 1674–1687. https://doi.org/10.1242/dev.133728.

Fan, L., Deng, S., Sui, X., Liu, M., Cheng, S., Wang, Y., et al. (2018). Constitutive activation of β-catenin in ameloblasts leads to incisor enamel hypomineralization. *Journal of Molecular Histology, 49*, 499–507. https://doi.org/10.1007/s10735-018-9788-x.

Feldman, P. R., Fiebig, K. M., Piwko, C., Mints, B. M., Brown, D., Cahan, D. J., et al. (2021). Safety and efficacy of ALRV5XR in women with androgenetic alopecia or telogen effluvium: A randomised, double-blinded, placebo-controlled clinical trial. *EClinicalMedicine, 37*, 100978. https://doi.org/10.1016/j.eclinm.2021.100978.

Fu, N. Y., Nolan, E., Lindeman, G. J., & Visvader, J. E. (2020). Stem cells and the differentiation hierarchy in mammary gland development. *Physiological Reviews, 100*, 489–523. https://doi.org/10.1152/physrev.00040.2018.

Gay, D., Plikus, M. V., Lee, I., Treffeisen, E., Wang, A., & Cotsarelis, G. (2020). Cutaneous epithelial stem cells. *Principles of Tissue Engineering*. INC. https://doi.org/10.1016/b978-0-12-818422-6.00071-x.

Gong, L., Xu, X. G., & Li, Y. H. (2018). Embryonic-like regenerative phenomenon: Wound-induced hair follicle neogenesis. *Regenerative Medicine*, *13*, 729–739. https://doi.org/10.2217/rme-2018-0028.

Greco, V., Chen, T., Rendl, M., Schober, M., Pasolli, H. A., Stokes, N., et al. (2009). A two-step mechanism for stem cell activation during hair regeneration. *Cell Stem Cell*, *4*, 155–169. https://doi.org/10.1016/J.STEM.2008.12.009/ATTACHMENT/7A26AEC5-9DB1-47F5-B8A6-E9A8028513B9/MMC1.PDF.

Gritli-Linde, A., Hallberg, K., Harfe, B. D., Reyahi, A., Kannius-Janson, M., Nilsson, J., et al. (2007). Abnormal hair development and apparent follicular transformation to mammary gland in the absence of hedgehog signaling. *Developmental Cell*, *12*, 99–112. https://doi.org/10.1016/J.DEVCEL.2006.12.006/ATTACHMENT/79DEC185-459D-4F86-B90B-0D613DBF9FA3/MMC1.PDF.

Harada, H., & Ohshima, H. (2004). New perspectives on tooth development and the dental stem cell niche. *Archives of Histology and Cytology*, *67*, 1–11.

Harada, H., Toyono, T., Toyoshima, K., & Ohuchi, H. (2002). FGF10 maintains stem cell population during mouse incisor development. *Connective Tissue Research*, *43*, 201–204. https://doi.org/10.1080/713713524.

Hardcastle, Z., Mo, R., Hui, C. C., & Sharpe, P. T. (1998). The shh signalling pathway in tooth development: Defects in Gli2 and Gli3 mutants. *Development*, *125*, 2803–2811. https://doi.org/10.1242/DEV.125.15.2803.

Harshuk-Shabso, S., Dressler, H., Niehrs, C., Aamar, E., & Enshell-Seijffers, D. (2020). Fgf and Wnt signaling interaction in the mesenchymal niche regulates the murine hair cycle clock. *Nature Communications*, *111*(11), 1–14. https://doi.org/10.1038/s41467-020-18643-x.

Hatsell, S. J., & Cowin, P. (2006). Gli3-mediated repression of hedgehog targets is required for normal mammary development. *Development*, *133*, 3661–3670. https://doi.org/10.1242/DEV.02542.

Hatsell, S., & Frost, A. R. (2007). Hedgehog signaling in mammary gland development and breast cancer. *Journal of Mammary Gland Biology and Neoplasia*, *12*, 163–173. https://doi.org/10.1007/S10911-007-9048-2/TABLES/1.

Hosoya, A., Shalehin, N., Takebe, H., Shimo, T., & Irie, K. (2020). Sonic hedgehog signaling and tooth development. *International Journal of Molecular Sciences*, *21*, 1–12. https://doi.org/10.3390/ijms21051587.

Hsu, Y.-C., Li, L., & Fuchs, E. (2014). Emerging interactions between skin stem cells and their niches. *Nature Medicine*, *20*, 847–856. https://doi.org/10.1038/nm.3643.

Hsu, Y. C., Pasolli, H. A., & Fuchs, E. (2011). Dynamics between stem cells, niche, and progeny in the hair follicle. *Cell*, *144*, 92–105. https://doi.org/10.1016/j.cell.2010.11.049.

Hu, J. K., Du, W., Shelton, S. J., Oldham, M. C., Dipersio, C. M., & Klein, O. D. (2017). A FAK-YAP-mTOR signaling axis regulates stem cell-based tissue renewal in mice. *Cell Stem Cell*, *21*, 91–106. https://doi.org/10.1016/j.stem.2017.03.023.A.

Huang, X., Li, Z., Liu, A., Liu, X., Guo, H., Wu, M., et al. (2021). Microenvironment influences odontogenic mesenchymal stem cells mediated dental pulp regeneration. *Frontiers in Physiology*, *12*, 1–11. https://doi.org/10.3389/fphys.2021.656588.

Hui, M., Cazet, A., Nair, R., Watkins, D. N., O'Toole, S. A., & Swarbrick, A. (2013). The hedgehog signalling pathway in breast development, carcinogenesis and cancer therapy. *Breast Cancer Research*, *15*, 1–14. https://doi.org/10.1186/BCR3401/FIGURES/2.

Inman, J. L., Robertson, C., Mott, J. D., & Bissell, M. J. (2015). Mammary gland development: Cell fate specification, stem cells and the microenvironment. *Development, 142*, 1028–1042. https://doi.org/10.1242/dev.087643.

Ishikawa, Y., Ida-Yonemochi, H., Saito, K., Nakatomi, M., & Ohshima, H. (2021). The sonic hedgehog–patched–Gli signaling pathway maintains dental epithelial and pulp stem/progenitor cells and regulates the function of odontoblasts. *Frontiers in Dental Medicine, 2*, 1–11. https://doi.org/10.3389/fdmed.2021.651334.

Ito, M., Yang, Z., Andl, T., Cui, C., Kim, N., Millar, S. E., et al. (2007). Wnt-dependent de novo hair follicle regeneration in adult mouse skin after wounding. *Nature, 447*, 316–320. https://doi.org/10.1038/nature05766.

Jernvall, J., Keranen, S. V. E., & Thesleff, I. (2002). Evolutionary modification of development in mammalian teeth: Quantifying gene expression patterns and topography. *Proceedings of the National Academy of Sciences, 97*, 14444–14448. https://doi.org/10.1073/pnas.97.26.14444.

Jheon, A. H., Prochazkova, M., Meng, B., Wen, T., Lim, Y. J., Naveau, A., et al. (2016). Inhibition of notch signaling during mouse incisor renewal leads to enamel defects. *Journal of Bone and Mineral Research, 31*, 152–162. https://doi.org/10.1002/jbmr.2591.

Jimenez-Rojo, L., Pagella, P., Harada, H., & Mitsiadis, T. A. (2019). Dental epithelial stem cells as a source for mammary gland regeneration and Milk producing cells in vivo. *Cell, 8*, 1–14. https://doi.org/10.3390/cells8101302.

Jing, J., Feng, J., Li, J., Zhao, H., Ho, T. V., He, J., et al. (2021). Reciprocal interaction between mesenchymal stem cells and transit amplifying cells regulates tissue homeostasis. *eLife, 10*, 1–18. https://doi.org/10.7554/eLife.59459.

Joshi, P. A., Waterhouse, P. D., Kannan, N., Narala, S., Fang, H., Di Grappa, M. A., et al. (2015). RANK signaling amplifies WNT-responsive mammary progenitors through R-SPONDIN1. *Stem Cell Reports, 5*, 31–44. https://doi.org/10.1016/j.stemcr.2015.05.012.

Juuri, E., Isaksson, S., Jussila, M., Heikinheimo, K., & Thesleff, I. (2013). Expression of the stem cell marker, SOX2, in ameloblastoma and dental epithelium. *European Journal of Oral Sciences, 121*, 509–516. https://doi.org/10.1111/eos.12095.

Juuri, E., Saito, K., Ahtiainen, L., Seidel, K., Tummers, M., Hochedlinger, K., et al. (2012). Sox2 + stem cells contribute to all epithelial lineages of the tooth via Sfrp5 + progenitors. *Developmental Cell, 23*, 317–328. https://doi.org/10.1016/j.devcel.2012.05.012.

Keränen, S. V. E., Kettunen, P., Åberg, T., Thesleff, I., & Jernvall, J. (1999). Gene expression patterns associated with suppression of odontogenesis in mouse and vole diastema regions. *Development Genes and Evolution, 209*, 495–506. https://doi.org/10.1007/s004270050282.

Kessenbrock, K., Smith, P., Steenbeek, S. C., Pervolarakis, N., Kumar, R., Minami, Y., et al. (2017). Diverse regulation of mammary epithelial growth and branching morphogenesis through noncanonical Wnt signaling. *Proceedings of the National Academy of Sciences of the United States of America, 114*, 3121–3126. https://doi.org/10.1073/pnas.1701464114.

Kim, J. Y., Ohn, J., Yoon, J. S., Kang, B. M., Park, M., Kim, S., et al. (2019). Priming mobilization of hair follicle stem cells triggers permanent loss of regeneration after alkylating chemotherapy. *Nature Communications, 10*. https://doi.org/10.1038/s41467-019-11665-0.

Kim, R., Yu, T., Li, J., Prochazka, J., Sharir, A., Green, J. B. A., et al. (2021). Early perturbation of Wnt signaling reveals patterning and invagination-evagination control points in molar tooth development. *Development, 148*. https://doi.org/10.1242/DEV.199685.

Kollar, E. J., & Fisher, C. (1980). Tooth induction in chick epithelium: Expression of quiescent genes for enamel synthesis. *Science, 207*, 993–995. https://doi.org/10.1126/SCIENCE.7352302.

Komiya, Y., & Habas, R. (2008). Wnt secretion and extra-cellular regulators. *Organogenesis, 4*, 68–75.

Kratochwil, K. (1969). Organ specificity in mesenchymal induction demonstrated in the embryonic development of the mammary gland of the mouse. *Developmental Biology, 20*, 46–71. https://doi.org/10.1016/0012-1606(69)90004-9.

Krivanek, J., Soldatov, R. A., Kastriti, M. E., Chontorotzea, T., Herdina, A. N., Petersen, J., et al. (2020). Dental cell type atlas reveals stem and differentiated cell types in mouse and human teeth. *Nature Communications, 111*(11), 1–18. https://doi.org/https://doi.org/10.1038/s41467-020-18512-7.

Kuang-Hsien Hu, J., Mushegyan, V., & Klein, O. D. (2014). On the cutting edge of organ renewal: Identification, regulation and evolution of incisor stem cells. *Genesis, 52*, 79. https://doi.org/10.1002/DVG.22732.

Laurikkala, J., Mikkola, M., Mustonen, T., Åberg, T., Koppinen, P., Pispa, J., et al. (2001). TNF signaling via the ligand-receptor pair ectodysplasin and edar controls the function of epithelial signaling centers and is regulated by Wnt and activin during tooth organogenesis. *Developmental Biology, 229*, 443–455. https://doi.org/10.1006/DBIO.2000.9955.

Lee, M. J., Kim, E. J., Otsu, K., Harada, H., & Jung, H. S. (2016). Sox2 contributes to tooth development via Wnt signaling. *Cell and Tissue Research, 365*, 77–84. https://doi.org/10.1007/s00441-016-2363-4.

Lee, J. H., Kim, T. S., Yang, T. H., Koo, B. K., Oh, S. P., Lee, K. P., et al. (2008). A crucial role of WW45 in developing epithelial tissues in the mouse. *The EMBO Journal, 27*, 1231–1242. https://doi.org/10.1038/emboj.2008.63.

Lee, M. Y., Racine, V., Jagadpramana, P., Sun, L., Yu, W., Du, T., et al. (2011). Ectodermal influx and cell hypertrophy provide early growth for all murine mammary rudiments, and are differentially regulated among them by Gli3. *PLoS One, 6*. https://doi.org/10.1371/journal.pone.0026242.

Li, J., Chatzeli, L., Panousopoulou, E., Tucker, A. S., & Green, J. B. A. (2016). Epithelial stratification and placode invagination are separable functions in early morphogenesis of the molar tooth. *Development, 143*, 670–681. https://doi.org/10.1242/dev.130187.

Li, C. Y., Hu, J., Lu, H., Lan, J., Du, W., Galicia, N., et al. (2016). αE-catenin inhibits YAP/TAZ activity to regulate signalling Centre formation during tooth development. *Nature Communications, 71*(7), 1–13. https://doi.org/10.1038/ncomms12133.

Li, L., Tang, Q., Wang, A., & Chen, Y. P. (2019). Regrowing a tooth: In vitro and in vivo approaches. *Current Opinion in Cell Biology, 61*, 126–131. https://doi.org/10.1016/j.ceb.2019.08.002.

Lilja, A. M., Rodilla, V., Huyghe, M., Hannezo, E., Landragin, C., Renaud, O., et al. (2018). Clonal analysis of Notch1-expressing cells reveals the existence of unipotent stem cells that retain long-term plasticity in the embryonic mammary gland. *Nature Cell Biology, 20*, 677–687. https://doi.org/10.1038/s41556-018-0108-1.

Lim, C. H., Sun, Q., Ratti, K., Lee, S. H., Zheng, Y., Takeo, M., et al. (2018). Hedgehog stimulates hair follicle neogenesis by creating inductive dermis during murine skin wound healing. *Nature Communications, 9*. https://doi.org/10.1038/s41467-018-07142-9.

Lim, X., Tan, S. H., Yu, K. L., Lim, S. B. H., & Nusse, R. (2016). Axin2 marks quiescent hair follicle bulge stem cells that are maintained by autocrine Wnt/β-catenin signaling. *Proceedings of the National Academy of Sciences of the United States of America, 113*, E1498–E1505. https://doi.org/10.1073/pnas.1601599113.

Lin, M., Leimeister, C., Gessler, M., & Kopan, R. (2000). Activation of the notch pathway in the hair cortex leads to aberrant differentiation of the adjacent hair-shaft layers. *Development, 2432*, 2421–2432.

Lu, X., Yang, J., Zhao, S., & Liu, S. (2019). Advances of Wnt signalling pathway in dental development and potential clinical application. *Organogenesis, 15*, 101–110. https://doi.org/10.1080/15476278.2019.1656996.

Lumsden, A. G. S. (1988). Spatial organization of the epithelium and the role of neural crest cells in the initiation of the mammalian tooth germ. *Development, 103*, 155–169. https://doi.org/10.1242/DEV.103.SUPPLEMENT.155.

Macias, H., & Hinck, L. (2012). Mammary gland development. *Wiley Interdisciplinary Reviews: Developmental Biology, 1*, 533. https://doi.org/10.1002/WDEV.35.

Mayer, J. A., Foley, J., De La Cruz, D., Chuong, C. M., & Widelitz, R. (2008). Conversion of the nipple to hair-bearing epithelia by lowering bone morphogenetic protein pathway activity at the dermal-epidermal Interface. *The American Journal of Pathology, 173*, 1339. https://doi.org/10.2353/AJPATH.2008.070920.

Michno, K., Boras-Granic, K., Mill, P., Hui, C. C., & Hamel, P. A. (2003). Shh expression is required for embryonic hair follicle but not mammary gland development. *Developmental Biology, 264*, 153–165. https://doi.org/10.1016/S0012-1606(03)00401-9.

Mikkola, M. L., & Millar, S. E. (2006). The mammary bud as a skin appendage: Unique and shared aspects of development. *Journal of Mammary Gland Biology and Neoplasia, 11*, 187–203. https://doi.org/10.1007/s10911-006-9029-x.

Mill, P., Mo, R., Fu, H., Grachtchouk, M., Kim, P. C. W., Dlugosz, A. A., et al. (2003). Sonic hedgehog dependent activation of Gli2 is essential for embryonic hair follicle development. *Genes & Development, 17*, 282. https://doi.org/10.1101/GAD.1038103.

Millar, S. E. (2002). Molecular mechanisms regulating hair follicle development. *The Journal of Investigative Dermatology, 118*, 216–225. https://doi.org/10.1046/j.0022-202x.2001.01670.x.

Mina, M., & Kollar, E. J. (1987). The induction of odontogenesis in non-dental mesenchyme combined with early murine mandibular arch epithelium. *Archives of Oral Biology, 32*, 123–127. https://doi.org/10.1016/0003-9969(87)90055-0.

Mitsiadis, T. A., Catón, J., Pagella, P., Orsini, G., & Jimenez-Rojo, L. (2017). Monitoring notch signaling-associated activation of stem cell niches within injured dental pulp. *Frontiers in Physiology, 8*, 1–9. https://doi.org/10.3389/fphys.2017.00372.

Mitsiadis, T. A., Chéraud, Y., Sharpe, P., & Fontaine-Pérus, J. (2003). Development of teeth in chick embryos after mouse neural crest transplantations. *Proceedings of the National Academy of Sciences of the United States of America, 100*, 6541–6545. https://doi.org/10.1073/pnas.1137104100.

Mitsiadis, T. A., Graf, D., Luder, H., Gridley, T., & Bluteau, G. (2010). BMPs and FGFs target notch signalling via jagged 2 to regulate tooth morphogenesis and cytodifferentiation. *Development, 137*, 3025–3035. https://doi.org/10.1242/dev.049528.

Mitsiadis, T. A., Regaudiat, L., & Gridley, T. (2005). Role of the notch signalling pathway in tooth morphogenesis. *Archives of Oral Biology, 50*, 137–140. https://doi.org/10.1016/J.ARCHORALBIO.2004.10.006.

Monkkonen, T., Landua, J. D., Visbal, A. P., & Lewis, M. T. (2017). Epithelial and non-epithelial Ptch1 play opposing roles to regulate proliferation and morphogenesis of the mouse mammary gland. *Development, 144*, 1317–1327. https://doi.org/10.1242/dev.140434.

Müller-Röver, S., Handjiski, B., Van Der Veen, C., Eichmüller, S., Foitzik, K., McKay, I. A., et al. (2001). A comprehensive guide for the accurate classification of murine hair follicles in distinct hair cycle stages. *The Journal of Investigative Dermatology, 117*, 3–15. https://doi.org/10.1046/J.0022-202X.2001.01377.X.

Murashima-Suginami, A., Kiso, H., Tokita, Y., Mihara, E., Nambu, Y., Uozumi, R., et al. (2021). Anti-USAG-1 therapy for tooth regeneration through enhanced BMP signaling. *Science Advances*, 7, 1–11. https://doi.org/10.1126/sciadv.abf1798.

Nakatomi, M., Ida-Yonemochi, H., Nakatomi, C., Saito, K., Kenmotsu, S., Maas, R. L., et al. (2018). Msx2 prevents stratified squamous epithelium formation in the enamel organ. *Journal of Dental Research, 97*, 1355–1364. https://doi.org/10.1177/0022034518777746.

Närhi, K., Järvinen, E., Birchmeier, W., Taketo, M. M., Mikkola, M. L., & Thesleff, I. (2008). Sustained epithelial beta-catenin activity induces precocious hair development but disrupts hair follicle down-growth and hair shaft formation. *Development*, *135*, 1019–1028. https://doi.org/10.1242/DEV.016550.

Nguyen, T. M., & Aragona, M. (2021). Regulation of tissue architecture and stem cell dynamics to sustain homeostasis and repair in the skin epidermis. *Seminars in Cell & Developmental Biology*. https://doi.org/10.1016/J.SEMCDB.2021.09.008.

Nusse, R. (2008). Wnt signaling and stem cell control. *Cell Research*, *185*(18), 523–527. https://doi.org/10.1038/cr.2008.47.

Nusse, R., & Clevers, H. (2017). Wnt/β-catenin signaling, disease, and emerging therapeutic modalities. *Cell*, *169*, 985–999. https://doi.org/10.1016/J.CELL.2017.05.016.

Okolowsky, N., Furth, P. A., & Hamel, P. A. (2014). Oestrogen receptor-alpha regulates non-canonical hedgehog-signalling in the mammary gland. *Developmental Biology*, *391*, 219–229. https://doi.org/10.1016/J.YDBIO.2014.04.007.

Oliver, R. F. (1966). Regeneration of dermal papillae in rat vibrissae. *The Journal of Investigative Dermatology*, *47*, 496–497. https://doi.org/10.1038/JID.1966.175.

Oshima, M., Ogawa, M., & Tsuji, T. (2017). Functional tooth regeneration, in: Organ regeneration: 3D stem cell culture & Manipulationn. *Methods in Molecular Biology*. https://doi.org/10.1007/978-1-4939-6949-4.

Panciera, T., Azzolin, L., Fujimura, A., Di Biagio, D., Frasson, C., Bresolin, S., et al. (2016). Induction of expandable tissue-specific stem/progenitor cells through transient expression of YAP/TAZ. *Cell Stem Cell*, *19*, 725–737. https://doi.org/10.1016/j.stem.2016.08.009.

Paus, R., Haslam, I. S., Sharov, A. A., & Botchkarev, V. A. (2013). Pathobiology of chemotherapy-induced hair loss. *The Lancet Oncology*, *14*, e50–e59. https://doi.org/10.1016/S1470-2045(12)70553-3.

Peled, A., Sarig, O., Samuelov, L., Bertolini, M., Ziv, L., Weissglas-Volkov, D., et al. (2016). Mutations in TSPEAR, encoding a regulator of notch signaling, affect tooth and hair follicle morphogenesis. *PLoS Genetics*, *12*, 1–17. https://doi.org/10.1371/journal.pgen.1006369.

Phoon, Y. P., Chivukula, I. V., Tsoi, Y. L., Kanatani, S., Uhlén, P., Kuiper, R., et al. (2020). Notch activation in the mouse mammary luminal lineage leads to ductal hyperplasia and altered partitioning of luminal cell subtypes. *Experimental Cell Research*, *395*. https://doi.org/10.1016/j.yexcr.2020.112156.

Pispa, J., & Thesleff, I. (2003). Mechanisms of ectodermal organogenesis. *Developmental Biology*, *262*, 195–205. https://doi.org/10.1016/S0012-1606(03)00325-7.

Plikus, M. V., Mayer, J., De Cruz, D., Baker, R. E., Maini, P. K., Maxson, R., et al. (2009). Cyclic dermal BMP signaling regulates stem cell activation during hair regeneration. *Nature*, *451*, 340–344. https://doi.org/10.1038/nature06457.Cyclic.

Prochazkova, M., Chavez, M. G., Prochazka, J., Felfy, H., Mushegyan, V., & Klein, O. D. (2015). Embryonic versus adult stem cells. *Stem Cell Biol. Tissue Eng. Dent. Sci.*, *249–262*. https://doi.org/10.1016/B978-0-12-397157-9.00020-5.

Raafat, A., Goldhar, A. S., Klauzinska, M., Xu, K., Amirjazil, I., Mccurdy, D., et al. (2012). Development of the mouse mammary gland. *Journal of Cellular Physiology*, *226*, 1940–1952. https://doi.org/10.1002/jcp.22526.Expression.

Rajaram, R. D., Buric, D., Caikovski, M., Ayyanan, A., Rougemont, J., Shan, J., et al. (2015). Progesterone and Wnt4 control mammary stem cells via myoepithelial crosstalk. *The EMBO Journal*, *34*, 641–652. https://doi.org/10.15252/EMBJ.201490434.

Rangel, M. C., Bertolette, D., Castro, N. P., Klauzinska, M., Cuttitta, F., & Salomon, D. S. (2016). Developmental signaling pathways regulating mammary stem cells and contributing to the etiology of triple-negative breast cancer. *Breast Cancer Research and Treatment*, *156*, 211–226. https://doi.org/10.1007/s10549-016-3746-7.

Rishikaysh, P., Dev, K., Diaz, D., Shaikh Qureshi, W. M., Filip, S., & Mokry, J. (2014). Signaling involved in hair follicle morphogenesis and development. *International Journal of Molecular Sciences*, *15*, 1647–1670. https://doi.org/10.3390/ijms15011647.

Rognoni, E., & Walko, G. (2019). The roles of YAP/TAZ and the hippo pathway in healthy and diseased skin. *Cell*, *8*, 1–29.

Rognoni, E., & Watt, F. M. (2018). Skin cell heterogeneity in development, wound healing, and cancer. *Trends in Cell Biology*, *0*. https://doi.org/10.1016/j.tcb.2018.05.002.

Saito, K., Michon, F., Yamada, A., Inuzuka, H., Yamaguchi, S., Fukumoto, E., et al. (2020). Sox21 regulates Anapc10 expression and determines the fate of ectodermal. *The Organ, iScience 23*, 101329. https://doi.org/10.1016/j.isci.2020.101329.

Sanz-Navarro, M., Seidel, K., Sun, Z., Bertonnier-Brouty, L., Amendt, B. A., Klein, O. D., et al. (2018). Plasticity within the niche ensures the maintenance of a Sox2+ stem cell population in the mouse incisor. *Development*, *145*. https://doi.org/10.1242/DEV.155929.

Sarkar, L., & Sharpe, P. T. (1999). Expression of Wnt signalling pathway genes during tooth development. *Mechanisms of Development*, *85*, 197–200. https://doi.org/10.1016/S0925-4773(99)00095-7.

Schlegelmilch, K., Mohseni, M., Kirak, O., Pruszak, J., Rodriguez, J. R., Zhou, D., et al. (2011). Yap1 acts downstream of α-catenin to control epidermal proliferation. *Cell*, *144*, 782–795. https://doi.org/10.1016/j.cell.2011.02.031.

Seidel, K., Marangoni, P., Tang, C., Houshmand, B., Du, W., Maas, R. L., et al. (2017). Resolving stem and progenitor cells in the adult mouse incisor through gene coexpression analysis. *eLife*, *6*, e24712. https://doi.org/10.7554/eLife.24712.

Sengel, P. (1976). Morphogenesis of skin. In *Developmental and cell biology, series* (p. 3). https://doi.org/10.1136/bmj.1.6004.290-a.

Sharir, A., Marangoni, P., Zilionis, R., Wan, M., Wald, T., Hu, J. K., et al. (2019). A large pool of actively cycling progenitors orchestrates self-renewal and injury repair of an ectodermal appendage. *Nature Cell Biology*, *21*, 1102–1112. https://doi.org/10.1038/s41556-019-0378-2.

Sizemore, G., Balakrishnan, S., Hammer, A., Thies, K., Trimboli, A., Wallace, J., et al. (2017). Stromal PTEN inhibits the expansion of mammary epithelial stem cells through Jagged-1. *Physiology & Behavior*, *176*, 139–148. https://doi.org/10.1159/000444169. Carotid.

Skibinski, A., Breindel, J. L., Prat, A., Galván, P., Smith, E., Rolfs, A., et al. (2014). The hippo transducer TAZ interacts with the SWI/SNF complex to regulate breast epithelial lineage commitment. *Cell Reports*, *6*, 1059–1072. https://doi.org/10.1016/j.celrep.2014.02.038.

Slepicka, P. F., Somasundara, A. V. H., & dos Santos, C. O. (2021). The molecular basis of mammary gland development and epithelial differentiation. *Seminars in Cell & Developmental Biology*, *114*, 93–112. https://doi.org/10.1016/J.SEMCDB.2020.09.014.

Son, M. J., Jeong, J. K., Kwon, Y., Ryu, J. S., Mun, S. J., Kim, H. J., et al. (2018). A novel and safe small molecule enhances hair follicle regeneration by facilitating metabolic reprogramming. *Experimental & Molecular Medicine*, *50*. https://doi.org/10.1038/s12276-018-0185-z.

Steinhart, Z., & Angers, S. (2018). Wnt signaling in development and tissue homeostasis. *Development*, *145*. https://doi.org/10.1242/DEV.146589.

Suen, W. J., Li, S. T., & Yang, L. T. (2020). Hes1 regulates anagen initiation and hair follicle regeneration through modulation of hedgehog signaling. *Stem Cells*, *38*, 301–314. https://doi.org/10.1002/stem.3117.

Sun, X., Are, A., Annusver, K., Sivan, U., Jacob, T., Dalessandri, T., et al. (2020). Coordinated hedgehog signaling induces new hair follicles in adult skin. *eLife*, *9*. https://doi.org/10.7554/eLife.46756.

Takeo, M., Asakawa, K., Toyoshima, K., Ogawa, M., & Tong, J. (2021). Expansion and characterization of epithelial stem cells with potential for cyclical hair regeneration. *Scientific Reports*. https://doi.org/10.1038/s41598-020-80624-3.

Tao, Y., Mis, M., Blazer, L., Ustav Jnr, M., Steinhart, Z., Chidiac, R., et al. (2019). Tailored tetravalent antibodies potently and specifically activate wnt/frizzled pathways in cells, organoids and mice. *eLife*, *8*. https://doi.org/10.7554/ELIFE.46134.

Totaro, A., Castellan, M., Battilana, G., Zanconato, F., Azzolin, L., Giulitti, S., et al. (2017). YAP/TAZ link cell mechanics to notch signalling to control epidermal stem cell fate. *Nature Communications*, *8*, 1–13. https://doi.org/10.1038/ncomms15206.

Tripurani, S. K., Wang, Y., Fan, Y. X., Rahimi, M., Wong, L., Lee, M. H., et al. (2018). Suppression of Wnt/β-catenin signaling by EGF receptor is required for hair follicle development. *Molecular Biology of the Cell*, *29*, 2784–2799. https://doi.org/10.1091/mbc.E18-08-0488.

Uyttendaele, H., Panteleyev, A. A., De Berker, D., Tobin, D. T., & Christiano, A. M. (2004). Activation of Notch1 in the hair follicle leads to cell-fate switch and Mohawk alopecia. *Differentiation*, *72*, 396–409. https://doi.org/10.1111/J.1432-0436.2004.07208006.X.

Varelas, X. (2014). The hippo pathway effectors TAZ and YAP in development, homeostasis and disease. *Development*, *141*, 1614–1626. https://doi.org/10.1242/dev.102376.

Volponi, A. A., Zaugg, L. K., Neves, V., Liu, Y., & Sharpe, P. T. (2018). Tooth repair and regeneration. *Current Oral Health Reports*, *5*, 295–303.

Walker, J. V., Zhuang, H., Singer, D., Illsley, C. S., Kok, W. L., Sivaraj, K. K., et al. (2019). Transit amplifying cells coordinate mouse incisor mesenchymal stem cell activation. *Nature Communications*, *10*. https://doi.org/10.1038/s41467-019-11611-0.

Wang, D., Cai, C., Dong, X., Yu, Q. C., Zhang, X. O., Yang, L., et al. (2015). Identification of multipotent mammary stemcells by protein C receptor expression. *Nature*, *517*, 81–84. https://doi.org/10.1038/nature13851.

Whyte, J. L., Smith, A. A., & Helms, J. A. (2012). Wnt signaling and injury repair. *Cold Spring Harbor Perspectives in Biology*, *4*. https://doi.org/10.1101/CSHPERSPECT.A008078.

Willems, A., & Sinclair, R. (2021). Alopecias in humans: Biology, pathomechanisms and emerging therapies. *Veterinary Dermatology*. https://doi.org/10.1111/VDE.13014.

Wu, P., Zhang, Y., Xing, Y., Xu, W., Guo, H., Deng, F., et al. (2019). The balance of Bmp6 and Wnt10b regulates the telogen-anagen transition of hair follicles. *Cell Communication and Signaling: CCS*, *17*, 1–10. https://doi.org/10.1186/s12964-019-0330-x.

Yoshizaki, K., Fukumoto, S., Bikle, D. D., & Oda, Y. (2020). Transcriptional regulation of dental epithelial cell fate. *International Journal of Molecular Sciences*, *21*, 1–14. https://doi.org/10.3390/ijms21238952.

Yoshizaki, K., Hu, L., Nguyen, T., Sakai, K., Ishikawa, M., Takahashi, I., et al. (2017). Mediator 1 contributes to enamel mineralization as a coactivator for Notch1 signaling and stimulates transcription of the alkaline phosphatase gene. *The Journal of Biological Chemistry*, *292*, 13531–13540. https://doi.org/10.1074/jbc.M117.780866.

Yu, T., & Klein, O. D. (2020). Molecular and cellular mechanisms of tooth development, homeostasis and repair. *Development*, *147*. https://doi.org/10.1242/DEV.184754.

Yu, Q. C., Verheyen, E. M., & Zeng, Y. A. (2016). Mammary development and breast cancer: A Wnt perspective. *Cancers (Basel).*, *8*, 1–26. https://doi.org/10.3390/cancers8070065.

Zeng, Y. A., & Nusse, R. (2010). Wnt proteins are self-renewal factors for mammary stem cells and promote their long-term expansion in culture. *Cell Stem Cell*, *6*, 568–577. https://doi.org/10.1016/j.stem.2010.03.020.

Zhang, Y. V., Cheong, J., Ciapurin, N., McDermitt, D. J., & Tumbar, T. (2009). Distinct self-renewal and differentiation phases in the niche of infrequently dividing hair follicle stem cells. *Cell Stem Cell*, *5*, 267–278. https://doi.org/10.1016/J.STEM.2009.06.004.

Zhang, Y., Li, F., Song, Y., Sheng, X., Ren, F., Xiong, K., et al. (2016). Numb and Numbl act to determine mammary myoepithelial cell fate, maintain epithelial identity, and support lactogenesis. *The FASEB Journal*, *30*, 3474–3478. https://doi.org/10.1096/fj.201600387R.

Zhang, H., Pasolli, H. A., & Fuchs, E. (2011). Yes-associated protein (YAP) transcriptional coactivator functions in balancing growth and differentiation in skin. *Proceedings of the National Academy of Sciences of the United States of America*, *108*, 2270–2275. https://doi.org/10.1073/pnas.1019603108.

Zhao, C., Cai, S., Shin, K., Lim, A., Kalisky, T., Lu, W. J., et al. (2017). Stromal Gli2 activity coordinates a niche signaling program for mammary epithelial stem cells. *Science*, *80*, 356. https://doi.org/10.1126/science.aal3485.

Zhao, H., Feng, J., Seidel, K., Shi, S., Klein, O., Sharpe, P., et al. (2014). Secretion of shh by a neurovascular bundle niche supports mesenchymal stem cell homeostasis in the adult mouse incisor. *Cell Stem Cell*, *14*, 160–173. https://doi.org/10.1016/J.STEM.2013.12.013.

Zheng, L., Rui, C., Zhang, H., Chen, J., Jia, X., & Xiao, Y. (2019). Sonic hedgehog signaling in epithelial tissue development. *Regenerative Medicine Research*, 7, 3. https://doi.org/10.1051/RMR/190004.

Zhu, X., Zhao, P., Liu, Y., Zhang, X., Fu, J., Ivy Yu, H. M., et al. (2013). Intra-epithelial requirement of canonical Wnt signaling for tooth morphogenesis. *The Journal of Biological Chemistry*, *288*, 12080–12089. https://doi.org/10.1074/JBC.M113.462473.

FSC
www.fsc.org
MIX
Papier | Fördert gute Waldnutzung
FSC® C083411

Kolibri 360 GmbH
Ferdinand-Jühlke-Straße 7
99095 Erfurt, Deutschland
produktsicherheit@kolibri360.de